3D Stem Cell Culture: Stem Cell Biology Essentials

3D Stem Cell Culture: Stem Cell Biology Essentials

Editor: Royce Fortin

STATES
ACADEMIC PRESS
www.statesacademicpress.com

www.statesacademicpress.com

STATES
ACADEMIC PRESS

Cataloging-in-Publication Data

3D stem cell culture : stem cell biology essentials / edited by Royce Fortin.
p. cm.
Includes bibliographical references and index.
ISBN 978-1-63989-794-0
1. Stem cells. 2. Cell culture. 3. Cultures (Biology). 4. Cytology--Technique. I. Fortin, Royce.
QH588.S83 A17 2023
616.027 74--dc23

States Academic Press,
109 South 5th Street,
Brooklyn, NY 11249, USA

ISBN 978-1-63989-794-0 (Hardback)

Contents

Preface

A 3D stem cell culture is an artificially created environment that allows biological cells to grow or interact with surrounding extracellular environment in three dimensions. It can be classified into two types of methods, namely, scaffold techniques and scaffold-free techniques. Solid scaffolds, hydrogels, and other materials are used in scaffold techniques. Recent studies have explored the potentiality of human CD34+ stem cells generated in-vitro agarose gel 3D model to understand the process of bone ossification. Scaffold-free techniques include the use of low adhesion plates, hanging drop plates, micropatterned surfaces, rotating bioreactors, magnetic levitation, and magnetic 3D bioprinting. This book presents recent researches on 3D stem cell culture. Scientists and students actively engaged in the field of stem cell biology will find it full of crucial and unexplored concepts.

All of the data presented henceforth, was collaborated in the wake of recent advancements in the field. The aim of this book is to present the diversified developments from across the globe in a comprehensible manner. The opinions expressed in each chapter belong solely to the contributing authors. Their interpretations of the topics are the integral part of this book, which I have carefully compiled for a better understanding of the readers.

At the end, I would like to thank all those who dedicated their time and efforts for the successful completion of this book. I also wish to convey my gratitude towards my friends and family who supported me at every step.

Editor

1

Characterization of In Vitro 3D Cell Model Developed from Human Hepatocellular Carcinoma (HepG2) Cell Line

Martina Štampar [1,2], Barbara Breznik [1], Metka Filipič [1,2] and Bojana Žegura [1,2,*]

1 Department of Genetic Toxicology and Cancer Biology, National Institute of Biology, 1000 Ljubljana, Slovenia; martina.stampar@nib.si (M.Š.); barbara.breznik@nib.si (B.B.); metka.filipic@nib.si (M.F.)

2 Jozef Stefan International Postgraduate School, 1000 Ljubljana, Slovenia

* Correspondence: Bojana.zegura@nib.si

Abstract: In genetic toxicology, there is a trend against the increased use of in vivo models as highlighted by the 3R strategy, thus encouraging the development and implementation of alternative models. Two-dimensional (2D) hepatic cell models, which are generally used for studying the adverse effects of chemicals and consumer products, are prone to giving misleading results. On the other hand, newly developed hepatic three-dimensional (3D) cell models provide an attractive alternative, which, due to improved cell interactions and a higher level of liver-specific functions, including metabolic enzymes, reflect in vivo conditions more accurately. We developed an in vitro 3D cell model from the human hepatocellular carcinoma (HepG2) cell line. The spheroids were cultured under static conditions and characterised by monitoring their growth, morphology, and cell viability during the time of cultivation. A time-dependent suppression of cell division was observed. Cell cycle analysis showed time-dependent accumulation of cells in the G0/G1 phase. Moreover, time-dependent downregulation of proliferation markers was shown at the mRNA level. Genes encoding hepatic markers, metabolic phase I/II enzymes, were time-dependently deregulated compared to monolayers. New knowledge on the characteristics of the 3D cell model is of great importance for its further development and application in the safety assessment of chemicals, food products, and complex mixtures.

Keywords: 3D cell model; HepG2; cell proliferation; cell cycle; gene expression

---

1. Introduction

In recent years, considerable efforts have been made to develop hepatic in vitro 3D cell models with higher predictability for detecting the genotoxic effects of chemicals and environmental particular liver, which is the main target organ of chemical activation and detoxification processes [1,2]. In toxicology, it is nowadays recommended to use alternative in vitro models for the implementation of the 3R (reduce, refine, and replace) strategy as well as considering the inaccurate prediction of animal models due to inter-species variability. This demands an urgent need for advanced, robust, cost, and time-efficient in vitro models for the safety assessment [3,4]. It is widely recognised that human primary hepatocytes (PHH) are the golden standard for studying metabolism and toxicity, as they are of human origin and express metabolic enzymes that are relevant for human metabolism of xenobiotics [5–9]. However, there are several shortcomings related to the application of PHH such as limited availability, the complexity of isolation and culturing of cells as their expansion in culture is not possible, short life span, rapid dedifferentiation, loss of many hepatocyte functions and hepatic phenotype when cultured in monolayer cultures, high costs associated with the performance of the experiments, and the most important—genetic and metabolic differences due to polymorphism of donors. Due to the above-mentioned facts, PHH are not suitable for routine use for genotoxicity testing [5,10–12]. Alternatively, hepatic carcinoma-derived cell lines, such as HepG2, C3A, HepaRG, HuH6, and many others, are frequently used in genotoxicity studies, due to their unlimited growth, availability, and high reproducibility of results. These cell lines [13–16] have several phenotypic characteristics and some functional properties of liver cells [17,18] and therefore, represent an effective compromise between the ease of culturing and the expression of several key enzymes involved in xenobiotic metabolism [19–21]. However, the major limitations of hepatic cells cultured in two (2D) dimensions are the low expression of CYP450 enzymes, xenobiotic receptors, and phase II enzymes [5,22] and thus, inadequate expression of liver cell function in vivo. Consequently, it is highly plausible that such cell models give inaccurate and false-positive results [23].

In drug development and hepatotoxicity research, there is a high demand for establishing new, reliable, and uniformed models [24] with higher predictability for the consequences of human exposure. The report from the Workshop on Genotoxicity Testing (IWGT) recommends focusing on the development of 3D models, which better reflect the in vivo conditions and are capable of creating the complex microenvironment with the purpose of advancing our understanding of complex biological phenomena [4]. Another important advantage of 3D cell models compared to 2D cell models is that they enable prolonged exposures, due to their increased stability as they retain high cell viability and morphology over the period of several weeks [25]. Many static and perfused techniques for culturing 3D cell models (spheroids) are available, such as non-adhesive surfaces, hanging drop cultures, spinner flasks, bioreactors, and micro-moulding [9,26,27]. Each of these techniques offers various advantages; however, many are technically challenging, expensive, and need appropriate facilities [28]. The most important advantages of culturing cells in the form of 3D are increased cell–cell and cell–matrix interactions and higher expression of liver-specific functions (albumin content, urea synthesis, and expression of cytochromes), thus providing a physiologically more relevant model in vivo [29–35]. In recent years, hepatic 3D cell models are, in addition to being used for drug development, also applied for studying genotoxic effects, such as chromosomal instability [28] and DNA damage [32,35,36] induced by various chemicals and environmental samples [37]. Although 3D cell models are superior to 2D cell models, the lack of standardisation of 3D cell culture protocols and insufficient characterisation of the spheroids prevent the integration of 3D cell culture models into the field of toxicology research, particularly due to variability in structure and composition of the formed spheroids [38,39].

The aim of the present study was to characterise the HepG2 cells grown in 3D conformation in terms of hepatic properties and the mRNA expression profile of selected genes coding for cell proliferation, drug-metabolising enzymes, transporters, and liver-specific factors. The HepG2 spheroids were formed by the forced floating method from an initial cell density of 6000 and 3000 cells/spheroid. As the age of the spheroids can influence the outcome and sensitivity of the cell model [40], spheroid growth, quality parameters (surface area and perimeter), and viability of cells were monitored over a period of 6 days and 12 days, respectively. The viability of cells in spheroids was determined by the Live/Dead staining using confocal microscopy, while the proliferation of cells (KI67 marker) in spheroids of different age and cell cycle analysis was assessed with flow cytometry. The mRNA expression of selected hepatic markers, genes involved in cell proliferation, and those involved in phase I/II metabolism in spheroids with an initial density of 3000 cells/spheroid was assessed using real-time quantitative PCR (Fluidigm).

2. Materials and Methods

2.1. Chemicals

Minimum essential medium eagle (MEME), penicillin/streptomycin, Na-pyruvate, non-essential amino acids (NEAA), L-glutamine, $NaHCO_3$, dimethylsulphoxide (DMSO), methylcellulose, propidium iodide

(PI), and fluorescein diacetate (FDA) were purchased from Sigma (St. Louis, MO, USA). Trypsin-EDTA (0.25%), foetal bovine serum (FBS), and TRIzol® reagent were obtained from Gibco (Praisley, Scotland, UK). Hoechst 33258 dye was obtained from Invitrogen (Waltham, MA, USA). Phosphate-buffered saline (PBS), methanol, and ethanol were purchased from PAA Laboratories (Dartmouth, NH, USA). Triton X-100 was obtained from Fisher Sciences (Massachusetts, NJ, USA), while the high capacity cDNA archive kit, TaqMan Universal PCR Master Mix (4440038), and TaqMan Gene Expression Assays were from Applied Biosystems (Massachusetts, NJ, USA). The PreAmp GrandMasterMix (TA05-50) was obtained from TATAA Biocenter AB (Göteborg, Sweden). GE 48.48 Dynamic Array Sample & Assay loading Reagent Kit—10 IFCs (85000821) and 48.48 Dynamic Array: Gene expression chip were obtained from Fluidigm (South San Francisco). Anti-Ki-67-FITC (130-107-586) antibodies and REA Control (I)-FITC (131-104-611) were obtained from Miltenyi Biotec (Bergisch Gladbach, Germany).

2.2. *Cell Culture and Formation of 3D Spheroids*

The HepG2 cell line was obtained from the ATCC cell bank (HB-8065™) and was grown in MEME media supplemented with 10% FBS, 1% NEAA, 100 IU/mL pen/strep, 0.1 g/mL NaHCO3, 0.1 g/mL Na-pyruvate, and 2 mM L-glutamine at 37 °C in 5% CO_2 atmosphere. The spheroids were developed by the forced floating method described in Štampar et al. (2019) [35] using a growth medium supplemented with 4% methylcellulose [41]. Two culture conditions for spheroids formation were used. The spheroids with an initial density of 3000 cells/spheroid and 6000 cells/spheroid were seeded and grown for 12 and 4 days, respectively, depending on the specific endpoint measured. The culture media was changed every 2–3 days to obtain the optimal growth of spheroids.

2.3. *Monitoring the Growth and Morphology of Spheroids*

The surface area of at least 10 spheroids with the initial density of 3000 cells/spheroid and 6000 cells/spheroid was monitored during the time of cultivation (up to 12 and 4 days, respectively). The surface area (mm^2) of each spheroid was monitored microscopically every day and determined by planimetry at 40× magnification using the NIS elements software 4.13 v (Nikon Instruments, Melville, NY, USA) connected with the Ti Eclipse inverted microscope (Nikon, Japan). Evaluation of the spheroid growth was determined three times independently.

2.4. *Quantification and Viability Determination of the Whole Spheroid by Live/Dead Staining*

At least three spheroids for each condition (3000 and 6000 cells/spheroid) were stained and monitored during the time of cultivation. The culture media were replaced with FBS-free media supplemented with FDA (8 μg/mL) and incubated for 1 h in darkness at 37 °C in a 5% CO_2 atmosphere. After staining, the cells were washed with PBS. Subsequently, PI (20 μg/mL) was added and incubated in the dark for an additional 5 min. Following the incubation, the staining mixture was washed with PBS and substituted with fresh serum-free MEME media (100 μL). The staining was performed according to [37] with modifications. The Z-stacks images of single spheroids were captured using the confocal laser scanning microscope Leica TCS SP8 at 100x magnification. Z-stacks of optical sections were captured across the entire spheroid thickness using excitation and emission (PI: 493/636 nm, FDA: 488/530 nm) settings for simultaneous dual-channel recordings; approximately 50 Z-stacks per spheroid were taken. Z-stacks were processed and analysed using the Leica confocal software and presented as a "maximum intensity projection image" gallery. Three independent experiments were performed (N = 3). The quantification of Z-stacks was proceeded in the program Image-Pro 10 (Media Cybernetics, Inc., Rockville, MD, USA), where at least 20 Z-stacks per spheroid were quantified. The percentage of dead cells in the spheroid was calculated as a ratio between the spheroid area and the number of dead cells.

2.5. Analyses of Cell Cycle and Cell Proliferation by Flow Cytometry

The analyses of cell cycle and cell proliferation in spheroids were performed with flow cytometry. The cells from spheroids with an initial density of 3000 and 6000/cells were collected during the time of cultivation for 18 and 4 days, respectively. For the analysis, 30 spheroids per sample were collected and dissociated with enzymatic digestion and mechanical degradation, as described by Štampar et al. (2019). The obtained single-cell suspension was washed (1x PBS), fixed in ethanol, and kept at −20 °C until the analysis (for details see Hercog et al. (2019 [42]; 2020 [37])). After the fixation step, the cells were washed in cold PBS, centrifuged, and labelled with anti-KI67-FITC (50-fold diluted antibodies in 1% BSA). Subsequently, the cells were washed with BSA and PBS and afterwards stained with Hoechst 33258 dye (diluted in 0.1% Triton X-100 1:500). The cell analyses were performed on a MACSQuant Analyzer 10 (Miltenyi Biotech, Bergisch Gladbach, Germany), where the FITC intensity, corresponding to the proliferation marker KI67+, was detected in the B1 (525/550 nm) channel and the analysis of the cell cycle was proceeded in the V1 (450/550 nm) channel. Rea-FITC control (Miltenyi Biotec, Bergisch Gladbach, Germany) was used to exclude unspecific binding. The experiments were repeated in three independent biological repetitions, where each time, 2.5×10^4 single cells per experimental point were recorded. The obtained data were analysed and the graphics were prepared in FlowJo software V10 (Becton Dickinson, Franklin Lakes, NJ, USA). The cell cycle distribution of the solvent control and treated samples and KI67-positive cells for each day was compared to the first day of the measurements. The statistical analysis for cell cycle was performed by the two-way ANOVA with Bonferroni's multiple comparisons test, alpha 0.05%, while the statistical significance of KI67-positive cells was assessed by one-way ANOVA with Dunnett's multiple comparisons test, α0.05, both using GraphPad Prism V6 (GraphPad Software, San Diego, CA, USA).

2.6. The qPCR Analyses of the Expression of Selected Genes

For the gene expression analysis, only spheroids with a density of 3000 cells/spheroid were collected during the time of cultivation (every two to three days, starting with day three of cultivation). The basal expression of the selected genes was determined in HepG2 monolayer cultures (2D) cultured for 2 days (this is usually the time when in 2D, the gene expression is evaluated after the exposure to various compounds; 24 h for cells to attach to the surface of the plastic and 24 h of subsequent exposure) and in spheroids (3D) cultured for 3, 5, 7, 10, 12, 14, and 17 days. Total mRNA was isolated from one T25 plate in the case of 2D and from a pool of 25 spheroids per sample in the case of 3D, using the TRIzol reagent. The quality and the concentration of total mRNA were measured with a NanoDrop 1000 spectrophotometer (Thermo Fischer Scientific, Wilmington, DE, USA), while degradation was checked by gel-electrophoresis (BioRad Power PAC 3000 and UVP Chem Studio PLUS, Analytik Jena AG, Upland, CA, USA). The cDNA High Capacity Archive Kit was applied for the reverse transcription of 1 μg of total mRNA per sample. The expression of selected genes was quantified by the qPCR on 48.48 Dynamic Array™ IFC method, where TaqMan Universal PCR Master Mix and the pre-amplificated (TATAA PreAmp GrandMasterMix, Tataa Biocenter, Gothenburg, Sweden) Taqman Gene Expression Assays were used. To eliminate the effects of inhibition and to assess the performance of a primer set, a series of 5-fold dilutions of each target gene was analysed. The qPCR experiments were performed on 48.48 Dynamic Array™ IFC chips for gene expression on the BioMark HD machine system (Fluidigm, UK) and data were analysed with an open web program QuantGenious [43]. Experiments were repeated three times independently, each time in two replicates. The difference (2D versus 3D) in gene expression greater than 1.5-fold was considered as up/downregulation (relative expression >1.5 or <0.66, respectively).

The selected genes: *ALB* (albumin, Hs00910225_m1); *AFP* (α-fetoprotein), Hs00173490_m1; *ALDH3A1* (aldehyde dehydrogenase 3 family member A1), Hs00964880_m1; *TOP2A* (topoisomerase

2-α), Hs01032137_m1; *PCNA* (Proliferating cell nuclear antigen), Hs00427214_g1; *KI67* (cellular marker for proliferation), Hs01032443_m1; *CCND1* (encodes the cyclin D1 protein), Hs00765553_m1; *CDKN1A* (Cyclin Dependent Kinase Inhibitor 1A), Hs00355782_m1; *CYP1A2* (cytochrome P450 family 1 subfamily A member 2), Hs00167927_m1; *CYP1A1* (cytochrome P450 family 1 subfamily A member 1), Hs01054797_g1; *CYP3A4* (cytochrome P450 family 3 subfamily A member 4), Hs02514989_s1; *AHR* (aryl hydrocarbon receptor), Hs00169233_m1; *UGT2B7* (UDP glucuronosyltransferase family 2 member B7), Hs00426592_m1; *UGT1A1* (UDP glucuronosyltransferase 1 family, polypeptide A1), Hs02511055_s1; *SULT1C2* (sulfotransferase family 1C member 2), Hs00602560_m1; *SULT1B1* (sulfotransferase family 1B member 1), Hs00234899_m1; *NAT1* (N-acetyltransferase 1), Hs02511243_s1; *NAT2* (N-acetyltransferase 2), Hs01854954_s1; *HIF1α* (Hypoxia-inducible factor 1-alpha), Hs00153153_m1; *BBC3* (p53 upregulated modulator of apoptosis), Hs00248075_m1 were pre-amplificated. In all experiments, *GAPDH* (Human Endogenous Control, Hs99999905_m1) and *HPRT1* (Hypoxanthine phosphoribosyltransferase 1), Hs02800695_m1) were used as reference genes.

The different tests were performed between the 2 days old monolayer and spheroids on day 3 and all subsequent days with the two-way ANOVA (Dunnett's test) by multiple unpaired t-test analysis using the Sidak–Bonferroni method ($^+ p < 0.05$, $^* p < 0.05$, respectively). The statistical significance within days was conducted by the two-way ANOVA, considering Tukey's multiple comparisons test († $p < 0.05$).

3. Results and Discussion

The 2D cultures traditionally used for studying the genotoxic effects of chemicals have several limitations, which consequently lead to misleading results. This has become even more evident with the development of in vitro models that enable the growth of cells in three dimensions (3D), which is physiologically more similar to in vivo conditions [44]. However, before 3D cell models can be integrated for genotoxicity testing research, there is a need for the development and subsequent standardisation of robust models that accurately predict the possible effects of studied compounds [45].

To our knowledge, the present study is the first where a comprehensive characterisation of the HepG2 3D cell model was performed. Based on previous results, two initial cell densities of 3000 and 6000 cells per spheroid were selected for the development of 3D cell models [35]. The HepG2 spheroids were formed by the forced floating method and cultured under static conditions for several days. During the time of cultivation, the spheroids were characterised by measuring the surface area, and viability of cells in spheroids. The characterisation is a very important step in the development of 3D cell models as the obtained data provide the information on whether the model is comparable to the liver in vivo conditions and is thus, more accurate than traditional monolayer culture.

3.1. The Effects on Growth and Morphological Changes over Time

The growth of at least 10 spheroids with the initial density of 3000 cells/spheroid (Figure S1A,B) and 6000 cells/spheroid (Figure S1C,D) was daily measured from 24 h to 7 days. The surface area was monitored by light microscopy and planimetry. The size of the spheroids varied according to the number of cells seeded in each well and increased with time of incubation. The spheroids with an initial density of 3000 cells/spheroid (Figure S1A) grew time-dependently from the first day of seeding. At the end of the cultivation (7 days), the surface area (0.39 ± 0.01 mm^2) increased 95% compared to the surface area at 24 h (0.2 ± 0.02 mm^2) (Figure S1B). In contrast to this, the spheroids with an initial density of 6000 cells/spheroid (Figure S1C) grew slower and the surface area after 7 days of cultivation increased for only 20% (0.56 ± 0.03 mm^2) compared to 24 h (0.45 ± 0.05 mm^2) (Figure S1D). At both initial densities, the spheroids maintained uniform spherical shape over the course of the culturing. These results show a steady growth of spheroids during the time of culture, which is in line with other studies on HepG2 spheroids [18,32]. The increase in the average surface area was higher at the lower initial cell density, meaning that the cells proliferated at a higher rate compared to spheroids with higher initial density, which was also reported by Lee et al. (2009) [46]. Based on these results, we concluded that 3000 cells per well was the more optimal density for the formation of spheroids that could be used for long-term exposures; therefore, this density was selected for further characterisation of the spheroids.

3.2. Determination of Live/Dead Cells in Spheroids over Cultivation Time

Live/Dead staining enabled us to investigate at which initial density and time of cultivation the viability of cells started to decrease and consequently, form the necrotic core. A very well-known limitation of aged spheroids is the formation of a necrotic core, which results from the accumulation of metabolic waste products and insufficient diffusion of oxygen/nutrients into the centre of the spheroid starting at a diameter above 200 to 500 μm, as reported by Nath and Devi (2016) [47]. The visual analysis of HepG2 spheroids by confocal microscopy verified the time-resolved viability of the cells in a 3D culture, which was more or less stable over the cultivation time. In Figure 1A,B, representative spheroids from day 3 until day 12 and day 6 in the case of 3000 and 6000 cells/spheroid, respectively, stained with FDA (live) and PI (dead) are shown. Although the percentage of PI-positive cells in spheroids with an initial density of 3000 cells/spheroid increased from 10 days of cultivation onwards, we noticed that only a few dead cells were observed in the centre of the spheroids, meaning that no necrotic core was formed. The quantification of PI-positive cells representing dead cells confirmed a time-dependent increase in non-viable cells that was significant after 10 and 12 days, reaching 14.5% and 18.9% on average (Figure 1C,D), respectively. In spheroids with an initial density of 6000 cells/spheroid, a higher percentage of dead cells compared to 3000 cells/spheroid was observed already after 3 days of cultivation that was 8.4% and 4.0% on average, respectively. Correlating these results with the spheroid growth data, the spheroids with an initial density of 3000 cells/spheroid gradually increased in size and stayed uniformly spherical with limited degrees of necrosis up to day 12. However, in larger spheroids, small patches of death cells started to occur approximately on day 4 and by day 6, reached 16.5% dead cells on average with the visible formation of a necrotic core. Spheroids usually consist of three main zones—an outer proliferating rim, a quiescent viable zone, and an inner necrotic core that can develop due to lower diffusion of nutrients and oxygen forming hypoxic conditions [48–50]. In cancer research, the zones in larger tumour spheroids with a necrotic core resemble the cellular heterogeneity of solid in vivo tumours [51] and the necrosis occurring in the centre of the spheroid is a desirable characteristic as it mimics in vivo conditions. However, in genetic toxicology, where a model has to recapitulate an in vivo-like liver microenvironment, this is an undesirable characteristic. There are only a few studies so far that specifically determined the size or time at which hepatic spheroids develop necrosis that is associated with the cell type, cell number, and culture conditions [32,52]; therefore, the present study contributes new knowledge on the formation of a necrotic core in spheroids with time of cultivation. Previously, in encapsulated 3D HepG2 aggregates, no necrotic core was observed up to three days of culturing, while with prolonged cultivation, the thickness of aggregates disabled the determination of necrotic cells in the centre of the 3D cell model [19]. Elje et al. reported [32] a small necrotic core in HepG2 3D spheroids developed in hanging drops after 1 day of culturing, which was more represented as separate dead cells; however, the viability was stable over time and the cells were cultured for up to 21 days. Furthermore, in hepatic C3A spheroids with initial density of 2500 cells/spheroid, small patches of cell death started to occur approximately at day 14 and, by day 18, a necrotic core was formed [52]. Altogether, these studies, including ours, show that necrosis occurs in the core of the spheroid with time of cultivation and depends on the cell type and 3D conformation.

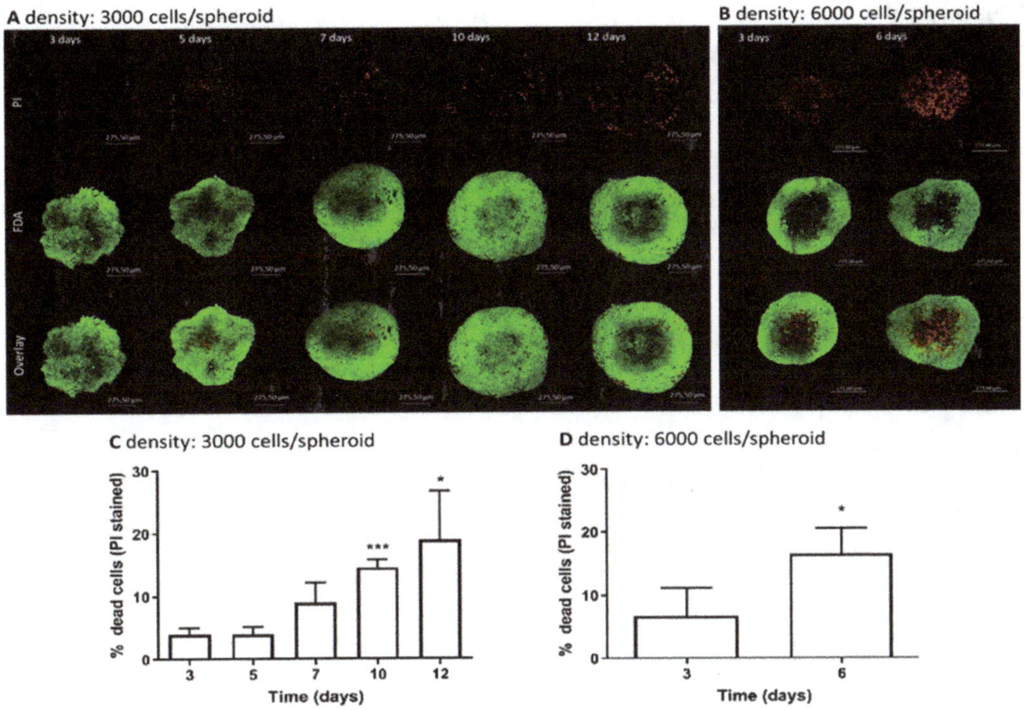

Figure 1. Images of Live/Dead-stained spheroids(**A**,**B**) and the quantification of the percentage of PI-positive cells in images of spheroids (**C**,**D**) captured over the period of cultivation at different densities (**A**–**C**) 3000 cells/spheroid and (**B**–**D**) 6000 cells/spheroid. The cells in spheroids were stained with FDA (green, live cells) and PI (red, dead cells). The Z-stacks were obtained using a confocal microscope at 100× magnification (N = 3). A "maximum intensity projection image" of the spheroid was generated from 50 Z-stacks images (**A**,**B**). Z-stacks were quantified with the Image-Pro 10 software (**C**,**D**) where at least 20 Z-stacks per spheroid were measured and the percentage of dead cells in the spheroid was calculated (N = 3). The statistical significance was calculated with the Student t-test, with alpha 0.05, * $p < 0.05$, ** $p < 0.01$, *** $p < 0.001$, **** $p < 0.0001$.

3.3. *Distribution of Cells within the Cell Cycle and Cell Proliferation during the Time of Cultivation*

Previous studies on HepG2 3D models reported a strong decrease in cell proliferation, which is associated with a time-dependent cell differentiation process [32,53]. In our study, the proliferation of cells and their distribution within the cell cycle of spheroids with the initial density of 3000 and 6000 cells/spheroid were determined by flow cytometry over the time of cultivation, with simultaneous detection of the fluorescent signals of FITC corresponding to the proliferation marker KI67 and Hoechst 33258 corresponding to the cell cycle distribution. It is well known that KI67 protein is present during all active phases of the cell cycle (G1, S, G2, and M), and it is absent from the resting cells (G0) [54,55]. Therefore, the novel approach of simultaneous staining allowed us to further distinguish the distribution of cells in the G0 (non-proliferating) and G1 (proliferating) phase. The results showed that the overall number of proliferating cells and the number of proliferating cells within the G0/G1 phase decreased over the time of cultivation, which is clearly the consequence of the time-dependent increase in non-proliferating cells (G0) within the G0/G1 peak (Figure 2A–C). The percentage of proliferating cells gradually declined with the time reaching 50% decrease after approximately 7 and 2 days of cultivation at an initial density of 3000 (Figure 2A) and 6000 (Figure 2B) cells/spheroid, respectively. At higher cell density, we noticed that already after 24 h, only 65.5% of cells proliferated and by day 4, only 28% of cells expressed KI67 protein, meaning that at day 4, the majority of cells did not proliferate. On the contrary, at lower initial cell density, the decrease in cell proliferation was slower. After 3 days of spheroid cultivation, 82% of cells proliferated, while with further cultivation, the proliferation rate decreased to approximately 68%, 54%, and 13% after 5, 7, and 18 days, respectively. The same trend was observed for proliferating cells within the G0/G1 phase. Similarly, a decrease in the number of proliferating cells in HepG2 spheroids and spheroids cultured in hydrogels was reported by Ramaiahgari et al. (2014) [31] and Lee et al. (2009) [46]. Our results clearly showed that spheroids with lower initial density maintain the proliferation of cells at a higher rate and over the longer period compared to the spheroids with higher initial density, meaning that spheroids with lower initial density are more suitable for long-term exposures.

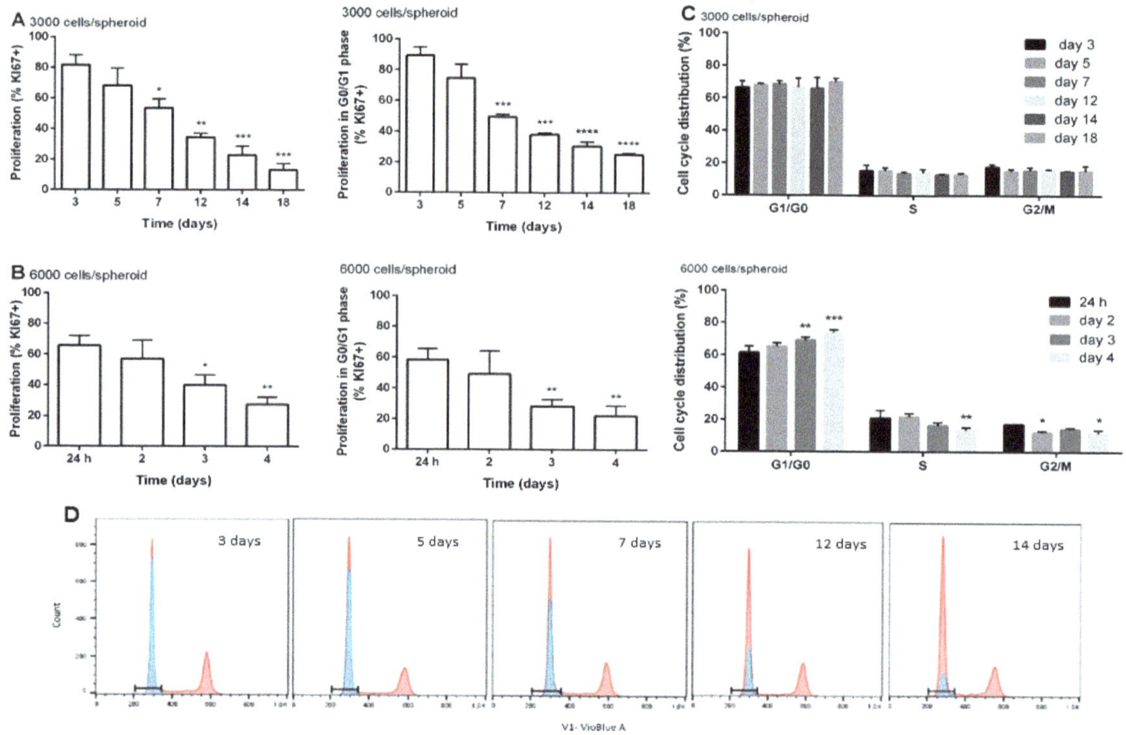

Figure 2. The proliferation of cells and the distribution of cells in different phases of the cell cycle during the time of cultivation. Percentage of KI67-positive cells and the percentage of KI67-positive cells within the G0/G1 phase of cell cycle overtime at density 6000 cells/spheroid (**A**). Percentage of KI67-positive cells and the percentage of KI67-positive cells within the G0/G1 phase of cell cycle overtime at density 3000 cells/spheroid (**B**). Distribution of cells in different phases of the cell cycle over the period of cultivation (**C**). Representative overlays of simultaneous staining with Hoechst for the cell cycle (red) and anti-KI67 antibody (blue) (**D**). The results are presented as the mean ± SD (N = 3). The statistical analysis was conducted in GraphPad Prism 6, by the two-way ANOVA using the Bonferroni multiple comparisons test, * $p < 0.05$, ** $p < 0.01$, *** $p < 0.001$, **** $p < 0.0001$.

The cell cycle duration determines the unique doubling time of the cells [26] that is coordinately controlled by cyclin-dependent kinases and their cyclin partners, whose levels fluctuate throughout the phases of the cell cycle [56] (Figure 2C). In the G1 phase of the cell cycle, crucial decisions on DNA replication and completion of the cell division are made [57]. From the obtained results, we can see that at the initial density of 3000 cells/spheroid, the distribution of cells by phases does not change over time of cultivation. At all measured time points during the time of cultivation, approximately 67.7 ± 3.4% of proliferating cells were in the G1 phase, 14 ± 1.8% of cells were in the S phase, and 15.8 ± 1.7% were in the G2 phase, meaning that the cell cycle was not disturbed during 18 days, only the ratio of proliferating/non-proliferating cells in G0/G1 cells changed. On the contrary, in spheroids with the initial density of 6000 cells/spheroid, a slight increase in proliferating cells in the G1 phase over the time of cultivation was observed, with 61.3% of cells in G1 after 24 h reaching up to 73.9% after 4 days, with a concurrent decrease in cells in the S and G2 phase from 21% and 17% after 24 h to 13.8% and 11.77% after 4 days, respectively (Figure 2D), suggesting cell cycle arrest and thus, the accumulation of cells in the G0/G1 phase with a time of spheroid cultivation. The cell cycle arrest can lead to the inhibition of cell proliferation and/or apoptosis [56], which is in line with the obtained data of the KI67 proliferation marker. These results again show that spheroids with lower initial density maintain a higher rate of cell proliferation compared to spheroids with higher initial density.

3.4. Gene Expression in Spheroids

For the first time, analysis of the expression of genes involved in the proliferation (*TOP2A*, *PCNA*, *KI67*, *CCND1*, and *CDKN1A*), apoptosis (*BBC3*), genes of hepatic markers (*AFP* and *ALB*), and gene a transcription factor encoding aryl hydrocarbon receptor (*AHR*), phase I (*CYP1A1*, *CYP1A2*, and *CYP3A4*), and phase II (*UGT1A1*, *UGT2B7*, *SULT1B1*, *SULT1C2*, *NAT1*, and *NAT2*) xenobiotic-metabolising enzymes was studied in HepG2 spheroids. Gene expression was evaluated in spheroids formed from the lower initial density, namely 3000 cells/spheroid that were cultured for 3, 5, 7, 10, 12, 14, and 17 days. Data for the individual genes were compared to the expression of genes from the monolayer culture at the age of 2 days and are presented as the ratio between 2D (at 2 days of culturing) and 3D at the corresponding time point.

3.4.1. Expression of Genes Involved in Cell Proliferation

Cell proliferation can be regulated by several factors, such as mitogens, growth factors, and survival factors [58]. We studied the genes *PCNA*, *KI67*, and *TOP2A* over the time of culturing, as the expression of the genes involved in the process of cell proliferation may be affected by the time the cells are in culture. A proliferating cell nuclear antigen (*PCNA*), a protein with an important role in DNA replication, has many cell cycle-dependent properties and its absence leads to cell cycle arrest in the S and G2/M phase [59]. Another important proliferation marker KI67 encodes a nuclear antigen during the G1, S, and G2–M phases of proliferating cells, meaning that it is present during all active phases of the cell cycle, except the G0 phase [60,61]. The expression of *TOP2α* is cell cycle-dependent and encodes DNA topoisomerase, which is an enzyme that controls and alters the topologic states of DNA during transcription [60,62]. The results showed that all three studied genes, *TOP2A*, *PCNA*, and *KI67*, were time-dependently downregulated compared to 2 days old monolayer culture (Figure 3A). The major shift in the downregulation of the proliferation markers occurred approximately at day seven.

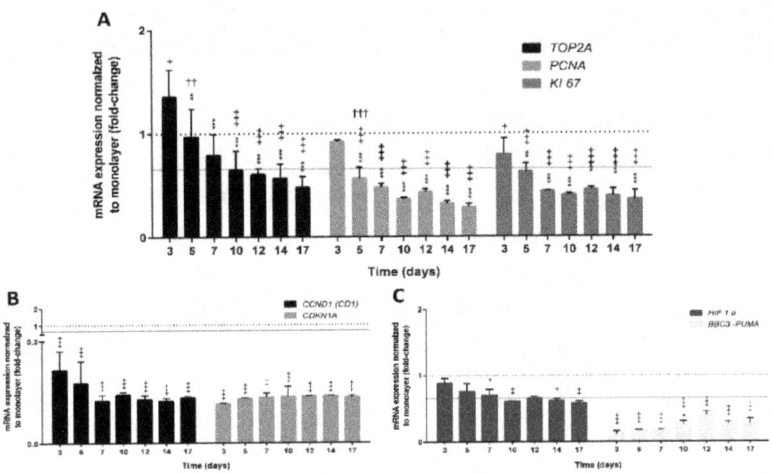

Figure 3. Relative expression of genes involved in the proliferation process (**A**,**B**) and apoptosis (**C**) over time from 3 to 17 days. The significant difference between (i) the monolayer culture (2D) and the spheroids (3D) (+ $p < 0.05$, ++ $p < 0.01$, +++ $p < 0.001$); (ii) the first day (at day 3) of measurement and all subsequent days (* $p < 0.05$, ** $p < 0.01$, *** $p < 0.001$); and (iii) within days (†† $p < 0.01$, ††† $p < 0.001$) was calculated in GraphPad Prism 6, by two-way ANOVA considering Dunnett's multiple comparisons test. Results are presented as the mean ± SD (N = 3). The dotted line denotes the expression of the corresponding gene in monolayer culture (1-fold change), the grey line indicates up- or downregulation of genes with the threshold set at 1.5-fold, which is more than 1.5 or less than 0.66 relative expression, respectively.

The changes in *KI67* on the mRNA level showed the same trend as the data obtained with flow cytometry, where a decrease in KI67-positive cells over the time of cultivation, with approximately 50% of proliferating cells, was detected at day seven. As described before, HepG2 cells accumulated in the G1/G0 phase, due to arrested division in the G0 phase, which is especially noticeable at the initial density of 6000 cells/spheroid. The low proliferation rate in the 3D model can therefore be effectively utilised for studying the effects of long term exposure to various compounds, which is not feasible in a 2D model as the confluence limits the duration of culturing [31,63]. A decrease in cell proliferation in different 3D cell models overtime was reported in several studies with HepG2, using extra-cellular matrix-based hydrogel [31] and spheroids prepared by the hanging drop method [32], and HepG2/C3A, culturing in rotating bioreactors [30]. The occurrence of reduced proliferation over the time of cultivation may also have an impact on the decreased expression of cyclin protein coding genes that control cell cycle progression [64]. In our study, the expression of cyclin protein coding gene *CCND1* was downregulated and a decrease in the first seven days was noticed from a −4.76-fold change after 3 days of cultivation to −8.33-fold change after 7 days of cultivation (Figure 3B). The *CCND1* encodes the cyclin D1 protein, whose elevated expression has been reported in many human tumours and correlates with increased cell proliferation and differentiation due to the reduced G1/S transition [65]. The expression of *CDKN1A*, which is the cell cycle-related gene responsible for the G2/M checkpoint [66], was downregulated (−9.09-fold at day 3) compared to monolayer culture. We noticed a slight increase in *CDKN1A* expression with time of incubation reaching the maximal level at day 7 (−7.14-fold) (Figure 3B). After the seventh day, the expression of both cyclin protein coding genesremained unchanged, which, compared to 2D monolayer culture, indicates the non-proliferating differentiated phenotype of HepG2 spheroids as described by Hiemstra et al. (2019) [67] and by Ramaiahgari et al. (2014) [31].

Altogether, the results suggest that the expression of gene markers related to the proliferation and division of cells decreased over time. Similarly, the expression of *BBC3*, a gene related to apoptosis, was downregulated in spheroids at the age of 3 days (−8.7-fold) compared to monolayer culture; however, with further incubation, the expression of *BBC3* increased, reaching the highest level after day 12 (−2.56-fold change) and then, declining again (Figure 3C). The pro-apoptotic protein BBC3 interacts with anti-apoptotic Bcl-2 family members, resulting in mitochondria-induced apoptosis and cell death through the caspase cascade [68]. It was reported that the diffusion of oxygen into the centre of spheroids is difficult if a diameter is larger than 200 μm (larger spheroids), which consequently causes hypoxia in the core [49]. Live/Dead stained spheroids (density 3000 cells/spheroid) showed no signs of necrosis in the centre; however, in spheroids with a higher initial density (6000 cells/spheroid), necrosis was observed at day 6. This could be related to the hypoxia as reported by Ramaiahgari et al. (2014) [31] (Figure 3). We further investigated the transcription level of the *HIF1α* gene, one of the hypoxia-inducible factors, which mediates cellular adaptation to hypoxic conditions [69]. The obtained results showed that at lower initial density (3000 cells/spheroid) during 17-day cultivation, the *HIF1α* was downregulated compared to monolayer culture and the changes were significantly different from day 7 onwards with no further time-dependent deregulations (Figure 3C), which indicates that no excessive hypoxia was present in HepG2 spheroids up to 17 days.

3.4.2. Expression of Genes Encoding Hepatic Markers

Long-term cultivation of HepG2 cells in 3D conformation led to the enhancement of the expression of liver-specific markers, as shown for alpha-fetoprotein (*AFP*) and albumin (*ALB*) compared to monolayer culture (Figure 4). Alpha-fetoprotein is involved in pleiotropic activities affecting the processes of cell differentiation and growth regulation [70]. Compared to 2D, *AFP* expression in spheroids was 4.57-fold higher, which was statistically significant, at the age of 3 days and reached the maximal mRNA level at day 7 (6.84-fold). With further cultivation, the expression of *AFP* did not change much. Similarly, the expression of *ALB* in 3D significantly differed from the expression in 2D and was 4.95-fold upregulated at day 7. The *ALB* expression gradually increased and reached the maximal level at day 10 (9.56-fold) with a slightly decrease after 12 and 17 days (7.67-fold and

8.03-fold, respectively). Albumin is a stable protein and has a serum half-life of 20 days. Its synthesis is typically regulated on the transcriptional level [71]. Previously, the elevated expression of albumin in HepG2 spheroids compared to monolayer cultures was reported at mRNA [22,35] and protein [31] levels. Altogether, these studies show that hepatic functions are strongly enhanced in 3D systems compared to the 2D monolayer cultures [15,18,19,25,38,53].

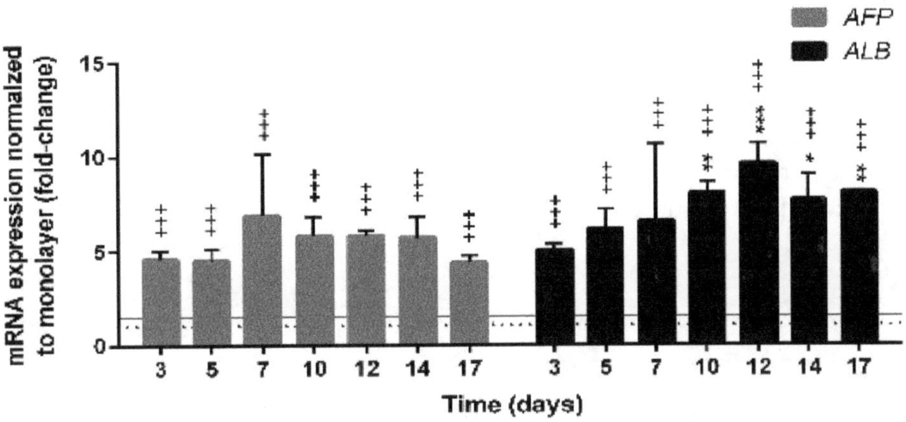

Figure 4. Monitoring the mRNA expression of hepatic markers over time. The significant difference between (i) the monolayer culture (2D) and the spheroids (3D) (+++ $p < 0.001$) and (ii) the first day (at day 3) of measurement and all subsequent days (* $p < 0.05$, ** $p < 0.01$, *** $p < 0.001$); was calculated in GraphPad Prism 6, by the two-way ANOVA considering Dunnett's multiple comparisons test. Results are presented as the mean ± SD (N = 3). The dotted line denotes the expression of the corresponding gene in monolayer culture (1-fold change); the grey line indicates up- or downregulation of genes with the threshold set at 1.5-fold, which is more than 1.5 or less than 0.66 relative expression, respectively.

3.4.3. Expression of Genes Involved in the Xenobiotic Metabolism

Biotransformation of xenobiotic substances is divided into phase I and II reactions, where phase I reactions include the transformation of a parent compound to more polar metabolite(s), while phase II biotransformation results in metabolic inactivation by conjugating reactions including glucuronidation, sulfation, acetylation, methylation, glutathione, and amino acid conjugation. In general, the respective conjugates are more hydrophilic than the parent compounds and are excreted from the organism [72,73]. The most important enzymes from phase I reactions belong to the superfamily CYP450 and play an important role in cellular metabolism and homeostasis, and detoxification and metabolic activation of xenobiotic compounds into reactive metabolites [74,75]. In general, hepatocellular carcinoma monolayer cell cultures exhibit low expression of drug-metabolising enzymes including CYPs and therefore, do not represent ideal alternative systems to human hepatocytes for drug metabolism and hepatotoxicity testing [76]. Recently, several published studies have described the higher expression of CYP enzymes in hepatic cells cultured in 3D conformation [6,18,21,25,28,31,35,77]. Moreover, it has been reported that HepG2 spheroids developed by the forced floating method are metabolically competent, thus expressing various CYP450 enzymes, and are sensitive for detecting the genotoxic effects of indirect-acting genotoxic compounds [35]. In the present study, we evaluated the gene expression of *CYP1A1*, *CYP1A2*, *CYP3A4*, *AHR*, and *ALDH3A1* in 3D spheroids at the age of 3, 5, 7, 10, 12, 14, and 17 days, and compared to the expression of monolayer culture at the age of 2 days (Figure 5). The results revealed that all studied CYPs were clearly upregulated in a 3D model compared to 2D (Figure 5A). The basal mRNA level of *CYP1A1* was approximately 2-fold higher than in monolayer culture and remained at the same level throughout the whole duration of the spheroid cultivation. Similarly, the expression of *CYP1A2* remained constant from day 5 onwards. *CYP3A4*, the major hepatic CYP contributing to the metabolism of more than 50% of xenobiotic substances [76], was reported not to be expressed in HepG2 cells grown in monolayer [20], which strongly limits the use of HepG2 cells for the assessment of the drug metabolism. As already previously shown [35,67,78], the results of the present study confirmed the significant upregulation of *CYP3A4* from day 3 onwards (2.16-fold) with a gradual increase in the mRNA level, reaching the highest level at day 12 (5.76-fold). Altogether, these results demonstrate that HepG2 cells grown in 3D conformation show differentiation into more metabolically competent cells when compared to monolayer cultures. The expression pattern of metabolic genes closer to primary hepatocytes has also been reported for several hepatocellular carcinoma cell lines grown under certain prolonged culturing conditions. For instance, the confluent growth of Huh7 cells resulted in a cell phenotype change and an increase in the *CYP3A4* mRNA level, protein content, and activity after 4 weeks in confluent culture [79], while differentiated HepaRG cells expressed higher levels of CYPs compared to non-differentiated [80,81]. Additionally, HepG2 cells grown in 3D conformation expressed an increased level of *CYP1A1*, *CYP1A2*, and *CYP3A4* genes [35], while HepG2/C3A cells grown in 3D under dynamic conditions for 3 weeks showed elevated expression of *CYP1A1* and *CYP3A4* [77] compared to 2D cultures.

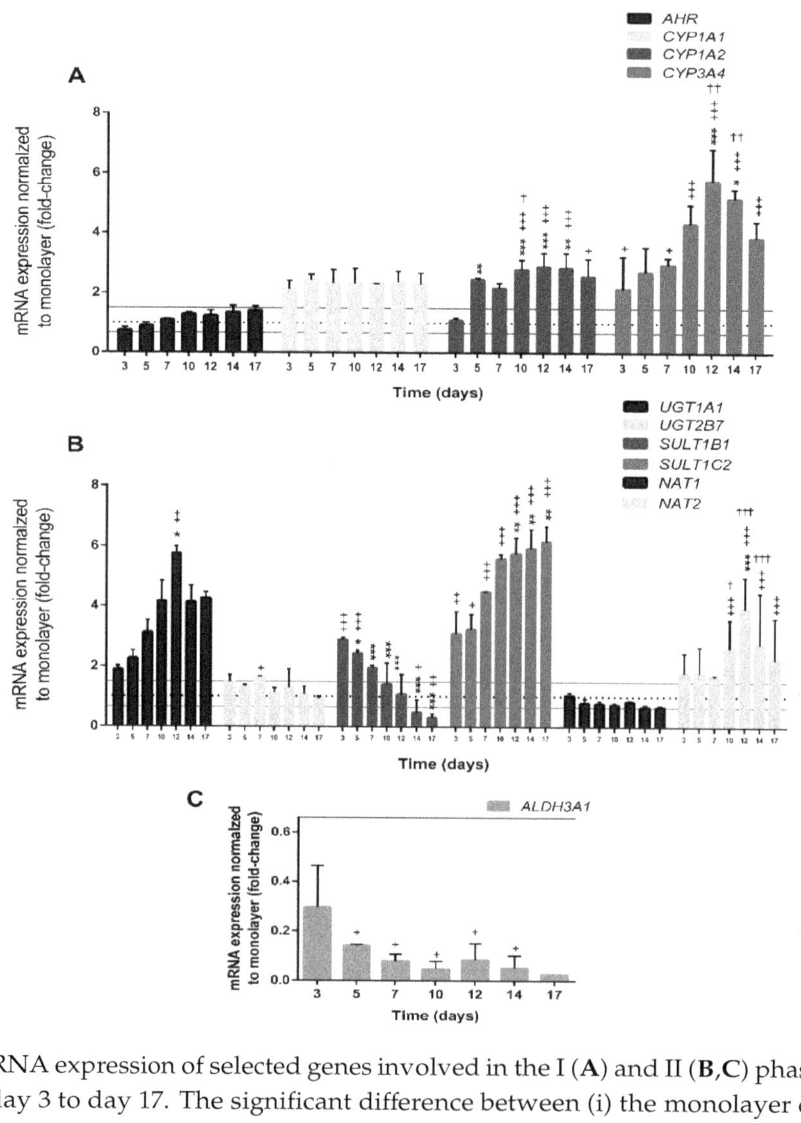

Figure 5. The mRNA expression of selected genes involved in the I (**A**) and II (**B**,**C**) phase of metabolism over time from day 3 to day 17. The significant difference between (i) the monolayer culture (2D) and the spheroids (3D) (+ $p < 0.05$, ++ $p < 0.01$, +++ $p < 0.001$); (ii) the first day (at day 3) of measurement and all subsequent days (* $p < 0.05$, ** $p < 0.01$, *** $p < 0.001$); and (iii) within days († $p < 0.05$, †† $p < 0.01$, ††† $p < 0.001$) was calculated in GraphPad Prism 6, by two-way ANOVA considering Dunnett's multiple comparisons test. Results are presented as the mean ± SD (N = 3). The dotted line denotes the expression of the corresponding gene in monolayer culture (1-fold change), the grey line indicates up- or downregulation of genes with the threshold set at 1.5-fold, which is more than 1.5 or less than 0.66 relative expression, respectively.

Another phase I enzyme ALDH3A1 (Figure 5C), a member of aldehyde dehydrogenase 3 family that catalyses the aliphatic and aromatic aldehydes to the corresponding acids [82,83], was downregulated already in 3 days old spheroids (−3.45-fold) and with further cultivation, the expression of *ALDH3A1* dropped sharply (−12.8-fold and −20-fold at 10 and 14 days, respectively). Low basal expression of *ALDH3A1* in the normal liver was reported by Muzio et al. (2012) [84], while induced levels were determined in liver, colon, bladder, and lung [82,85,86]. The results showing low gene expression are in line with the decreased proliferation of HepG2 cells in spheroids, as it is known that the activation of ALDH3A1 stimulates the proliferation [84], thus the low mRNA level of *ALDH3A1* corroborates decreased cell proliferation and arrested cell cycle determined by the flow cytometry. The expression of drug-metabolising enzymes including *CYP1A1, CYP1A2*, and *ALDH3A1* is regulated by the activation of nuclear factors such as aryl hydrocarbon receptors (AhR) [87]. The basal expression of the *AHR* gene in HepG2 spheroids at the age of 3 days did not significantly differ from the expression in monolayer culture; however, with further incubation, it slowly increased over the time of incubation, reaching the maximal level (1.41-fold) at 17 days. Altogether, the obtained results are in line with the findings of Ramaiahgari et al. (2014) [31], Shah et al. (2018) [28], and Štampar et al. (2019 [35]; 2020 [77]), who described that HepG2 spheroids express higher mRNA levels of phase I metabolic enzymes, which is an important physiological function of hepatic cells in vivo [88].

The enzymes from phase II drug metabolism represent the detoxification step of xenobiotic compounds with the main pathway by the formation of glucuronide conjugates by glucuronosyltransferases (UGTs). In HepG2 spheroids (Figure 5B), the basal mRNA level of *UGT1A1* compared to monolayer culture showed great upregulation from day 3 onwards that was time-dependent, reaching a maximal level at day 12 (5.74-fold). Subsequently, the *UGT1A1* mRNA level decreased (4.13-fold at day 14) and remained constant during further cultivation. Time-dependent elevation of *UGT1A1* expression in HepG2 spheroids was already previously reported by Ramaiahgari et al. (2014) [31] and Štampar et al. (2019) [35]. For the gene *UGT2B7*, slight but not significant upregulation was detected at day 3 (1.49-fold); however, its expression did not increase with time. Previously, it has been demonstrated that dynamic HepG2/C3A spheroids expressed an increased mRNA level of *UGT2B7* after three weeks of culturing [77]. In addition to glucuronidation, sulfonation is very important in the biotransformation of xenobiotics [89], where sulfotransferases (SULTs) catalyse the sulfate conjugation of a variety of exogenous chemicals and endobiotics using 3′-phosphoadenosine-5′-phosphosulfate as the donor [90]. On the other hand, in the reaction, pro-carcinogens are converted into highly reactive intermediates that can act as chemical mutagens and carcinogens by covalently binding to DNA [91]. The results showed that the mRNA level of *SULT1B1* was significantly upregulated (2.89-fold) already after 3 days of culturing; however, the expression was downregulated after prolonged cultivation (−2.13 fold and −3.33-fold at 12 and 17 days, respectively). On the contrary, *SULT1C2* was significantly elevated already at day 3 (3.09-fold) and the expression sharply increased with time (6.16-fold by day 17). Another important group of phase II enzymes is *N*-acetyltransferases (NAT) that catalyse the activation of aromatic and heterocyclic amines via *O*-acetylation, while *N*-acetylation of the parent amines is considered a detoxification step [73]. The gene encoding *NAT1* was not importantly deregulated in HepG2 spheroids—at most, it was slightly downregulated (−1.5-fold at day 17). In contrast, the expression of *NAT2* was significantly elevated with the highest expression observed at day 12 (3.95-fold). Transcriptomic analyses revealed that in dynamic HepG2/C3A spheroids, *NAT1* was not importantly deregulated when compared to 2D culture, while *NAT2* was significantly downregulated in 25 days old culture [77]. NATs are substrate-specific and have distinct tissue distribution, where NAT1 has a ubiquitous tissue distribution and its expression is related to cancers, while NAT2 activity has been described in the liver, colon, and intestinal epithelium [73]. In the present study, many crucial genes that are involved in the activation and detoxification of xenobiotic substances and are in HepG2 monolayer cultures expressed at a very low rate, or are even not detectable, were clearly expressed in HepG2 3D spheroids.

In a study on HepG2 spheroids, Elenberger et al. (2018), based on ultrastructural and organo-typic functional investigations, identified clearly different phases of HepG2 spheroids, including an early phase (day 3 to 6), mid-stage phase (day 6 to 12), and late phase (day 15 to 18), all showing significant differences in cell-to-cell interactions, specialised microstructures such as the formation of bile canaliculi, and metabolic activities including albumin and urea secretion. Similarly, the results of the present study suggested three stages of spheroid formation; the early stage at the age of 3–6 days, mid-stage at the age of 7–12 days, and the late stage at the age of >14 days. In HepG2 spheroids, the expression of hepatic markers and metabolic genes from phases I and II changed over time of cultivation with important changes observed by day 7. The highest mRNA levels for the majority of metabolic genes were noticeable at 10 to 12 days, which, with further cultivation, gradually declined, suggesting that the aforementioned cultivation time is long enough for HepG2 cell differentiation and the development of a metabolically competent cell model with the quantitative and qualitative expression of phase I and II metabolic enzymes compared to a 2D cell model. In addition to metabolic activity, we noticed an important change in cell proliferation at day 7 with a slight increase in cell death, which turned out to be significant after 10 days of cultivation. Taken together, our study demonstrated that the spheroidal age needs to be considered as an important parameter in the development of spheroid-based in vitro models.

4. Conclusions

In vitro, 3D cell models compared to traditional monolayer cultures better resemble the cell organisation of tissues and organs and thus, more accurately mimic the in vivo microenvironment. In recent years, they have also become a promising tool in the field of genetic toxicology in order to reduce, replace, and refine animal experiments. Nevertheless, before 3D models can be routinely used for genotoxicity assessment, they have to be comprehensively characterised and growth conditions need to be optimised to allow for the reproducibility and comparability of the results. Furthermore, systematic characterisation allows us to identify all the crucial advantages and disadvantages, which is of high importance for the further use of the new 3D models. To our knowledge, the present study is the first where a 3D HepG2 cell model was systematically characterised and standardised including advanced cell cycle and proliferation analysis by flow cytometry, and gene expressions. In the present work, we developed uniform 3D HepG2 spheroids of similar size and shape grown under static conditions. The influence of spheroidal age on cell proliferation and metabolic status was studied over a 17-day cultivation period to gain a deeper understanding of the morphological and physiological characteristics of HepG2 spheroids. Based on new knowledge obtained within our study, we can conclude that the initial cell density for the formation of spheroids is very important in order to obtain spheroids with viable dividing cells, which is a prerequisite for studying the adverse geno-/toxic effects, as a division of cells is necessary for damage to be incorporated into DNA. Compared to the 2D monolayer cultures, HepG2 spheroids showed a time-dependent reduction in cell proliferation with cell division arrested in the G0/G1 phase of the cell cycle. Moreover, the spheroids revealed increased liver-specific functions and demonstrated strong physiological relevance concerning gene expression of hepatic markers and metabolic enzymes, in particular for sulfotransferases in phase II, thus indicating differentiation into more metabolically competent cells; this, however, has to be further confirmed at the protein level. We believe that the 3D HepG2 cell model with characterised cell growth and proliferation, as well as known expression of hepatic markers and metabolic enzymes, will contribute to a more reliable assessment of genotoxic activity of chemicals, due to its higher physiological relevance for human exposure and may, therefore, provide an alternative to animal models, which comply with the 3Rs policy to reduce in vivo testing.

Supplementary Materials

Figure S1: The growth and morphology of spheroids (planimetry) monitored during 7 days of cultivation. The surface area of spheroid size was measured every 24 h (A–B: initial density of 3000 cells/spheroid and C–D: initial density of 6000 cells/spheroid). The images were taken using an inverted microscope at 40× magnification

(N = 3). Results are presented as the mean ± SD (N = 10). The statistical analysis was performed in GraphPad Prism 6, by the one-way ANOVA using the Dunnett's multiple comparisons tests, ** $p < 0.01$, *** $p < 0.001$.

Author Contributions: Conceptualization: B.Ž., M.Š.; methodology: M.Š., B.B., B.Ž.; software: M.Š., B.B.; validation: M.Š., B.Ž., B.B.; formal analysis: M.Š., B.Ž., B.B.; investigation: M.Š., B.B., B.Ž.; resources: M.F., B.Ž.; data curation: M.Š., B.Ž.; writing—original draft: M.Š., B.Ž.; writing—review and editing: M.F., B.B.; visualization: M.Š., B.B., B.Ž.; supervision: B.Ž.; project administration: B.Ž., M.F.; funding acquisition: M.F., B.Ž. All authors have read and agreed to the published version of the manuscript.

References

1. Thummel, K. Factors influencing drug metabolism. In *xPharm: The Comprehensive Pharmacology Reference*; Elservier: Amsterdam, The Netherlands, 2007; pp. 1–18. ISBN 9780080552323.
2. Croom, E. Metabolism of xenobiotics of human environments. In *Progress in Molecular Biology and Translational Science*; Elsevier: Amsterdam, The Netherlands, 2012; Volume 112, pp. 31–88.
3. Corvi, R.; Madia, F. In vitro genotoxicity testing–Can the performance be enhanced? *Food Chem. Toxicol.* **2017**, *106*, 600–608. [CrossRef] [PubMed]
4. Pfuhler, S.; Van Benthem, J.; Curren, R.; Doak, S.H.; Dusinska, M.; Hayashi, M.; Heflich, R.H.; Kidd, D.; Kirkland, D.; Luan, Y.; et al. Use of in vitro 3D tissue models in genotoxicity testing: Strategic fit, validation status and way forward. Report of the working group from the 7th International Workshop on Genotoxicity Testing (IWGT). *Mutat. Res. Genet. Toxicol. Environ. Mutagen.* **2020′** *850–851*, 503135. [CrossRef] [PubMed]
5. Kyffin, J.A.; Sharma, P.; Leedale, J.A.; Colley, H.; Murdoch, C.; Mistry, P.; Webb, S.D. Impact of cell types and culture methods on the functionality of in vitro liver systems—A review of cell systems for hepatotoxicity assessment. *Toxicol. In Vitr.* **2018**, *48*, 262–275. [CrossRef] [PubMed]
6. Godoy, P.; Hewitt, N.J.; Albrecht, U.; Andersen, M.E.; Ansari, N.; Bhattacharya, S.; Bode, J.G.; Bolleyn, J.; Borner, C.; Böttger, J.; et al. Recent advances in 2D and 3D in vitro systems using primary hepatocytes, alternative hepatocyte sources and non-parenchymal liver cells and their use in investigating mechanisms of hepatotoxicity, cell signaling and ADME. *Arch. Toxicol.* **2013** *87*, 1315–1530. [CrossRef] [PubMed]
7. Harris, A.J.; Dial, S.L.; A Casciano, D. Comparison of basal gene expression profiles and effects of hepatocarcinogens on gene expression in cultured primary human hepatocytes and HepG2 cells. *Mutat. Res. Mol. Mech. Mutagen.* **2004**, *549*, 79–99. [CrossRef] [PubMed]
8. Lecluyse, E.L. Human hepatocyte culture systems for the in vitro evaluation of cytochrome P450 expression and regulation. *Eur. J. Pharm. Sci.* **2001** *13*, 343–368. [CrossRef]
9. Fey, S.J.; Wrzesinski, K. Determination of Drug Toxicity Using 3D Spheroids Constructed From an Immortal Human Hepatocyte Cell Line. *Toxicol. Sci.* **2012** *127*, 403–411. [CrossRef]
10. Wang, Z.; Luo, X.; Anene-Nzelu, C.; Yu, Y.; Hong, X.; Singh, N.H.; Xia, L.; Liu, S.; Yu, H. HepaRG culture in tethered spheroids as an in vitro three-dimensional model for drug safety screening. *J. Appl. Toxicol.* **2015**, *35*, 909–917. [CrossRef]
11. Otto, M.; Hansen, S.H.; Dalgaard, L.; Dubois, J.; Badolo, L. Development of an in vitro assay for the investigation of metabolism-induced drug hepatotoxicity. *Cell Biol. Toxicol.* **2008**, *24*, 87–99. [CrossRef]
12. Drewitz, M.; Helbling, M.; Fried, N.; Bieri, M.; Moritz, W.; Lichtenberg, J.; Kelm, J.M. Towards automated production and drug sensitivity testing using scaffold-free spherical tumor microtissues. *Biotechnol. J.* **2011**, *6*, 1488–1496. [CrossRef]
13. Guillouzo, A.; Corlu, A.; Aninat, C.; Glaise, D.; Morel, F.; Guguen-Guillouzo, C. The human hepatoma HepaRG cells: A highly differentiated model for studies of liver metabolism and toxicity of xenobiotics. *Chem. Biol. Interact.* **2007**, *168*, 66–73. [CrossRef] [PubMed]
14. Donato, M.; Tolosa, L.; Gómez-Lechón, M.J. Culture and Functional Characterization of Human Hepatoma HepG2 Cells. In *Protocols in In Vitro Hepatocyte Research*; Springer: New York, NY, USA, 2015; pp. 77–93, ISBN 9781493920747.

15. Wrzesinski, K.; Fey, S.J. After trypsinisation, 3D spheroids of C3A hepatocytes need 18 days to re-establish similar levels of key physiological functions to those seen in the liver. *Toxicol. Res.* **2013**, *2*, 123–135. [CrossRef]
16. Waldherr, M.; Lundt, N.; Klaas, M.; Betzold, S.; Wurdack, M.; Baumann, V.; Estrecho, E.; Nalitov, A.; Cherotchenko, E.; Cai, H.; et al. Observation of bosonic condensation in a hybrid monolayer MoSe2-GaAs microcavity. *Nat. Commun.* **2018**, *9*, 1–6. [CrossRef] [PubMed]
17. Duret, C.; Gerbal-Chaloin, S.; Ramos, J.; Fabre, J.-M.; Jacquet, E.; Navarro, F.; Blanc, P.; Cunha, A.S.; Maurel, P.; Daujat-Chavanieu, M.; et al. Isolation, Characterization, and Differentiation to Hepatocyte-Like Cells of Nonparenchymal Epithelial Cells from Adult Human Liver. *STEM CELLS* **2007**, *25*, 1779–1790. [CrossRef] [PubMed]
18. Hurrell, T.; Lilley, K.S.; Cromarty, A.D. Proteomic responses of HepG2 cell monolayers and 3D spheroids to selected hepatotoxins. *Toxicol. Lett.* **2019**, *300*, 40–50. [CrossRef]
19. Bazou, D.; Coakley, W.; Hayes, A.; Jackson, S. Long-term viability and proliferation of alginate-encapsulated 3-D HepG2 aggregates formed in an ultrasound trap. *Toxicol. Vitr.* **2008**, *22*, 1321–1331. [CrossRef]
20. Wilkening, S.; Stahl, F.; Bader, A. Comparison of primary human hepatocytes and hepatoma cell line HepG2 with regard to their biotransformation properties. *Drug Metab. Dispos.* **2003**, *31*, 1035–1042. [CrossRef]
21. Nakamura, K.; Kato, N.; Aizawa, K.; Mizutani, R.; Yamauchi, J.; Tanoue, A. Expression of albumin and cytochrome P450 enzymes in HepG2 cells cultured with a nanotechnology-based culture plate with microfabricated scaffold. *J. Toxicol. Sci.* **2011**, *36*, 625–633. [CrossRef]
22. Chang, T.T.; Hughes-Fulford, M. Monolayer and Spheroid Culture of Human Liver Hepatocellular Carcinoma Cell Line Cells Demonstrate Distinct Global Gene Expression Patterns and Functional Phenotypes. *Tissue Eng. Part A* **2009**, *15*, 559–567. [CrossRef]
23. Gerets, H.H.J.; Tilmant, K.; Gerin, B.; Chanteux, H.; Depelchin, B.O.; Dhalluin, S.; Atienzar, F.A. Characterization of primary human hepatocytes, HepG2 cells, and HepaRG cells at the mRNA level and CYP activity in response to inducers and their predictivity for the detection of human hepatotoxins. *Cell Biol. Toxicol.* **2012**, *28*, 69–87. [CrossRef]
24. Hughes, B. Industry concern over EU hepatotoxicity guidance. *Nat. Rev. Drug Discov.* **2008**, *7*, 719. [CrossRef] [PubMed]
25. Bell, C.C.; Hendriks, D.F.G.; Moro, S.M.L.; Ellis, E.; Walsh, J.; Renblom, A.; Puigvert, L.F.; Dankers, A.C.A.; Jacobs, F.; Snoeys, J.; et al. Characterization of primary human hepatocyte spheroids as a model system for drug-induced liver injury, liver function and disease. *Sci. Rep.* **2016**, *6*, 25187. [CrossRef] [PubMed]
26. Ravi, M.; Paramesh, V.; Kaviya, S.; Anuradha, E.; Solomon, F.P. 3D Cell Culture Systems: Advantages and Applications. *J. Cell. Physiol.* **2015**, *230*, 16–26. [CrossRef] [PubMed]
27. Breslin, S.; O'Driscoll, L. Three-dimensional cell culture: The missing link in drug discovery. *Drug Discov. Today* **2013**, *18*, 240–249. [CrossRef] [PubMed]
28. Shah, U.-K.; Mallia, J.D.O.; Singh, N.; Chapman, K.E.; Doak, S.H.; Jenkins, G. A three-dimensional in vitro HepG2 cells liver spheroid model for genotoxicity studies. *Mutat. Res. Genet. Toxicol. Environ. Mutagen.* **2018**, *825*, 51–58. [CrossRef] [PubMed]
29. Li, C.-L.; Tian, T.; Nan, K.-J.; Zhao, N.; Guo, Y.H.; Cui, J.; Wang, J.; Zhang, W.G. Survival advantages of multicellular spheroids vs. monolayers of HepG2 cells in vitro. *Oncol. Rep.* **2008**, *20*, 1465–1471. [CrossRef]
30. Aucamp, J.; Calitz, C.; Bronkhorst, A.J.; Wrzesinski, K.; Hamman, J.; Gouws, C.; Pretorius, P.J. Cell-free DNA in a three-dimensional spheroid cell culture model: A preliminary study. *Int. J. Biochem. Cell Biol.* **2017**, *89*, 182–192. [CrossRef]
31. Ramaiahgari, S.C.; Braver, M.W.D.; Herpers, B.; Terpstra, V.; Commandeur, J.N.; Van De Water, B.; Price, L.S. A 3D in vitro model of differentiated HepG2 cell spheroids with improved liver-like properties for repeated dose high-throughput toxicity studies. *Arch. Toxicol.* **2014**, *88*, 1083–1095. [CrossRef]
32. Elje, E.; Hesler, M.; Rundén-Pran, E.; Mann, P.; Mariussen, E.; Wagner, S.; Dusinska, M.; Kohl, Y. The comet assay applied to HepG2 liver spheroids. *Mutat. Res. Genet. Toxicol. Environ. Mutagen.* **2019**, *845*, 403033. [CrossRef]
33. Loessner, D.; Stok, K.S.; Lutolf, M.P.; Hutmacher, D.W.; Clements, J.A.; Rizzi, S.C. Bioengineered 3D platform to explore cell–ECM interactions and drug resistance of epithelial ovarian cancer cells. *Biomaterials* **2010**, *31*, 8494–8506. [CrossRef]
34. Wrzesinski, K. From 2D to 3D—A New Dimension for Modelling the Effect of Natural Products on Human Tissue. *Curr. Pharm. Des.* **2015**, *21*, 5605–5616. [CrossRef] [PubMed]

35. Štampar, M.; Tomc, J.; Filipič, M.; Žegura, B. Development of in vitro 3D cell model from hepatocellular carcinoma (HepG2) cell line and its application for genotoxicity testing. *Arch. Toxicol.* **2019**, *93*, 3321–3333. [CrossRef] [PubMed]
36. Mandon, M.; Huet, S.; Dubreil, E.; Fessard, V.; Le Hégarat, L. Three-dimensional HepaRG spheroids as a liver model to study human genotoxicity in vitro with the single cell gel electrophoresis assay. *Sci. Rep.* **2019**, *9*, 10548. [CrossRef] [PubMed]
37. Hercog, K.; Štampar, M.; Štern, A.; Filipič, M.; Žegura, B. Application of advanced HepG2 3D cell model for studying genotoxic activity of cyanobacterial toxin cylindrospermopsin. *Environ. Pollut.* **2020**, *265*, 114965. [CrossRef]
38. Eilenberger, C.; Rothbauer, M.; Ehmoser, E.-K.; Ertl, P.; Küpcü, S. Effect of Spheroidal Age on Sorafenib Diffusivity and Toxicity in a 3D HepG2 Spheroid Model. *Sci. Rep.* **2019**, *9*, 4863. [CrossRef]
39. Vinken, M.; Hengstler, J.G. Characterization of hepatocyte-based in vitro systems for reliable toxicity testing. *Arch. Toxicol.* **2018**, *92*, 2981–2986. [CrossRef]
40. Gong, X.; Lin, C.; Cheng, J.; Su, J.; Zhao, H.; Liu, T.; Wen, X.; Zhao, P. Generation of Multicellular Tumor Spheroids with Microwell-Based Agarose Scaffolds for Drug Testing. *PLoS ONE* **2015**, *10*, e0130348. [CrossRef]
41. Breznik, B.; Motaln, H.; Vittori, M.; Rotter, A.; Turnšek, T.L. Mesenchymal stem cells differentially affect the invasion of distinct glioblastoma cell lines. *Oncotarget* **2017**, *8*, 25482–25499. [CrossRef]
42. Hercog, K.; Maisanaba, S.; Filipič, M.; Sollner-Dolenc, M.; Kač, L.; Zegura, B. Genotoxic activity of bisphenol A and its analogues bisphenol S, bisphenol F and bisphenol AF and their mixtures in human hepatocellular carcinoma (HepG2) cells. *Sci. Total Environ.* **2019**, *687*, 267–276. [CrossRef]
43. Baebler, Š.; Svalina, M.; Petek, M.; Stare, K.; Rotter, A.; Pompe-Novak, M.; Gruden, K. quantGenius: Implementation of a decision support system for qPCR-based gene quantification. *BMC Bioinform.* **2017**, *18*, 276. [CrossRef]
44. Witte, R.P.; Kao, W.J. Keratinocyte-fibroblast paracrine interaction: The effects of substrate and culture condition. *Biomaterials* **2005**, *26*, 3673–3682. [CrossRef] [PubMed]
45. Bokhari, M.; Carnachan, R.J.; Cameron, N.R.; Przyborski, S.A. Culture of HepG2 liver cells on three dimensional polystyrene scaffolds enhances cell structure and function during toxicological challenge. *J. Anat.* **2007**, *211*, 567–576. [CrossRef] [PubMed]
46. Lee, C.-M.; Pohl, J.; Morgan, E.T. Dual Mechanisms of CYP3A Protein Regulation by Proinflammatory Cytokine Stimulation in Primary Hepatocyte Cultures. *Drug Metab. Dispos.* **2009**, *37*, 865–872. [CrossRef] [PubMed]
47. Nath, S.; Devi, G.R. Three-dimensional culture systems in cancer research: Focus on tumor spheroid model. *Pharmacol. Ther.* **2016**, *163*, 94–108. [CrossRef] [PubMed]
48. Mehta, G.; Hsiao, A.Y.; Ingram, M.; Luker, G.D.; Takayama, S. Opportunities and challenges for use of tumor spheroids as models to test drug delivery and efficacy. *J. Control. Release* **2012**, *164*, 192–204. [CrossRef]
49. Asthana, A.; Kisaalita, W.S. Microtissue size and hypoxia in HTS with 3D cultures. *Drug Discov. Today* **2012**, *17*, 810–817. [CrossRef]
50. Wrzesinski, K.; Rogowska-Wrzesinska, A.; Kanlaya, R.; Borkowski, K.; Schwämmle, V.; Dai, J.; Joensen, K.E.; Wojdyla, K.; Carvalho, V.B.; Fey, S.J. The Cultural Divide: Exponential Growth in Classical 2D and Metabolic Equilibrium in 3D Environments. *PLoS ONE* **2014**, *9*, e106973. [CrossRef]
51. Zanoni, M.; Piccinini, F.; Arienti, C.; Zamagni, A.; Santi, S.; Polico, R.; Bevilacqua, A.; Tesei, A. 3D tumor spheroid models for in vitro therapeutic screening: A systematic approach to enhance the biological relevance of data obtained. *Sci. Rep.* **2016**, *6*, 19103. [CrossRef]
52. Gaskell, H.; Sharma, P.; Colley, H.E.; Murdoch, C.; Williams, D.P.; Webb, S.D. Characterization of a functional C3A liver spheroid model. *Toxicol. Res.* **2016**, *5*, 1053–1065. [CrossRef]
53. Luckert, C.; Schulz, C.; Lehmann, N.; Thomas, M.; Hofmann, U.; Hammad, S.; Hengstler, J.G.; Braeuning, A.; Lampen, A.; Hessel, S. Comparative analysis of 3D culture methods on human HepG2 cells. *Arch. Toxicol.* **2017**, *91*, 393–406. [CrossRef]
54. Li, L.T.; Jiang, G.; Chen, Q.; Zheng, J.N. Ki67 is a promising molecular target in the diagnosis of cancer (Review). *Mol. Med. Rep.* **2015**, *11*, 1566–1572. [CrossRef] [PubMed]
55. Gerlach, C.; Sakkab, D.Y.; Scholzen, T.; Daßler, R.; Alison, M.R.; Gerdes, J. Ki-67 expression during rat liver regeneration after partial hepatectomy. *Hepatology* **1997**, *26*, 573–578. [CrossRef] [PubMed]

56. Wang, W.; Heideman, L.; Chung, C.S.; Pelling, J.C.; Koehler, K.J.; Birt, D.F. Cell-cycle arrest at G2/M and growth inhibition by apigenin in human colon carcinoma cell lines. *Mol. Carcinog.* **2000**, *28*, 102–110. [CrossRef]
57. Bartek, J.; Lukas, J. Mammalian G1- and S-phase checkpoints in response to DNA damage. *Curr. Opin. Cell Biol.* **2001**, *13*, 738–747. [CrossRef]
58. Alberts, B.; Johnson, A.; Lewis, J.; Raff, M.; Roberts, K.; Walter, P. Extracellular Control of Cell Division, Cell Growth, and Apoptosis. *Mol. Biol. Cell* **2002**, *91*, 40–53.
59. Dillehay, K.L.; Seibel, W.L.; Zhao, D.; Lu, S.; Dong, Z. Target validation and structure–activity analysis of a series of novel PCNA inhibitors. *Pharmacol. Res. Perspect.* **2015**, *3*, e00115. [CrossRef] [PubMed]
60. Yang, C.; Su, H.; Liao, X.; Han, C.; Yu, T.; Zhu, G.; Wang, X.; Winkler, C.A.; O'Brien, S.J.; Peng, T. Marker of proliferation Ki-67 expression is associated with transforming growth factor beta 1 and can predict the prognosis of patients with hepatic B virus-related hepatocellular carcinoma. *Cancer Manag. Res.* **2018**, *10*, 679–696. [CrossRef]
61. Zhang, D.; Liu, E.; Kang, J.; Yang, X.; Liu, H. MiR-3613-3p affects cell proliferation and cell cycle in hepatocellular carcinoma. *Oncotarget* **2017**, *8*, 93014–93028. [CrossRef]
62. Neubauer, E.; Wirtz, R.M.; Kaemmerer, D.; Athelogou, M.; Schmidt, L.; Sänger, J.; Lupp, A. Comparative evaluation of three proliferation markers, Ki-67, TOP2A, and RacGAP1, in bronchopulmonary neuroendocrine neoplasms: Issues and prospects. *Oncotarget* **2016**, *7*, 41959–41973. [CrossRef]
63. Ramaiahgari, S.C.; Waidyanatha, S.; Dixon, D.; DeVito, M.J.; Paules, R.S.; Ferguson, S.S. From the Cover: Three-Dimensional (3D) HepaRG Spheroid Model With Physiologically Relevant Xenobiotic Metabolism Competence and Hepatocyte Functionality for Liver Toxicity Screening. *Toxicol. Sci.* **2017**, *159*, 124–136. [CrossRef]
64. Duronio, R.J.; Xiong, Y. Signaling Pathways that Control Cell Proliferation. *Cold Spring Harb. Perspect. Biol.* **2013**, *5*, a008904. [CrossRef] [PubMed]
65. Snoj, N.; Dinh, P.; Bedard, P.; Sotiriou, C. Molecular Biology of Breast Cancer. In *Molecular Pathology: The Molecular Basis of Human Disease*; Elsevier Inc.: Amsterdam, The Netherlands, 2009; pp. 501–517, ISBN 9780123744197.
66. Tamura, R.E.; De Vasconcellos, J.F.; Sarkar, D.; Libermann, A.T.; Fisher, P.B.; Zerbini, L.F. GADD45 Proteins: Central Players in Tumorigenesis. *Curr. Mol. Med.* **2012**, *12*, 634–651. [CrossRef] [PubMed]
67. Hiemstra, S.; Ramaiahgari, S.C.; Wink, S.; Callegaro, G.; Coonen, M.; Meerman, J.; Jennen, D.; Van Den Nieuwendijk, K.; Dankers, A.; Snoeys, J.; et al. High-throughput confocal imaging of differentiated 3D liver-like spheroid cellular stress response reporters for identification of drug-induced liver injury liability. *Arch. Toxicol.* **2019**, *93*, 2895–2911. [CrossRef] [PubMed]
68. Nakano, K.; Vousden, K.H. PUMA, a Novel Proapoptotic Gene, Is Induced by p53. *Mol. Cell* **2001**, *7*, 683–694. [CrossRef]
69. Mathieu, J.; Zhou, W.; Xing, Y.; Sperber, H.; Ferreccio, A.; Agoston, Z.; Kuppusamy, K.T.; Moon, R.T.; Ruohola-Baker, H. Hypoxia-Inducible Factors Have Distinct and Stage-Specific Roles during Reprogramming of Human Cells to Pluripotency. *Cell Stem Cell* **2014**, *14*, 592–605. [CrossRef] [PubMed]
70. Li, M.; Li, H.; Li, C.; Wang, S.; Jiang, W.; Liu, Z.; Zhou, S.; Liu, X.; McNutt, M.A.; Li, G. Alpha-fetoprotein: A new member of intracellular signal molecules in regulation of the PI3K/AKT signaling in human hepatoma cell lines. *Int. J. Cancer* **2011**, *128*, 524–532. [CrossRef] [PubMed]
71. Friedrichs, B. Th. Peters. Jr.: All about Albumin. Biochemistry, Genetics, and Medical Applications. XX and 432 pages, numerous figures and tables. Academic Press, Inc., San Diego, California, 1996. Price: 85.00 US $. *Food/Nahrung* **1997**, *41*, 382. [CrossRef]
72. Anzenbacher, P.; Anzenbacherová, E. Cytochromes P450 and metabolism of xenobiotics. *Cell. Mol. Life Sci.* **2001**, *58*, 737–747. [CrossRef]
73. Jancova, P.; Anzenbacher, P.; Anzenbacherova, E. Phase II drug metabolizing enzymes Activity and expression of detoxification enzymes: Effect of age, obesity induction and administration of plant extracts View project microflora View project PHASE II DRUG METABOLIZING ENZYMES. *Biomed. Pap. Med. Fac. Univ. Palacky Olomouc Czech Repub.* **2010**, *154*, 103–116. [CrossRef]
74. Watkins, P.B. Role of Cytochromes P45O in Drug Metabolism and Hepatotoxicity. *Semin. Liver Dis.* **1990**, *10*, 235–250. [CrossRef]
75. Manikandan, P.; Nagini, S. Cytochrome P450 Structure, Function and Clinical Significance: A Review. *Curr. Drug Targets* **2018**, *19*, 38–54. [CrossRef] [PubMed]

76. Donato, M.; Jover, R.; Gómez-Lechón, M. Hepatic Cell Lines for Drug Hepatotoxicity Testing: Limitations and Strategies to Upgrade their Metabolic Competence by Gene Engineering. *Curr. Drug Metab.* **2013**, *14*, 946–968. [CrossRef]
77. Štampar, M.; Frandsen, H.S.; Rogowska-Wrzesinska, A.; Wrzesinski, K.; Filipič, M.; Žegura, B. Hepatocellular carcinoma (HepG2/C3A) cell-based 3D model for genotoxicity testing of chemicals. *Sci. Total. Environ.* **2020**, 143255. [CrossRef] [PubMed]
78. Takahashi, Y.; Hori, Y.; Yamamoto, T.; Urashima, T.; Ohara, Y.; Tanaka, H. 3D spheroid cultures improve the metabolic gene expression profiles of HepaRG cells. *Biosci. Rep.* **2015**, *35*, e00208. [CrossRef] [PubMed]
79. Sivertsson, L.; Ek, M.; Darnell, M.; Edebert, I.; Ingelman-Sundberg, M.; Neve, E.P.A. CYP3A4 Catalytic Activity Is Induced in Confluent Huh7 Hepatoma Cells. *Drug Metab. Dispos.* **2010**, *38*, 995–1002. [CrossRef] [PubMed]
80. Kanebratt, K.P.; Andersson, T.B. Evaluation of HepaRG Cells as an in Vitro Model for Human Drug Metabolism Studies. *Drug Metab. Dispos.* **2008**, *36*, 1444–1452. [CrossRef] [PubMed]
81. Gunness, P.; Mueller, D.; Shevchenko, V.; Heinzle, E.; Ingelman-Sundberg, M.; Noor, F. 3D Organotypic Cultures of Human HepaRG Cells: A Tool for In Vitro Toxicity Studies. *Toxicol. Sci.* **2013**, *133*, 67–78. [CrossRef]
82. Sotiropoulou, M.; Pappas, P.; Marselos, M. Effects of 3-methylcholanthrene and aspirin co-administration on ALDH3A1 in HepG2 cells. *Chem. Biol. Interact.* **2001**, *130–132*, 235–245. [CrossRef]
83. Lindahl, R. Aldehyde Dehydrogenases and Their Role in Carcinogenesis. *Crit. Rev. Biochem. Mol. Biol.* **1992**, *27*, 283–335. [CrossRef]
84. Muzio, G.; Maggiora, M.; Paiuzzi, E.; Oraldi, M.; Canuto, R. Aldehyde dehydrogenases and cell proliferation. *Free Radic. Biol. Med.* **2012**, *52*, 735–746. [CrossRef]
85. Holmes, R.S.; Hempel, J. Comparative studies of vertebrate aldehyde dehydrogenase 3: Sequences, structures, phylogeny and evolution. Evidence for a mammalian origin for the ALDH3A1 gene. *Chem. Biol. Interact.* **2011**, *191*, 113–121. [CrossRef] [PubMed]
86. Liu, H.; Cui, B.; Xu, Y.; Hu, C.; Liu, Y.; Qu, G.; Li, D.; Wu, Y.; Zhang, D.; Quan, S.; et al. Ethyl carbamate induces cell death through its effects on multiple metabolic pathways. *Chem. Biol. Interact.* **2017**, *277*, 21–32. [CrossRef] [PubMed]
87. Hahn, M.E. The aryl hydrocarbon receptor: A comparative perspective. *Comp. Biochem. Physiol. Part C Pharmacol. Toxicol. Endocrinol.* **1998**, *121*, 23–53. [CrossRef]
88. Snykers, S.; De Kock, J.; Rogiers, V.; Vanhaecke, T. In Vitro Differentiation of Embryonic and Adult Stem Cells into Hepatocytes: State of the Art. *Stem Cells* **2009**, *27*, 577–605. [CrossRef] [PubMed]
89. Gamage, N.; Barnett, A.; Hempel, N.; Duggleby, R.G.; Windmill, K.F.; Martin, J.L.; McManus, M.E. Human Sulfotransferases and Their Role in Chemical Metabolism. *Toxicol. Sci.* **2006**, *90*, 5–22. [CrossRef] [PubMed]
90. Altmann, B.; Ahrens, R.; Welle, A.; Dinglreiter, H.; Schneider, M.; Schober, A. Microstructuring of multiwell plates for three-dimensional cell culture applications by ultrasonic embossing. *Biomed. Microdevices* **2012**, *14*, 291–301. [CrossRef] [PubMed]
91. Song, N.R. Bioactivation of benzylic and allylic alcohols via sulfo-conjugation. *Chem. Biol. Interact.* **1998**, *109*, 221–235. [CrossRef]

2

A Scaffold-Free 3-D Co-Culture Mimics the Major Features of the Reverse Warburg Effect In Vitro

Florian Keller [1,2], Roman Bruch [1], Richard Schneider [3], Julia Meier-Hubberten [3], Mathias Hafner [1,2] and Rüdiger Rudolf [1,2,*]

1 Institute of Molecular and Cell Biology, Mannheim University of Applied Sciences, 68163 Mannheim, Germany; f.keller@hs-mannheim.de (F.K.); r.bruch@hs-mannheim.de (R.B.); m.hafner@hs-mannheim.de (M.H.)

2 Institute of Medical Technology, Medical Faculty Mannheim of Heidelberg University and Mannheim University of Applied Sciences, 68167 Mannheim, Germany

3 TIP Oncology, Merck Healthcare KGaA, 64289 Darmstadt, Germany; richard.schneider@merckgroup.com (R.S.); julia.meier-hubberten@merckgroup.com (J.M.-H.)

* Correspondence: r.rudolf@hs-mannheim.de

Abstract: Most tumors consume large amounts of glucose. Concepts to explain the mechanisms that mediate the achievement of this metabolic need have proposed a switch of the tumor mass to aerobic glycolysis. Depending on whether primarily tumor or stroma cells undergo such a commutation, the terms 'Warburg effect' or 'reverse Warburg effect' were coined to describe the underlying biological phenomena. However, current in vitro systems relying on 2-D culture, single cell-type spheroids, or basal-membrane extract (BME/Matrigel)-containing 3-D structures do not thoroughly reflect these processes. Here, we aimed to establish a BME/Matrigel-free 3-D microarray cancer model to recapitulate the metabolic interplay between cancer and stromal cells that allows mechanistic analyses and drug testing. Human HT-29 colon cancer and CCD-1137Sk fibroblast cells were used in mono- and co-cultures as 2-D monolayers, spheroids, and in a cell-chip format. Metabolic patterns were studied with immunofluorescence and confocal microscopy. In chip-based co-cultures, HT-29 cells showed facilitated 3-D growth and increased levels of hexokinase-2, TP53-induced glycolysis and apoptosis regulator (TIGAR), lactate dehydrogenase, and: translocase of outer mitochondrial membrane 20 (TOMM20), when compared with HT-29 mono-cultures. Fibroblasts co-cultured with HT-29 cells expressed higher levels of mono-carboxylate transporter 4, hexokinase-2, microtubule-associated proteins 1A/1B light chain 3, and ubiquitin-binding protein p62 than in fibroblast mono-cultures, in both 2-D cultures and chips. Tetramethylrhodamin-methylester (TMRM) live-cell imaging of chip co-cultures revealed a higher mitochondrial potential in cancer cells than in fibroblasts. The findings demonstrate a crosstalk between cancer cells and fibroblasts that affects cellular growth and metabolism. Chip-based 3-D co-cultures of cancer cells and fibroblasts mimicked features of the reverse Warburg effect.

Keywords: reverse Warburg effect; fibroblasts; MCT4; LC3; mitochondria; optical tissue clearing

---

1. Introduction

The human colon adenocarcinoma cell line HT-29 is frequently used as a model for colorectal cancer. Colorectal cancer is one of the most frequent malignancies in the Western world. Although annual rates declined over the last decades, in 2017, the life-time risk of disease in the USA was at 4.6% and 4.2% for men and women, respectively, with a death rate of roughly 38% [1]. Upon surgical removal of colorectal adenoma, 3% of patients develop colorectal cancer within a median follow-up of 7.7 years [2].

Thus, effective treatment options are still needed. To decrease the median development time and cost, efficient pre-clinical screening methods are being investigated by researchers and developers worldwide [3].

Like in many other malignancies, colorectal cancer also exhibits an extraordinarily high demand of energy, which is due to excessive metabolic activity and proliferation [4]. Therefore, total energy intake and macronutrient consumption serve as risk indicators [5]. Furthermore, the characteristic exuberant glucose uptake of colorectal lesions has been regularly exploited for diagnostic purposes in clinical molecular imaging by using ^{18}F-fluorodeoxyglucose as a biomarker to identify metastatic sites [6]. However, metabolic interactions within tumors are complex and involve heterogeneous tumor cell populations ranging from anabolic to catabolic forms that must be considered for the establishment of reliable cancer models [7]. Furthermore, stromal components, such as mesenchymal cells, might also participate in the metabolic interplay [8]. While the increased overall glucose consumption in a tumor indicates an altered metabolic activity with a tendency towards aerobic glycolysis [9], it is debated if all cells in a tumor undergo such a metabolic switch, or whether this occurs primarily for either tumor or stromal cells. According to the Warburg hypothesis, it is the cancer cells that do so [10]. However, this is in conflict with the observation that in situ, tumor cells often display intact mitochondria [11]. Thus, a reverse Warburg hypothesis was formulated that proposes the occurrence of a signaling from the cancer cell to the stromal fibroblast, which modulates the metabolic behavior in the fibroblast and induces its shift to aerobic glycolysis [12]. This may lead to the production of reactive oxygen species and a general change in redox status. Accordingly, extensive alterations within tumor tissue sites [13], such as angiogenesis, migration, acidification, and hypoxia, are the consequence [14].

While cancer cells eventually enter a neoplastic state through reduced restrictions of proliferative potential [15], a metabolic shift of fibroblasts can bring these cells from a suppressive to a cancer-supportive state [16]. These stromal cells are usually referred to as cancer-associated fibroblasts (CAFs) and exhibit a strongly altered expression profile [17], including overexpression of smooth muscle actin or vimentin [18] and downregulation of caveolin 1. In addition, CAFs show enhanced autophagic activity, leading to higher expression levels of microtubule-associated proteins 1A/1B light chain 3 (MAP1LC3, short LC3) and ubiquitin-binding protein p62 (P62), both markers for autophagosome formation [19], and a loss of mitochondria through mitophagy [20]. This further favors a shift to aerobic glycolysis and the accumulation of short carbon-rich metabolites like lactate, which can be shuttled to surrounding cancer cells [21]. Therefore, mono-carboxylate transporters (MCT) play an important role [22], with MCT4 and MCT1 being principal exporters and importers, respectively. While MCT4 is often upregulated in CAFs, MCT1 expression is typically abundant in tumor cells [23]. The increased lactate levels can enhance metabolic activity in oxidative cancer cells [24,25], and the metabolic plasticity of tumor cell mitochondria is essential for adequate processing of nutrients available on a non-regular basis [26]. Such metabolic interactions can be instrumental for tumorigenesis, drug resistance, and tumor relapse [24].

Current metabolic models often either originate from primary tissue biopsies [27] or include xenograft tissues [28] relying on Matrigel or basal membrane extract (BME) as the supporting material [29]. However, a major issue here is that approaches based on primary cells are time consuming, expensive, and prone to low reproducibility due to the intrinsic variability of the patient material itself. On the other hand, commonly used two-dimensional mono-cultures often lack important components of intercellular signaling, which are crucial for drug testing [30]. Therefore, Matrigel, as a naturally derived hydrogel scaffold for 3-D cell cultures that are more homogeneous compared to biopsies or xenografts, has been the gold standard for a wide range of cell types [31]. These hydrogels provide biophysical cell-adhesive properties and can contribute to cell viability and growth [32], but they also contain an ill-defined mixture of growth factors that influences cellular behavior in an unpredictable way due to batch-to-batch variances [33], potentially obliterating the examination of delicate alterations in short-term cultures.

We aimed to model the reverse Warburg effect using 2-D and 3-D co-cultures of human HT-29 colon cancer cells and human CCD-1137Sk fibroblasts. While Dynarray cell chip microarray cultures reflected major aspects of the reverse Warburg effect, 2-D cultures partially reproduced these findings but lacked effects on HT-29 cells. Conversely, spheroid co-cultures did not show enhanced growth, suggesting a lack of an efficient metabolic advantage under this condition.

2. Materials and Methods

2.1. Cell Culture

CCD-1137Sk human fibroblasts were kept in Iscove's Modified Dulbecco's Medium (Capricorn; IMDM-A; Lot #CP18-2245) supplemented with 10% fetal bovine serum (Capricorn; FBS-12B; Lot #CP16-1422), and 1% penicillin/streptomycin (Capricorn; PS-B; Lot #CP18-2207). HT-29 human colon cancer cells were cultivated in McCoy's 5a media (Capricorn; MCC-A; Lot #CP19-2689), supplemented with 10% FBS-12B and 1% PS-B. The media for MDA-MB-231 human breast cancer cells consisted of Dulbecco's Modified Eagle Medium (Capricorn; DMEM-HPA; Lot #CP18-2096) supplemented with 10% FBS-12B, 1% PS-B, and 1% Minimum Essential Medium Nonessential Amino Acids (Capricorn; NEAA-B; Lot #CP17-1726). All cell lines were passaged twice per week with a seeding density of 1×10^6 cells/T75 flask. Two-dimensional cell culture analysis was based on cover slips (12 mm; VWR; ECN 631-1577; Lot #43395 819) placed in cell culture dishes (Greiner; PS; 664 160) before seeding a total of 1.2×10^6 cells in appropriate ratios and culturing for four days.

2.2. Spheroid Formation

To analyze growth behavior, the spheroid formation was performed with a total of 6000 cells seeded per well within different mono- and co-culture ratios using 96-well cell-repellent microplates (Greiner; 650970; Lot #E2001347). The BME dependency of the spheroid metabolism was analyzed by setting 250 HT-29 or 4000 CCD-1137Sk cells in a mono-culture condition. Additionally, 250 cancer cells were co-cultivated together with 500 fibroblasts. This ensured a more homogeneous spheroid mean diameter distribution around 300 µm after four days, preventing size-dependent effects from influencing the cellular biology approach. Appropriate amounts for 2.5% or 10% BME dissolved in media were added immediately after the culture suspensions. Cell aggregation was enhanced by centrifugation for 6 min at 500 rcf.

2.3. Dynarray Preparation

For Dynarray (300MICRONS; Dynarrays MCA-C300-300-PC) seeding, the microarrays were sterilized with 100%, 70%, 50%, and 30% isopropanol for 1 min, respectively, before washing twice in sterile deionized water. Then, 100 µg/mL Type-I collagen (Roche; rat tail Collagen; 11179179001) dissolved in 150 µL of sterile deionized water was incubated for 1 h on top of the cavities at room temperature for coating the microarray. The cancer cells and fibroblast cells were seeded at appropriate ratios to result in a density of 4000 cells per cavity. The Dynarrays were placed in cell culture dishes and after 4 h at 37 °C and 5% CO_2, 10 mL of media were added. After fixation in 4% PFA/PBS, Dynarrays were either stained, cleared, and imaged as whole mounts or orthogonal Dynarray slices were made. In brief, chips were embedded in 2% agarose (Roche; Agarose MP; 11 388 983 001) dissolved in Tris-acetate-EDTA (TAE) buffer (0.04 M tris base, 0.02 M acetic acid, 0.01 M EDTA) and 50-µm-thick slices were collected using a vibratome (Leica Biosystems; Vibratome Model VT1000S). To label the cells with a CellTracker, mono-type suspensions of either HT-29 or CCD-1137Sk cells were incubated for 1 h in fetal bovine serum (FBS)-free media (37 °C/5% CO_2) and washed with PBS directly before co-seeding. For live-cell analysis, Dynarray cultures were incubated in Tetramethylrhodamin-methylester (TMRM)-containing complete media (30 min/37 °C), washed 3 times with PBS, and transferred into 25 mM -(2-hydroxyethyl)-1-piperazineethanesulfonic acid (HEPES) buffer to ensure steady pH conditions throughout the imaging process.

2.4. Antibodies and Dyes

Primary antibody incubation was performed in a static incubation (ON/4 °C) for 2-D and on a roller mixer (ON/37 °C) for 3-D samples. Secondary antibodies were used for 1 h at room temperature for flat and as the primary antibodies for spheroid and Dynarray preparation. Antibody or dye details and concentrations were as follows: Anti-Carcino Embyonic Antigen CD66e (CEA; Thermofisher scientific; MIC0101; Lot #VD2969991; dilution 1:250), Anti-Type-I Collagen (Collagen 1; Rockland; 600-401-103-0.5; Lot #41250; dilution 1:100), Anti-Type-IV Collagen (Coll4; Rockland; 600-401-106-0.5; Lot #40995; dilution 1:100), CellTracker Deep Red (CTdr; Lifetechnologies; C34565; Lot #1781143; stock 2 mM; dilution 1:1000), CellTracker Green (CTg; Lifetechnologies; C7025; Lot #1913919; stock 10 mM; dilution 1:1000), 4′,6-Diamidino-2-Phenylindole (DAPI; Roche; 10236276001; Lot #28114320; 1 mg/mL; dilution 1:1000), Anti-Hexokinase II (HK-2; abcam; ab209847; Lot #GR3220265-2; dilution 1:100), Anti- Microtubule-associated proteins 1A/1B Light Chain 3 (LC3; CellSignalling; 3868S; Lot #11; dilution 1:200), Anti-Lactate dehydrogenase (LDH; abcam; ab47010; Lot #GR3307117-1; dilution 1:1000), Anti-Monocarboxylate Transporter 1 (MCT1; Sigma-Aldrich; Anti-SLC16A1; Lot #C114705; dilution 1:500), Anti-Monocarboxylate Transporter 4 (MCT4; SantaCruz Biotechnology; sc-376140; Lot #D3018; dilution 1:400), Anti-Ubiquitin-binding protein p62 protein (P62; Progen; GP62-C; Lot #703241-03; dilution 1:100), Anti-Succinate Dehydrogenase (SDH; Thermofisher scientific; 459200; VC296863; dilution 1:200), Anti-TP53-induced glycolysis and apoptosis regulator (TIGAR; abcam; ab37910; Lot #GR3210648-2; dilution 1:500); Anti-Tetramethylrhodamine (TMRM; Invitrogen; T-668; 500 μM; dilution 1:2000), Anti-Translocase of outer mitochondrial membrane 20 (TOMM20; Sigma-Aldrich; HPA011562-100UL; Lot #4450; dilution 1:400), Goat α-mouse AF 488 (M; Invitrogen; A11001; Lot #1834337; dilution 1:1000), Donkey α-rabbit AF 488 (R; Invitrogen; A21206; Lot #176375; dilution 1:1000), Goat α-guinea pig AF 555 (GP; Invitrogen; A21435; Lot #2015338; dilution 1:1000), and Goat α-rabbit AF 647 (R; Invitrogen; A21246; Lot #55002A; dilution 1:1000).

2.5. Data Acquisition

Brightfield (Zeiss Axiovert 25/objective CP-ACHROMAT, 5x/0.12Ph0) and confocal (Leica SP8, objectives HC PL FLUOTAR 10 × /0.30, HC PL APO 20 × /0.75 IMM CORR and HC PL APO CS2 63 × /1.2 W CORR) microscopy and the evaluation of spheroid size [34] were conducted as published earlier [35]. For confocal analysis, pictures with a resolution of 1024 × 1024 pixels w×ere taken with 1-μm z-steps for 3-D stacks. Whole-mount staining was enhanced by a clearing procedure based on the glycerol clearing protocol as published [36]. Briefly, samples were immersed in an aqueous glycerol solution with a refractive index of 1.459 containing 88% of glycerol (ON/37 °C) on a roller mixer. Microscopy was performed in the same solution. Therefore, samples were fixed in 4% PFA/PBS, treated with 2% Triton X-100 (2 min/RT), incubated in penetration buffer (0.2% Triton X-100, 0.3 M glycine, 20% DMSO in PBS/30 min/37 °C), blocked with bovine albumin serum (1% BSA, 0.2% Triton X-100, 10% DMSO in PBS/2 h/37 °C), and stained with antibodies incubated in appropriate buffer (0.2% Tween 20, 10 μg/mL heparin, 1% BSA, 5% DMSO in PBS/ON/37 °C). For intermediate washing steps, a buffer containing 0.2% Tween 20 and 10 μg/mL heparin in PBS was used.

2.6. Data Analysis

Two-dimensional image analysis and assembly were done with ImageJ software (v1.48v) and based on signal intensities in split channels. In 2-D, regions of interest were set based on: 4′,6-diamidino-2-phenylindol (DAPI) and the carcinoembryonic antigen (CEA) channel threshold.

For 3-D image analysis and corresponding data visualization, a dedicated Python script was established. Each stack was divided into a set of small non-overlapping regions with x, y, and z dimensions of 32, 32, and 16 voxels, respectively. To minimize the influence of background signals, the images were first segmented based on their nuclei channel. Therefore, the DAPI signal of an image volume was smoothed using three-dimensional Gaussian blur and then segmented using a threshold value calculated by Otsu's method [37]. To also include adjacent regions outside the nuclei, high sigma values of 4, 4, and 0 for the x, y, and z dimension were used for Gaussian smoothing. An exemplary sum-projection of a segmentation including the visualization of the regions' borders can be seen in Figure 6C. Only regions with more than 50% foreground based on segmentation were included. For the remaining regions, the average foreground intensity of the voxels was calculated for DAPI, CEA, and the corresponding marker channel. For visualization, composite results of 5 cavities for each culture type were assembled.

Statistics were calculated with GraphPad Prism 7. For direct comparison of two data points within the immunofluorescence evaluation, significance was defined based on the *p*-values (* $p \leq 0.05$; ** $p \leq 0.01$; *** $p \leq 0.001$; **** $p \leq 0.0001$) in a multiple comparison t-test analysis without assumptions. Normal distribution and homoscedasticity were tested with the Kolmogorov–Smirnov procedure. For testing of co-localization, the Coloc2 algorithm of ImageJ was used with a PSF of 3.0 and 10 randomizations. The Pearson's correlation (R) above threshold was evaluated. Furthermore, one-way ANOVA with Holm–Sidak multiple comparison was performed for the growth curve comparisons. Significance was defined based on *p*-values (* $p \leq 0.05$; ** $p \leq 0.01$; *** $p \leq 0.001$; **** $p \leq 0.0001$).

3. Results

3.1. MCT4 as well as Markers for Glycolysis and Autophagy are Upregulated in CCD-1337Sk Fibroblasts upon Co-Culture with HT-29 Cells

First, we investigated whether co-cultures of HT-29 cells and CCD-1137Sk fibroblasts exhibited an altered expression of lactate transporters as compared to the respective mono-cultures. Therefore, 2-D mono- and co-cultures were set up and cultured at a confluency of up to 80% for four days. Then, immunofluorescence staining was first done for MCT1 and MCT4, as markers for lactate influx and efflux, respectively. While both MCT1 and MCT4 were expressed more strongly in HT-29 tumor cells than in fibroblasts, only MCT4 increased significantly in the fibroblasts upon co-cultivation (Figure 1A–C). Conversely, MCT1 remained low in fibroblast cells under all conditions (Figure 1A–C). Discrimination between HT-29 and CCD-1137Sk cells was done on the basis of three criteria. First, HT-29 consistently grew in dense islets, both in mono- and co-culture, whereas CCD-1137Sk typically showed a spindle-shaped morphology and grew in between the HT-29 islets in the co-cultures. Second, the molecular markers, carcinogen embryonic antigen (CEA) [38] and collagen 4 (Coll4) [39], were primarily expressed in either HT-29 or CCD-1137Sk cells, respectively (Figure S1). These served as additional confirming features for the cell-type selection. Finally, DAPI staining of CCD-1137Sk cell nuclei was mostly darker than that of HT-29 cells and showed a more elongated and larger area. This trait was also used for the later analyses of the 3-D data sets.

Next, the effects of the co-culturing of HT-29 and CCD-1137Sk cells on their metabolic profiles were addressed. Therefore, immunofluorescence staining of the same cultures as those mentioned in Figure 1 was done for hexokinase-2 (HK-2), lactate dehydrogenase (LDH), TP53-induced glycolysis and apoptosis regulator (TIGAR), succinate dehydrogenase (SDH), and translocase of outer mitochondrial membrane 20 (TOMM20), as markers for glucose breakdown, pyruvate-lactate metabolism, negative glycolysis regulation, oxidative phosphorylation, and mitochondrial content, respectively (Figure 2A–C) [4].

While HT-29 cells did not show any significant change in any of these markers, CCD-1137Sk cells displayed altered expression levels consistent with an upregulation of glycolysis and a downregulation of oxidative phosphorylation. Indeed, HK-2 went up, whereas TIGAR and TOMM20 decreased. LDH and SDH remained unaltered under these conditions.

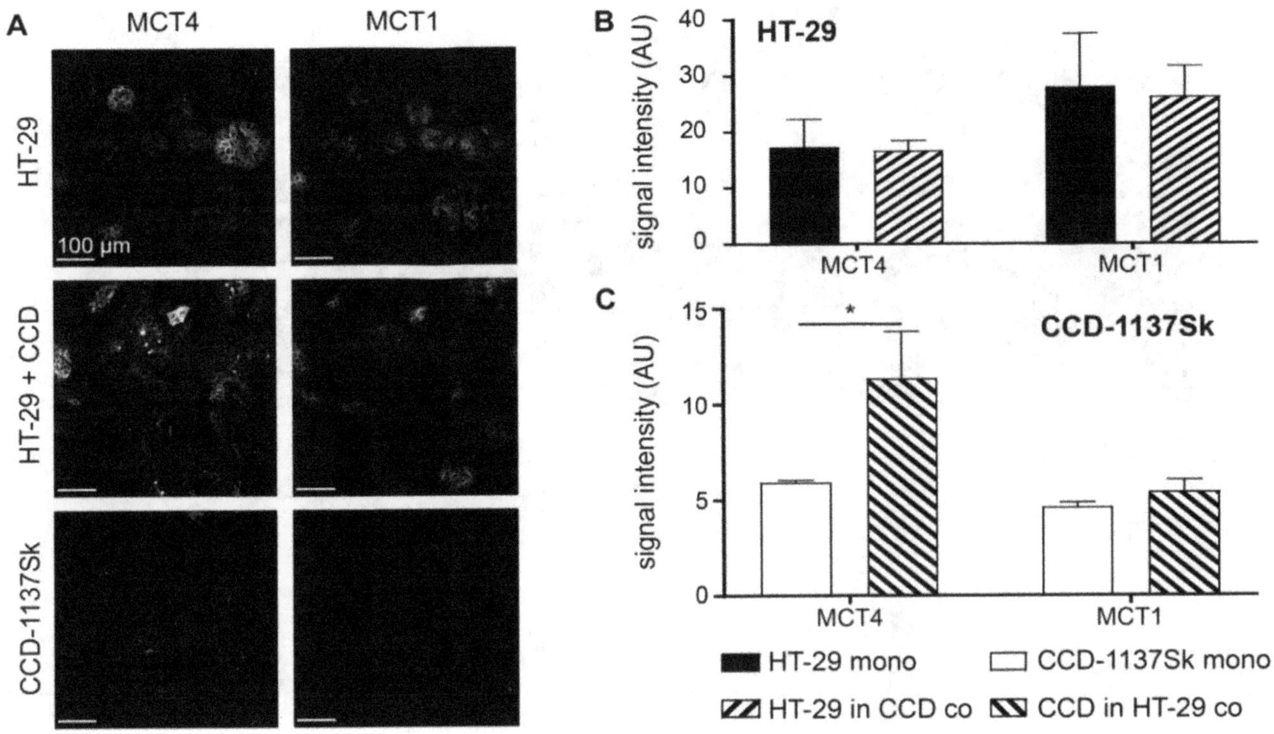

Figure 1. Monolayer co-cultures of HT-29 and CCD-1137Sk show enhanced expression of mono-carboxylate transporters, MCT4 in fibroblasts. HT-29 and CCD-1137Sk cells were either seeded alone or in co-culture and grown to a sub-confluent state for four days. Then, cells were fixed and stained with diamidino-2-phenylindol (DAPI) as well as antibodies against MCT4 and MCT1 as markers for nuclei, lactate export, and lactate import, respectively (**A**–**C**). (**A**) Representative confocal images of fluorescence staining for markers and cultures as indicated. (**B**) and (**C**) Graphs show quantitative analysis of the fluorescence intensity values for markers and cell type as indicated. Mean + SEM (n = 3 experiments; * $p < 0.05$).

The data on lactate transporters and metabolic markers were consistent with a co-culture-induced switch of the CCD-1137Sk cells towards a catabolic phenotype. To further corroborate this possibility, the expression of markers for autophagy, namely LC3 and P62, were assessed under mono- and co-culture conditions. As shown in Figure 3, both proteins were upregulated in CCD-113Sk in co-culture with HT-29 cells. Conversely, no alteration of LC3 or P62 levels was found in HT-29 cells upon co-culturing with fibroblasts.

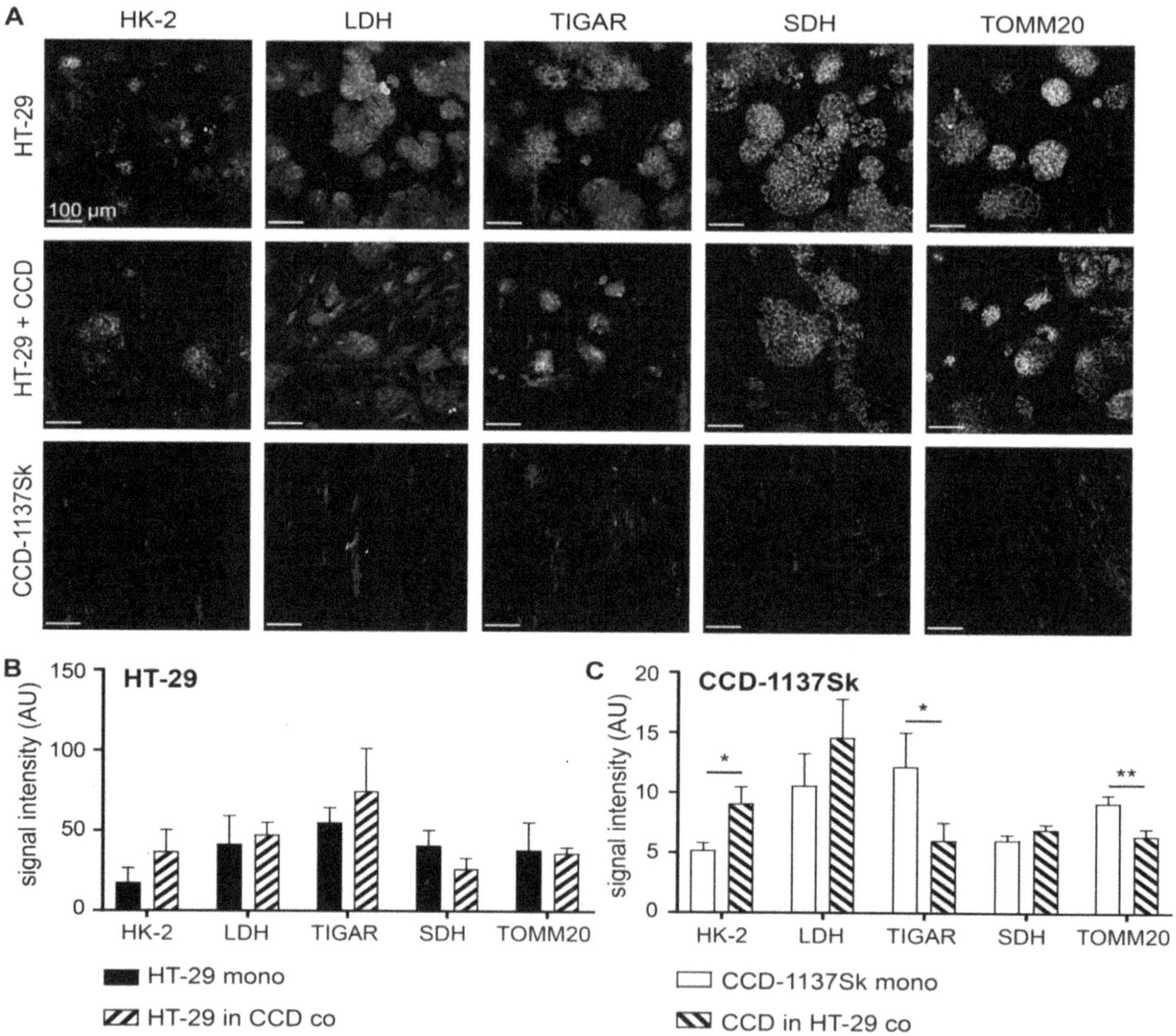

Figure 2. Fibroblasts in monolayer co-cultures of HT-29 and CCD-1137Sk show enhanced glycolytic and reduced oxidative phosphorylation markers. HT-29 and CCD-1137Sk cells were either seeded alone or in co-culture and grown to a sub-confluent state for four days. Then, cells were fixed and stained with DAPI as well as antibodies against HK-2, lactate dehydrogenase (LDH), TP53-induced glycolysis and apoptosis regulator (TIGAR), succinate dehydrogenase (SDH), and: translocase of outer mitochondrial membrane 20 (TOMM20) as markers for nuclei, glucose breakdown, pyruvate-lactate metabolism, negative glycolysis regulation, oxidative phosphorylation, and mitochondrial content, respectively (**A–C**). (**A**) Representative confocal images of fluorescence staining for markers and cultures as indicated. B and (**C**) Graphs show quantitative analysis of fluorescence intensity values for markers and cell types as indicated. Mean + SEM ($n = 3$ experiments; * $p < 0.05$, ** $p < 0.01$).

Finally, to investigate whether the observed modifications of fibroblast metabolism were exclusively related to the co-culture with HT-29 cells, analogous experiments were carried out with the human breast cancer cell line MDA-MB-231 in mono- and co-culture with CCD-1137Sk fibroblasts. As shown in Supplementary Material Figures S2–S4, the overall results were essentially the same: Upon co-culture with MDA-MB-231 cells, CCD-1137Sk cells showed enhanced levels of MCT4, HK-2, LC3, and P62 and reduced presence of TIGAR and TOMM20. In contrast, the breast cancer cells did not show any such alteration.

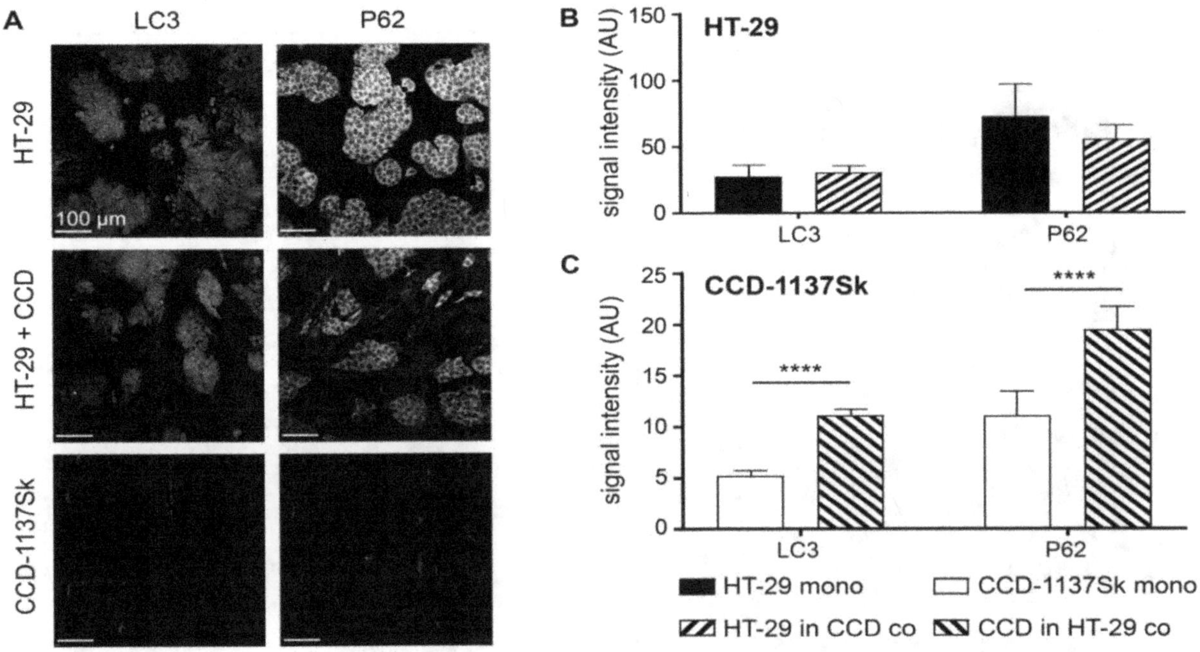

Figure 3. Fibroblasts in monolayer co-cultures of HT-29 and CCD-1137Sk display enhanced levels of autophagy markers. HT-29 and CCD-1137Sk cells were either seeded alone or in co-culture and grown to a sub-confluent state for four days. Then, cells were fixed and stained with DAPI as well as antibodies against LC3 and P62 as markers for nuclei and autophagy (**A–C**). (**A**) Representative confocal images of fluorescence staining for markers and cultures as indicated. (**B**) and (**C**) Graphs showing quantitative analysis of fluorescence intensity values for markers and cell type as indicated. Mean + SEM ($n = 3$ experiments; **** $p < 0.0001$).

3.2. Addition of CCD-1137Sk Fibroblasts does not Alter Growth of HT-29 Spheroids

Although 2-D cultures are ideal for quantitative analyses, they often lack gradients of drugs, oxygen, waste products, and nutrients as they are present in vivo. Therefore, we next established three-dimensional spheroid co-cultures with CCD-1137Sk fibroblasts and HT-29 colon cancer cells. To test the best conditions for a homogeneous distribution of both cell types in spheroids, different ratios of cancer cells and fibroblasts were either seeded together or sequentially in ultra-low attachment plates and allowed to grow for up to 30 days. Every seeding procedure resulted in rather round spheroids. Since the condensation of the first cell population into a spheroid was typically accomplished within two days (Figure 4A), the secondary seeding in the sequential protocols was performed on that day. Size measurements of the mean diameters revealed that neither the different tested cell-line ratios, nor seeding sequences altered the assembly or growth of spheroids (Figure 4B). Indeed, all spheroids with cancer cells alone or cancer cells plus fibroblasts showed a very similar behavior: After an initial condensation phase, which led to a minimal mean diameter of spheroids at two to three days, they followed an asymptotic growth curve and reached a maximal spheroid diameter of approximately 1000 µm (Figure 4B). Thus, CCD-1137Sk fibroblasts did not contribute to any major change in the growth of HT-29 spheroids. Accordingly, fibroblast mono-culture spheroids remained small and did not increase in size over time (Figure 4B). An examination of the distribution of Coll4 and CEA proteins on optical slices from co-culture spheroids suggested that fibroblasts remained rather confined to small islets (Figure 4C, arrowheads). Thus, it appeared that there was limited cross-talk between both cell types, suggesting that the spheroid model was not ideal to study mutual interactions between HT-29 cancer cells and CCD-1137Sk fibroblasts.

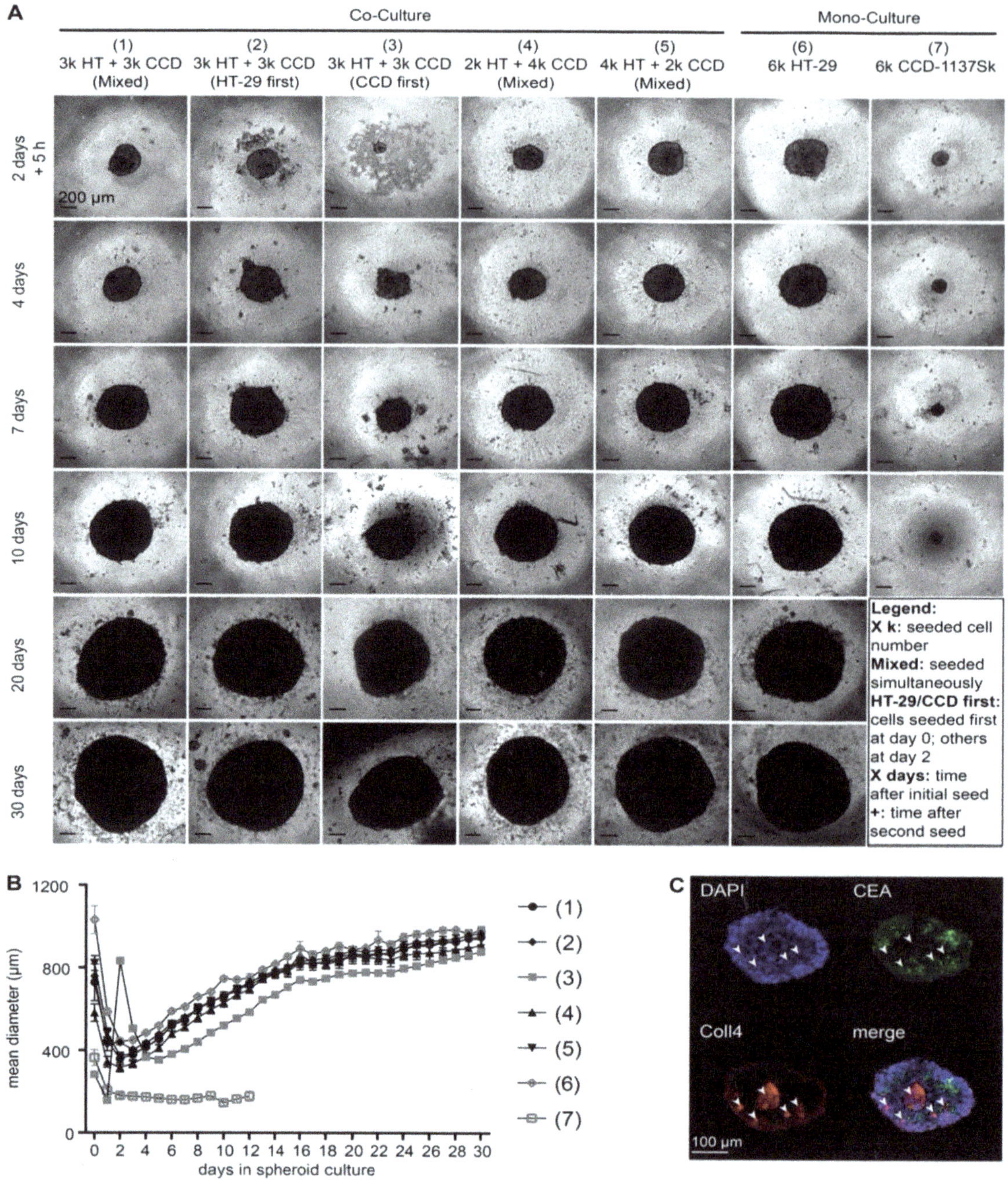

Figure 4. Growth of HT-29 spheroids is unaffected in the presence of CCD-1137Sk cells. Human HT-29 or CCD-1137Sk cells were seeded into ultra-low attachment plates to form spheroids, following different seeding orders (mixed, HT-29 first or CCD-1137Sk first) and cell numbers (ranging from 2000 to 6000 cells per type) as indicated. From 0 to 30 days in culture, brightfield images were taken to perform spheroid-size analysis based on the mean diameter (**A–B**). Additionally, co-culture spheroids with 250 HT-29 cells together with simultaneously seeded 500 CCD-1137Sk cells were fixed after four days and stained with DAPI as well as with anti-CEA and anti-Coll4 antibodies as markers for nuclei, HT-29 cells, and fibroblasts, respectively (**C**). (**A**) Representative brightfield images of HT-29 and CCD-1137Sk mono- and co-culture spheroids seeded in different ratios and orders, as depicted. (**B**) Graph shows a quantitative analysis of the mean diameters of mono- and co-culture spheroids at different time-points as indicated. Mean + SEM (n = 5–9 experiments). (**C**) Representative optical slice of DAPI (blue), anti-CEA (green), and anti-Coll4 signals (red). Arrowheads, exemplary clusters of Coll4-positive areas. Bottom right panel shows an overlay image of all three channels.

3.3. Growth of CCD-1137Sk and HT-29 Co-Cultures is Enhanced in Dynarray Chips

Reasoning that the spherical arrangement with its observed strict spatial separation between HT-29 and CCD-1137Sk cells might have contributed to the lack of an effect of co-culturing on spheroid growth, we switched to chip-based 3-D Dynarrays. These have cavities of 300 μm in diameter and depth and their inner surfaces can be coated with extracellular-matrix components or gels. Furthermore, the walls of the cavities are porous, allowing the passage of nutrients. HT-29 and CCD-1137Sk cells were either seeded as mono- or co-cultures into collagen-1-coated Dynarray chips and cellular growth was assessed. In all cases, 4000 cells were seeded per cavity, i.e., either 4000 HT-29 or CCD-1137Sk cells in the mono-cultures or 2000 cells of each type for the co-cultures. In this format, the fibroblast marker collagen 1 was widely distributed throughout the co-cultures (Figure 5A), suggesting a better mixing of both cell types than in spheroids. MCT4 fluorescence signals were comparable in both HT-29 and CCD1137-Sk cell areas of the co-culture chips (Figure 5A). Furthermore, the growth behavior of HT-29 cells in Dynarray chips was influenced by the presence or absence of fibroblasts. While HT-29 cells adhered in the mono-culture to the inner chip walls and grew in only few flat layers (Figure 5B), co-cultures with fibroblasts filled out the entire chip cavities and even often grew beyond (Figure 5B). Total cell counts increased in co-cultures by 55% as compared to HT-29 mono-cultures after four days in culture (Figure 5C). This demonstrates that the growth behavior of HT-29 and/or CCD-1137Sk cells can be modulated by 3-D co-culturing and that this is dependent on the type of 3-D culture system.

3.4. Fluorescence Staining of CEA, Coll4, and DAPI Shows Different Intensities in HT-29 and CCD-1137Sk Cells

To assess the potential effects of co-culturing on metabolic markers in Dynarray cultures of HT-29 and CCD-1137Sk cells, we sought cell-type-specific features that could later serve as discriminators. Thus, in analogy to the 2-D cultures, chips were stained with DAPI, anti-CEA, and anti-Coll4, and were then visualized by 3-D confocal microscopy. This revealed a high density of nuclei (Figure 6A), impairing an instance segmentation of individual cells. Yet, a correlative quantitative voxel analysis could be implemented to determine the relative staining intensities. Therefore, 3-D image stacks were subdivided into voxel cubes, as shown in Figure 6C, and the average fluorescence intensity per cube was determined after nuclei segmentation and background subtraction. This showed the following principal features. First, in mono-cultures, HT-29 cells showed brighter CEA and darker Coll4 signals than CCD-1137Sk cells (Figure 6B). In this and all following graphs, a red vertical line marks the CEA brightness range that contains 95% of all CCD-1137Sk voxel cubes in mono-culture. Second, also in mono-cultures, DAPI fluorescence was much brighter in HT-29 cells than in CCD-1137Sk cells (Figure 6B, color code). Third, in co-cultures, the Coll4 vs. CEA distribution looked more similar to the HT-29 mono-culture than to the CCD-1137Sk mono-culture distribution (Figure 6B), suggesting that most cells in co-cultures were HT-29 cells. However, a limited number of signals in the lower left quadrant of the co-culture distribution showed a combination of low DAPI and CEA signals on the one hand and higher Coll4 signals on the other hand, suggesting that these could indicate the fibroblast population (Figure 6B). Altogether, a picture emerged that using this kind of voxel-based analysis, the two cell populations could be differentially assessed with respect to their marker expression, at least under the assumption that in co-cultures the marker expression did not alter in a confounding manner. However, it also appeared that either the fibroblast population in the co-cultures was rather small and/or that signals from HT-29 cells were preferentially detected in co-cultures using this approach. In the following, we used CEA and DAPI staining as discriminators to address the expression of metabolic markers in the two cell types.

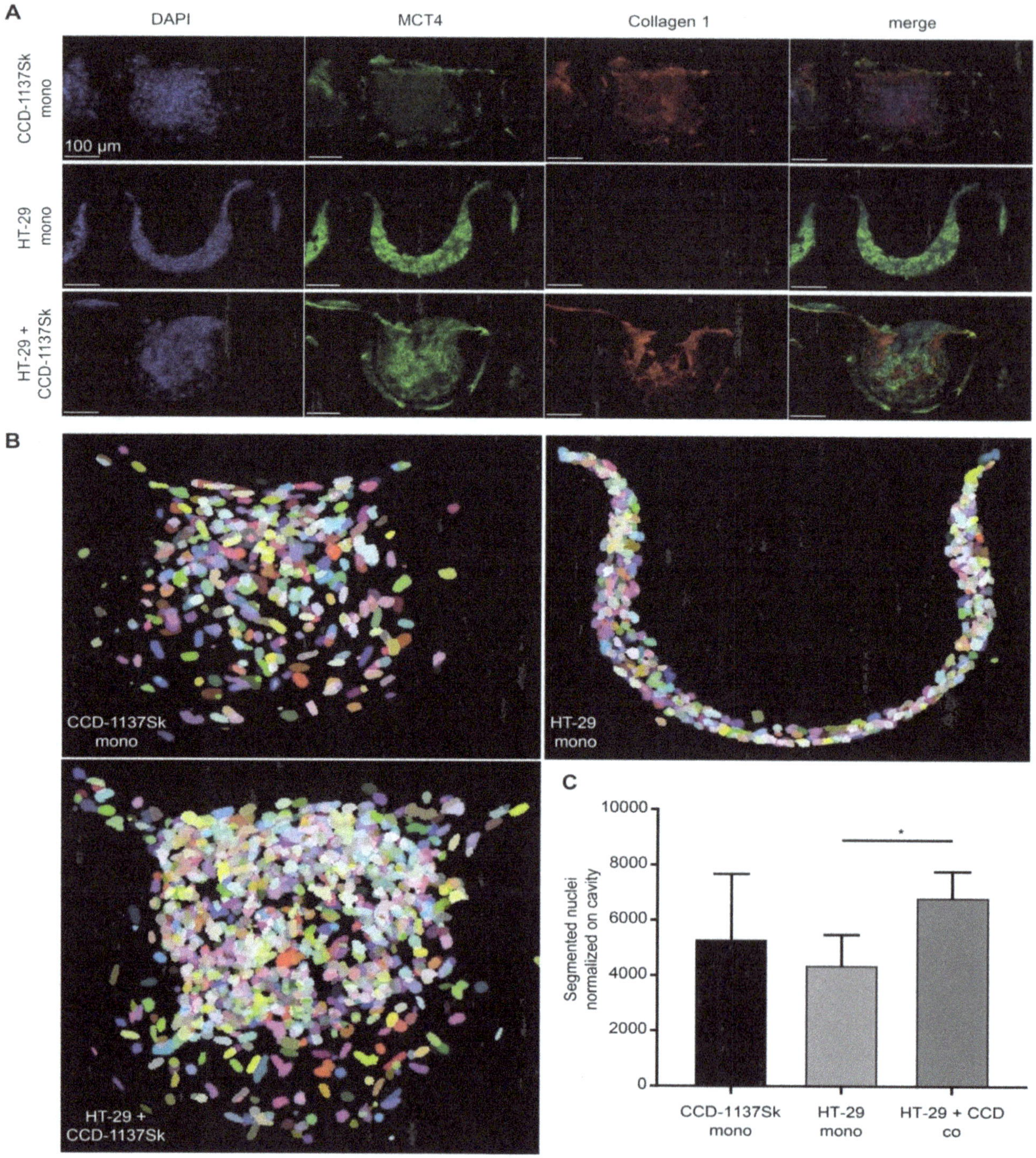

Figure 5. Chip-based 3-D cultures reveal morphological and metabolic changes upon co-cultivation of HT-29 and CCD-1137Sk cells. A total of 4000 cells per chip cavity were seeded to yield either mono- or co-cultures of HT-29 and CCD-1137Sk cells. In the case of co-cultures, 2000 HT-29 cells were co-seeded with 2000 CCD-1137Sk cells per well. After four days, chips were fixed, transversally sectioned, and then stained with DAPI as well as antibodies against MCT4 and collagen 1 as markers for nuclei, lactate export, and extracellular matrix (ECM), respectively. Subsequently, sections were imaged with confocal microscopy. (**A**) Representative maximum-z projections. (**B**) Representative display of segmented nuclei from different culture conditions as indicated. Pseudo-colors were chosen to discriminate between individual nuclei. They do not indicate different cell types. (**C**) Quantitative analysis of nuclear counts. Mean + SEM (n = 3 experiments).

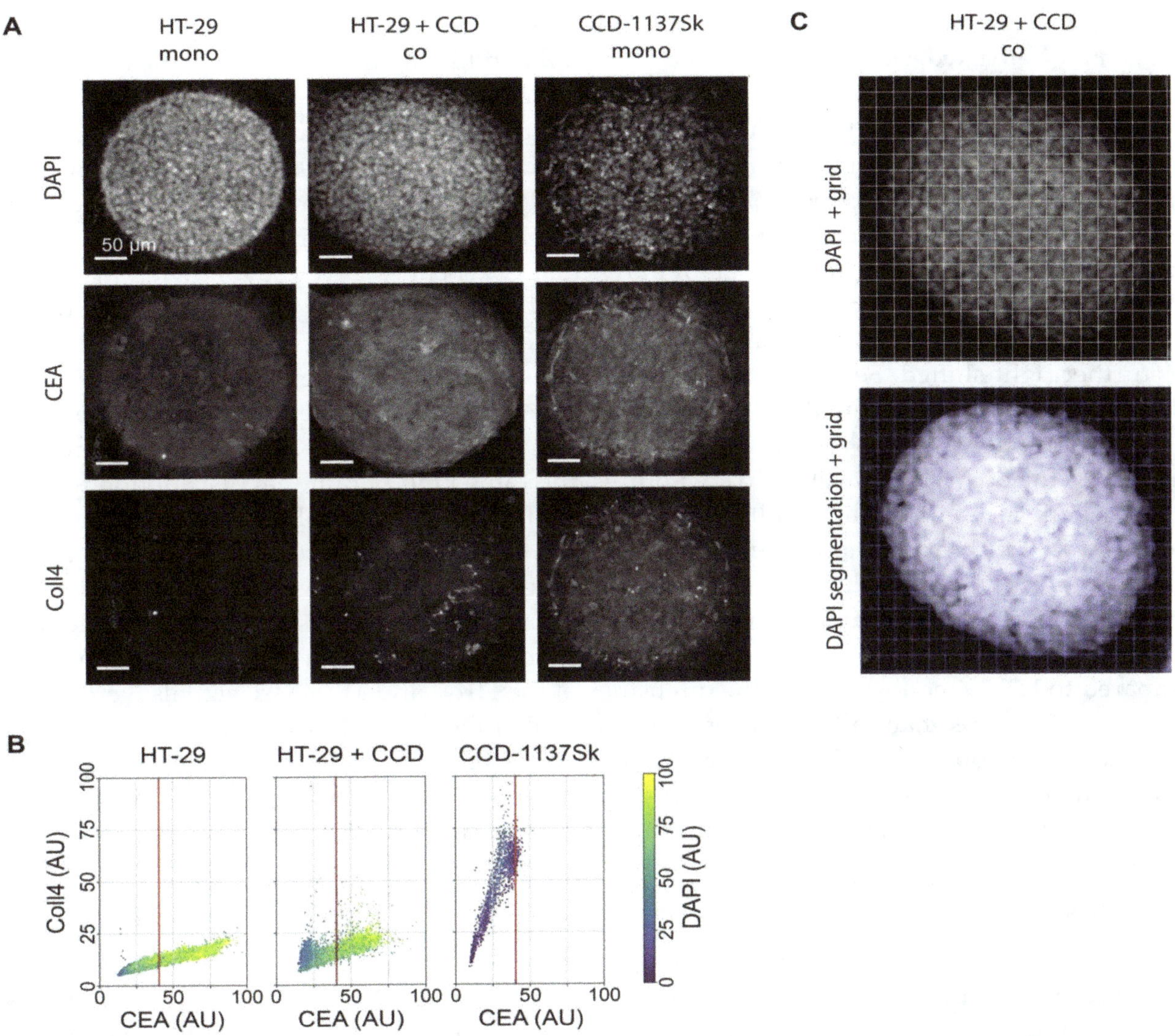

Figure 6. DAPI, CEA, and Coll4 staining jointly discriminate HT-29 and CCD-1137Sk cell populations in 3-D Dynarray co-cultures. A total of 4000 cells per chip cavity were seeded to yield either mono- or co-cultures of HT-29 and CCD-1137Sk cells. In case of co-cultures, 2000 HT-29 cells were co-seeded with 2000 CCD-1137Sk cells per well. After four days, chips were fixed, cleared, and stained with DAPI and antibodies against CEA and Coll4 as markers for nuclei, cancer cells, and fibroblasts, respectively. Subsequently, samples were imaged with 3-Dconfocal microscopy. (**A**) Representative sum-z projections. (**B**) Scatter plots showing the anti-Coll4 intensity distribution as a function of the anti-CEA intensity for HT-29 mono-cultures (left), HT-29 + CCD-1137Sk co-cultures (middle), and CCD-1137Sk mono-cultures (right). Each dot shows the average intensity of voxel cubes in a total of five Dynarray cavities. Red vertical line, CEA brightness range that contains 95% of all CCD-1137Sk voxel cubes in mono-culture. (**C**) Exemplary images showing the voxel-cube grid used for the quantification of the average fluorescence intensities. Upper panel, raw sum-z projection of DAPI signal; lower panel, sum-z projection upon nuclei segmentation and background subtraction.

3.5. Fluorescence Signals of Metabolic Marker Proteins Show Differential Regulation in Dynarray Co-Cultures

Given that the voxel-cube analysis was promising to distinguish relative marker-protein expression levels in 3-D cultures, we wanted to assess a potential metabolic cross-talk between HT-29 and CCD-1137Sk cells in chips. Therefore, Dynarrays were seeded with mono-cultures of HT-29 or

CCD-1137Sk cells or with co-cultures of both cell lines as described before and grown for four days. Then, chips were fixed, cleared, and stained with DAPI and anti-CEA to segregate the cell types from each other. In addition, the samples were labeled with antibodies against metabolic marker proteins HK-2, LDH, TIGAR, or TOMM20. Three-dimensional confocal microscopy of stained specimens was done, and image stacks were quantitatively analyzed using the voxel-cube approach. As already evident from simple sum-z projections (Figure 7A), all metabolic-marker proteins exhibited different intensities, within a given culture, between the two cell types, and in the mono- vs. co-culture condition. The quantitative analysis of all individual voxel cubes gave the following major results: First, comparing the mono-cultures of HT-29 and CCD-1137Sk, CEA signals always extended more to the brighter value range for HT-29 than for CCD-1137Sk cells; in co-cultures, the CEA brightness was further right-shifted. Second, for the markers HK-2 and TIGAR, the fluorescence distribution in HT-29 mono-cultures was bimodal with a low- and a high-intensity cluster, suggesting the presence of two subpopulations of cells with either a low or high abundance of HK-2 and TIGAR; this did not correlate to the expression of CEA. In co-cultures, the same kind of bimodal distribution was found, but here, both CEA and the metabolic marker protein fluorescence signals were higher, leading to an overall shift of the data cloud to the upper-right part of the graphs; this was more pronounced for HK-2 than for TIGAR. Third, CCD-1137Sk cells in mono-cultures showed no bimodal distribution for any of the markers but rather displayed a continuous range of fluorescence intensities for all four tested markers. However, while TIGAR and LDH reached high values in CCD-1137Sk mono-cultures compared to HT-29 mono-cultures, the intensities of the HK-2 and TOMM20 signals were similar or lower in CCD-113Sk than in HT-29 cells. Finally, while LDH and TOMM20 were apparently going up in the HT-29 population of co-cultures (right from the red line), TIGAR and LDH were seemingly downregulated in CCD-1137Sk cells upon co-cultivation. Though, the latter effect might have also been caused by a potential skew towards HT-29 cells of the analysis method.

3.6. Autophagy is Enhanced in Dynarray Co-Cultures of HT-29 and CCD-1137Sk Cells

Next, the voxel-cube approach was used to assess the potential regulation of autophagy in 3-D chip cultures. Therefore, mono- and co-cultures of HT-29 and CCD-1137Sk cells in Dynarrays were fixed after four days, cleared, and stained with DAPI and anti-CEA as before. To study the presence of autophagy, samples were co-labeled with anti-LC3 or anti-P62. Although a first look on the sum-z projections suggested that the fluorescence intensity of both LC3 and P62 was higher in co-cultures than in mono-cultures of either HT-29 or CCD-1137Sk (Figure 8A), the voxel-cube analysis did not confirm this. Rather, it seemed that the brightness distributions of both LC3 (Figure 8B) and P62 (Figure 8C) were rather similar when comparing HT-29 mono-cultures and co-cultures. For HT-29 mono- and co-cultures, LC3 signals showed a bimodal distribution, similar to the ones observed for HK-2 and TIGAR. Conversely, CCD-1137Sk cells in mono-culture again showed simple continuous ranges of LC3 and P62 fluorescence intensities. Since the lack of higher LC3 and P62 values in co-cultures upon voxel-cube analysis did not reflect the sum-z projections well and there was suspicion that the voxel-cube approach might have a skew towards HT-29 signals, the data were additionally analyzed by fluorescence intensity measurements per μm^3 regardless of cell type. As depicted in Figure 8D, this corroborated the qualitative impression of the sum-z projections. Indeed, using this approach, LC3 and P62 signals were highest in the co-cultures and lowest in the CCD-1137Sk mono-cultures. Of course, this did not allow a cell-type-specific statement, but in consideration of the whole picture, a major contribution of fibroblasts to the high autophagy marker values in co-cultures appeared likely.

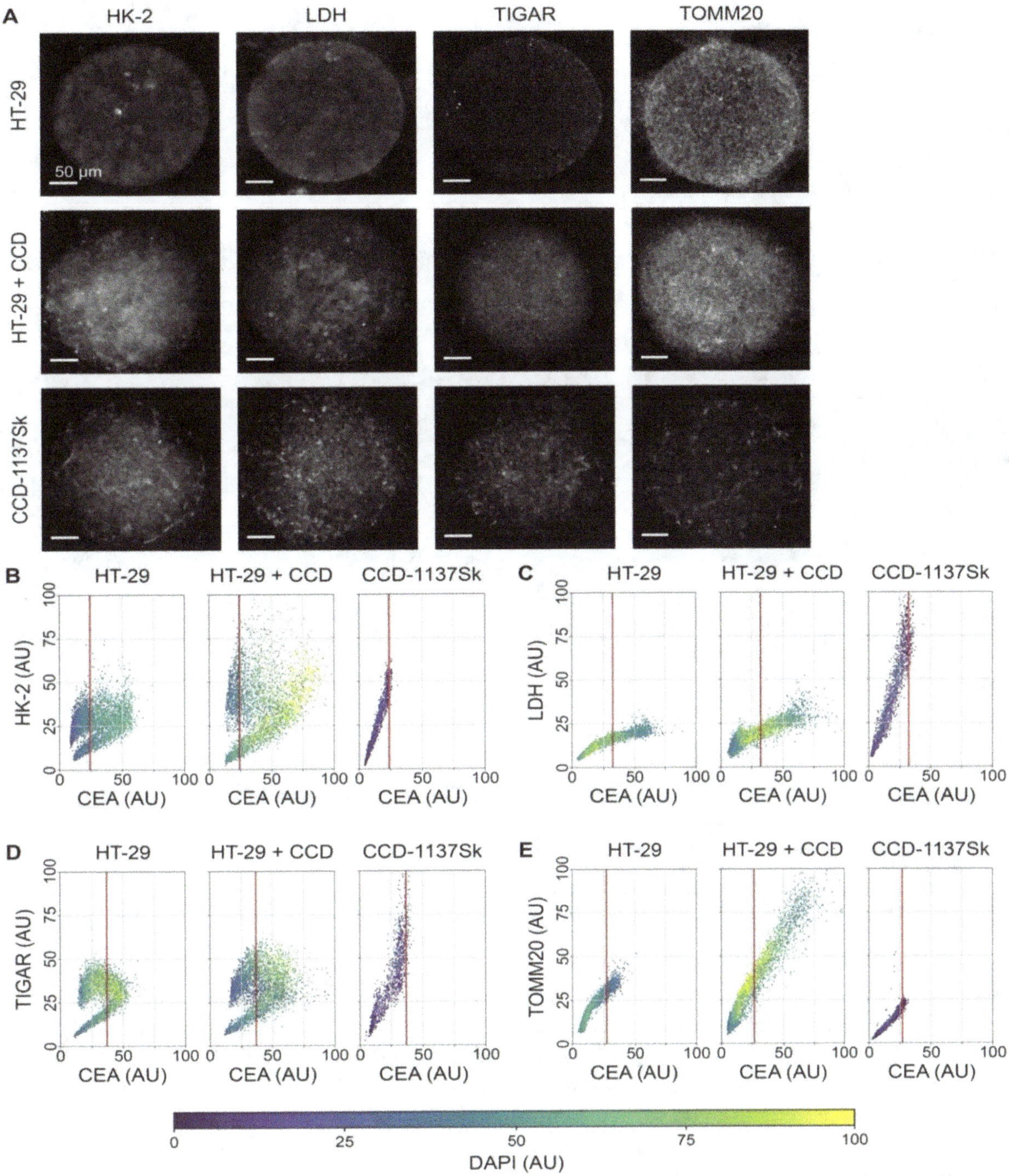

Figure 7. The expression of relevant metabolic marker proteins is altered upon co-culturing of HT-29 and CCD-1137Sk cells in Dynarrays. A total of 4000 cells per chip cavity were seeded to yield either mono- or co-cultures of HT-29 and CCD-1137Sk cells. In the case of co-cultures, 2000 HT-29 cells were co-seeded with 2000 CCD-1137Sk cells per well. After four days, chips were fixed, cleared, and stained with DAPI and antibodies against CEA as well as HK-2, LDH, TIGAR, or TOMM20 as markers for nuclei, cancer cells, and different metabolic activities, respectively. Then, samples were imaged with 3-D confocal microscopy. (**A**) Representative sum-z projections. (**B–E**) Scatter plots showing the fluorescence intensity distributions of anti-HK-2 (**B**), anti-LDH (**C**), anti-TIGAR (**D**), or anti-TOMM20 (**E**) as a function of the anti-CEA intensity for HT-29 mono-cultures (left), HT-29 + CCD-1137Sk co-cultures (middle), and CCD-1137Sk mono-cultures (right). Each dot shows the average intensity of a voxel cube in a total of five Dynarray cavities. Red vertical line, CEA brightness range that contains 95% of all CCD-1137Sk voxel cubes in mono-culture.

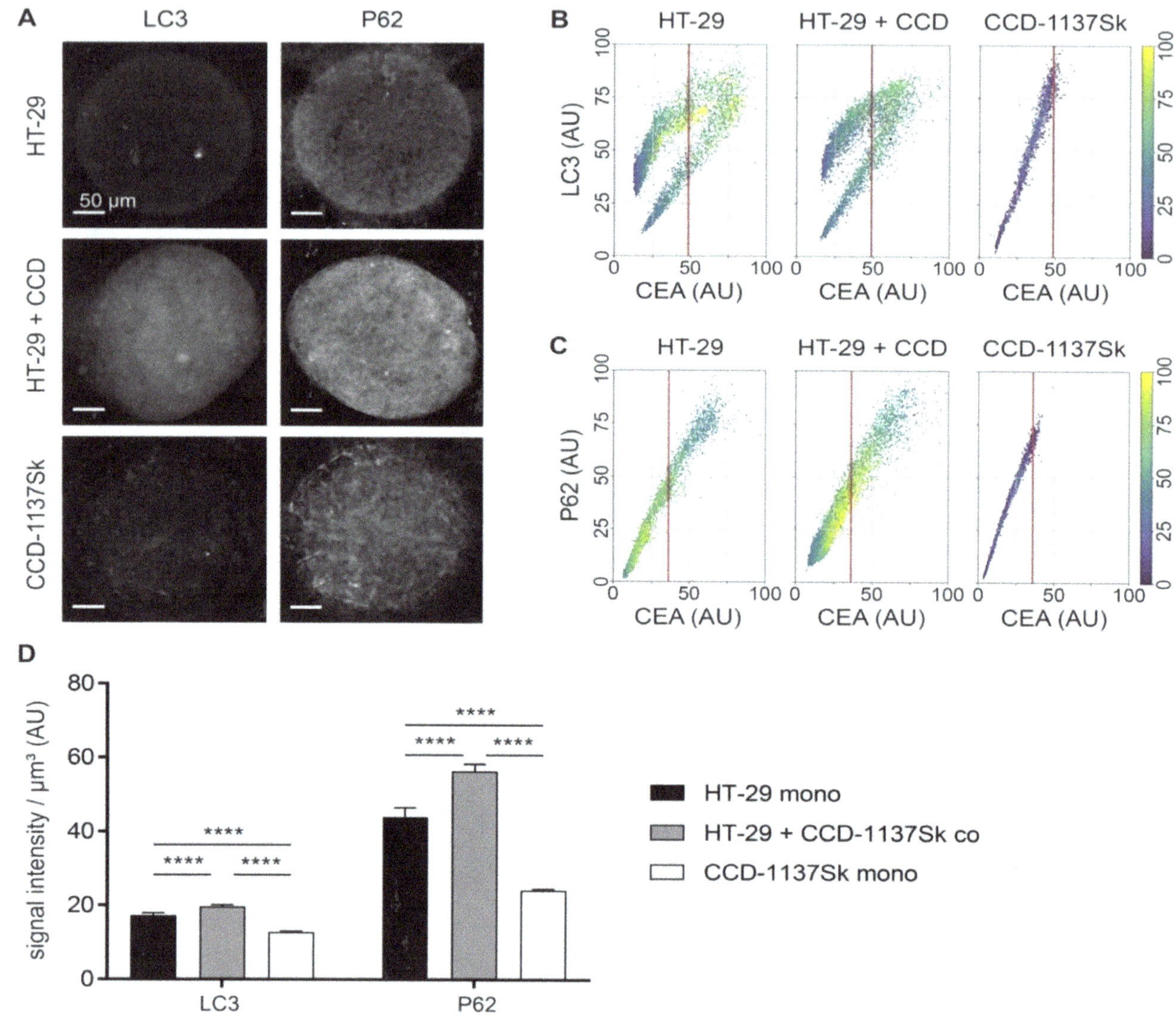

Figure 8. Co-culturing of HT-29 and CCD-1137Sk cells in Dynarrays leads to a rise of autophagic markers. A total of 4000 cells per chip cavity were seeded to yield either mono- or co-cultures of HT-29 and CCD-1137Sk cells. In the case of co-cultures, 2000 HT-29 cells were co-seeded with 2000 CCD-1137Sk cells per well. After four days, chips were fixed, cleared, and stained with DAPI and antibodies against CEA as well as LC3 or P62 as markers for nuclei, cancer cells, and autophagy, respectively. Then, samples were imaged with 3-D confocal microscopy. (**A**) Representative sum-z projections. (**B**,**C**) Scatter plots showing fluorescence intensity distributions of anti-LC3 (**B**) or anti-P62 signals (**C**) as a function of anti-CEA intensity for HT-29 mono-cultures (left), HT-29 + CCD-1137Sk co-cultures (middle), and CCD-1137Sk mono-cultures (right). Each dot shows the average intensity of a voxel cube of a total of five Dynarray cavities. Red vertical line, CEA brightness range that contains 95 % of all CCD-1137Sk voxel cubes in mono-culture. (**D**) LC3/P62 fluorescence intensity per μm^3. Mean + SEM ($n = 5$ experiments).

3.7. In Microarray Chip Co-Cultures, CCD-1137Sk Fibroblasts Show Low Mitochondrial Membrane Potential

Cells undergoing Warburg- or reverse Warburg-based metabolic switches are expected to vary greatly with respect to mitochondrial activity. Thus, to test the presence of active mitochondria, live cell imaging was performed using the red-fluorescent vital dye TMRM. This accumulates in mitochondria depending on their membrane potential. To be able to differentiate between the cell types in live cell imaging, CCD-1137Sk and HT-29 cells were loaded with either CellTracker DeepRed (CTdr) or CellTracker Green (CTg), respectively. Then, they were co-cultured in Dynarray chips for five days. Next, 30 min after loading with TMRM, 3-D live cell confocal microscopy was performed. At this point, it is necessary to say that in the control experiments, the CellTracker staining itself also enhanced LC3 expression primarily in CCD-1137Sk cells of HT-29 co-cultures (data not shown). Therefore, we cannot completely exclude that the TMRM data shown here were partially affected by the experimental procedure itself. However, as observed upon immunostaining against collagen 1 and CEA (see Figure 5A), the CellTracker signals also revealed clusters of fibroblasts interspersed within the HT-29 colony (Figure 9A). While CTg-positive HT-29 cells exhibited variable but mostly bright TMRM fluorescence, CTdr-positive CCD-1137Sk cells were almost completely devoid of such signals. Co-localization analysis of TMRM with the labeling intensity for each CellTracker dye using Pearson's coefficient measurements confirmed this observation (Figure 9B). This suggests that HT-29 cancer cells were present in different states of mitochondrial activity, while fibroblasts were mostly lacking a consistent mitochondrial membrane potential. Moreover, additional immunostaining of co-culture chips showed that regions outside the circular entry of the chip cavities, which were primarily populated by fibroblasts, were strongly positive for MCT4 but not for MCT1 (Figure 10). Conversely, the circular body within the 3-D cavities, which was composed of mixed HT-29 and CCD-1137Sk populations, exhibited variable amounts of MCT1 and MCT4, showing a prevalence for either one or the other kind of transporter.

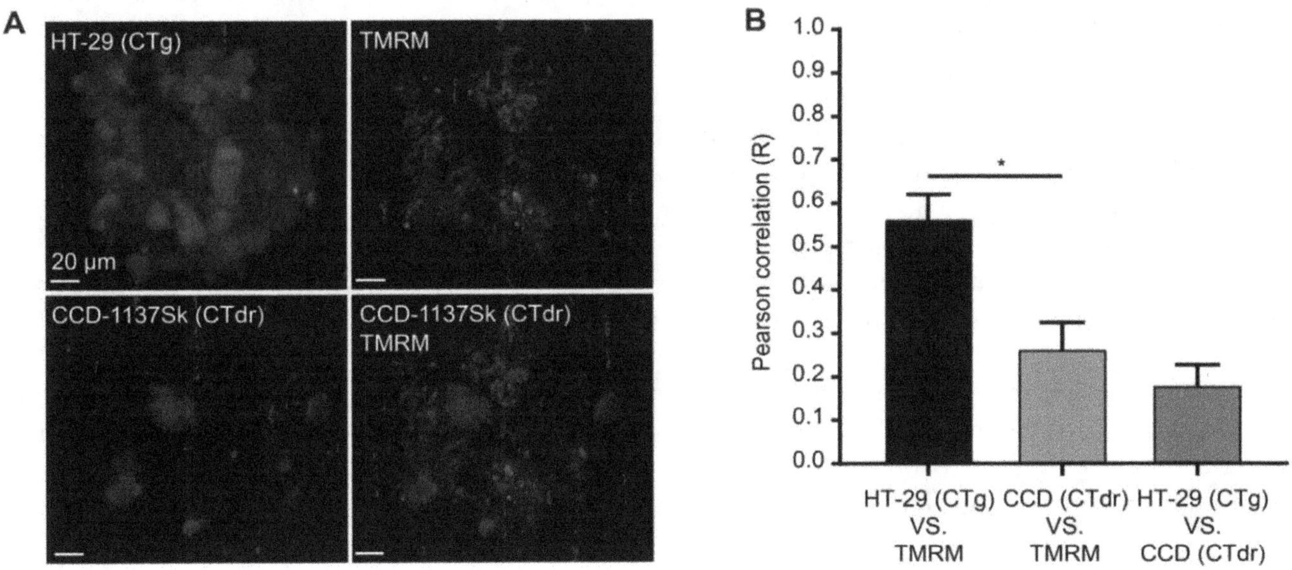

Figure 9. Chip-based 3-D cultures reveal a low mitochondrial membrane potential in fibroblasts upon co-cultivation of HT-29 and CCD-1137Sk cells. Human HT-29 and CCD-1137Sk cells were live stained with CTg and CTdr, respectively, and then co-seeded in Dynarray chips for four days. To evaluate the mitochondrial membrane potential, TMRM was analyzed in 3-D live cell imaging. (**A**) Representative maximum-z projections of fluorescence signals for CTg, CTdr, and TMRM, as indicated. (**B**) Quantitative analysis of Pearson's correlation between CT-dyes and TMRM as a measure of cell-type-specific TMRM fluorescence intensity. Mean + SEM (n = 3 experiments).

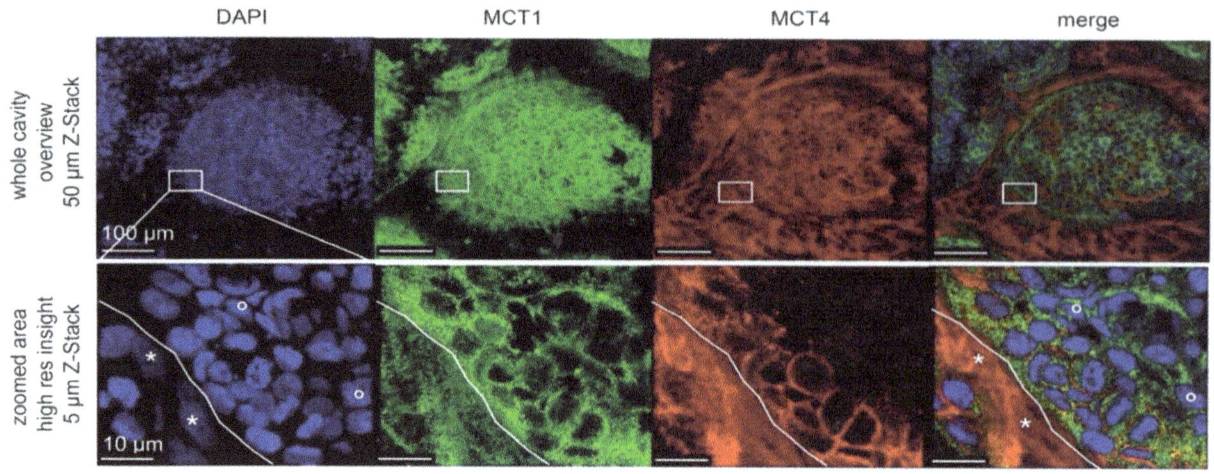

Figure 10. High-resolution confocal images reveal differential cell-type-specific expression of lactate shuttle proteins upon co-cultivation of HT-29 and CCD-1137Sk cells. Per chip cavity, 2000 HT-29 and 2000 CCD-1137Sk cells were co-seeded. After four days, cultures were fixed, cleared, and then stained with DAPI as well as antibodies against MCT4 and MCT1 as markers for nuclei, lactate export, and lactate import, respectively. Subsequently, samples were imaged with confocal microscopy. Shown are representative maximum-z projections with the indicated thickness. Lower panels were taken from the same specimen region as boxed in the upper panels. White solid lines in the lower panels indicate the border between a zone primarily populated by CCD-1137Sk (lower left) and HT-29 cells (upper right). Asterisks and open circles point to typically dim and large nuclei of CCD-1137Sk vs. bright and small nuclei of HT-29 cells.

4. Discussion

Since a few years ago, mutual interactions between cancer and stromal cells have emerged as major contributors to tumor metabolism [40]. Paracrine and autocrine signaling of growth factors, proteases, inhibitors, and extracellular matrix components might lead to widespread cross-talk between cancer cells and their surroundings. This influences the metabolism of all cells involved [41] and has become a widely studied target for cancer treatment. However, cell model systems with desired properties in terms of both fast and easy usability, as well as sufficient physiological complexity, including effects within tumor populations of cancer and stromal cells for drug treatment investigation, are rare.

Current in vitro model systems for cancer studies mostly employ either cell lines or primary tumor cells and these are either cultured on planar surfaces or, using different methods, in 3-D. Regarding the choice of 'cell line' versus 'primary cells', the latter appears superior for personalized medicine [42] and for a higher physiological relevance [43]. However, since primary cells are derived from a limited amount of tumor material and since the genotype and phenotype are variable not only between any two patients but also within the tumor population of a given patient, primary cells exhibit major inherent variability in terms of availability and genetic constitution [44]. Therefore, while in vitro models derived from primary cells might be advantageous for personalized medicine and advanced stages of drug testing, the use of cell lines can still be an option for the study of more fundamental cell biological phenomena and principles of cellular signaling due to their unlimited availability and relative phenotypic robustness. With respect to culturing in 2-D or 3-D, the former is still widely used due to its unbeatable applicability and analytical amenability, in particular, in the context of image-based assays [45]. Nevertheless, signaling processes mediated by cell–cell contact or paracrine mechanisms might be reliably addressed in such adherent cell cultures. However, since contact via the entire cell surface as well as the establishment of physiological cell organization, of gradients of drugs, oxygen, waste products etc. can only be obtained in 3-D cell cultures, the latter has increasingly gained importance in recent years. Therefore, spheroids and, particularly, tumor organoids have become key to in vitro cancer studies [46]. Spheroids are mostly based on the self-assembly of immortalized cells by impairing their settlement on any non-cellular surface. In contrast, tumor organoids involve the use of primary tumor-derived cells and their embedding in hydrogels formed by extracellular matrix extracts, typically Matrigel, BME, or similar. Models of intermediate complexity use Matrigel in combination with cell lines, in particular, for obtaining 3-D cultures from cell lines that otherwise hardly form such cultures [35,47]. Although Matrigel/BME-based tumor organoids have been seen as a 'gold standard' for in vitro cancer research, the use of Matrigel also has important drawbacks. For example, there is a considerable lot-to-lot variability, Matrigel is derived from tumor tissue, and it inherently adds specific physicochemical matrix properties [48]. Furthermore, it has a significant autofluorescence in the green wavelength range [49,50], rendering its use in marker-specific microscopic analyses less efficient. We compared 3-D cultures made with and without BME and experienced issues of autofluorescence. On top, BME-based cultures were much less regular and condensed, leading to an increased size of 3-D data stacks, an issue that requires more efforts on data acquisition time, storage, and analysis (Figure S5A–C). Most worrying, though, were the differences observed in terms of marker expression. For example, in CCD-1137Sk mono-cultures, the presence of BME increased the fluorescence signals obtained for CEA and decreased those for LC3 (Figure S5D).

Recapitulating, different model systems might be superior to others for certain aspects, but the choice of an appropriate condition will likely depend strongly on the research topic. In this context, here, we used yet another assay system, i.e., the Dynarray chip. From a purely technological point of view, this can be an interesting option for mechanistic studies where both intrinsic cell–cell interactions as well as a defined surface composition should be guaranteed. These chips feature equally spaced and equally sized cavities made by an optically transparent polymer, which can be coated with any material of interest, e.g., Matrigel, BME, collagen, or also just poly-L-lysine, to enhance interaction between the chip and cells. We chose collagen 1 to allow for a defined support and to avoid potentially interfering signaling molecules that might come with Matrigel/BME. Indeed, at least in 2-D cultures, coating with collagen 1 did not affect the expression of any of the tested markers at the level of immunofluorescence (Figure S6). In Dynarray co-cultures of HT-29 or MDA-MB-231 cells with CCD-1137Sk fibroblasts, the latter were typically making most of the contacts to the collagen surface and the morphology of these peripheral cells was often spindle shaped, forming bridges between the chip wall and the mass of the co-culture. Conversely, in the centers of the cavities, cells were better mixed, although some segregation of fibroblasts and tumor cells into nests was also observed, like in the spheroids. Altogether, this mediated a high homogeneity of 3-D cultures between different cavities and different chips. Another technical advantage of the chips over the spheroid cultures was their spatial regularity, which eased automated microscopic acquisition and subsequent data analysis.

Regarding the metabolic interactions, the 2-D cultures and Dynarrays revealed good consensus for fibroblasts. Indeed, in both systems, data were consistent with a higher level of HK-2 and autophagy markers as well as a loss of TIGAR and TOMM20 in CCD-1137Sk cells upon co-culturing with HT-29. With respect to the metabolic profile of HT-29 cells, 2-D cultures did not reveal any significant change, while in Dynarrays, the data showed an upregulation of HK-2, TIGAR, LDH, and TOMM20. This could indicate that the signaling interplay in the 3-D cultures was more extensive, leading to a clearer response of the HT-29 cells. Furthermore, the lactate export carrier MCT4 was upregulated in fibroblasts in 2-D upon co-cultivation and it showed a prevalence over MCT1 in the fibroblasts of the Dynarray cultures. The mitochondrial membrane potential, as measured by the TMRM fluorescence intensity, was mostly low in fibroblasts. In contrast, the TMRM signal was higher but variable from cell to cell in HT-29 cells. Finally, chip-based 3-D co-cultures of HT-29 and CCD-1137Sk cells grew more robustly than the corresponding mono-cultures, although it was not possible to unequivocally attribute this to one of the two cell types. Altogether, these data pointed to a metabolic cross-talk between cancer and fibroblast cells in co-cultures, which primarily affected the metabolic behavior of the fibroblasts, and in Dynarrays, HT-29 cells as well. This result is most consistent with a scenario in which HT-29 cells induce CCD-1137Sk cells to switch to a glycolytic and autophagic state, to lose mitochondrial activity, and to gain lactate export capabilities. Conversely, HT-29 cells apparently remained mixed in terms of mitochondrial activity and the expression of lactate importers and exporters, suggesting the presence of both catabolic and anabolic states. This was nicely corroborated by the bimodal expression patterns of HK-2, TIGAR, and LC3. Similar features, likely responsible for the differential susceptibility to drug treatment, were previously observed in mouse models and tumor tissue [51,52]. In particular, metabolic heterogeneity of malignant cells [53] and bad prognosis together with decreased overall survival upon enhanced MCT4 expression of tumors are fitting to the present study [54]. Finally, HT-29 in Dynarray co-cultures also gained in HK-2, TIGAR, LDH, and TOMM20, consistent with a combination of high glucose and lactate metabolism capability [26] that could jointly fuel the production of pyruvate and oxidative phosphorylation.

While the metabolic interaction patterns were observed in monolayer and chip-based co-cultures, no effect of co-culturing was evident on spheroid growth. This might be explained by three not mutually exclusive scenarios: First, fibroblasts remained rather isolated and encapsulated in spheroid co-cultures while they were more loosely interspersed in the chip-based co-cultures. Thus, one might speculate that the distribution of paracrine signals between the tumor and fibroblast cells was not comparable between the two culture conditions. Second, it might be that the relatively large spheroids of up to 1000 μm in diameter created more profound gradients of oxygen, nutrients, and waste products than the 300-μm-wide Dynarray-chip cultures. This might have intrinsically changed the basic metabolic profile of these cells or their propensity to react to each other. Finally, it cannot be completely excluded that the measurement of the spheroid diameter was insufficient to detect metabolic alterations induced upon co-cultivation. For this reason, the seeding densities in experiments focusing on the metabolic aspects of spheroid cultures had to be altered in a way that ensured a homogeneous size distribution as compared to the 300-μm Dynarray cavities. More analyses should be undertaken to reveal changes in the metabolic state of the cells. However, even if these existed, they were apparently so subtle that they had no robust effect on cell growth as observed in the chip format.

5. Conclusions

In summary, this work described different cellular co-culture models to mimic mutual metabolic interactions between HT-29 cancer and CCD-1137Sk fibroblast cells. In particular, the differential expression patterns of MCT1, MCT4, HK-2, TIGAR, TOMM20, LC3, and P62 in HT-29 and CCD-1137Sk cells, as well as the low mitochondrial membrane potential in fibroblasts as observed in mono-layer 2-D and chip-based 3-D co-cultures were consistent with reverse Warburg-type interactions. Together with the additional synergistic effects of co-culturing on cell proliferation in Dynarray chips, these features were comparable to those observed in tumor tissue. In contrast, spheroid co-cultures did not show such effects. In the future, the presented models might be employed for addressing basic molecular signaling and cell physiology aspects, as well as for testing metabolic modifiers, in combination with anticancer drugs for the treatment of neoplastic disorders.

Supplementary Materials

Figure S1: CEA can serve as a cancer-cell marker in 2-D mono- and co-cultures of HT-29, MDA-MB-231, and CCD-1137Sk cells. Figure S2: Monolayer co-cultures of MDA-MB-231 and CCD-1137Sk show enhanced expression of MCT4 in fibroblasts. Figure S3: Fibroblasts in monolayer co-cultures of MDA-MB-231 and CCD-1137Sk show enhanced glycolytic and reduced oxidative phosphorylation markers. Figure S4: Fibroblasts in monolayer co-cultures of MDA-MB-231 and CCD-1137Sk display enhanced levels of autophagy markers. Figure S5: Culture-embedding in BME affects growth and fluorescence properties of spheroids. Figure S6: Surface coating with Collagen 1 does not alter the expression patterns of several metabolic markers.

Author Contributions: Conceptualization, F.K. and R.R.; methodology, F.K. and R.B.; validation, F.K.; formal analysis, F.K. and R.B.; investigation, F.K.; resources, R.S. and J.M.-H.; writing—original draft preparation, F.K.; writing—review and editing, F.K., R.S., J.M.-H., M.H., R.R.; supervision, M.H., R.R.; project administration, R.S., J.M.-H., M.H., R.R.; funding acquisition, M.H., R.R. All authors have read and agreed to the published version of the manuscript.

References

1. Siegel, R.L.; Miller, K.D.; Fedewa, S.A.; Ahnen, D.J.; Meester, R.G.S.; Barzi, A.; Jemal, A. Colorectal cancer statistics. *CA Cancer J. Clin.* **2017**, *67*, 177–193. [CrossRef] [PubMed]
2. Løberg, M.; Kalager, M.; Holme, Ø.; Hoff, G.; Adami, H.-O.; Bretthauer, M. Long-term colorectal-cancer mortality after adenoma removal. *N. Engl. J. Med.* **2014**, *371*, 799–807. [CrossRef] [PubMed]
3. Sensi, F.; D'Angelo, E.; D'Aronco, S.; Molinaro, R.; Agostini, M. Preclinical three-dimensional colorectal cancer model: The next generation of in vitro drug efficacy evaluation. *J. Cell. Physiol.* **2018**, *234*, 181–191. [CrossRef] [PubMed]
4. Wilde, L.; Roche, M.; Domingo-Vidal, M.; Tanson, K.; Philp, N.; Curry, J.; Martinez-Outschoorn, U. Metabolic coupling and the Reverse Warburg Effect in cancer: Implications for novel biomarker and anticancer agent development. *Semin. Oncol.* **2017**, *44*, 198–203. [CrossRef]
5. Sun, Z.; Liu, L.; Wang, P.P.; Roebothan, B.; Zhao, J.; Dicks, E.; Cotterchio, M.; Buehler, S.; Campbell, P.T.; McLaughlin, J.R.; et al. Association of total energy intake and macronutrient consumption with colorectal cancer risk: Results from a large population-based case-control study in Newfoundland and Labrador and Ontario, Canada. *Nutr. J.* **2012**, *11*, 18. [CrossRef]
6. Benz, M.R.; Czernin, J.; Dry, S.M.; Tap, W.D.; Allen-Auerbach, M.S.; Elashoff, D.; Phelps, M.E.; Weber, W.A.; Eilber, F.C. Quantitative F18-fluorodeoxyglucose positron emission tomography accurately characterizes peripheral nerve sheath tumors as malignant or benign. *Cancer* **2010**, *116*, 451–458. [CrossRef]
7. Xiao, Y.; Kim, D.; Dura, B.; Zhang, K.; Yan, R.; Li, H.; Han, E.; Ip, J.; Zou, P.; Liu, J.; et al. Ex vivo Dynamics of Human Glioblastoma Cells in a Microvasculature-on-a-Chip System Correlates with Tumor Heterogeneity and Subtypes. *Adv. Sci.* **2019**, *6*, 1801531. [CrossRef]
8. Martinez-Outschoorn, U.E.; Lisanti, M.P.; Sotgia, F. Catabolic cancer-associated fibroblasts transfer energy and biomass to anabolic cancer cells, fueling tumor growth. *Semin. Cancer Biol.* **2014**, *25*, 47–60. [CrossRef]
9. Gatenby, R.A.; Gillies, R.J. Why do cancers have high aerobic glycolysis? *Nat. Rev. Cancer* **2004**, *4*, 891–899. [CrossRef]
10. Otto, A.M. Warburg effect(s)-a biographical sketch of Otto Warburg and his impacts on tumor metabolism. *Cancer Metab.* **2016**, *4*, 5. [CrossRef]
11. Vyas, S.; Zaganjor, E.; Haigis, M.C. Mitochondria and Cancer. *Cell* **2016**, *166*, 555–566. [CrossRef] [PubMed]
12. Martinez-Outschoorn, U.E.; Pavlides, S.; Whitaker-Menezes, D.; Daumer, K.M.; Milliman, J.N.; Chiavarina, B.; Migneco, G.; Witkiewicz, A.K.; Martinez-Cantarin, M.P.; Flomenberg, N.; et al. Tumor cells induce the cancer associated fibroblast phenotype via caveolin-1 degradation: Implications for breast cancer and DCIS therapy with autophagy inhibitors. *Cell Cycle* **2010**, *9*, 2423–2433. [CrossRef] [PubMed]
13. Chiavarina, B.; Whitaker-Menezes, D.; Migneco, G.; Martinez-Outschoorn, U.E.; Pavlides, S.; Howell, A.; Tanowitz, H.B.; Casimiro, M.C.; Wang, C.; Pestell, R.G.; et al. HIF1-alpha functions as a tumor promoter in cancer associated fibroblasts, and as a tumor suppressor in breast cancer cells: Autophagy drives compartment-specific oncogenesis. *Cell Cycle* **2010**, *9*, 3534–3551. [CrossRef] [PubMed]
14. Wallace, D.C. Mitochondria and cancer. *Nat. Rev. Cancer* **2012**, *12*, 685–698. [CrossRef]
15. Porporato, P.E.; Filigheddu, N.; Pedro, J.M.B.-S.; Kroemer, G.; Galluzzi, L. Mitochondrial metabolism and cancer. *Cell Res.* **2018**, *28*, 265–280. [CrossRef]
16. Costa, A.; Kieffer, Y.; Scholer-Dahirel, A.; Pelon, F.; Bourachot, B.; Cardon, M.; Sirven, P.; Magagna, I.; Fuhrmann, L.; Bernard, C.; et al. Fibroblast Heterogeneity and Immunosuppressive Environment in Human Breast Cancer. *Cancer Cell* **2018**, *33*, 463–479.e10. [CrossRef]
17. Augsten, M. Cancer-Associated Fibroblasts as Another Polarized Cell Type of the Tumor Microenvironment. *Front. Oncol.* **2014**, *4*, 820. [CrossRef]
18. Shiga, K.; Hara, M.; Nagasaki, T.; Sato, T.; Takahashi, H.; Takeyama, H. Cancer-Associated Fibroblasts: Their Characteristics and Their Roles in Tumor Growth. *Cancers* **2015**, *7*, 2443–2458. [CrossRef]
19. Klionsky, D.J.; Abdelmohsen, K.; Abe, A.; Abedin, M.J.; Abeliovich, H.; Zorzano, A.; Zughaier, S.M. Guidelines for the use and interpretation of assays for monitoring autophagy (3rd edition). *Autophagy* **2016**, *12*, 1–222. [CrossRef]

20. Pavlides, S.; Vera, I.; Gandara, R.; Sneddon, S.; Pestell, R.G.; Mercier, I.; Martinez-Outschoorn, U.E.; Whitaker-Menezes, D.; Howell, A.; Sotgia, F.; et al. Warburg meets autophagy: Cancer-associated fibroblasts accelerate tumor growth and metastasis via oxidative stress, mitophagy, and aerobic glycolysis. *Antioxid. Redox Signal.* **2012**, *16*, 1264–1284. [CrossRef]
21. Wu, Y.; Yao, J.; Xie, J.; Liu, Z.; Zhou, Y.; Pan, H.; Han, W. The role of autophagy in colitis-associated colorectal cancer. *Signal Transduct. Target. Ther.* **2018**, *3*, 31. [CrossRef] [PubMed]
22. Park, S.J.; Smith, C.P.; Wilbur, R.R.; Cain, C.P.; Kallu, S.R.; Valasapalli, S.; Sahoo, A.; Guda, M.R.; Tsung, A.J.; Velpula, K.K. An overview of MCT1 and MCT4 in GBM: Small molecule transporters with large implications. *Am. J. Cancer Res.* **2018**, *8*, 1967–1976. [PubMed]
23. Whitaker-Menezes, D.; Martinez-Outschoorn, U.E.; Lin, Z.; Ertel, A.; Flomenberg, N.; Witkiewicz, A.K.; Birbe, R.C.; Howell, A.; Pavlides, S.; Gandara, R.; et al. Evidence for a stromal-epithelial "lactate shuttle" in human tumors: MCT4 is a marker of oxidative stress in cancer-associated fibroblasts. *Cell Cycle* **2011**, *10*, 1772–1783. [CrossRef] [PubMed]
24. Fu, Y.; Liu, S.; Yin, S.; Niu, W.; Xiong, W.; Tan, M.; Li, G.; Zhou, M. The reverse Warburg effect is likely to be an Achilles' heel of cancer that can be exploited for cancer therapy. *Oncotarget* **2017**, *8*, 57813–57825. [CrossRef] [PubMed]
25. Pérez-Escuredo, J.; Dadhich, R.K.; Dhup, S.; Cacace, A.; Van Hée, V.F.; De Saedeleer, C.J.; Sboarina, M.; Rodriguez, F.; Fontenille, M.J.; Brisson, L.; et al. Lactate promotes glutamine uptake and metabolism in oxidative cancer cells. *Cell Cycle* **2016**, *15*, 72–83. [CrossRef] [PubMed]
26. Cannino, G.; Ciscato, F.; Masgras, I.; Sánchez-Martín, C.; Rasola, A. Metabolic plasticity of tumor cell mitochondria. *Front. Oncol.* **2018**, *8*. [CrossRef]
27. Katt, M.E.; Placone, A.L.; Wong, A.D.; Xu, Z.S.; Searson, P.C. In Vitro Tumor Models: Advantages, Disadvantages, Variables, and Selecting the Right Platform. *Front. Bioeng. Biotechnol.* **2016**, *4*, 12. [CrossRef]
28. Georges, L.M.C.; de Wever, O.; Galván, J.A.; Dawson, H.; Lugli, A.; Demetter, P.; Zlobec, I. Cell Line Derived Xenograft Mouse Models Are a Suitable in vivo Model for Studying Tumor Budding in Colorectal Cancer. *Front. Med.* **2019**, *6*, 139. [CrossRef]
29. Benton, G.; Kleinman, H.K.; George, J.; Arnaoutova, I. Multiple uses of basement membrane-like matrix (BME/Matrigel) in vitro and in vivo with cancer cells. *Int. J. Cancer* **2011**, *128*, 1751–1757. [CrossRef]
30. Riedl, A.; Schlederer, M.; Pudelko, K.; Stadler, M.; Walter, S.; Unterleuthner, D.; Unger, C.; Kramer, N.; Hengstschläger, M.; Kenner, L.; et al. Comparison of cancer cells in 2D vs. 3D culture reveals differences in AKT-mTOR-S6K signaling and drug responses. *J. Cell Sci.* **2017**, *130*, 203–218. [CrossRef]
31. Fang, Y.; Eglen, R.M. Three-Dimensional Cell Cultures in Drug Discovery and Development. *SLAS Discov.* **2017**, *22*, 456–472. [CrossRef] [PubMed]
32. Nunes, A.S.; Barros, A.S.; Costa, E.C.; Moreira, A.F.; Correia, I.J. 3D tumor spheroids as in vitro models to mimic in vivo human solid tumors resistance to therapeutic drugs. *Biotechnol. Bioeng.* **2019**, *116*, 206–226. [CrossRef] [PubMed]
33. Li, Y.; Kumacheva, E. Hydrogel microenvironments for cancer spheroid growth and drug screening. *Sci. Adv.* **2018**, *4*, 1–11. [CrossRef]
34. Chen, W.; Wong, C.; Vosburgh, E.; Levine, A.J.; Foran, D.J.; Xu, E.Y. High-throughput image analysis of tumor spheroids: A user-friendly software application to measure the size of spheroids automatically and accurately. *J. Vis. Exp.* **2014**. [CrossRef] [PubMed]
35. Keller, F.; Rudolf, R.; Hafner, M. Towards optimized breast cancer 3D spheroid mono- and co-culture models for pharmacological research and screening. *J. Cell. Biotechnol.* **2019**, *5*, 89–101. [CrossRef]
36. Nürnberg, E.; Vitacolonna, M.; Klicks, J.; von Molitor, E.; Cesetti, T.; Keller, F.; Bruch, R.; Ertongur-Fauth, T.; Riedel, K.; Scholz, P.; et al. Routine Optical Clearing of 3D-Cell Cultures: Simplicity Forward. *Front. Mol. Biosci.* **2020**, *7*, 441. [CrossRef]
37. Smith, P.; Reid, D.B.; Environment, C.; Palo, L.; Alto, P.; Smith, P.L. Otsu_1979_otsu_method. *IEEE Trans. Syst. Man Cybern.* **1979**, *9*, 62–66. [CrossRef]
38. Tiernan, J.P.; Perry, S.L.; Verghese, E.T.; West, N.P.; Yeluri, S.; Jayne, D.G.; Hughes, T.A. Carcinoembryonic antigen is the preferred biomarker for in vivo colorectal cancer targeting. *Br. J. Cancer* **2013**, *108*, 662–667. [CrossRef]
39. Pastor-Pareja, J.C. Atypical basement membranes and basement membrane diversity—What is normal anyway? *J. Cell Sci.* **2020**, *133*. [CrossRef]

40. da Cunha, B.R.; Domingos, C.; Stefanini, A.C.B.; Henrique, T.; Polachini, G.M.; Castelo-Branco, P.; Tajara, E.H. Cellular Interactions in the Tumor Microenvironment: The Role of Secretome. *J. Cancer* **2019**, *10*, 4574–4587. [CrossRef]
41. Mueller, M.M.; Fusenig, N.E. Friends or foes—Bipolar effects of the tumour stroma in cancer. *Nat. Rev. Cancer* **2004**, *4*, 839–849. [CrossRef] [PubMed]
42. Ginsburg, G.S.; Phillips, K.A. Precision medicine: From science to value. *Health Aff.* **2018**, *37*, 694–701. [CrossRef] [PubMed]
43. Pan, C.; Kumar, C.; Bohl, S.; Klingmueller, U.; Mann, M. Comparative proteomic phenotyping of cell lines and primary cells to assess preservation of cell type-specific functions. *Mol. Cell. Proteom.* **2009**, *8*, 443–450. [CrossRef] [PubMed]
44. Sasaki, N.; Clevers, H. Studying cellular heterogeneity and drug sensitivity in colorectal cancer using organoid technology. *Curr. Opin. Genet. Dev.* **2018**, *52*, 117–122. [CrossRef]
45. Kapałczyńska, M.; Kolenda, T.; Przybyła, W.; Zajączkowska, M.; Teresiak, A.; Filas, V.; Ibbs, M.; Bliźniak, R.; Łuczewski, Ł.; Lamperska, K. 2D and 3D cell cultures—A comparison of different types of cancer cell cultures. *Arch. Med. Sci.* **2018**, *14*, 910–919. [CrossRef]
46. Chaicharoenaudomrung, N.; Kunhorm, P.; Noisa, P. Three-dimensional cell culture systems as an in vitro platform for cancer and stem cell modeling. *World J. Stem Cells* **2019**, *11*, 1065–1083. [CrossRef]
47. Rustamov, V.; Keller, F.; Klicks, J.; Hafner, M.; Rudolf, R. Bone Sialoprotein Shows Enhanced Expression in Early, High-Proliferation Stages of Three-Dimensional Spheroid Cell Cultures of Breast Cancer Cell Line MDA-MB-231. *Front. Oncol.* **2019**, *9*, 36. [CrossRef]
48. Cruz-Acuña, R.; García, A.J. Engineered materials to model human intestinal development and cancer using organoids. *Exp. Cell Res.* **2019**, *337*, 109–114. [CrossRef]
49. Shen, Y.; Zhu, D.; Lu, W.; Liu, B.; Li, Y.; Cao, S. The characteristics of intrinsic fluorescence of type I collagen influenced by collagenase I. *Appl. Sci.* **2018**, *8*, 1947. [CrossRef]
50. Solanki, A.; King, D.; Thibault, G.; Wang, L.; Gibbs, S.L. Quantification of fluorophore distribution and therapeutic response in matched in vivo and ex vivo pancreatic cancer model systems. *PLoS ONE* **2020**, *15*, 1–18. [CrossRef]
51. Sharick, J.T.; Jeffery, J.J.; Karim, M.R.; Walsh, C.M.; Esbona, K.; Cook, R.S.; Skala, M.C. Cellular Metabolic Heterogeneity In Vivo Is Recapitulated in Tumor Organoids. *Neoplasia* **2019**, *21*, 615–626. [CrossRef] [PubMed]
52. Blank, A.; Roberts, D.E.; Dawson, H.; Zlobec, I.; Lugli, A. Tumor Heterogeneity in Primary Colorectal Cancer and Corresponding Metastases. Does the Apple Fall Far From the Tree? *Front. Med.* **2018**, *5*, 234. [CrossRef] [PubMed]
53. Schcolnik-Cabrera, A.; Chávez-Blanco, A.; Domínguez-Gómez, G.; Dueñas-González, A. Understanding tumor anabolism and patient catabolism in cancer-associated cachexia. *Am. J. Cancer Res.* **2017**, *7*, 1107–1135. [PubMed]
54. Bovenzi, C.D.; Hamilton, J.; Tassone, P.; Johnson, J.; Cognetti, D.M.; Luginbuhl, A.; Keane, W.M.; Zhan, T.; Tuluc, M.; Bar-Ad, V.; et al. Prognostic Indications of Elevated MCT4 and CD147 across Cancer Types: A Meta-Analysis. *Biomed Res. Int.* **2015**, *2015*, 242437. [CrossRef]

3

Innovative Visualization and Quantification of Extracellular Vesicles Interaction with and Incorporation in Target Cells in 3D Microenvironments

Enrico Ragni [1,†], Silvia Palombella [2,†], Silvia Lopa [2,†], Giuseppe Talò [2], Carlotta Perucca Orfei [1], Paola De Luca [1], Matteo Moretti [2,3,*,‡] and Laura de Girolamo [1,*,‡]

1 IRCCS Istituto Ortopedico Galeazzi, Laboratorio di Biotecnologie Applicate all'Ortopedia, Via R. Galeazzi 4, I-20161 Milano, Italy; enrico.ragni@grupposandonato.it (E.R.); carlotta.perucca@grupposandonato.it (C.P.O.); deluca.paola@grupposandonato.it (P.D.L.)

2 IRCCS Istituto Ortopedico Galeazzi, Cell and Tissue Engineering Laboratory, Via R. Galeazzi 4, I-20161 Milano, Italy; silvia.palombella@grupposandonato.it (S.P.); silvia.lopa@grupposandonato.it (S.L.); giuseppe.talo@grupposandonato.it (G.T.)

3 Regenerative Medicine Technologies Laboratory, Ente Ospedaliero Cantonale (EOC), 6900 Lugano, Switzerland

* Correspondence: matteo.moretti@grupposandonato.it (M.M.); laura.degirolamo@grupposandonato.it (L.d.G.)

† These authors contributed equally to this work.

‡ These authors contributed equally to this work.

Abstract: Extracellular vesicles (EVs) showed therapeutic properties in several applications, many in regenerative medicine. A clear example is in the treatment of osteoarthritis (OA), where adipose-derived mesenchymal stem cells (ASCs)-EVs were able to promote regeneration and reduce inflammation in both synovia and cartilage. A still obscure issue is the effective ability of EVs to be internalized by target cells, rather than simply bound to the extracellular matrix (ECM) or plasma membrane, since the current detection or imaging technologies cannot fully decipher it due to technical limitations. In the present study, human articular chondrocytes (ACHs) and fibroblast-like synoviocytes (FLSs) isolated from the same OA patients were cocultured in 2D as well as in 3D conditions with fluorescently labeled ASC-EVs, and analyzed by flow cytometry or confocal microscopy, respectively. In contrast with conventional 2D, in 3D cultures, confocal microscopy allowed a clear detection of the tridimensional morphology of the cells and thus an accurate discrimination of EV interaction with the external and/or internal cell environment. In both 2D and 3D conditions, FLSs were more efficient in interacting with ASC-EVs and 3D imaging demonstrated a faster uptake process. The removal of the hyaluronic acid component from the ECM of both cell types reduced their interaction with ASC-EVs only in the 2D system, showing that 2D and 3D conditions can yield different outcomes when investigating events where ECM plays a key role. These results indicate that studying EVs binding and uptake both in 2D and 3D guarantees a more precise and complementary characterization of the molecular mechanisms involved in the process. The implementation of this strategy can become a valuable tool not only for basic research, but also for release assays and potency prediction for clinical EV batches.

Keywords: mesenchymal stem cells; extracellular vesicles; orthopedics; osteoarthritis; extracellular matrix; flow cytometry; confocal microscopy; microfluidics

1. Introduction

Extracellular vesicles (EVs) are nanoparticles delimited by a lipid bilayer that are naturally released from almost all cell types [1]. Particles differ in both size and origin, either endosomal or plasma membrane-derived [2]. Endosomal EVs are generally smaller (50–150 nM) and referred to as exosomes, whereas plasma-membrane EVs are larger (>100 nM) and defined as microvesicles. The difference in their origin suggested a differential fingerprint, for both cargo and functions [2], although, to date, no molecular signatures have been defined to sharply distinguish these two categories of nanoparticles. Therefore, due to the absence of a consensus on specific markers for EV subtypes, they have recently been divided into small (<200 nM) and large (>200 nM) particles, superseding the classical exosome and microvesicle definition, with a more general definition based on either substantial overlap for some determinants in their surface biochemical composition (presence of CD63/CD81/CD9, etc.) or cell of origin [3]. In living organisms, EVs act as intercellular messengers to transfer biological signals such as nucleic acids, proteins and lipids [4]. Due to this function of cargo transfer, EVs have been recently studied as therapeutics for the delivery of both natural and engineered loads [5]. In this perspective, EVs, especially when collected from mesenchymal stem cells (MSCs), showed innate therapeutic properties in the frame of regenerative medicine [6]. Notably, pre-clinical studies shed light on the potential of MSC-EVs for several pathologies affecting the liver, heart, kidney, brain, bone, and joints [6]. In particular, in osteoarthritis (OA), characterized by both degenerative and inflammatory processes, MSC-EVs were able to promote the regeneration of joint components, such as synovium and cartilage, and to reduce pain [7] through the delivery of their cargo enriched in anti-inflammatory and trophic factors and miRNAs [8].

Mandatory to understanding and ameliorating EV for potential future clinical use is the correct visualization and quantification of their uptake at the cell, tissue, and organ levels in explorative and pre-clinical settings. At present, the majority of available data has been generated in vitro, although some major issues have still to be faced to consistently translate knowledge in clinics. In this frame, regardless of the protocol used, studies concerning EV internalization rely mainly on the detection of fluorescent signals in target cells or tissues incubated with pre-labeled EVs. Among the compounds for labeling EVs, membrane-specific fluorescent dyes (PKH67, PKH26, etc.) [9] or membrane-permeable carboxyfluorescein diacetate succimidyl ester (CFSE) [10] are the most used. Except for very expensive and rare high-resolution instrumentation, in the majority of laboratories, the available instrumentation allows fluorescence detection through conventional flow cytometry and/or fluorescent or confocal microscopy. In our experience, despite giving valuable information such as the ability of EVs from adipose-derived MSCs to interact with fibroblast-like synoviocytes (FLSs) isolated from OA patients [11], these techniques do not allow an exact understanding of the level of interaction between EVs and cells. This may be ascribed to the fact that recipient cells are subjected to trypsin digestion and multiple washing steps, possibly hampering EV-cell interaction and consequently the detection of the fluorescent signal [12]. Moreover, while by flow cytometry it is impossible to discern internalized EVs from those simply attached to the cell surface or bound to the plasma membrane, still this is very difficult also by confocal microscopy on standard monolayer cultures. This is due to the reduced z-axis thickness of target cells grown in an unnatural 2D environment on plastic cell culture surfaces, which makes it difficult to clearly discriminate between the extracellular matrix, plasma membrane, and cytoplasm [13].

Due to the above-mentioned reasons, there is an urgent need to establish advanced models to deeply characterize the interaction of EVs with single cells. In fact, the development of new tools able to discriminate and quantify EV adhesion vs internalization is mandatory to study molecular mechanisms in basic research and establish release assays to characterize alternative EV batches for regenerative medicine purposes. In this context, 3D culture models, like the ones that can be established in microfluidic devices, represent an invaluable tool. Culturing cells in a 3D environment that resembles their natural niche allows indeed to study their behavior in an in vivo-like milieu where cells may acquire a more natural and tridimensional morphology. In this scenario, confocal microscopy could make possible to discriminate between cells with internalized EVs and cells that have interacted only superficially with EVs, a crucial issue for biodistribution and potential enhancement of EV uptake beside simple interaction with target cells, for both basic research and clinical applications. Among the most diverse approaches to establish a 3D culture, microfluidics provides several advantages, such as the low number of samples required and the possibility to establish spatially confined cocultures, creating compartmentalized environments that can simulate human tissues, organs, or pathology features [14].

This work focused on the technical advantage of studying the interaction and internalization of MSC-derived EVs in donor-matched FLSs and articular chondrocytes (ACHs) isolated from OA patients in a 3D physiological-like microenvironment. 3D confocal microscopy data were compared with those obtained through conventional flow cytometry on cells grown in 2D on traditional plastic surfaces. Finally, the role of the hyaluronic acid (HA)-based matrix in EV docking was dissected to confirm HA as a crucial factor in the trafficking of EVs in target cells. We hypothesized that 2D and 3D approaches should be combined to achieve complementary data on cell interaction with EVs and that this strategy could represent a valuable tool for both basic types of research on EV-cell molecular mechanisms and clinical translation for therapeutic EV batches choice.

2. Materials and Methods

2.1. Ethics Statement

This work was completed at IRCCS Istituto Ortopedico Galeazzi. Institutional Review Board approval (San Raffaele Hospital Ethics Committee approval on date 8 March 2018, registered under number 6/int/2018) was obtained before the beginning of the study. Sampling was performed after the procurement of patient informed consent (CI_REGAIN_adulto_v2) and following the 1964 Helsinki declaration and its later amendments.

2.2. Adipose Tissue-Derived Mesenchymal Stem Cell (ASC) Isolation and Culture

Adipose waste material was collected from three female donors (54 ± 8 years old) undergoing liposuction. Tissue was processed as previously reported [11]. Briefly, after type I collagenase (Worthington Biochemical Co., Lakewood, NJ, USA) digestion at 37 °C for 30 min, tissues were filtered (100 µM cell strainer) and centrifuged at 1000× *g* (5 min, RT). Pellets were suspended in DMEM + 10% FBS and seeded at 5×10^3 cells/cm^2 (37 °C, 5% CO_2, 95% humidity).

2.3. ASC Characterization by Flow Cytometry

ASCs at passage three were analyzed by flow cytometry with a CytoFLEX flow cytometer (Beckman Coulter, Fullerton, CA, USA), collecting at least 10,000 events. Antibodies used to confirm ASC phenotype [15] were: anti-CD44-PE (Cat# 130-110-293), CD90-FITC (clone REA897), CD105-PerCP-Vio700 (clone REA794), CD45-PE Vio770 (clone REA747) (Miltenyi Biotec, Bergisch Gladbach, Germany). Doublets were removed from analysis gating events on FSC-H and FSC-A plot.

2.4. EV Production

ASCs at passage three and 90% confluence were washed twice with PBS, and DMEM without FBS was added. After 48 h, supernatants were collected and serially centrifuged at 376× *g*, 1000× *g*, 2000× *g* and twice at 4000× *g* to remove floating cells and debris. When fluorescent EVs were needed, the supernatant was labeled with 10 µM CFSE (Sigma-Aldrich, Milan, Italy) for 1 h at 37 °C. Eventually, EVs or CFSE-labeled EVs were collected by ultracentrifugation at 100,000× *g* for 3 h at 4 °C. No more than 25 µL supernatant were left and pellets were washed with 25 mL PBS to remove excess dye and final pellets, again with no more than 25 µL supernatant, were suspended in PBS, 100 µL per 25 mL of initial culture supernatant. Initial CFSE concentration at this step was 1:4000 reduced. EVs from the three ASCs isolates were pooled for incorporation studies. To confirm a lack of major protein contamination, the number of particles was related to total protein amount and EV batches considered of good purity when falling in the 10^8 to 10^{10} particle/μg protein range, as described in [16].

2.5. EV Characterization by Flow Cytometry

CFSE-EVs were analyzed by flow cytometry with a CytoFLEX flow cytometer calibrated with FITC-fluorescent microbeads to allow the detection of fluorescent particles as small as 100 nM, as previously reported [11]. Calibration standards were 160, 200, 240, and 500 nM sizes (Biocytex, Marseille, France). EVs were 1:10,000 diluted in PBS and 100 μL stained with anti-CD63-APC (clone H5C-6) and CD81-APC (clone 5A6) (Biolegend, San Diego, CA, USA) antibodies for 30 min at 4 °C. Gains were: FSC = 106, SSC = 61, FITC = 272, PE = 116, and PC7 = 371. After incubation, samples were diluted with PBS to 1000 μL before analysis. At least 10,000 events were collected. CFSE-EVs were first compared in the FITC channel with a PBS+CFSE sample used as background signal, to gate only stained EVs. Only events falling in this gate were used to analyze Ab unstained and stained CFSE-EVs and cytograms in the APC channel overlaid to detected positive particles. This strategy of initial FITC gating allowed to remove from the analysis Ab aggregates that, being APC labeled, do not give autofluorescence spillover in the FITC channel.

2.6. EV Characterization by Transmission Electron Microscopy

Purified EVs were blotted on Formvar carbon-coated grids for 10 min and drops removed. After negative stain (2% uranyl acetate aqueous suspension, 10 min), the grid was dried at RT. TALOS L120C transmission electron microscope (Thermo Fisher Scientific, Waltham, MA, USA) at 120 kV was used to examine the samples.

2.7. EV Characterization by Nanoparticle Tracking Analysis (NTA)

Nanosight LM10-HS system (NanoSight Ltd., Amesbury, UK) was used to visualize purified EVs that were 1:100 diluted in PBS. Three 30 s recordings were performed for each sample. Dedicated software provided both the concentration measurements and the high-resolution particle size distribution profiles.

2.8. ACHs and FLSs Isolation and Culture

Donor-matched FLSs and ACHs were isolated from synovial membranes and full-thickness cartilage explants obtained from three male donors (67 ± 5 yo) undergoing hip replacement. Synovial membranes were treated as previously described [11]. Briefly, minced tissues were enzymatically digested (37 °C, 3 h) by 0.25% w/v type I collagenase (Worthington Biochemical Co., Freehold, NJ, USA). Digested samples were filtered through a 100 μM cell strainer and centrifuged (376× *g*, 5 min). Pellets were suspended in DMEM supplemented with 10% FBS and FLSs selected for plastic adherence. Cartilage explants were processed as previously described [17,18]. Briefly, minced tissues from the femoral head were incubated overnight at 37 °C with 0.15% w/v type II collagenase (Worthington Biochemical Corporation). After centrifugation, cells were seeded in DMEM supplemented with 10% FBS, and ACHs selected for plastic adherence. Both ACHs and FLSs were maintained in an incubator at 37 °C in a humidified atmosphere with 5% CO_2 and used for the following experiments at the same passage for donor-matched cells.

2.9. Determination of EV Interaction with ACHs and FLSs in 2D Cultures by Flow Cytometry

ACHs and FLSs were seeded in duplicate each at 10,000 cells/cm^2 in 24-well plates. After 24 h (time 0 for interaction assessment), the medium was removed and replaced with either DMEM + 10% FBS (ultracentrifuged for 24 h at 100,000× *g* to remove endogenous EVs), or DMEM + 10% FBS + CFSE-labeled EVs (ratio of 50,000 EVs per seeded cell). Due to average 1:50 purified CFSE-EVs dilution in culture medium, residual CFSE concentration during EVs uptake was below 0.1 nM, greatly inferior to the recommended concentration range (0.1–10 μM) for labeling and therefore avoiding a specific and EV-independent cell staining. After 1, 6, and 24 h, the medium was removed and cells detached to be compared with cells detached at time 0 before EV supplementation. ACHs and FLSs were analyzed with a CytoFLEX flow cytometer and fluorescence in the FITC channel detected. To adjust for technical differences in the number of seeded cells or cell growth between seeding and flow cytometer analysis leading to slightly different EVs to cells ratios, the fluorescence values were normalized per counted events in the cell gate under SSC-A vs FSC-A under the assumption of a linear and inverse proportion between fluorescence intensity and EVs to cell ratio (e.g., double cell number leading to half mean fluorescence intensity and vice versa). The assumption is based on the working conditions that are far below the saturation EVs to cells ratio (100,000 to 1) previously demonstrated in [11].

2.10. Hyaluronic Acid (HA) Matrix Removal and Determination of EV Interaction in 2D Cultures by Flow Cytometry

To remove HA matrix before seeding in 24-well plates, after trypsin detachment, both ACHs and FLSs were split in two, and one aliquot was treated for 10 min at 37 °C with 10 U/mL hyaluronidase (Sigma-Aldrich), whereas the other aliquot was incubated for the same time without the enzyme (mock). After two washes, treated cells were supplemented with 1 mM 4-methylumbelliferone (4-MU) (Sigma-Aldrich), an HA synthesis inhibitor, in the complete medium, and eventually, both hyaluronidase + 4-MU treated and untreated cells were seeded as previously described. 4-MU was left all night in treated cells. The following day, CFSE-labeled EVs (ratio of 50,000 EVs per seeded cell) were added in both treated and untreated cells and 4-MU was maintained in treated cells. Flow cytometry was performed at 0 (before EV addition) and 24 h and 4-MU treated (±EVs) *vs* untreated (±EVs) samples compared. Fluorescence values were normalized by detected events under SSC-A vs FSC-A plot.

2.11. Assessment of CFSE-Labeled EV Transfer to ACHs and FLSs by Microscopy

As per paragraph 2.9., ACHs and FLSs were seeded in 8-wells multi-chamber glass slides and CFSE-labeled EVs (ratio of 50,000 EVs per seeded cell) were supplemented the day after. Before seeding, both cell types were labeled with CellTracker™ Deep Red Dye (Invitrogen, Eugene, OR, USA) following the manufacturer's instructions. Briefly, after detachment, 150,000 cells were suspended in serum-free medium and the dye solution was added at a final concentration of 500 nM. Cells were then incubated for 30 min and washed once before being counted and seeded. Images were taken at 0 and 24 h after EV addition.

2.12. Design and Fabrication of the Microfluidic Chip

The microfluidic chip was designed using Solid Edge ST9 CAD Software (Siemens AG, Texas, USA). The device comprised two gel compartments to inject fibrin-embedded tissue-specific cells, which are ACHs and FLSs, divided by a central channel for the injection of medium and EVs. The width and the height of the channels were 1 mM and 150 μM, respectively, for a total length of 7 mM. The three channels were separated by regularly spaced trapezoidal posts (basis: 0.07 mm/0.3 mm; height: 0.2 mM; distance: 0.1 mM) (Figure 1).

The mold for the microfluidic chip was generated in acrylic resin using a DLP 3D Printer (ASIGA, Sydney, Australia). Microfluidic chips were then produced by replica molding using PDMS SYLGARD® 182 Silicone Elastomer (Dow Corning, Michigan, USA), poured on 3D printed molds and cured at 65 °C for 120 min in oven, finally punched to obtain the ports (gel compartments: 1 mM Ø for inlet/outlet; central channel: 1 mM Ø for inlet, 2 mM Ø for outlet) and bonded to a glass slide through plasma bonding using a Plasma Cleaner (Harrick Plasma Cleaners, New York, USA).

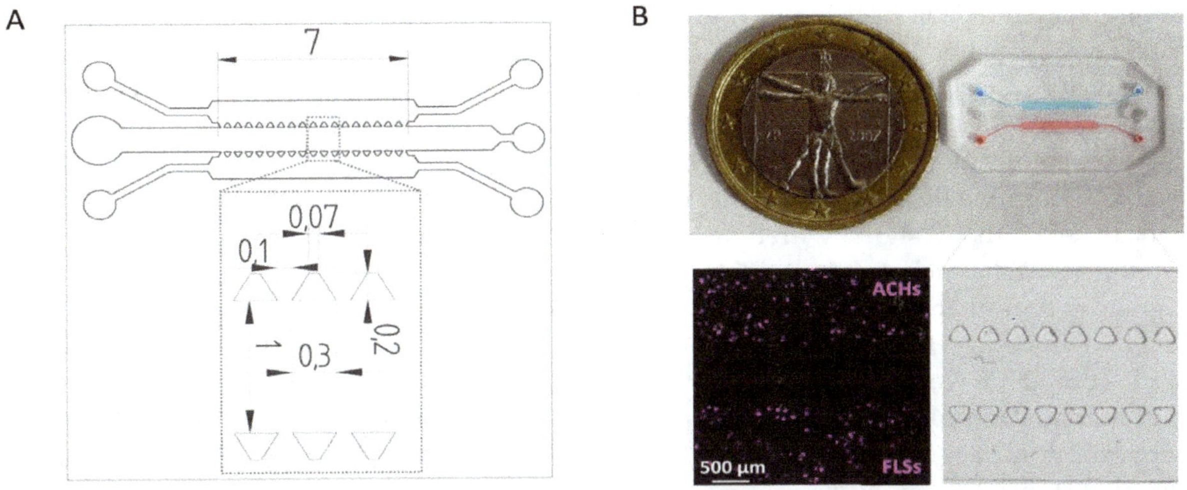

Figure 1. Design and fabrication of the microfluidic device. (**A**) Technical drawing of the microfluidic device with dimensions of trapezoidal posts and channels expressed in mm. (**B**) Top-view of the microfluidic device with ACH (blue) and FLS (red) compartments. The inset shows a stereomicroscope image of the microfluidic chip features. Representative picture showing ACHs and FLSs embedded in fibrin gels after staining.

2.13. *Cell Culture in the Microfluidic Chip*

Before seeding in the microfluidic device, ACHs and FLSs were stained with CellTracker™ Deep Red Dye (Invitrogen, Eugene, Oregon, USA). For HA removal experiments, cells were treated with hyaluronidase before staining, as previously described. For fibrin gel embedding, cells were suspended at 3×10^6 cells/mL in human thrombin (4 UI/mL diluted in 40 mM $CaCl_2$, Tisseel kit, Baxter), mixed 1:1 with human fibrinogen (20 mg/mL diluted in PBS, Sigma-Aldrich), and injected in the respective microfluidic compartments. The chips were incubated at RT for 7 min to let the gels polymerize. This procedure resulted in a final cell concentration of 1.5×10^6 cells/mL in 10 mg/mL fibrin gels.

Complete medium with or without 4-MU depending on the experimental groups was injected in the central channel of the chip after gel polymerization. After culturing chips for 24 h at 37 °C and 5% CO_2, 50,000 EVs per cell were injected in the central channel with or without 4-MU depending on the experimental setting. Chips were fixed at the desired time points (1, 6, and 24 h after EV injection) with 2% paraformaldehyde for 15 min at RT. Chips not injected with EVs were also fixed and used to define background fluorescence signals. Pictures were taken by a confocal microscope (Leica SP8) acquiring a z-stack of 150 µM corresponding to the entire channel height. Each sliced picture was taken at a distance of 1 µM.

2.14. *Quantification of Internalized EVs*

To evaluate the interaction between cells and EVs, image analysis was performed by ImageJ software. We developed a macro script to automatize the analysis process using as input two-channel z-stack images acquired by confocal microscopy. For each stack image, the macro performed four counting steps: 1. count of the total number of cells and generate ROIs of the cell areas, 2. count of the number of cells with internalized EVs, 3. count of the number of cells interacting with EVs (either on the surface or internalized), 4. evaluation of the signal intensity (gray value) of the EV channel, using the cell ROIs generated in step 1. In particular, for the first phase, a "Z project" of the maximum signal intensity was generated from the cell channel and the background was removed using the command "Subtract Background". After applying the "Watershed" filter to separate close cells, the "Analyze particles" tool (minimum cell size: 80 µm^2) was used to count the total number of cells. The gray value for the signal of the cell tracker was 8 bit sampled corresponding to 255 values of gray intensity. The threshold was set to include as much signal as possible (from 15 to 255), after subtracting the dark background, to include the whole cell. The selected threshold was the result of an empirical iterative process, where different independent expert biologists analyzed sample images in order to optimize the value. For the second phase, the ImageJ plugin "Colocalization" (https://imagej.nih.gov/ij/plugins/colocalization.html), was used to generate an image of the colocalized points, representing the EVs internalized by the cells. On this image, the "Analyze particles" tool was used to determine the number of cells with internalized EVs, considering positive only the cells that colocalized with EV signal with a minimum size of 5 μM^2. Subsequently, the function "AND" between the stack channel of cells and the stack channel of EVs was used to evaluate the presence of close signals in both channels. Boolean "AND" function is similar to Colocalization but less restrictive allowing to include also the cell with EVs on the membrane. The resulting image, after applying the "Z project" command, was examined with the "Analyze Particles" tool with a threshold size of 64 μM^2 (80% on the minimum cell size) determining the number of cells interacting with the EVs, regardless if present on the surface and/or internalized. Furthermore, the "Measure" tool was used to evaluate the EV signal strength in the ROIs generated in the first step. This section has been performed in the Z Project of the EV channel to evaluate all the EVs near or inside the cells. After that, a new stack was obtained subtracting (Image Calculator, Subtract) to EVs channel stack the colocalized point stack, to evaluate only the signal strength only on the cell surface. Also, in this case, the "Measure" command was used on the Z Project of the resulting stack using the same ROIs generated in step 1.

2.15. Statistics

Statistical analysis was performed using GraphPad Prism Software version 5 (GraphPad, San Diego, CA, US). Normal data distribution was assessed by the Kolmogorov–Smirnov normality test. Student's *t*-test was used to compare data. Level of significance was set at p-value < 0.05.

3. Results

3.1. ASCs and ASC-EVs Showed Typical Features Confirming Their Identity

ASCs homogeneously expressed mesenchymal antigens CD44, CD90 and CD105, being negative for hematopoietic marker CD45 (Figure 2A), confirming their identity. As revealed by NTA, purified ASC-EVs mostly ranged between 50 nM and 400 nM (Figure 2B), with the majority of particles resulting in smaller than 200 nM. EV size was further confirmed by transmission electron microscopy (Figure 2B). A calibrated flow cytometer able to identify particles as small as 100 nM [8,11] confirmed the NTA dimensional range (Figure 2C). The absence of EVs major protein contamination, that can influence size determination with particle aggregates, was confirmed by protein analysis of EV batches that resulted to have $0.33 \times 10^9 \pm 0.12$ particles/µg protein (N = 4, mean ± SEM), falling in the 10^8 to 10^{10} range claimed for pure preparations [16]. Flow cytometry also confirmed the solid (>90%) presence of the EV markers CD63 and CD81 (Figure 2C).

3.2. 2D Experiments Showed a Higher Interaction of ASC-EVs with FLSs Than ACHs

ACHs and FLSs grown on 2D surface were incubated for 24 h with CFSE-labeled EVs, and interaction kinetics scored at 1, 6, and 24 h. Flow cytometry demonstrated the fluorescence gain over time in both cell lines. At 24 h, both ACHs and FLSs resulted to be CFSE positive, with a complete shift of the fluorescence peak (Figure 3A,B). EVs: ACHs of 50,000:1 did not lead to uptake saturation, since halving the cell number, and therefore doubling the available EVs per cells, resulted in a two-fold increase in detected fluorescence (data not shown). Fluorescence microscopy further supported the detection of EV interaction with the whole cell population (Figure 3C,D), with particles and particle clusters in the cytoplasm (Figure 3E) indicating a specific and EV-related process rather than unspecific labeling, given by free or protein aggregate-bound and released CFSE remnant, that typically results in a diffuse staining. FLSs did not show a linear time-dependent gain of signal, due to a flection in the fluorescence increment curve between 6 and 24 h. The same trend was observed also in ACHs, suggesting a conserved mechanism (Figure 4). Eventually, comparing flow cytometry fluorescence plots, the interaction between EVs and FLSs resulted statistically higher than with ACHs at all time points (Figure 4).

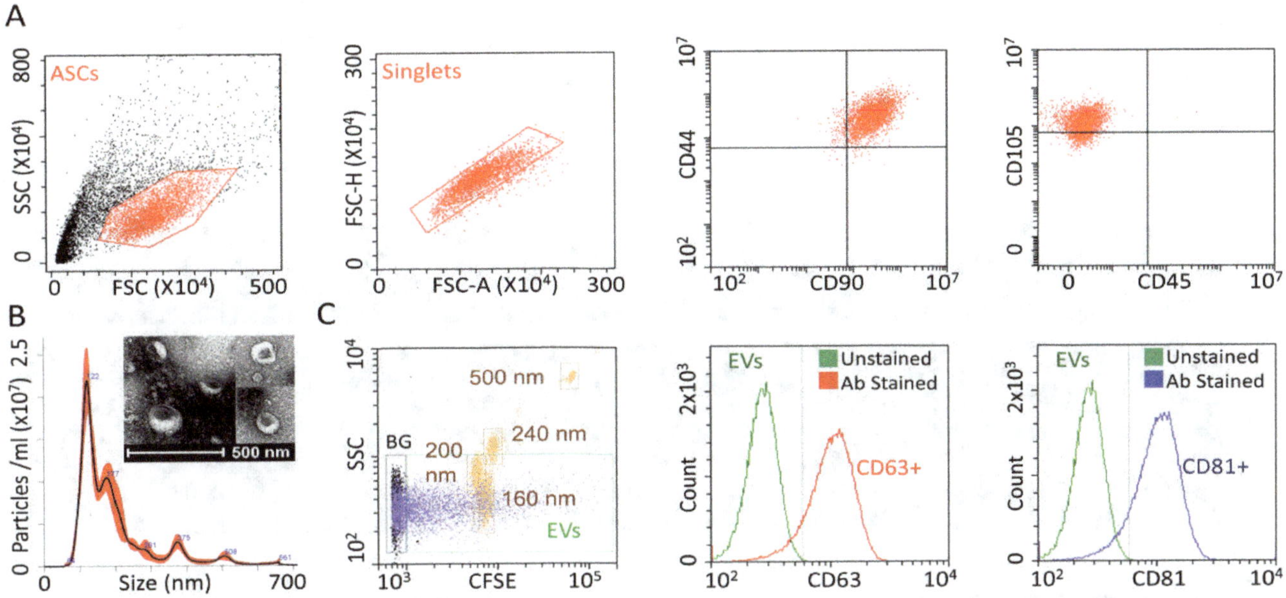

Figure 2. ASC and ASC-EV phenotype. (**A**) After FSC/SSC gating and singlets definition, ASCs resulted positive for CD90/44/105 MSC markers and negative for CD45 determinant. Representative cytograms of a single population are shown. (**B**) The size distribution of nanoparticles by NanoSight particle-tracking analysis and transmission electron micrographs of ASC-EVs showing particles with characteristic cup-shaped morphology (**C**) The first panel shows the merged cytograms of FITC-fluorescent nanometric beads (160 nM, 200 nM, 240 nM and 500 nM, squared in brown) used for flow cytometer calibration and CFSE-fluorescently labeled ASC-EVs (total population in violet, CFSE positive events squared in green as EVs), both gated with respect to PBS+CFSE background (total population in black and squared in black as BG), and confirmed instrument sensitivity to identify ASC-EVs as small as below 200 nm. Events gated in the EVs trace were used for assessing CD63/81 staining shown in the last two panels made by merging unstained and Ab stained (CD63 in red and CD81 in blue) CFSE-EVs cytograms. Both EV markers were detected on the majority of isolated ASC-EVs. One representative donor is shown.

3.3. *3D Culture Imaging Allowed to Discriminate Internalized ASC-EVs Both in ACHs and FLSs*

After being embedded in 3D fibrin gel, ACHs and FLSs were incubated with EVs and interaction at different time points was detected by confocal microscopy. Similar to 2D cultures, pictures of 3D cultures confirmed that after 24 h EV fluorescence was clearly associated with both ACHs and FLSs, including cell prolongations (Figure 5A). The analysis of the intensity of total EV fluorescence, regardless of its cell location, demonstrated an increase over time for both ACHs and FLSs. Differently from monolayer cultures where the fluorescence peaks were compact at all time points, in 3D cultures at 6 and 24 h the fluorescence intensity spectra were more distributed (Figure 5B). Supporting 2D data, the saturation mechanism and the higher EV interaction with FLSs compared to ACHs were confirmed, although not reaching statistical significance (Figure 5C). A deeper examination of the images allowed us to discriminate cells with internalized particles (Figure 5D), thus providing more information than classical 2D cultures. As reported in Figure 5E, the number of interacting (inside + outside EVs) ACHs and FLSs increased over time, reaching a plateau phase (around 80% in both cell types) 6 h after injection. Notably, although EVs started to interact at 1 h, incorporation was not detected at this time point. At 6 h, 24% and 42% of ACHs and FLSs resulted in CFSE positive inside their cytoplasm, respectively. As per total fluorescence and total interaction, the incorporation at 24 h reached a plateau for both ACHs (34% of cells with internal fluorescence) and FLSs (37%).

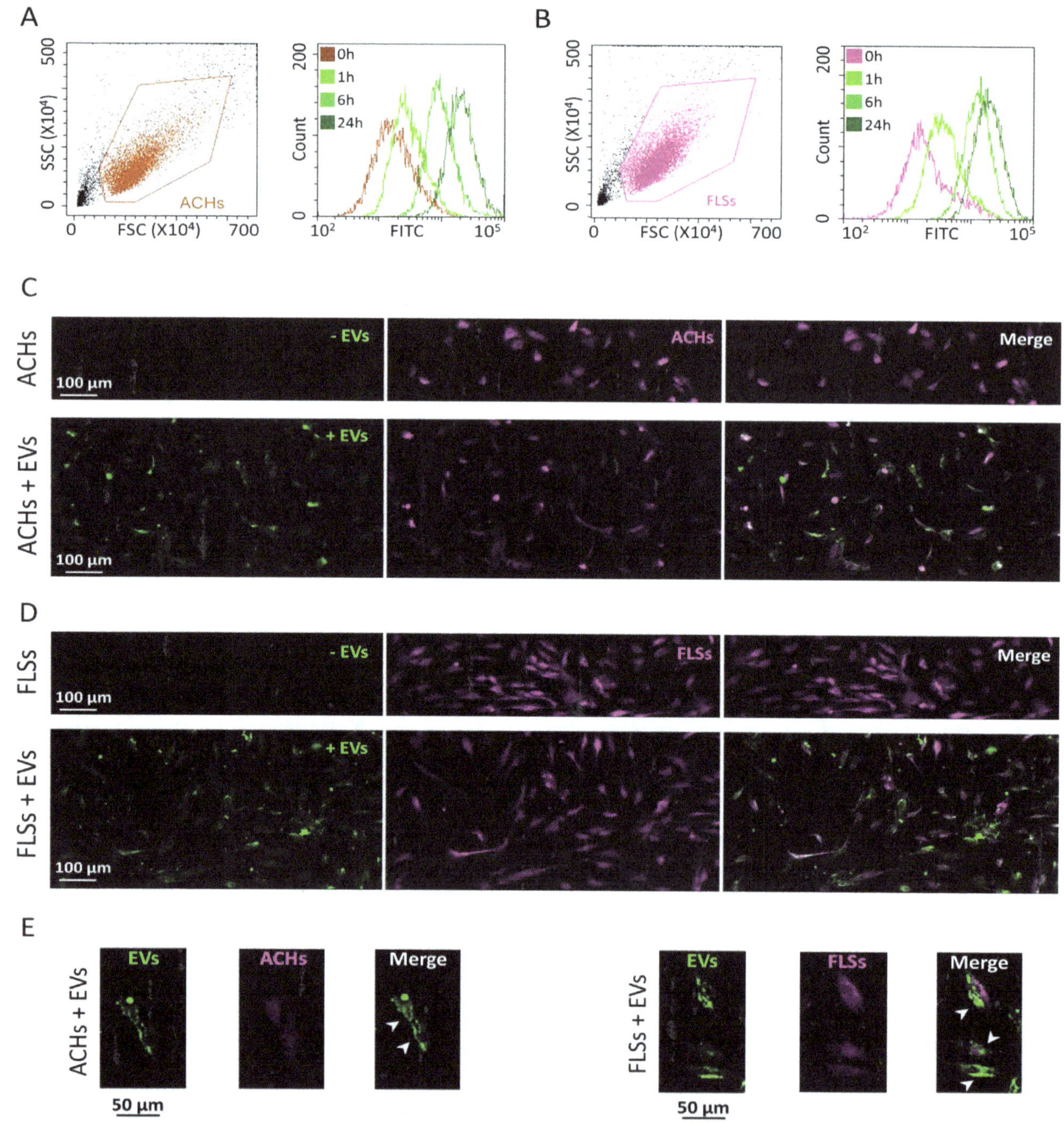

Figure 3. ASC-EV interaction with ACHs and FLSs over time in 2D. (**A**,**B**) Cytograms at time 0, 1-6-24 h for ACHs and FLSs incubated with ASC-EVs showing an increase in FITC-channel intensity signal over time. One representative donor is shown. (**C**,**D**) Images at time 0 and after 24 h of ACHs and FLSs grown in 2D and incubated with ASC-EVs. EVs were CFSE-stained and are shown in green, whereas cells were stained with CellTracker™ Deep Red Dye and are shown in violet. One representative donor is shown. (**E**) Magnification of ACHs and FLSs interaction with ASC-EVs that clearly stain in a punctuate fashion the cytoplasm and perinuclear area of single cells, as indicated in the merge pictures by the white arrows.

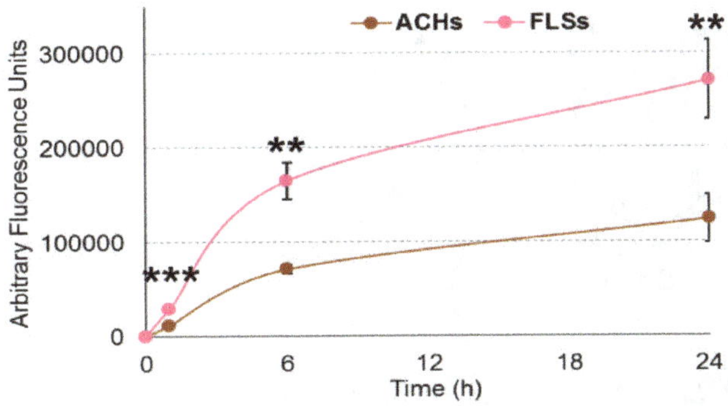

Figure 4. Differential ASC-EV interaction with ACHs and FLSs over time in 2D. Fluorescence intensities in the FITC channel for ACHs and FLSs over time. N = 3, values are shown as mean ± SE Significance for p-value < 0.05. ** for p-value < 0.01 and *** for p-value < 0.001.

3.4. *Hyaluronic Acid Removal Affected ASC-EV Interaction with Cells in 2D but Not in 3D Cultures*

Cells treated with hyaluronidase and maintained in complete medium containing hyaluronan synthesis inhibitor 4-methylumbelliferone (4-MU) were co-incubated with EVs. Fluorescence was detected at 24 h with either flow cytometry for 2D cultures or confocal microscopy for 3D microfluidic devices. Flow cytometry confirmed the significant (p-value < 0.05) reduction (42 ± 7%) in fluorescence in FLSs (Figure 6A). The diminution was also observed in ACHs with a similar percentage (47 ± 1%) (Figure 6B).

When analyzed in the 3D-based model, both ACHs and FLSs treated with 4-MU showed the presence of a fluorescent signal outside the cell membrane (Figure 7A and Supplementary File 1). The analysis of signal intensity did not result in decreased interaction between ASC-EVs and cells compared to untreated cells (Figure 7B). Consistently, 4-MU treatment did not alter the percentage of both ACHs and SFLs interacting with or incorporating EVs (Figure 7C).

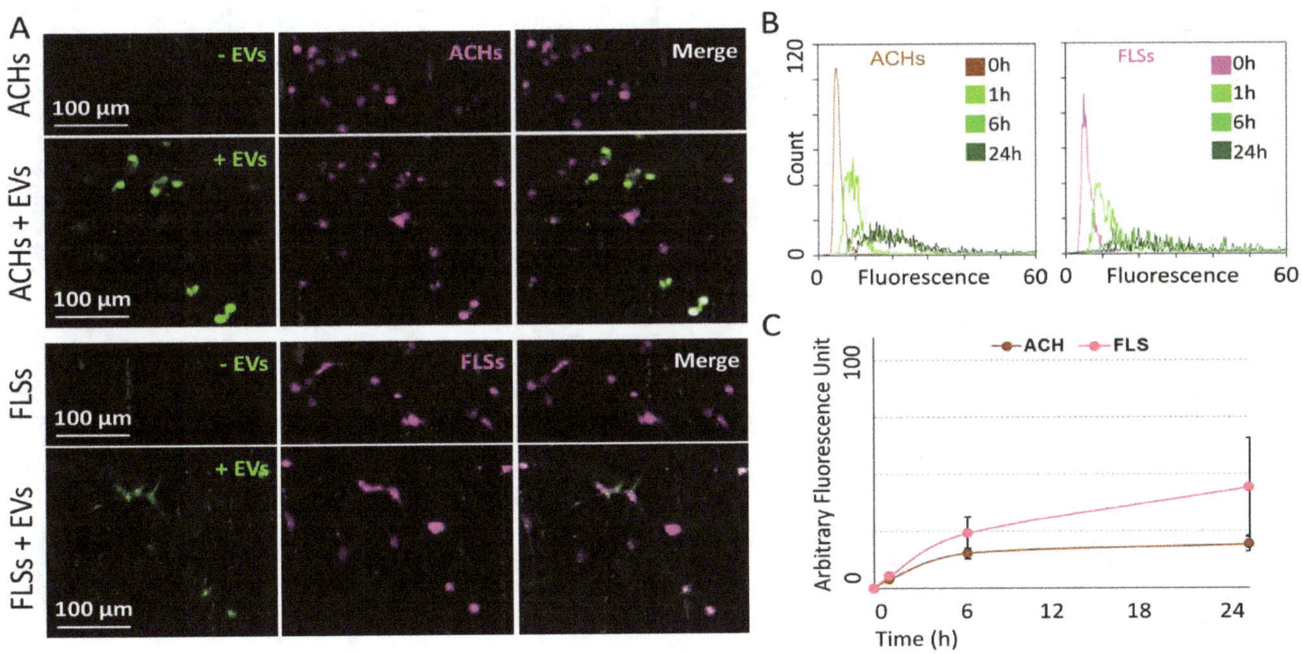

Figure 5. *Cont.*

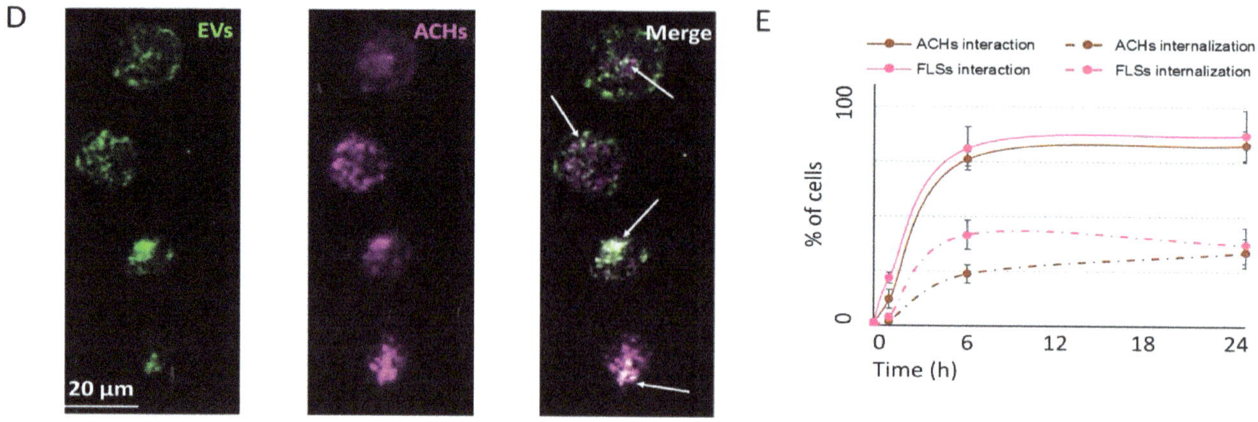

Figure 5. ASC-EV interaction with and incorporation in ACHs and FLSs over time in 3D microfluidic devices. (**A**) Pictures at time 0 and after 24 h of ACHs and FLSs grown in 3D chips and incubated with ASC-EVs. EVs were CFSE-stained and are shown in green, whereas cells were stained with CellTracker™ Deep Red Dye and are shown in violet. The merged pictures clearly show the overlapping of the signal of ASC-EVs with cells. One representative donor is shown. (**B**) Green fluorescence intensity at different time points of ACHs and FLSs incubated or not with ASC-EVs inside the 3D microfluidic device. (**C**) The trend over time of mean green fluorescence of ACHs and FLSs incubated with ASC-EVs. N = 3, values are shown as mean ± SE. Significance for p-value < 0.05. (**D**) Representative pictures of ASC-EVs (green) incubated with ACHs (violet) after 24 h in 3D microfluidic device. EVs clearly surrounded ACHs and were also internalized, as indicated in the merge pictures by the white arrows. (**E**) Trend over time of the number of cells either interacting with ASC-EVs regardless outside or inside (continuous line) or with internalization (dashed line). N = 3, values are shown as mean ± SE. Significance for p-value < 0.05.

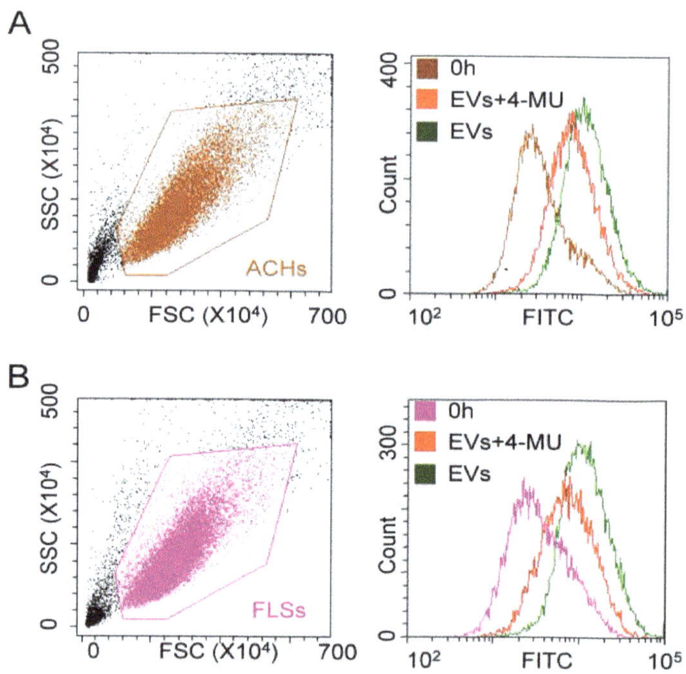

Figure 6. HA removal effect on ASC-EV interaction with ACHs and FLSs in 2D. (**A**,**B**) Cytograms showing fluorescence at time 0 and after 24 h for both ACHs and FLSs incubated with ASC-EVs, in presence and absence of HA-synthesis inhibitor 4-MU. HA removal reduces fluorescence increase at 24 h. 4-MU treated samples at time 0 are not shown for clarity, since no differences with control samples at time 0 emerged. One representative donor is shown.

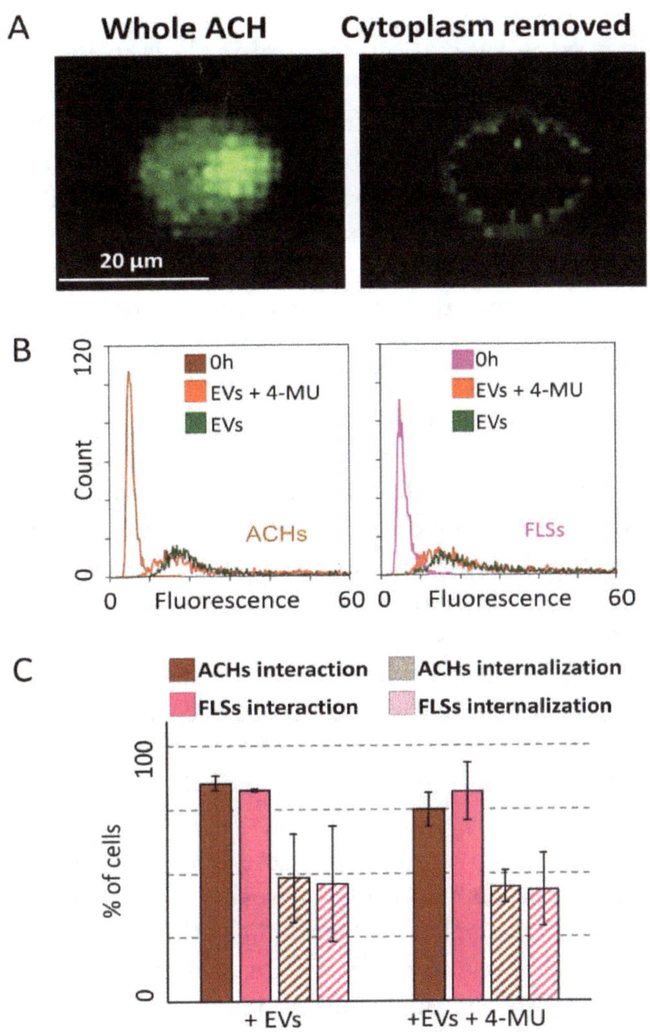

Figure 7. HA removal effect on ASC-EV interaction with and incorporation in ACHs and FLSs in 3D microfluidic devices. (**A**) Image of a selected ACHs before and after removal of fluorescence signals associated with the cytoplasm. The presence of EVs outside the cell body is clear. (**B**) EV fluorescence intensity at time 0 and 24 h of ACHs and FLSs treated or not with 4-MU inside the 3D microfluidic device. 4-MU treated samples at time 0 are not shown for clarity, since no differences with control samples at time 0 emerged. N = 3 (**C**) ASC-EVs interacting with or internalized in ACHs or FLSs with or without treatment with 4-MU at 24 h. N = 3, values are shown as mean ± SE. Significance for p-value < 0.05.

4. Discussion

In this work, classical flow cytometry on 2D cultures and confocal microscopy for 3D cultures were used to identify and dissect the interactions between ASC-EVs and ACHs and/or FLSs by combining the possible complementary information given by these two different systems. Before performing any experiment, the identity of ASCs was confirmed by assessing the expression of standard surface markers [15] and ASC-EVs were characterized obtaining data consistent with the definition of "small EVs" that was recently released by the International Society for Extracellular Vesicles, in place of the old term "exosomes" [3]. For the first time, a clear distinction between adhesion of ASC-EVs on the cell surface and cell uptake was demonstrated through the use of a 3D microfluidic device. Moreover, the role of the extracellular matrix, and in particular of HA, in implementing EV docking was taken into account. The identification of a combined strategy of investigation would indeed be useful for basic research as well as for release assays and potency prediction for clinical EV batches.

In the last years, EV biogenesis started to be deciphered by a large body of evidence [3]. On the contrary, our understanding of the EV-cell interaction mode is in its infancy. This is a crucial key point since EVs, to exert their function, must firstly interact with recipient cells, being known that a preferential bind to specific target cell types occurs in vivo and, presumably, in humans [5]. In several in vivo studies that systemically administered EVs of different origins, a large proportion of injected particles accumulated in the spleen and liver [19–21]. Therefore, the biodistribution issue emerged as a pivotal aspect for safety and efficacy of EVs, to successfully improve targeted EV delivery. Together with internalization and cargo release, the ability of EVs to bind specific cell types could be used and modified to produce more efficient EVs, such as engineered drug delivery platforms to desired sites of action. In order to investigate and shed light on this issue, a profound knowledge of the molecular mechanisms behind EV-cell binding is mandatory, together with the development of dedicated and advanced technical platforms for detection and analysis.

To date, several molecules have been postulated to direct EV-cell recognition, like lipids, glycans, laminin or fibronectin via integrins [22–25]. These results were mainly obtained through conventional detection techniques that are currently used to quantitate EVs and visualize their uptake, including flow cytometers and confocal microscopes. Although useful milestones and mandatory for analysis, the aforementioned tools have some limitations. Flow cytometry can detect the fluorescence increase in target cells when incubated with labeled EVs. In this work, by flow cytometry, FLSs grown on 2D surfaces confirmed the previously observed time-dependent increase in fluorescence together with the saturation over time [11]. Similar kinetics was scored for donor-matched ACHs, confirming a conserved EV-cell interaction mechanism that was present also in unrelated cell types as bladder cancer [12] or lymphoma cells [26]. Due to the capacity of the flow cytometer to compare samples by median fluorescence intensity, for the first time in our study FLSs and ACHs were compared. The results showed that FLSs had a higher ratio of interaction with fluorescent particles with respect to ACHs. Under the paradigm of biodistribution, this result might suggest FLSs as better interactors when using ASC-EVs as therapeutic vehicles, and synovia as the preferential target tissue compared to cartilage in OA patients. Although these findings would need further confirmation, they could pave the way to a more accurate interpretation of the mechanisms of action of the so called orthobiologics, which include MSCs and their products.

Despite the importance of these results, flow cytometry cannot provide any data able to discern between adhesion of EVs on the extracellular matrix/plasma membrane and their eventual internalization, a crucial parameter to predict EV efficacy. Although with some pitfalls, confocal microscopy is commonly used to assess this parameter. In fact, confocal microscopy on 2D cultures requires very experienced operators able to discriminate between surface or internal localization of fluorescence signals. This is due to the reduced height of analyzed cells, leading also to the lack of or reduced signal on the cell surface adherent to the plastic surface [13]. On the contrary, the combination of confocal microscopy with 3D culture allows the clear detection of single cells embedded in a hydrogel-based matrix, with cell morphology resembling the in vivo condition, facilitating the recognition of fluorescent signals inside cells [27]. Here, fibrin was chosen to provide cells with a natural 3D matrix due to its ease-of-use in microfluidic applications and high biocompatibility [28]. Beyond these properties, fibrin matrix also shows some microstructural features similar to collagen [29], a major component of synovial and chondral extracellular matrix, thus being suitable to generate an in vivo-mimicking microenvironment for both synovial fibroblasts and articular chondrocytes. In our study, in both FLSs and ACHs, the results obtained in 3D cultures confirmed the flow cytometry findings on 2D cells in terms of kinetics of interaction with EVs and the superior ability of FLSs to recruit ASC-EVs. Most importantly, 3D experiments showed that only a portion of target cells were able to internalize EVs, and that at early time points FLSs again performed better than ACHs. This crucial parameter was impossible to be determined with flow cytometry, since all the cells resulted EV-positive as observed in 3D pictures when analyzing the total fluorescent signal instead of the fine-tuned localization. The distinction between efficient internalization and solely surface adhesion of

EVs is of fundamental importance, making microfluidic devices compatible with hydrogel-based 3D culture a valuable tool for all those studies and future clinical tests aimed at confirming and comparing internalization of naïve or engineered EVs-therapeutic products. The future understanding of the technical or biological reasons why only a portion of cells internalized the ASC-EVs and the underlying mechanism, in the view of clinical translation, is mandatory to modify the uptake and the fusion mechanisms and thus improve this percentage. A possible technical explanation might be a partial injury of the cells during the chip preparation, although viability before fibrin embedding always resulted higher than 95%. Also, on a biological point of view, the shift from a 2D culture condition to 3D might have turned some cells to a resting state in order to adapt to the new environment, and therefore delay their intracellular metabolism and uptake kinetics. Again, the possibility to grow cells in an in vivo-like microenvironment will make microfluidics a useful tool for these approaches.

In view of ameliorating EV uptake by dissecting molecular interactions, recently it has been proposed that hyaluronic acid of the extracellular matrix may be involved in EV docking [11,30]. This may be an important way of interaction between those cells enriched in HA-based ECM, such as FLSs and ACHs [31,32], and CD44 (HA ligand) decorated EVs (as those released by ASCs and MSCs in general) [11,33]. In the orthopedic field, CD44-coated EVs were found bound to the soluble HA fibrils in both healthy and OA synovial fluids [34], suggesting HA molecules as rails for the movement of both naturally occurring or therapeutically injected EVs towards other HA-enriched structures. In this perspective, we recently demonstrated in 2D cultures that either blocking CD44 availability on ASC-EVs or removing HA from FLSs surface resulted in a reduction of EV-cell interaction [11]. These led to the hypothesis of the HA matrix as a "sponge" that increases the local concentration of EVs around the plasma membrane, thus facilitating its passive uptake or binding to specific receptors. Nevertheless, the repeal of this mechanism did not completely abolish EV interaction, suggesting the involvement of passive diffusion together with the presence of other ligand-receptor mechanisms, e.g., CD44 presence on FLSs and ACHs surface [35] together with HA coating of MSC-EVs [36]. Here, the 2D model confirmed the role of the HA matrix of EV-cell interaction for ACHs, as already shown for FLSs, suggesting a conserved mechanism of interaction [11]. The HA production was also demonstrated for both ACHs [37] and FLSs [38] in 3D systems, suggesting a conserved function, and in ACHs the expression of the major HA synthase *HAS2* was shown to be comparable between 2D cultures and 3D fibrin-based matrix similar to the one used in this report. Nevertheless, aware of possible subtle differences in HA amount between the two cell culture systems herein analyzed that do not exactly reproduce the conditions used in the mentioned reports, the 3D model clearly showed that even after the removal of the hyaluronan coating, a fluorescent signal could be detected around the cells. This reinforced the hypothesis of other molecules involved in EV docking on the cell surface. In contrast with the 2D culture, in the 3D system, after HA removal, we did not observe a diminution of the total fluorescence of cells. The difference between 2D and 3D may be due to the considerable diversity between the two experimental settings. In 2D cultures, cells are covered with a liquid phase, where EVs rapidly move by Brownian motion, and therefore the time of interaction and possibly docking with the cell surface is reduced. To increase this possibility, the HA coating is crucial for reducing EV motion by acting as flypaper and at the same time increasing their concentration along the cell surface. Conversely, in the 3D model, the fibrin gel naturally acts itself like a sponge, retaining EVs in greater number and more closely than the liquid phase in 2D culture, thus not inducing a fluorescence reduction of 4-MU treated cells. These data reinforce the hypothesis that ECM, either cell-associated or tissue-structural, plays a key role during the process of EV–cell interaction by forming a "spongy" network, embedding vesicles.

5. Conclusions

Future studies will be needed to confirm the role of ECM in the uptake of EVs from cells and tissues of different origins. Given the findings of our study, we suggest a combined strategy of investigation which would take advantage of both the cytofluorimetric characterization of cells grown in 2D as well as of confocal microscopy analysis on 3D cultures. In particular, hydrogel-based 3D cultures mimicking the environment/organization of soft tissues and allowing a more physiological EV-cell interaction will allow for the dissection the array of molecular mechanisms underlying EV-cell interaction. This will provide useful insights to develop new approaches aimed to improve EV docking in target cells or reduce adhesion in those cell types that should not be treated. Moreover, the combination of 2D and 3D approaches will be useful for all the studies aimed not only at the characterization of the interaction between EVs and target cells, but also at confirming the internalization of shuttled cargo.

Author Contributions: Conceptualization, E.R. and S.L.; Data curation, G.T.; Formal analysis, E.R., S.P., S.L. and G.T.; Funding acquisition, M.M. and L.d.G.; Investigation, S.P. and C.P.O.; Methodology, E.R., S.P. and S.L.; Resources, G.T.; Software, G.T.; Supervision, M.M. and L.d.G.; Validation, C.P.O. and P.D.L.; Writing – original draft, E.R., S.P. and S.L.; Writing – review & editing, M.M. and L.d.G. All authors have read and agreed to the published version of the manuscript.

References

1. Lotvall, J.; Hill, A.F.; Hochberg, F.; Buzas, E.I.; Di Vizio, D.; Gardiner, C.; Gho, Y.S.; Kurochkin, I.V.; Mathivanan, S.; Quesenberry, P.; et al. Minimal experimental requirements for definition of extracellular vesicles and their functions: A position statement from the International Society for Extracellular Vesicles. *J. Extracell. Vesicles* **2014**, *3*, 26913. [CrossRef] [PubMed]
2. Cocucci, E.; Meldolesi, J. Ectosomes and exosomes: Shedding the confusion between extracellular vesicles. *Trends Cell Biol.* **2015**, *25*, 364–372. [CrossRef]
3. Thery, C.; Witwer, K.W.; Aikawa, E.; Alcaraz, M.J.; Anderson, J.D.; Andriantsitohaina, R.; Antoniou, A.; Arab, T.; Archer, F.; Atkin-Smith, G.K.; et al. Minimal information for studies of extracellular vesicles 2018 (MISEV2018): A position statement of the International Society for Extracellular Vesicles and update of the MISEV2014 guidelines. *J. Extracell. Vesicles* **2018**, *7*, 1535750. [CrossRef] [PubMed]
4. Doyle, L.M.; Wang, M.Z. Overview of Extracellular Vesicles, Their Origin, Composition, Purpose, and Methods for Exosome Isolation and Analysis. *Cells* **2019**, *8*, 727. [CrossRef]
5. Murphy, D.E.; de Jong, O.G.; Brouwer, M.; Wood, M.J.; Lavieu, G.; Schiffelers, R.M.; Vader, P. Extracellular vesicle-based therapeutics: Natural versus engineered targeting and trafficking. *Exp. Mol. Med.* **2019**, *51*, 1–12. [CrossRef] [PubMed]
6. Zhao, T.; Sun, F.; Liu, J.; Ding, T.; She, J.; Mao, F.; Xu, W.; Qian, H.; Yan, Y. Emerging Role of Mesenchymal Stem Cell-derived Exosomes in Regenerative Medicine. *Curr. Stem Cell Res.* **2019**, *14*, 482–494. [CrossRef]
7. Mianehsaz, E.; Mirzaei, H.R.; Mahjoubin-Tehran, M.; Rezaee, A.; Sahebnasagh, R.; Pourhanifeh, M.H.; Mirzaei, H.; Hamblin, M.R. Mesenchymal stem cell-derived exosomes: A new therapeutic approach to osteoarthritis? *Stem Cell Res.* **2019**, *10*, 340. [CrossRef]
8. Ragni, E.; Perucca Orfei, C.; De Luca, P.; Vigano, M.; Colombini, A.; Lugano, G.; Bollati, V.; de Girolamo, L. miR-22-5p and miR-29a-5p Are Reliable Reference Genes for Analyzing Extracellular Vesicle-Associated miRNAs in Adipose-Derived Mesenchymal Stem Cells and Are Stable under Inflammatory Priming Mimicking Osteoarthritis Condition. *Stem Cell Rev. Rep.* **2019**, *15*, 743–754. [CrossRef]
9. Carpintero-Fernandez, P.; Fafian-Labora, J.; O'Loghlen, A. Technical Advances to Study Extracellular Vesicles. *Front. Mol. Biosci.* **2017**, *4*, 79. [CrossRef] [PubMed]
10. Lee, T.S.; Kim, Y.; Zhang, W.; Song, I.H.; Tung, C.H. Facile metabolic glycan labeling strategy for exosome tracking. *Biochim. Biophys. Acta Gen. Subj.* **2018**, *1862*, 1091–1100. [CrossRef]
11. Ragni, E.; Perucca Orfei, C.; De Luca, P.; Lugano, G.; Vigano, M.; Colombini, A.; Valli, F.; Zacchetti, D.; Bollati, V.; de Girolamo, L. Interaction with hyaluronan matrix and miRNA cargo as contributors for in vitro potential of mesenchymal stem cell-derived extracellular vesicles in a model of human osteoarthritic synoviocytes. *Stem Cell Res.* **2019**, *10*, 109. [CrossRef] [PubMed]
12. Franzen, C.A.; Simms, P.E.; Van Huis, A.F.; Foreman, K.E.; Kuo, P.C.; Gupta, G.N. Characterization of uptake and internalization of exosomes by bladder cancer cells. *Biomed. Res. Int.* **2014**, *2014*, 619829. [CrossRef] [PubMed]
13. North, A.J. Seeing is believing? A beginners' guide to practical pitfalls in image acquisition. *J. Cell Biol.* **2006**, *172*, 9–18. [CrossRef] [PubMed]

14. Sackmann, E.K.; Fulton, A.L.; Beebe, D.J. The present and future role of microfluidics in biomedical research. *Nature* **2014**, *507*, 181–189. [CrossRef]
15. Lv, F.J.; Tuan, R.S.; Cheung, K.M.; Leung, V.Y. Concise review: The surface markers and identity of human mesenchymal stem cells. *Stem Cells* **2014**, *32*, 1408–1419. [CrossRef]
16. Webber, J.; Clayton, A. How pure are your vesicles? *J. Extracell. Vesicles* **2013**, *2*, 19861. [CrossRef]
17. Vigano, M.; Lugano, G.; Colombini, A.; De Luca, P.; Perucca Orfei, C.; Ragni, E.; de Girolamo, L. Evaluation of Different Seeding Methods for Cell-Seeded Collagen Matrix-Supported Autologous Chondrocyte Transplantation. *Joints* **2018**, *6*, 215–219. [CrossRef]
18. Lopa, S.; Piraino, F.; Kemp, R.J.; Di Caro, C.; Lovati, A.B.; Di Giancamillo, A.; Moroni, L.; Peretti, G.M.; Rasponi, M.; Moretti, M. Fabrication of multi-well chips for spheroid cultures and implantable constructs through rapid prototyping techniques. *Biotechnol. Bioeng.* **2015**, *112*, 1457–1471. [CrossRef]
19. Wiklander, O.P.; Nordin, J.Z.; O'Loughlin, A.; Gustafsson, Y.; Corso, G.; Mager, I.; Vader, P.; Lee, Y.; Sork, H.; Seow, Y.; et al. Extracellular vesicle in vivo biodistribution is determined by cell source, route of administration and targeting. *J. Extracell. Vesicles* **2015**, *4*, 26316. [CrossRef]
20. Gangadaran, P.; Li, X.J.; Lee, H.W.; Oh, J.M.; Kalimuthu, S.; Rajendran, R.L.; Son, S.H.; Baek, S.H.; Singh, T.D.; Zhu, L.; et al. A new bioluminescent reporter system to study the biodistribution of systematically injected tumor-derived bioluminescent extracellular vesicles in mice. *Oncotarget* **2017**, *8*, 109894–109914. [CrossRef]
21. Charoenviriyakul, C.; Takahashi, Y.; Morishita, M.; Matsumoto, A.; Nishikawa, M.; Takakura, Y. Cell type-specific and common characteristics of exosomes derived from mouse cell lines: Yield, physicochemical properties, and pharmacokinetics. *Eur. J. Pharm. Sci.* **2017**, *96*, 316–322. [CrossRef] [PubMed]
22. Flannagan, R.S.; Canton, J.; Furuya, W.; Glogauer, M.; Grinstein, S. The phosphatidylserine receptor TIM4 utilizes integrins as coreceptors to effect phagocytosis. *Mol. Biol. Cell* **2014**, *25*, 1511–1522. [CrossRef] [PubMed]
23. Mulcahy, L.A.; Pink, R.C.; Carter, D.R. Routes and mechanisms of extracellular vesicle uptake. *J. Extracell. Vesicles* **2014**, *3*, 24641. [CrossRef] [PubMed]
24. Mu, W.; Rana, S.; Zoller, M. Host matrix modulation by tumor exosomes promotes motility and invasiveness. *Neoplasia* **2013**, *15*, 875–887. [CrossRef] [PubMed]
25. Sung, B.H.; Ketova, T.; Hoshino, D.; Zijlstra, A.; Weaver, A.M. Directional cell movement through tissues is controlled by exosome secretion. *Nat. Commun.* **2015**, *6*, 7164. [CrossRef]
26. Hazan-Halevy, I.; Rosenblum, D.; Weinstein, S.; Bairey, O.; Raanani, P.; Peer, D. Cell-specific uptake of mantle cell lymphoma-derived exosomes by malignant and non-malignant B-lymphocytes. *Cancer Lett.* **2015**, *364*, 59–69. [CrossRef]
27. Rothbauer, M.; Zirath, H.; Ertl, P. Recent advances in microfluidic technologies for cell-to-cell interaction studies. *Lab Chip* **2018**, *18*, 249–270. [CrossRef]
28. Park, C.H.; Woo, K.M. Fibrin-Based Biomaterial Applications in Tissue Engineering and Regenerative Medicine. *Adv. Exp. Med. Biol.* **2018**, *1064*, 253–261.
29. Lai, V.K.; Frey, C.R.; Kerandi, A.M.; Lake, S.P.; Tranquillo, R.T.; Barocas, V.H. Microstructural and mechanical differences between digested collagen-fibrin co-gels and pure collagen and fibrin gels. *Acta Biomater.* **2012**, *8*, 4031–4042. [CrossRef]
30. Rilla, K.; Mustonen, A.M.; Arasu, U.T.; Harkonen, K.; Matilainen, J.; Nieminen, P. Extracellular vesicles are integral and functional components of the extracellular matrix. *Matrix Biol.* **2019**, *75–76*, 201–219. [CrossRef]
31. Rilla, K.; Tiihonen, R.; Kultti, A.; Tammi, M.; Tammi, R. Pericellular hyaluronan coat visualized in live cells with a fluorescent probe is scaffolded by plasma membrane protrusions. *J. Histochem. Cytochem.* **2008**, *56*, 901–910. [CrossRef] [PubMed]
32. Clarris, B.J.; Fraser, J.R. On the pericellular zone of some mammalian cells in vitro. *Exp. Cell Res.* **1968**, *49*, 181–193. [CrossRef]
33. Ramos, T.L.; Sanchez-Abarca, L.I.; Muntion, S.; Preciado, S.; Puig, N.; Lopez-Ruano, G.; Hernandez-Hernandez, A.; Redondo, A.; Ortega, R.; Rodriguez, C.; et al. MSC surface markers (CD44, CD73, and CD90) can identify human MSC-derived extracellular vesicles by conventional flow cytometry. *Cell Commun. Signal.* **2016**, *14*, 2. [CrossRef] [PubMed]

34. Malda, J.; Boere, J.; van de Lest, C.H.; van Weeren, P.; Wauben, M.H. Extracellular vesicle-new tool for joint repair and regeneration. *Nat. Rev. Rheumatol.* **2016**, *12*, 243–249. [CrossRef] [PubMed]
35. Al-Sharif, A.; Jamal, M.; Zhang, L.X.; Larson, K.; Schmidt, T.A.; Jay, G.D.; Elsaid, K.A. Lubricin/Proteoglycan 4 Binding to CD44 Receptor: A Mechanism of the Suppression of Proinflammatory Cytokine-Induced Synoviocyte Proliferation by Lubricin. *Arthritis Rheumatol.* **2015**, *67*, 1503–1513. [CrossRef]
36. Arasu, U.T.; Karna, R.; Harkonen, K.; Oikari, S.; Koistinen, A.; Kroger, H.; Qu, C.; Lammi, M.J.; Rilla, K. Human mesenchymal stem cells secrete hyaluronan-coated extracellular vesicles. *Matrix Biol.* **2017**, *64*, 54–68. [CrossRef]
37. Li, Z.; Yao, S.; Alini, M.; Grad, S. Different response of articular chondrocyte subpopulations to surface motion. *Osteoarthr. Cartil.* **2007**, *15*, 1034–1041. [CrossRef]
38. Uehara, K.; Hara, M.; Matsuo, T.; Namiki, G.; Watanabe, M.; Nomura, Y. Hyaluronic acid secretion by synoviocytes alters under cyclic compressive load in contracted collagen gels. *Cytotechnology* **2015**, *67*, 19–26. [CrossRef]

4

A 3D Cell Death Assay to Quantitatively Determine Ferroptosis in Spheroids

Robin Demuynck [1,2], Iuliia Efimova [1,3], Abraham Lin [4,5], Heidi Declercq [2,3,6,†] and Dmitri V. Krysko [1,3,7,*,†]

1 Cell Death Investigation and Therapy Laboratory, Department of Human Structure and Repair, Ghent University, 9000 Ghent, Belgium; Robin.Demuynck@UGent.be (R.D.); Iuliia.Efimova@UGent.be (I.E.)
2 Tissue Engineering and Biomaterials Group, Department of Human Structure and Repair, Ghent University, 9000 Ghent, Belgium; heidi.declercq1@kuleuven.be
3 Cancer Research Institute Ghent, 9000 Ghent, Belgium
4 Plasma, Laser Ablation and Surface Modelling Group, University of Antwerp, 2610 Wilrijk, Belgium; abraham.lin@uantwerpen.be
5 Center for Oncological Research, University of Antwerp, 2610 Wilrijk, Belgium
6 Tissue Engineering lab, Department of Development and Regeneration, KU Leuven, 8500 Kortrijk, Belgium
7 Department of Pathophysiology, Sechenov First Moscow State Medical University, 119146 Moscow, Russia
* Correspondence: dmitri.krysko@ugent.be
† These authors shared senior authorship.

Abstract: The failure of drug efficacy in clinical trials remains a big issue in cancer research. This is largely due to the limitations of two-dimensional (2D) cell cultures, the most used tool in drug screening. Nowadays, three-dimensional (3D) cultures, including spheroids, are acknowledged to be a better model of the in vivo environment, but detailed cell death assays for 3D cultures (including those for ferroptosis) are scarce. In this work, we show that a new cell death analysis method, named 3D Cell Death Assay (3DELTA), can efficiently determine different cell death types including ferroptosis and quantitatively assess cell death in tumour spheroids. Our method uses Sytox dyes as a cell death marker and Triton X-100, which efficiently permeabilizes all cells in spheroids, was used to establish 100% cell death. After optimization of Sytox concentration, Triton X-100 concentration and timing, we showed that the 3DELTA method was able to detect signals from all cells without the need to disaggregate spheroids. Moreover, in this work we demonstrated that 2D experiments cannot be extrapolated to 3D cultures as 3D cultures are less sensitive to cell death induction. In conclusion, 3DELTA is a more cost-effective way to identify and measure cell death type in 3D cultures, including spheroids.

Keywords: ferroptosis; spheroids; 3D cultures; cell death assay; cancer

---

1. Introduction

The failure rates of new therapies during translation from preclinical research to clinical trials remain a big issue in cancer research [1]. This is largely due to limitations of the two-dimensional (2D) cell cultures, the most commonly used models in drug screening. The 2D cultures do not properly resemble the structures found in vivo as they often lack cell extracellular matrix (ECM) and cell-cell interactions, thereby influencing cell death responses to therapy (in particular to apoptosis) [2,3]. In recent years, the focus of drug screenings has shifted towards the use of three-dimensional (3D) culture methods [4–6] as they more closely mimic the in vivo tumour microenvironment due to cell secretion of its own ECM and the formation of cell-cell interactions [7]. Therefore, 3D structures better resemble the structure and phenotype of the in vivo organ they are derived from [8–11]. Moreover,

3D structures, such as spheroids, better represent in vivo drug resistance, as they form a more dense barrier for the drugs to cross, compared to 2D systems [12], with the core of the structure being most protected against the drugs. Therefore, with regard to the responses to therapy, including to cell death, 3D structures are better test models for drug screening compared to 2D monolayers [13].

Spheroids are one of the most widely used 3D models and can be formed by seeding cells in nonadherent plates allowing for self-assembly (Figure 1A) [7]. In spheroids, the ECM can be produced by cells, which acts as an extra barrier for drugs to cross in order to induce tumour cell death [14–16]. Moreover, the compactness and stiffness of spheroids resemble the in vivo hypoxic tumour microenvironment [17], which can cause drug resistance through cell cycle arrest and downregulation of pro-cell death proteins (e.g., caspase-3 in apoptosis) [18]. Altogether, spheroids can be more easily formed in a high-throughput manner compared to other 3D models and better resemble in vivo conditions, thus making them an attractive model for more efficient drug screening and analysis of cell death responses.

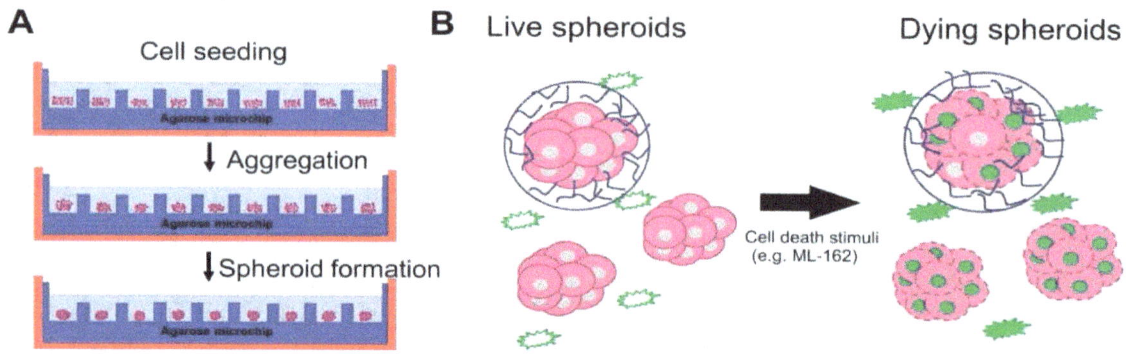

Figure 1. Principles of three-dimensional cell death assay (3DELTA). (**A**) Spheroids are formed by seeding cells on top of an agarose microwell chip. This chip is made by pouring liquid agarose on top of a polydimethylsiloxane (PDMS) mould. Cells will form clusters, leading to the formation of spheroids. (**B**) Sytox Green is nonfluorescent outside of viable cells. When cells are permeabilized at the end of the cell death process and the plasma membrane ruptures, Sytox Green will bind to the DNA and emit green fluorescence, which can then be measured.

Of interest, in the past, cancer cell death was perceived to be dichotomic with only two ways for a cell to die—in a regulated way (apoptosis, controlled via caspase-activation and cytochrome c release [19]) or in an uncontrolled way (accidental necrosis) [20,21]. In line with this concept, necrosis was seen as a form of unprogrammed cell death from cellular damage, which could be induced by different acute physical-chemical injuries (e.g., heat shock, detergents). However, in the past few decades, research has indicated more types of regulated cell death besides apoptosis, which are assembled under the umbrella term 'regulated necrosis' [22,23]. One example of regulated necrosis is called ferroptosis, a newly discovered pathway in cell death, which may be a potential alternative therapeutic pathway to overcome apoptosis-resistant cancer. Ferroptosis can be induced through blockage of Glutathione Peroxidase 4 (GPX4), an important inhibitor of ferroptotic cell death [24–26].

Of note, cell death has been thoroughly studied in 2D cultures, and despite the acceptance that 3D cultures including spheroids are better models for the in vivo environment, not many cell death assays exist for 3D cultures, even less for ferroptosis as the translation of the assay between culture systems is not straightforward. Most commonly used cell death assays in 2D include Calcein AM-propidium iodide (PI) stainings [27], immunohistochemical staining (for example Caspase-3 staining for apoptotic cells [7]), LDH- and ATP-based assays or flow cytometry for the detection of membrane and intracellular markers of cell death [28–30]. However, each of these assays present their own disadvantages for high-throughput cell death response screening in 3D cultures. For example, in Calcein AM-PI staining, PI ions can enter viable cells with a high membrane potential which in turn leads to false positive results [31]. For live/dead assays, such as Calcein AM-PI, the interior of the spheroids cannot be seen with ordinary fluorescence microscopy and require sectioning [28], a labour-extensive process. For immunohistochemical staining, sectioning is also required, making this method cumbersome when monitoring over time is needed [32]. ATP may be used as an indicator for cell viability, but the signal received from ATP luminescence assays is not stable and therefore not reliable [33]. Moreover, as a preparation step, the cells need to be lysed to extract the ATP and measure luminescence, thereby restricting the assay to endpoint analysis. Flow cytometry is another endpoint assay that can only be performed on single cells. Thus, spheroids must be disaggregated, a process and technique requiring extensive training before proficient use [34].

New tools are being developed to overcome the current challenges with 3D analysis, but a simple, quantitative method of determining real-time cell death modalities in 3D cultures remains difficult [35]. Therefore, the aim of this study is to develop a reliable assay which will allow for quantitative determination of cell death type in spheroids. In this study, spheroids were formed by seeding tumour cells (L929 mouse fibrosarcoma and SK-OV-3 human ovarian cancer cells) on top of nonadherent agarose microwell chips and allowing them to self-assemble. The cell death assay was optimized using spheroids collected on day 1. To validate the identification of cell death type, ferroptosis was induced in spheroids collected on day 1 and 10, in the presence or absence of different cell death inhibitors (zVAD-fmk for apoptosis, necrostatin-1s (Nec-1) for necroptosis and ferrostatin-1 (Fer-1), deferoxamin (DFO) and α-tocopherol (α-Toc) for ferroptosis). In this work we demonstrated that this method, named the 3D Cell Death Assay (3DELTA), can efficiently and quantitatively determine different cell death types, including ferroptosis in tumour spheroids. 3DELTA will make it possible to investigate high-throughput cell death responses to therapy with tumour spheroids. The results from this assay and 3D model will be more representative of therapy response in the clinic, and improve the rate of successful drug translation into clinical trials.

2. Material and Methods

2.1. Cell Lines and Culture

L929sAhFas mouse fibrosarcoma cells were cultured in Dulbecco's Modified Eagle Medium (DMEM) Glutamax (Life Technologies, Carlsbad, CA, USA), 10% foetal calf serum (FCS, Life Technologies) and 1% Penicillin/Streptomycin (Life Technologies) at 37 °C and 5% CO_2.

SK-OV-3-luc-GFP human ovarian cancer cells (kindly provided by LECR-UGent) were cultured in DMEM Glutamax containing 10% FCS, 1% Penicillin/Streptomycin at 37 °C and 5% CO_2. In this study for convenience, the L929sAhFas and SK-OV-3-luc GFP cell lines have been abbreviated to L929 and SKOV, respectively.

2.2. Spheroid Formation

An Ultrapure Agarose solution (3% *w*/*v*, Life Technologies) was dissolved in sterile PBS and heated. After heating, it was poured on top of a negative polydimethylsiloxane (PDMS) mould which has a diameter of 18 mm and a height of 3 mm (NaMiFab, Ghent University, 9000 Ghent, Belgium). Each microwell contains 2865 pores with a diameter of 200 µm and a depth of 220 µm each. At room temperature, the solution solidified and the agarose microwell was separated from the mould. The microwell was then placed in a 12-well culture plate. A total of 500 µL of the cell suspension, containing 1×10^6 cells, was seeded onto the microwell, which resulted in approximately 349 cells per pore. Either L929 or SKOV cells were used for spheroid formation and spheroids were cultured at 37 °C in a humidified 5% CO_2-containing incubator. Medium was refreshed after 24 h of culture and afterwards every two days. Either day 1 and 10 spheroids (L929) or only day 1 spheroids (SKOV) were used.

2.3. *Sytox Dyes*

Sytox dyes intercalate with DNA with a high affinity [36]. When cells are permeabilised (i.e., the plasma membrane is ruptured) at the end of any cell death process, Sytox enters the cell and binds to the DNA causing a large increase in fluorescence (quantum yield of 0.5, Figure 1B) which can be measured via a fluorescence microplate reader and the Sytox signal can be correlated to the percentage of cell death. Both Sytox Green and Sytox Blue were used for L929 and SKOVs, respectively.

2.4. *Fluorescence Microscopy and Morphological Assessment of Spheroids*

Induced spheroids were evaluated using an inverted fluorescence microscope (Olympus IX81). For L929 spheroids, a GFP filter was used to visualize Sytox Green (λ_{ex} = 485 nm, λ_{em} = 520 nm) while for SKOV-spheroids, a DAPI filter was used to visualize Sytox Blue (λ_{ex} = 444 nm, λ_{em} = 480 nm).

The geometry of spheroids was analysed using Xcellence image software (Olympus). Diameter (d), area (A) and perimeter (p) were measured after 1, 2, 5, 7 and 10 days of culture and sphericity was calculated using Excellence image software. Volume was calculated using the formula $V = \left(\pi d^3\right)/6$. Images of 23–29 spheroids were analysed for diameter, circularity and volume.

2.5. *Identification of Optimal Sytox Concentration*

L929- and SKOV-spheroids were collected at day 1 and seeded in different densities (ranging from 3 to 120 spheroids per well) in a 96-well plate in duplicates. A 2D control was seeded as well in different densities (ranging from 500 to 20,000 cells per well). Cells were stained either immediately after seeding (spheroids) or after 24 h (2D culture) using different concentrations of Sytox Green (ranging from 1 to 5 μM) for L929-spheroids and Sytox Blue (ranging from 2.5 to 5 μM) for SKOV spheroids and after 2 h spheroids were permeabilised with 10 μL Triton X-100 (0.05% *v*/*v*, MERCK, Darmstadt, Germany). After 2 h, Sytox intensity was measured using the Tecan Spark® (Tecan, Männedorf, Switzerland) 20M microplate fluorescence reader (Sytox Green: λ_{ex} = 485 nm, λ_{em} = 535 nm; Sytox Blue: λ_{ex} = 430 nm, λ_{em} = 460 nm).

2.6. *Determining the Optimal Timing of Triton X-100 Treatment and Triton X-100 Concentration*

L929-spheroids were collected at day 1 and seeded in different densities (30, 80 and 120 spheroids per well) in a 96-well plate in triplicate. A 2D control was seeded as well in different densities corresponding to the densities of cells present in the spheroids (10,620, 28,000 and 48,000 cells per well). Cells were stained either immediately after seeding (spheroids) or after 24 h (2D culture) with the optimal Sytox Green concentration (found in the previous experiment, Invitrogen, Carlsbad, CA, USA) and permeabilised with different concentrations of Triton X-100 (0.05% *v*/*v*, 0.10% *v/v* and 0.25% *v*/*v*) in a volume of 10 μL added to the medium. At different timepoints (each hour, starting from 2 h), Sytox Green intensity was measured using the Tecan Spark microplate fluorescence reader.

2.7. *Validation of 3DELTA: Quantification of Ferroptotic Cell Death*

Spheroids were collected at day 1 and 10 (L929) and day 1 (SKOV) and seeded at a density of 80 spheroids per well in a 96-well plate in duplicate. A 2D control was seeded as well at a density of 10,000 cells per well. Cells were stained either immediately after seeding (spheroids) or after 24 h (2D culture) using the optimal Sytox Green (L929) or Sytox Blue (SKOV) concentration. An apoptosis inhibitor (zVAD-fmk, BACHEM, Budendorf, Switzerland), necroptosis inhibitor (Nec-1s, Abcam, Cambridge, UK) and multiple ferroptosis inhibitors (Fer-1, DFO and α-Toc, Sigma, Saint Louis, MO, USA) were added 30 min before cell death induction with 5 μM ML-162 (AOBIOUS, Gloucester, MA, USA), an inhibitor of GPX4 and thus an inducer of ferroptosis. After 24 h, Sytox intensity was measured using the Tecan Spark microplate fluorescence reader. Afterwards, cells were permeabilised with the optimal Triton X-100 concentration (found in the previous experiment) to obtain 100% of cell death. After 2 h of incubation, Sytox intensity was measured again.

2.8. Visualization of Cell Death Pattern

Sytox fluorescence intensities were measured using the Tecan Spark® 20M microplate multimode reader. Each well of the 96-well plate was scanned in a '5 × 5 filled pattern' for a total of 21 positions per well. A script was written in Matlab (R2018b, Mathworks, Natick, MA, USA) to extract the fluorescence intensity data from the Tecan output file and a heat map was constructed using the pseudocolour plot function. Shading of each pixel of the image was based on the measured fluorescent intensity at specific positions in the well. Corners of the image have zero values as these areas were not included in the Tecan Spark-defined scan pattern.

2.9. Statistical Analysis

The percentage of the cell death is calculated by the following formula:

$$\frac{average_{Sytox}[ML162] - average_{Sytox}[background]}{average_{\ Sytox}[Triton\ X - 100] - average_{Sytox}[background]} * 100\% \quad (1)$$

Cell death data are analysed as repeated measurements. Data from 3 independent experiments were analysed. Statistical analysis is performed using GraphPad Prism 8.0 software and are represented by mean ± SEM. Data are analysed with two-way ANOVA. *p*-values less than 0.05 are considered significant.

3. Results

3.1. Linear Correlation between Sytox Intensity and Cell Density

First, the L929 and SKOV spheroid geometry was characterized over time (Figure 2A). The diameter, sphericity and volume were measured and calculated for day 1, 2, 5, 7 and 10. L929 diameter increased from an average diameter of 150 ± 1.30 µm on day 1 to 180.83 ± 1.03 µm on day 10 and SKOV diameter increased from 130 ± 1.63 µm on day 1 to 148.44 ± 2.81 µm on day 10. Sphericity remained stable around 90% and 75% for L929 and SKOV spheroids, respectively. The volume increase over time of spheroids correlates with an initial compaction phase followed by the proliferative phase of cancerous cells [37].

In order to correctly extrapolate the percentage of cell death over time, a linear correlation between the cell number and Sytox intensity is required. Therefore, the optimal Sytox concentration was established by testing a range of concentrations from 1 to 5 µM for Sytox Green and from 2.5 to 5 µM for Sytox Blue. Of note, Sytox Blue has a lower signal-to-noise ratio than Sytox Green [38]. For 2D cultures, the optimal concentration were already determined as 3.3 µM for Sytox Green in a previous study [39] and 5 µM for Sytox Blue, which was also confirmed in this study (supplemental Figure S1A,E). For L929 spheroids (Figure 2B) the same concentration can be applied for Sytox Green, based on the linear correlation (R^2 = 0.9902, Figure 2C). In the case of SKOV spheroids (Figure 2E), it was observed that 4 µM of Sytox Blue had the highest linear correlation with cell density (R^2 = 0.9156, Figure 2F). These optimal concentrations were then used for all further experiments. An optimal spheroid density was determined from a range of 30 to 120 spheroids per well (Figure 2D,G). Of note, we measured a large increase of fluorescence when the number of spheroids per well was above 30. Thus, in the following experiments, spheroids were seeded at a density of 80 spheroids per well.

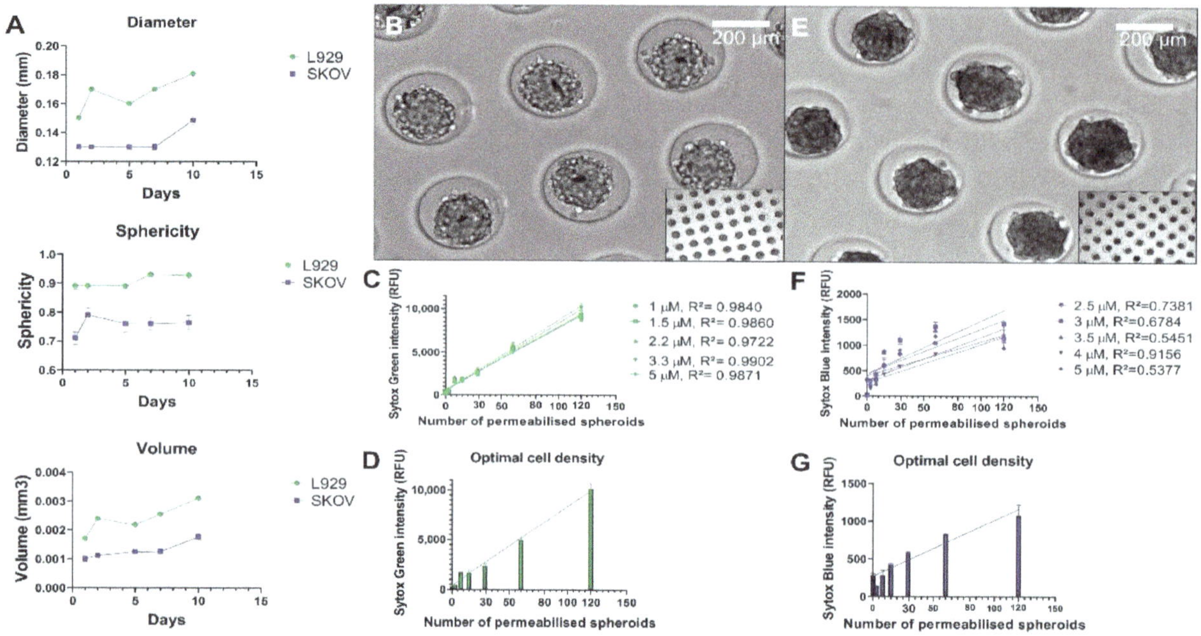

Figure 2. Characterization of the spheroids and analysis of linear correlation between Sytox intensity and cell density. (**A**) Diameter, sphericity and volume of L929 and SKOV spheroids. (**B**) Brightfield image of day 1 L929 spheroids showing globular spheroids that are not yet fully compacted. (**C**) Sytox Green intensity for different concentrations of Sytox Green was plotted out against different spheroid densities. Linear correlation was measured as R^2 and the highest correlation was found for Sytox Green 3.3 µM. (**D**) The increase of Sytox intensity for Sytox Green 3.3 µM is shown. Starting from 30 spheroids, a high increase in intensity was found until 120 spheroids. (**E**) Brightfield image of day 1 SKOV spheroids showing irregularly shaped spheroids that are quite compact. (**F**) Sytox Blue intensity for different concentrations of Sytox Blue was plotted out against different spheroid densities. Linear correlation was measured as R^2 and the highest correlation was found for Sytox Blue 4 µM. (**G**) The increase of Sytox intensity for Sytox Blue 4 µM is shown. Starting from 30 spheroids, a high increase in intensity was found until 120 spheroids.

3.2. Sytox Signal Penetrates All Cells in the Spheroids

In order to obtain the maximal intensity of the Sytox dyes, which corresponds to 100% cell membrane permeabilization and cell death, Triton X-100 was used to physically rupture the membrane. Previously in 2D cultures, Triton X-100 was used at a concentration of 0.05% (*v*/*v*) in order to permeabilize the cells and establish 100% cell death (39). However, spheroids maintain a structure of higher complexity than 2D cultures (Figures 1B and 2B,E). Therefore, optimization of spheroid permeabilization was performed. After staining with Sytox, spheroids were permeabilised using different concentrations of Triton X-100 (0.05%, 0.10% and 0.25%). To ensure that all cells undergo permeabilization after addition of Triton X-100, spheroids were disaggregated with trypsin. The fluorescence intensity of this 2D suspension was then compared to that of untrypsinized 3D spheroids.

Since the optimal gain setting is an inherent characteristic of fluorescence readers, we tested this first for our machine. The optimal gain setting was established by determining the gain with the highest linear correlation between Sytox Green intensity and cell density (Figure 3A). The highest correlation was found at gain 45 and 50 (R^2 = 1.000) and a gain 45 was used for the subsequent experiments. Next, top and bottom measurements were compared for spheroids (Figure 3B), which revealed that a more intense signal was received with less variation from top measurement, meaning that this setting was optimal for spheroid cultures. Furthermore, the incubation time with Triton X-100 required to permeabilize spheroids was tested and compared to 2D cultures (Figure 3C). Since the intensity of the Sytox signal did not change over time, this indicated that two hours was enough time to fully permeabilize spheroids with Triton X-100 at all concentrations.

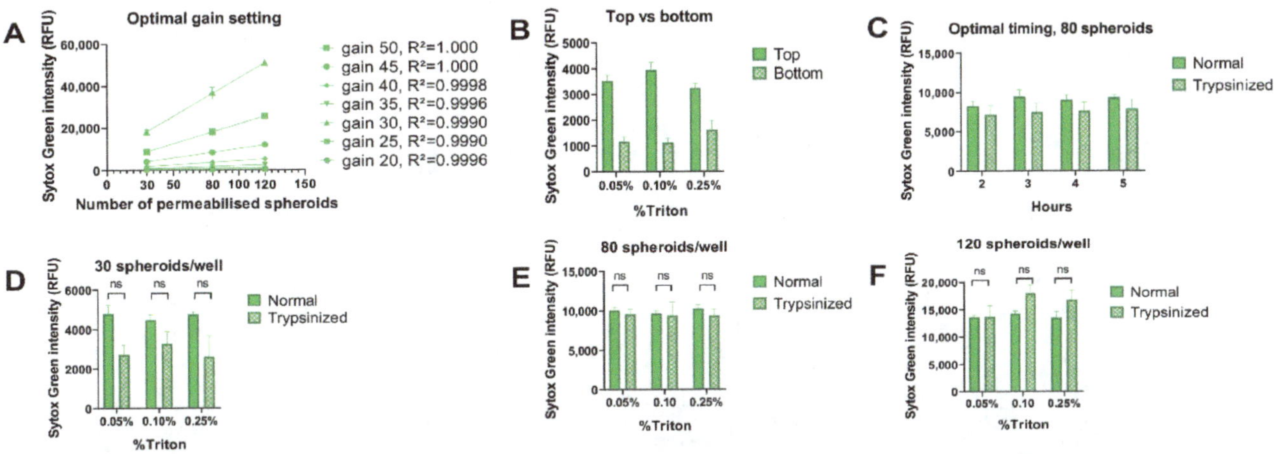

Figure 3. Triton X-100 efficiently permeabilizes spheroids. L929 spheroids were stained with 3.3 μM of Sytox Green and permeabilised with Triton X-100 in different concentrations (0.05%, 0.10% and 0.25%). Sytox Green intensity was measured using the Tecan Spark microplate reader. Either top or bottom measurement was used, and intensity was measured at different timepoints. (**A**) Optimal gain setting. The highest linearity is found at gain 45 and 50. (**B**) Comparison of top and bottom measurement. The most intense signal of Sytox Green with the lowest variation is found in top measurement. (**C**) Optimal timing for measurement. No differences were observed at the different timepoints. (**D–F**) Comparison of normal and trypsinized spheroids at different cell densities and with different Triton X-100 concentration. No significant differences in Sytox Green intensity were observed between normal and trypsinized. Furthermore, no significant differences were found between different Triton X-100 concentrations. Data are averages from three independent experiments ($n = 3$), each measured in triplicate; error bars = SEM. * $p < 0.05$, ** $p < 0.01$, *** $p < 0.001$, ns = not significant. RFU = relative fluorescent unit.

The comparison between standard and trypsinized spheroids was made for 30, 80 and 120 spheroids seeded per well; using different concentrations of Triton X-100: 0.05%, 0.10% or 0.25% (Figure 3D–F). No significant differences were found between different Triton X-100 concentrations and therefore, Triton X-100 0.05% (*v*/*v*) was used to permeabilize spheroids in all subsequent experiments. Importantly, no significant differences were observed between Sytox green intensity in standard spheroids and trypsinized spheroids ($p > 0.05$, $n = 3$). This indicates both that the cell death stain can penetrate into the core of intact spheroids and that the fluorescence emission can be detected.

3.3. Validation of 3DELTA: Quantification of Ferroptotic Cell Death

In order to validate cell death identification and quantification at the optimized measurement conditions for spheroids, ferroptotic cell death was induced with 5 μM ML-162, an inhibitor of GPX-4, and the 3DELTA method was performed. Different inhibitors of cell death modalities were added to confirm the type of cell death in spheroids—apoptosis (zVAD-fmk), necroptosis (Nec-1) and for ferroptosis (Fer-1, DFO and α-Toc) [40]. Visualization of cell death was performed for each well based on measured fluorescence intensities. A predefined scan pattern was used to measure Sytox intensity at specific points in the well, and a Matlab script was used to compile the data and construct heat maps. Based on the heat maps, the inhibitory effect of Fer-1, DFO and α-Toc and the distribution of cell death in each well are apparent (Figure 4A). Furthermore, it is clear that cell death induction is more efficient in day 1 spheroids compared to day 10 spheroids as well as inhibition of cell death.

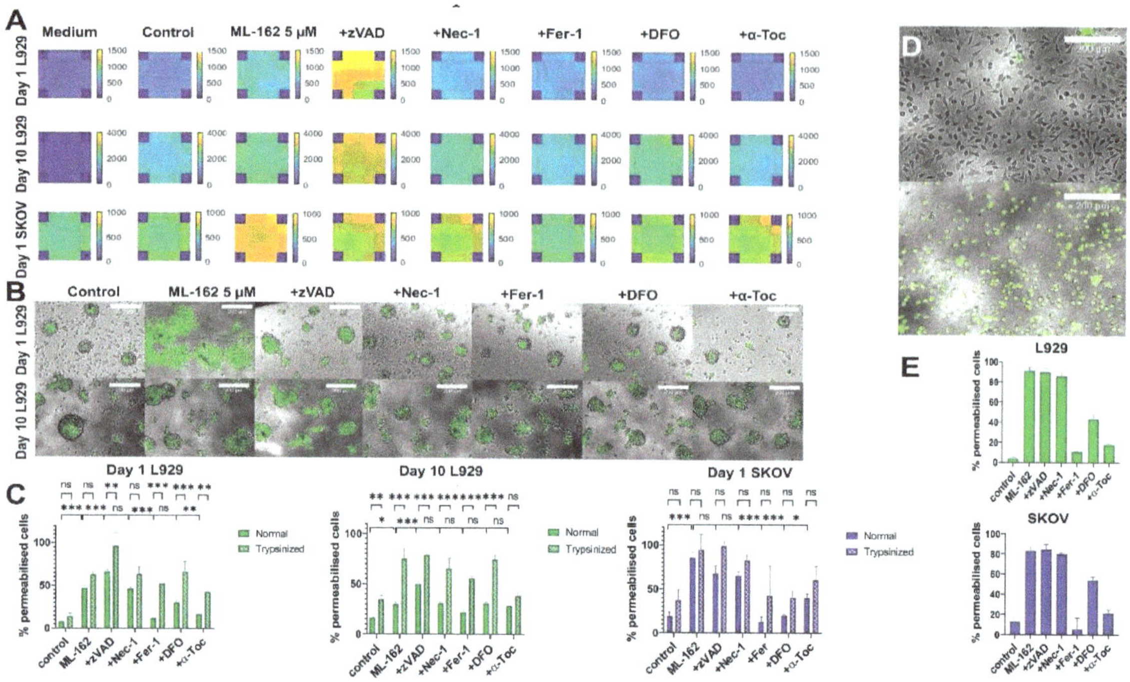

Figure 4. Validation of 3DELTA—quantification of ferroptotic cell death in spheroids. Ferroptosis was induced using 5 μM ML-162. Spheroids were stained with Sytox Green (L929) or Sytox Blue (SKOV) and cell death was measured after 24 hours as increase in fluorescence intensity using Tecan Spark microplate reader. Afterwards, spheroids were permeabilised with Triton X-100 0.05% (v/v). Fluorescence intensity was measured and used as 100% cell death. (**A**) Sytox was measured in a consistent, predefined pattern (21 positions per well), and heat maps of the entire well were generated in Matlab based on fluorescence intensities. Optimal gains were used for each cell line (gain 45, D1 L929; gain 45, D10 L929; gain 55, D1 SKOV). The corners shown here are zero values as these areas were not measured in the predefined Tecan Spark scan pattern. Experiments were performed in triplicate and representative images are shown here. (**B**) Fluorescence pictures of day 1 and 10 L929 spheroids. Control spheroids show cell death in their core which might be due to hypoxia leading to a necrotic core. Representative images are shown here. Scale bar = 200 μm (**C**) Quantification of cell death in normal and trypsinized spheroids. Data are averages from three independent experiments (n = 3), each measured in triplicate; error bars = SEM. * $p < 0.05$, ** $p < 0.01$, *** $p < 0.001$, ns = not significant. The first line represents the comparison between standard and trypsinized spheroids and the second line represents the comparison between control spheroids and spheroids induced with ML-162 and between induced spheroids and spheroids where inhibitors were added. (**D**) Brightfield image of control (upper panel) and induced (lower panel) L929 cells. All cells are stained with Sytox Green. (**E**) Quantification of cell death in L929 (upper panel) and SKOV (lower panel) 2D culture.

Interestingly, both day 1 and 10 spheroids showed cell death in the core in control spheroids (Figure 4B) which could be due to hypoxia leading to a necrotic core [17]. ML-162 induced a significant increase of cell death to around 50% compared to control L929 spheroids that were collected at day 1 (Figure 4A,C). However, the cell death response was decreased to around 30% in day 10 spheroids. The decrease in ferroptosis could be caused by an increase in complexity of spheroids over time and secretion of extracellular matrix [16]. For day 1 SKOV spheroids, approximately 85% of cell death was observed after stimulation with ML-162. The addition of ferroptosis inhibitors (i.e., Fer-1, DFO and α-Toc) to drug-induced spheroids significantly reduced cell death, indicating that the induced cell death followed the ferroptosis pathway. However, this was less prominent with DFO, compared to Fer-1 and α-Toc. The addition of zVAD-fmk, which inhibits apoptosis, appears to increase the amount of cell death and may be due to the induction of necroptosis via inhibition of caspase-8 [41]. It has previously been shown that zVAD-fmk can induce necroptosis via induction of tumour necrosis factor-α (TNFα) through the mitogen-activated kinases (MAPKs) pathway [42]. In day 10 L929 spheroids, inhibition of ferroptotic cell death was also less pronounced, which may be due to the more complex nature at day 10, preventing efficient diffusion of inhibitors towards the centre of the spheroids.

A comparison between normal and trypsinized spheroids was also made with the inhibitors (Figure 4C). Importantly, significant differences were found between the two conditions. These differences might be explained by the stress induced by the trypsinization procedure, which leads to dysregulated cell functions [43]. This was particularly apparent in day 10 spheroids where the cells were treated with collagenase-1 for 30 min. Collagenase-1 was used to disaggregate day 10 spheroids as trypsin could not efficiently degrade the ECM produced in this more advanced stage. Already in the control group, more cell death was observed, and here, the differences between trypsinized and nontrypsinized spheroids were even more pronounced. It also means that using trypsinized spheroids might give a wrong representation of cell death induction and lead to false positive or negative results. However, in day 1 SKOV spheroids, no significant differences between standard and trypsinized spheroids was found. In contrast, there was a large variation in the percentage of permeabilised cells in trypsinized SKOV spheroids, which may also contribute to false results in this case.

4. Discussion

High failure rates in drug development when entering into the clinic is a large healthcare problem. A major contributing factor is due to the poor models used in early target development which are mainly in 2D models. In this work, cell death was induced in spheroids which are more representative of the in vivo environment. We show that a new cell death analysis method (i.e., 3DELTA) can efficiently determine different cell death types including ferroptosis and quantitatively assess cell death in tumour spheroids. These results obtained from 3D cultures might be more relevant to the clinic and consequently will save time, money, and resources for drug development.

This work also strongly highlights the advantage of the 3DELTA method over existing cell death measurements [44], which require disaggregation of spheroids at one specific timepoint for analysis. Not only is the disaggregation process time-consuming, it also cannot be applied for kinetic measurements as it destroys the 3D environment. With the 3DELTA method, real-time kinetic and high-throughput measurements are possible if the machine is coupled to a CO_2 inlet and a heating control. Moreover, in this work it was shown that 2D experiments cannot be extrapolated towards 3D cultures as 3D cultures are less sensitive to cell death induction. In 2D cultures, ferroptosis induction leads to approximately 90% cell death (Figure 4E) while in day 1 spheroids this was already reduced to 50% and even to 30% in day 10 spheroids. Due to tumour heterogeneity, the responses between different tumour types might differ because of different sensitivity to cell death induction, which was also represented in this work as the SKOV cell line seemed to be more sensitive to ferroptosis [45]. These two cell lines have been chosen because they are extensively used in cell death research in 2D cultures [42,46,47]. Furthermore, use of two different cancer cell lines (murine fibrosarcoma and human ovarian cancer cell line) circumvents a lot of risks with cell type specificity of the DELTA assay. These data suggest that this method is also applicable to other cell lines. The application of 3D cultures is more representative to the in vivo situation as spheroids better resemble in vivo tumours. Moreover, ferroptosis will be an alternative strategy in cancer therapy, especially in apoptosis- and necroptosis-resistant cases [23,48]. Ferroptosis is induced via blockade of the system x_c cysteine/glutamate antiporter or glutathione peroxidase 4 (GPX4), resulting in a defective GSH-redox system where oxidized phosphatidylethanolamines are crucial for ferroptosis execution [26,49]. In the proposed 3DELTA method, it is possible to monitor ferroptotic cell death in time kinetics and without disaggregation of the spheroids. We were able to induce ferroptosis in 3D cancer models allowing the possibility of ferroptosis to be exploited in the future therapies. Of course, there are other important pathways which can contribute to cell death, such as autophagy. Autophagy often precedes cell death (e.g., apoptosis or ferroptosis) after which the cells will become Sytox positive [50,51]. Indeed, we have only used inhibitors for apoptosis, necroptosis and ferroptosis. However, this method can be easily adapted to include other inhibitors like autophagy inhibitors.

3DELTA proved to be a simple and efficient cell death assay to identify and quantify cell death modality in 3D cultures (i.e., spheroids) compared to the current state-of-the-art methods. It does not require the disaggregation of spheroids, making it less labour-extensive and allowing for potential real-time kinetic experiments. 3DELTA can be easily adapted to different stimuli and/or inhibitors. This is highly advantageous for high-throughput drug screenings in 96- and 384-well plates. However, several factors, summarized in Table 1, should be taken into consideration for the proper use of this assay. In summary, we have developed a more cost-effective way to measure cell death and determine its type in 3D cultures including spheroids.

Table 1. Troubleshooting.

Issue	Solution
Higher Sytox intensity in stimulated cells than with Triton X-100 leading to cell death higher than 100%	Resuspend more vigorously and repeat after 1 h Increase Triton X-100 concentration
Trypsinization incomplete for D10 spheroids	Include an incubation step with collagenase I instead of using trypsin
Spheroids disaggregate after 1 day during ferroptosis assay when using adherent plates	Use suspension plate instead of adherent plates to avoid cells adhering to plate
Intensity too high: overestimated	Use lower gains Optimize gain settings for new cell lines

Author Contributions: Conceptualization, H.D. and D.V.K.; data curation, R.D.; formal analysis, R.D. and A.L.; investigation, R.D.; methodology, R.D. and I.E.; software, A.L.; supervision, I.E., H.D. and D.V.K.; writing—original draft, R.D.; writing—review & editing, I.E., A.L., H.D. and D.V.K. All authors have read and agreed to the published version of the manuscript.

References

1. Witkiewicz Agnieszka, K.; Balaji, U.; Eslinger, C.; McMillan, E.; Conway, W.; Posner, B.; Mills, G.B.; O'Reilly, E.M.; Knudsen, E.S. Integrated Patient-Derived Models Delineate Individualized Therapeutic Vulnerabilities of Pancreatic Cancer. *Cell Rep.* **2016**, *16*, 2017–2031. [CrossRef]
2. Duval, K.; Grover, H.; Han, L.-H.; Mou, Y.; Pegoraro, A.F.; Fredberg, J.; Chen, Z. Modeling Physiological Events in 2D vs. 3D Cell Culture. *Physiology* **2017**, *32*, 266–277. [CrossRef]
3. Kapałczyńska, M.; Kolenda, T.; Przybyła, W.; Zajączkowska, M.; Teresiak, A.; Filas, V.; Ibbs, M.; Bliźniak, R.; Łuczewski, Ł.; Lamperska, K. 2D and 3D cell cultures a comparison of different types of cancer cell cultures. *Arch. Med. Sci.* **2018** *14*, 910–919.
4. Mittler, F.; Obeïd, P.; Rulina, A.V.; Haguet, V.; Gidrol, X.; Balakirev, M.Y. High-Content Monitoring of Drug Effects in a 3D Spheroid Model. *Front. Oncol.* **2017**, *7*, 293. [CrossRef] [PubMed]
5. Kessel, S.; Cribbes, S.; Dery, O.; Kuksin, D.; Sincoff, E.; Qiu, J.; Chan, L.L. High-Throughput 3D Tumor Spheroid Screening Method for Cancer Drug Discovery Using Celigo Image Cytometry. *SLAS Technol.* **2017**, *22*, 454–465. [CrossRef]
6. Liu, W.; Sun, M.; Han, K.; Wang, J. Large-Scale Antitumor Screening Based on Heterotypic 3D Tumors Using an Integrated Microfluidic Platform. *Anal. Chem.* **2019**. [CrossRef] [PubMed]
7. De Moor, L.; Merovci, I.; Baetens, S.; Verstraeten, J.; Kowalska, P.; Krysko, D.V.; De Vos, W.H.; Declercq, H. High-throughput fabrication of vascularized spheroids for bioprinting. *Biofabrication* **2018**, *10*, 035009. [CrossRef] [PubMed]
8. Drost, J.; Clevers, H. Organoids in cancer research. *Nat. Rev. Cancer* **2018**, *18*, 407–418. [CrossRef] [PubMed]

9. Kinoshita, T.; Higuchi, H.; Ayano-Kabashima-Niibe, S.G.; Hamamoto, Y.; Takaishi, H.; Kanai, T. Analysis of sensitivity and cell death pathways mediated by anti-cancer drugs using three-dimensional culture system. *Int. J. Cancer Res.* **2018**, *14*, 1–12.
10. Brodaczewska, K.K.; Bielecka, Z.F.; Maliszewska-Olejniczak, K.; Szczylik, C.; Porta, C.; Bartnik, E.; Czarnecka, A.M. Metastatic renal cell carcinoma cells growing in 3D on polyDlysine or laminin present a stemlike phenotype and drug resistance. *Oncol. Rep.* **2019**. [CrossRef]
11. Aref, A.R.; Campisi, M.; Ivanova, E.; Portell, A.; Larios, D.; Piel, B.P.; Mathur, N.; Zhou, C.; Coakley, R.V.; Bartles, A.; et al. 3D microfluidic ex vivo culture of organotypic tumor spheroids to model immune checkpoint blockade. *Lab Chip.* **2018**, *18*, 3129–3143. [CrossRef] [PubMed]
12. Ayuso, J.M.; Vitek, R.; Swick, A.D.; Skala, M.C.; Wisinski, K.B.; Kimple, R.J.; Lambert, P.F.; Beebe, D.J. Effects of culture method on response to EGFR therapy in head and neck squamous cell carcinoma cells. *Sci. Rep.* **2019**, *9*, 12480. [CrossRef] [PubMed]
13. Paskeviciute, M.; Petrikaite, V. Differences of statin activity in 2D and 3D pancreatic cancer cell cultures. *Drug Des. Dev. Ther.* **2017**, *11*, 3273–3280. [CrossRef] [PubMed]
14. Cui, X.; Hartanto, Y.; Zhang, H. Advances in multicellular spheroids formation. *J. R. Soc. Interface* **2017**, *14*, 20160877. [CrossRef] [PubMed]
15. McKenzie, A.J.; Hicks, S.R.; Svec, K.V.; Naughton, H.; Edmunds, Z.L.; Howe, A.K. The mechanical microenvironment regulates ovarian cancer cell morphology, migration, and spheroid disaggregation. *Sci. Rep.* **2018**, *8*, 7228. [CrossRef]
16. Rebelo, S.P.; Pinto, C.; Martins, T.R.; Harrer, N.; Estrada, M.F.; Loza-Alvarez, P.; Cabecadas, J.; Alves, P.M.; Gualda, E.J.; Sommergruber, W.; et al. 3D-3-culture: A tool to unveil macrophage plasticity in the tumour microenvironment. *Biomaterials* **2018**, *163*, 185–197. [CrossRef]
17. Leek, R.; Grimes, D.R.; Harris, A.L.; McIntyre, A. Methods: Using Three-Dimensional Culture (Spheroids) as an In Vitro Model of Tumour Hypoxia. *Adv. Exp. Med. Biol.* **2016**, *899*, 167–196.
18. Imamura, Y.; Mukohara, T.; Shimono, Y.; Funakoshi, Y.; Chayahara, N.; Toyoda, M.; Kiyota, N.; Takao, S.; Kono, S.; Nakatsura, T.; et al. Comparison of 2D- and 3D-culture models as drug-testing platforms in breast cancer. *Oncol. Rep.* **2015**, *33*, 1837–1843. [CrossRef]
19. Wallach, D.; Kang, T.-B. Programmed Cell Death in Immune Defense: Knowledge and Presumptions. *Immunity* **2018**, *49*, 19–32. [CrossRef]
20. Majno, G.; Joris, I. Apoptosis, oncosis, and necrosis. An overview of cell death. *Am. J. Pathol.* **1995**, *146*, 3–15.
21. Aaes, T.L.; Kaczmarek, A.; Delvaeye, T.; De Craene, B.; De Koker, S.; Heyndrickx, L.; Delrue, I.; Taminau, J.; Wiernicki, B.; De Groote, P.; et al. Vaccination with Necroptotic Cancer Cells Induces Efficient Anti-tumor Immunity. *Cell Rep.* **2016**, *15*, 274–287. [CrossRef] [PubMed]
22. Vanden Berghe, T.; Linkermann, A.; Jouan-Lanhouet, S.; Walczak, H.; Vandenabeele, P. Regulated necrosis: The expanding network of non-apoptotic cell death pathways. *Nat. Rev. Mol. Cell Biol.* **2014**, *15*, 135–147. [CrossRef] [PubMed]
23. Krysko, O.; Aaes, T.L.; Kagan, V.E.; D'Herde, K.; Bachert, C.; Leybaert, L.; Vandenabeele, P.; Krysko, D.V. Necroptotic cell death in anti-cancer therapy. *Immunol. Rev.* **2017**, *280*, 207–219. [CrossRef] [PubMed]
24. Friedmann Angeli, J.P.; Schneider, M.; Proneth, B.; Tyurina, Y.Y.; Tyurin, V.A.; Hammond, V.J.; Herbach, N.; Aichler, M.; Walch, A.; Eggenhofer, E.; et al. Inactivation of the ferroptosis regulator Gpx4 triggers acute renal failure in mice. *Nat. Cell Biol.* **2014**, *16*, 1180. [CrossRef]
25. Friedmann Angeli, J.P.; Krysko, D.V.; Conrad, M. Ferroptosis at the crossroads of cancer-acquired drug resistance and immune evasion. *Nat. Rev. Cancer* **2019**, *19*, 405–414. [CrossRef]
26. D'Herde, K.; Krysko, D.V. Ferroptosis: Oxidized PEs trigger death. *Nat. Chem. Biol.* **2017**, *13*, 4–5. [CrossRef] [PubMed]
27. Xiao, J.; Yao, R.; Xu, B.; Wen, H.; Zhong, J.; Li, D.; Zhou, Z.; Xu, J.; Wang, H. Inhibition of PDE4 Attenuates TNF-alpha-Triggered Cell Death Through Suppressing NF-kappaB and JNK Activation in HT-22 Neuronal Cells. *Cell Mol. Neurobiol.* **2019**. [CrossRef]
28. Costa, E.C.; Moreira, A.F.; de Melo-Diogo, D.; Gaspar, V.M.; Carvalho, M.P.; Correia, I.J. 3D tumor spheroids: An overview on the tools and techniques used for their analysis. *Biotechnol. Adv.* **2016**, *34*, 1427–1441. [CrossRef]

29. Gurunathan, S.; Park, J.H.; Han, J.W.; Kim, J.H. Comparative assessment of the apoptotic potential of silver nanoparticles synthesized by Bacillus tequilensis and Calocybe indica in MDA-MB-231 human breast cancer cells: Targeting p53 for anticancer therapy. *Int. J. Nanomed.* **2015**, *10*, 4203–4222. [CrossRef]
30. Krysko, D.V.; Vanden Berghe, T.; D'Herde, K.; Vandenabeele, P. Apoptosis and necrosis: Detection, discrimination and phagocytosis. *Methods* **2008**, *44*, 205–221. [CrossRef]
31. Kirchhoff, C.; Cypionka, H. Propidium ion enters viable cells with high membrane potential during live-dead staining. *J. Microbiol. Methods* **2017**, *142*, 79–82. [CrossRef]
32. Abdanipour, A.; Dadkhah, M.; Alipour, M.; Feizi, H. Effect of Ghrelin on Caspase 3 and Bcl2 Gene Expression in H2O2 Treated Rat's Bone Marrow Stromal Cells. *Adv. Pharm. Bull.* **2018**, *8*, 429–435. [CrossRef] [PubMed]
33. Veloria, J.R.; Devkota, A.K.; Cho, E.J.; Dalby, K.N. Optimization of a Luminescence-Based High-Throughput Screening Assay for Detecting Apyrase Activity. *SLAS Discov.* **2017**, *22*, 94–101. [CrossRef] [PubMed]
34. Grasser, U.; Bubel, M.; Sossong, D.; Oberringer, M.; Pohlemann, T.; Metzger, W. Dissociation of mono- and co-culture spheroids into single cells for subsequent flow cytometric analysis. *Ann. Anat.* **2018**, *216*, 1–8. [CrossRef] [PubMed]
35. Morgado Brajones, J.; Clouvel, G.; Dovillaire, G.; Levecq, X.; Lorenzo, C. Highly Sensitive Shack-Hartmann Wavefront Sensor: Application to Non-Transparent Tissue Mimic Imaging with Adaptive Light-Sheet Fluorescence Microscopy. *Methods Protoc.* **2019**. [CrossRef]
36. Thakur, S.; Cattoni, D.I.; Nollmann, M. The fluorescence properties and binding mechanism of SYTOX green, a bright, low photo-damage DNA intercalating agent. *Eur. Biophys. J.* **2015**, *44*, 337–348. [CrossRef] [PubMed]
37. Zhao, L.; Xiu, J.; Liu, Y.; Zhang, T.; Pan, W.; Zheng, X.; Zhang, X. A 3D Printed Hanging Drop Dripper for Tumor Spheroids Analysis Without Recovery. *Sci. Rep.* **2019**, *9*, 19717. [CrossRef] [PubMed]
38. Chiaraviglio, L.; Kirby, J.E. Evaluation of impermeant, DNA-binding dye fluorescence as a real-time readout of eukaryotic cell toxicity in a high throughput screening format. *Assay Drug Dev. Technol.* **2014**, *12*, 219–228. [CrossRef]
39. Grootjans, S.; Hassannia, B.; Delrue, I.; Goossens, V.; Wiernicki, B.; Dondelinger, Y.; Bertrand, M.J.M.; Krysko, D.V.; Vuylsteke, M.; Vandenabeele, P.; et al. A real-time fluorometric method for the simultaneous detection of cell death type and rate. *Nat. Protoc.* **2016**, *11*, 1444. [CrossRef]
40. Turubanova, V.D.; Balalaeva, I.V.; Mishchenko, T.A.; Catanzaro, E.; Alzeibak, R.; Peskova, N.N.; Efimova, I.; Bachert, C.; Mitroshina, E.V.; Krysko, O.; et al. Immunogenic cell death induced by a new photodynamic therapy based on photosens and photodithazine. *J. ImmunoTher. Cancer* **2019**, *7*, 350. [CrossRef]
41. Annibaldi, A.; Meier, P. Checkpoints in TNF-Induced Cell Death: Implications in Inflammation and Cancer. *Trends Mol. Med.* **2018**, *24*, 49–65. [CrossRef] [PubMed]
42. Wu, Y.T.; Tan, H.L.; Huang, Q.; Sun, X.J.; Zhu, X.; Shen, H.M. zVAD-induced necroptosis in L929 cells depends on autocrine production of TNFalpha mediated by the PKC-MAPKs-AP-1 pathway. *Cell Death Differ.* **2011**, *18*, 26–37. [CrossRef] [PubMed]
43. Huang, H.-L.; Hsing, H.-W.; Lai, T.-C.; Chen, Y.-W.; Lee, T.-R.; Chan, H.-T.; Lyu, P.-C.; Wu, C.-L.; Lu, Y.-C.; Lin, S.-T.; et al. Trypsin-induced proteome alteration during cell subculture in mammalian cells. *J. Biomed. Sci.* **2010**, *17*, 36. [CrossRef] [PubMed]
44. Rieseberg, M.; Kasper, C.; Reardon, K.F.; Scheper, T. Flow cytometry in biotechnology. *Appl. Microbiol. Biotechnol.* **2001**, *56*, 350–360. [CrossRef]
45. Dagogo-Jack, I.; Shaw, A.T. Tumour heterogeneity and resistance to cancer therapies. *Nat. Rev. Clin. Oncol.* **2018**, *15*, 81–94. [CrossRef]
46. Vanlangenakker, N.; Bertrand, M.J.M.; Bogaert, P.; Vandenabeele, P.; Vanden Berghe, T. TNF-induced necroptosis in L929 cells is tightly regulated by multiple TNFR1 complex I and II members. *Cell Death Dis.* **2011**, *2*, e230. [CrossRef]
47. Krysko, D.V.; Denecker, G.; Festjens, N.; Gabriels, S.; Parthoens, E.; D'Herde, K.; Vandenabeele, P. Macrophages use different internalization mechanisms to clear apoptotic and necrotic cells. *Cell Death Differ.* **2006**, *13*, 2011–2022. [CrossRef]
48. Hanahan, D.; Weinberg Robert, A. Hallmarks of Cancer: The Next Generation. *Cell* **2011**, *144*, 646–674. [CrossRef]
49. Dixon, S.J.; Lemberg, K.M.; Lamprecht, M.R.; Skouta, R.; Zaitsev, E.M.; Gleason, C.E.; Patel, D.N.; Bauer, A.J.; Cantley, A.M.; Yang, W.S.; et al. Ferroptosis: An iron-dependent form of nonapoptotic cell death. *Cell* **2012**, *149*, 1060–1072. [CrossRef]

50. Mariño, G.; Niso-Santano, M.; Baehrecke, E.H.; Kroemer, G. Self-consumption: The interplay of autophagy and apoptosis. *Nat. Rev. Mol. Cell Biol.* **2014**, *15*, 81–94. [CrossRef]
51. Zhou, Y.; Shen, Y.; Chen, C.; Sui, X.; Yang, J.; Wang, L.; Zhou, J. The crosstalk between autophagy and ferroptosis: What can we learn to target drug resistance in cancer? *Cancer Biol. Med.* **2019**, *16*, 630–646. [PubMed]

5

PP2A Deficiency Enhances Carcinogenesis of Lgr5^{+} Intestinal Stem Cells Both in Organoids and In Vivo

Yu-Ting Yen [1,2], May Chien [1,2], Yung-Chih Lai [2], Dao-Peng Chen [3], Cheng-Ming Chuong [2,4], Mien-Chie Hung [5,6] and Shih-Chieh Hung [1,2,7,*]

1 Drug Development Center, Institute of New Drug Development, China Medical University, Taichung 40402, Taiwan; d92449001@ntu.edu.tw (Y.-T.Y.); maymaychien2k12@gmail.com (M.C.)
2 Integrative Stem Cell Center, China Medical University Hospital, Taichung 40402, Taiwan; yungchihlai@gmail.com (Y.-C.L.); cmchuong@med.usc.edu (C.-M.C.)
3 Kim Forest Enterprise Co., Ltd., Taipei 22175, Taiwan; D96743@mail.cmuh.org.tw
4 Department of Pathology, Keck School of Medicine, University of Southern California, Los Angeles, CA 90033, USA
5 Center for Molecular Medicine and Graduate Institute of Cancer Biology, China Medical University, Taichung 40402, Taiwan; mhung77030@gmail.com
6 Cancer Biology Program, The University of Texas Graduate School of Biomedical Sciences, The University of Texas MD Anderson Cancer Center, Houston, TX 77030, USA
7 Department of Orthopaedics, China Medical University Hospital, Taichung 40402, Taiwan
* Correspondence: hung3340@gmail.com

Abstract: In most cancers, cellular origin and the contribution of intrinsic and extrinsic factors toward transformation remain elusive. Cell specific carcinogenesis models are currently unavailable. To investigate cellular origin in carcinogenesis, we developed a tumorigenesis model based on a combination of carcinogenesis and genetically engineered mouse models. We show in organoids that treatment of any of three carcinogens, DMBA, MNU, or PhIP, with protein phosphatase 2A (PP2A) knockout induced tumorigenesis in Lgr5^{+} intestinal lineage, but not in differentiated cells. These transformed cells increased in stem cell signature, were upregulated in EMT markers, and acquired tumorigenecity. A mechanistic approach demonstrated that tumorigenesis was dependent on Wnt, PI3K, and RAS-MAPK activation. In vivo combination with carcinogen and PP2A depletion also led to tumor formation. Using whole-exome sequencing, we demonstrate that these intestinal tumors display mutation landscape and core driver pathways resembling human intestinal tumor in The Cancer Genome Atlas (TCGA). These data provide a basis for understanding the interplay between extrinsic carcinogen and intrinsic genetic modification and suggest that PP2A functions as a tumor suppressor in intestine carcinogenesis.

Keywords: carcinogen; protein phosphatase 2A (PP2A); intestinal tumor; intestinal organoid; Lgr5^{+} crypt stem cell

---

1. Introduction

The cells of origin in most cancers have remained unknown. Chemical carcinogenesis mouse models recapitulating most of human cancers that are induced by exposure to environmental carcinogens [1], however, is difficult to be achieved in a cell-specific manner. Therefore, the current strategies to investigate the cellular origins of cancers are using genetically engineered mouse models (GEMMs), with either transgenic or conditionally targeted gene technologies to induce tumor in different cellular contexts [2]. Moreover, both models take a long time to develop cancer, limiting progress in the cancer research field.

The most applied animal model for studying intestinal tumorigenesis is based on activating mutations in the Wnt pathway, which relies on adenomatous polyposis coli (Apc) depletion [3] and beta-catenin (CTNNB1) activation [4], leading to beta-catenin stabilization and constitutive transcription of its down-stream genes. Recent progress in the understanding of the cell of origin of intestinal tumor was made using this model, although several inconsistencies were observed. After in vivo Apc depletion in leucine-rich-repeat containing G-protein-coupled receptor 5 $(Lgr5)^+$ crypt stem cells, tumor formation occurred within 3–5 weeks [3]. However, Apc depletion or being combined with $Kras^{G12D}$ mutation in progenitor and differentiated cells did not induce tumor formation [5,6]. However, tumor-initiating mutations can occur in $Lgr5^+$ crypt stem cells and in differentiated $Lgr5^-$ cells [4], indicating that the two hypotheses are not mutually exclusive. Ablation of $Lgr5^+$ cells in orthotopically transplanted tumors, generated by genetic modification in differentiated villus cells, suppressed tumor growth [7]. Interestingly, $Lgr5^+$ cells reappeared and tumors recurred when ablation was terminated 4 days later. The generation of $Lgr5^+$ cells from $Lgr5^-$ cells after $Lgr5^+$ ablation was also observed in the xenograft mouse model of human colon cancer stem cells (CSCs) [8]. However, the specific mechanism of $Lgr5^+$ cell generation from remaining $Lgr5^-$ cells remains unclear.

Aberrant activation of signal transduction pathway, a dynamic process involving an 'on/off' switch, can transform a normal cell to be malignant or further render cancer cells with the capacities for therapy resistance. Activating mutations in genes encoding kinases or signaling molecules, such as RAS and PI3K, switch on the signaling, continuously activating a survival and/or proliferation pathway, while activations of phosphatases, such as the serine/threonine phosphatase PP2A family, switch off the signaling [9]. Previous efforts through high-throughput screens of tyrosine kinome and tyrosine phosphatome have identified several driver or passenger mutations in a spectrum of malignancies, including intestinal tumor [10–12]. However, there are few if any studies focusing on the altered signalings driven by serine/threonine kinase mutations [13]. Human intestinal tumors contain active mutations in genes encoding proteins involved in the WNT, MAPK, TGF-beta, and PI3K pathways [14]. Ingenuity pathway analysis (IPA) of TCGA-COAD revealed PP2A complex and its subunits, such as PPP2R1A and PPP2CA, are intercalated among several driver mutation pathways (Figure S1A–C). Moreover, endogenous PP2A inhibitors, SET and CIP2A [15], are highly expressed in intestinal tumors in comparison to their matched normal tissue samples (Figure S1D). We showed that PP2A was suppressed in intestinal tumor stem cells (CSCs), thereby activating its substrate kinases to enhance survival under hypoxia and serum depletion [16], thereby increasing resistance to anti-angiogenesis therapy [17]. Our recent studies also demonstrated that reduced PP2A activity in colorectal and lung CSCs enhances suspension survival and induces tumor initiation [18], revealing the tumor suppressive role of PP2A [19]. Although higher numbers of Apc, p53, $Kras^{G12D}$, and Smad4 driver mutations may be required for human colorectal tumorigenesis, there are some intestinal tumors carry only one or no alteration in these driver mutations [20]. For example, gene fusions involving R-spondin 1 occurring in 10% of intestinal tumor are mutually exclusive with active Wnt signaling caused by *APC* or *CTNNB1* mutations [20]. The emerging novel intestinal tumorigenesis animal models should allow for elucidating the molecular mechanisms of these cancers.

Given that cancer is the product of complex interactions between the genetic and environmental predisposition factors, the combined use of chemical carcinogens that switch on kinases and GEMM with phosphatase deficiency is a logical approach for examining the complex interplay between genetic susceptibility and environmental exposure [21]. To investigate the cell origin of intestinal tumor, we first combined treatment with carcinogen 7,12-dimethylbenzanthracene (DMBA) that has previously been known to induce rodent s in the presence of 1,2-dimethyl-hydrazine [22] and PP2A inhibition via okadaic acid (OA) treatment or genetic deficiency. DMBA not only activates multiple mutations in different codons of ras [23] but also induces activation in other pathways, such as Notch [24], providing a screening approach for identifying key kinases or molecules. Besides DMBA, we also investigated the effects of *N*-methyl-*N*-nitrosourea (MNU) and 2-amino-1-methyl-6-phenylimidazo-[4,5-b]pyridine (PhIP) on intestinal carcinogenesis. MNU is one of the direct alkylating agents, which does not require metabolic activation for initiating carcinogenesis [25]. PhIP has received considerable attention because it has multi-organ targets and it was developed upon broiling of fish and meat [26]. Moreover, we established primary intestinal organoid models that recapitulate the rodent intestinal tumorigenesis paradigm [27]. These rodent intestinal tumorigenesis models are useful in the development of new strategies for targeting rodent intestinal CSCs and treatment of intestinal tumor.

2. Materials and Methods

2.1. Mouse Colonies

Ppp2r1a$^{flox/flox}$ mice, carrying conditional alleles with loxP sites flanking exon 5–6 of *Ppp2r1a*, were purchased from the Jackson Laboratory and crossed to *Lgr5-EGFP-CreERT2* or *Villin-Cre* mice to generate *Lgr5-EGFP-CreERT2; Ppp2r1a*$^{flox/flox}$ or *Villin-Cre; Ppp2r1a* $^{flox/flox}$ mice. NOD/SCID mice were purchased from Lasco Co., Ltd. (Taiwan). All animal studies and care of live animals were approved and performed following the guidelines made by the China Medical University Institutional Animal Care and Use Committee 2016-398-1; 2017-239.

2.2. Mouse Intestinal Organoid Cell Isolation, Culture, and Passage

Organoid culture was preformed according to a protocol modified from previously described methods [28]. In brief, the intestines were dissected, opened longitudinally and cut into small (2 mm) pieces. The tissues were rocked in dissociation reagent and incubated at room temperature (15–25 °C) for 15 min. The tissues were then mixed and filtered through a 70 μm sterile cell strainer. The crypts were collected by centrifugation at 140× *g* for 5 min at 4 °C. Approximately 500 crypts were suspended in 50 μL growth factor reduced phenol-free Matrigel (BD Biosciences, San Jose, CA, USA). Next, a 50 μL droplet of Matrigel/crypt mix was placed and polymerized in the center well of a 48-well plate. The basic culture medium (Dulbecco's modified Eagle's medium/F12 supplemented with penicillin/streptomycin), was supplemented with 50 ng/mL murine recombinant epidermal growth factor (EGF; Peprotech, Hamburg, Germany), Noggin (5% final volume) and R-spondin 1 (5% final volume) called "ENR" medium. Medium change was performed every 3–4 days. Each condition was examined in triplicate with multiple (>15) organoids in each sample. Each experiment was repeated twice.

2.3. Dysplasia Index

Histologic changes were scored blindly on the levels of four histological characteristics as previously described [27]: nuclear grade (enlarged nuclei with diffuse membrane irregularities and prominent nucleoli); stratification; mitoses and invasion (>2 foci). The dysplasia index was evaluated by all microscopic fields containing viable organoids with 5 fields per sample (*n*).

2.4. Primary Organoid Transplantation

For transplantation, cells from passage 7; day 50 organoid cultures were collected. Dissociated cells were pelleted by centrifugation and resuspended with Matrigel (50% Matrigel (BD), in a total volume of 100 μL containing indicated cell numbers and injected s.c. into NOD-SCID mice.

2.5. Immunofluorescence and Immunohistochemistry

Freshly isolated intestines were prepared according to a protocol modified from previously described methods [27]. The intestines were then applied for immunostaining. For immunostaining, the organoid cells were rinsed three times in ice-cold PBS. The organoid cells were spun down at 900 rpm for 10 min at 4 °C. Sections were deparaffinized and stained with H&E for the initial histology analysis. The immunofluorescence was performed on paraffin-embedded sections (5 μm). The permeabilized organoid cell samples were incubated with primary antibodies overnight at 4 °C. The samples were incubated with anti-PPP2R1A (GTX102206; GeneTex, Hsinchu City, Taiwan), anti-CK20 (GTX110600 Genetex), anti-Lgr5 (GTX50839 Genetex), anti-SMA (Abcam, ab5694, Cambridge, MA, USA), anti-beta-catenin (BD Transduction Labs, San Jose, CA, USA; 610154); the secondary antibodies used were DyLight® 650 Conjugated goat anti-rabbit (cat no. A120-101D5; Bethyl Laboratories Inc., Montgomery, TX, USA) and Goat anti-Rabbit IgG Antibody-FITC (Bethyl cat no. A120-101F) and DAPI (Molecular Probes) for 1 h at room temperature. The slides were mounted with SlowFade (SlowFade® AntiFade Kit, Molecular Probes, Waltham, MA, USA) followed by covering with a coverslip, and the edges were sealed to prevent drying. The specimens were examined with a Zeiss 710 Laser Scanning confocal microscope (Zeiss, Oberkochen, Germany).

Intestinal tissue was fixed and processed into paraffin blocks according to standard procedures. beta-catenin immunohistochemistry was performed as previously described [3]. Immunohistochemistry protocol hold as following: freshly isolated intestines were flushed with 10% formalin in PBS and fixed by incubation in a 10-fold excess of formalin overnight at room temperature. The formalin was removed and the intestines washed twice in PBS at room temperature. The intestines were then transferred to a tissue cassette and dehydrated by serial immersion in 20-fold volumes of 70, 96 and 100% EtOH for 2 h each at 4 °C. Excess ethanol was removed by incubation in xylene for 1.5 h room temperature and the cassettes then immersed in liquid paraffin (56 °C) overnight. Paraffin blocks were prepared using standard methods and 4μm tissue sections generated. These sections were de-waxed by immersion in xylene (2 × 5 min) and hydrated by serial immersion in 100% EtOH (2 × 1 min), 96% EtOH (2 × 1 min), 70% EtOH (2 × 1 min) and distilled water (2 × 1 min). Endogenous peroxidase activity was blocked by immersing the slides in peroxidase blocking buffer (0.040 M citric acid, 0.121 M disodium hydrogen phosphate, 0.030 M sodium azide, 1.5% hydrogen peroxide) for 15 min at room temperature. For beta-catenin, antigen retrieval involved 20 min boiling in Tris-EDTA pH 9.0, and blocking buffer (1% BSA in PBS) added to the slides for 30 min at room temperature. For beta-catenin (BD Transduction Labs, 610154), staining involved 1/100 dilution in blocking buffer (0.05% BSA in PBS) for 2 h at room temperature.

The slides were then rinsed in PBS and secondary antibody added (polymer HRP-labeled anti-mouse/rabbit, Envision) for 30 min at room temperature (Dako, Trappes, France). Slides were again washed in PBS and bound peroxidase detected by adding DAB substrate for 10 min at room temperature. Slides were then washed 2× in PBS and nuclei counterstained with Mayer's hematoxylin for 2 min, followed by two rinses in distilled water. Sections were dehydrated by serial immersion for 1 min each in 50 and 60% EtOH, followed by 2 min each in 70, 96, and 100% EtOH and xylene. Slides were mounted in mounting medium and a coverslip placed over the tissue section.

For immunohistochemistry (IHC) analysis, nuclei expressing beta-catenin after IHC staining were counted under 200× magnification. The Histological score (H-score) was determined based on the intensity and percentage of nucleus staining at each intensity [29], and calculated as follows: H-score = (nucleus showing highly beta-catenin expression) × 3 + (nucleus showing beta-catenin expression) × 2 + (nucleus showing weak beta-catenin expression) × 1.

2.6. Viral Infection of Organoids

For in vitro deficiency of the *Ppp2r1a*, organoid cultures containing floxed *Ppp2r1a* alleles were infected with adenovirus-encoding Cre recombinase (Ad-Cre) (Vector Biolabs, Philadelphia, PA, USA) at a titer of 100 multiplicity of infection (MOI) [27].

2.7. Tamoxifen Induction

Mice aged 6–8 weeks were injected intraperitoneally with a single 200 μL dose of tamoxifen in sunflower oil at 10 mg/mL.

2.8. Organoid Disaggregation, FACS, and Immunoblotting

Organoid cultures were recovered and dissociated from collagen gel by collagenase IV incubation, followed by incubation with 0.05% trypsin and EDTA. After extensive washing with 10% FBS, cells were filtered with 40-μm cell strainers (BD Falcon) Pellets were resuspended with FACS staining solution (5% FCS in PBS). Stringent wash was applied using ice-cold PBS, followed by isolation of $Lgr5^{-}EGFP^{+}$ cells using an FACSAria II (BD) [30]. For immunoblotting, the organoid cells were lysed in lysis buffer (1% Triton X-100, 150 mmol/L NaCl, 10 mmol/L Tris pH 7.4, 1 mmol/L EDTA pH 8.0, protease inhibitor cocktail) and then sonicated. The protein concentration was then measured. Next, equal amounts of protein (20 μg/well) were separated by SDS-polyacrylamide gel electrophoresis, transferred to nitrocellulose, and immunoblotted with primary antibodies. The membranes were blocked with CISblock buffer purchased from Cis-biotechnology, Taiwan. The following antibodies were used: anti-phospho-AKT (Ser-473), anti-AKT; anti-phospho-ERK1/2 (Thr-202/Tyr-204), anti-ERK, and anti-PP2A from Cell Signaling; anti-Lgr5 and anti-alkaline phosphate intestinal (Alpi) (Genetex, epitope C-terminus), beta-catenin (BD Transduction Labs, 610154), anti-Villin (Santa Cruz, Dallas, TX, USA), and anti-beta-actin and GAPDH (Sigma-Aldrich, St. Louis, MO, USA). Following the primary antibody incubation, the nitrocellulose membranes were incubated with secondary antibodies and visualized by ECL.

2.9. Antibody Arrays of Mouse AKT Pathway Phosphorylation

The RayBio™ Mouse AKT Pathway Phosphorylation Array Kit (cat. no. AAH-AKT1-2) was purchased and preformed according to a protocol modified from RayBiotech Inc. (Norcross, GA, USA). Briefly, the array membranes were blocked with blocking buffer for 30 min at room temperature. The membranes were then incubated with 2 mL of lysate prepared from organoid cultures with different treatments after normalization with equal amounts of protein. After extensive washing with wash buffer I (3 washings of 5 min each), and wash buffer II (3 washings of 5 min each) to remove unbound materials, the membranes were incubated with the Detection Antibody Cocktail for 1.5 to 2 h at room temperature, followed by wash with wash buffer I and II. Then the membranes were incubated with HRP-Anti-Rabbit IgG for 2 h at room temperature. The unbound HRP antibody was washed out with wash buffer I and II. Finally, each array membrane was exposed to X-ray film using a chemiluminescence detection system (Perkin Elmer, Wellesley, MA, USA).

2.10. Transcriptome Analysis

RNA was extracted from organoid culture using an RNeasy Kit (Qiagen, Hilden, Germany). RNA integrity was assessed using the RNA Nano6000 assay kit (Agilent Technologies, Santa Clara, CA, USA). For RNA-seq, library preparation and sequencing were performed by Novogene Technology. The output data (FASTQ files) were mapped to the target genome (TopHat v2.0.12), which can generate a database of splice junctions based on the gene model annotation file. HTSeq v0.6.1 was used to count the reads numbers mapped to each gene. Then the FPKM of each gene was calculated based on the length of the gene and reads count mapped to this gene. Differential expression heatmap results and biological variability were analyzed by ClustVis free web server [31] and gene set enrichment analysis (GSEA) [32], respectively. Data were submitted and approved by Gene Expression Omnibus (GEO; accession number GSE120241).

2.11. Whole-Exome Sequencing, Alignment, and Annotation

Exome sequences were captured with SureSelect^{XT} Mouse All Exon Kit (G7550E-001, Agilent, CA, USA) following the standard protocols. The products of exome capture should pass criteria: the length of fragments: 300 ± 30 bp and total amount: >600 ng. After exome capturing, the index-tagged samples were pooled and sequenced on Illumina HiSeq 2000. Burrows-Wheeler Alignment (v0.7.12) was employed to align reads to the reference genome (mm10) with default parameters. Aligned reads were sorted by picard-tools (v1.8). The duplicated reads were marked by picard-tools. Indel Realignment were performed with GenomeAnalysisTK (v3.5) using mm10 dbsnp database as known sites. Base quality score recalibration was also performed with GenomeAnalysisTK (v3.5) using mm10 dbsnp database. SNPs and indels were called by GenomeAnalysisTK HaplotypeCaller (v3.5), which used default parameters. Whole exome sequencing raw data was submitted to SRA database (SRA; http://trace.ncbi.nlm.nih.gov/Traces/sra/, accession number SRP162613)

2.12. Statistics

The p-values were determined using two-tailed Student's t-test (t groups) and One-way ANOVA (>2 groups). A p-value less than 0.05 was considered significant.

3. Results

3.1. Combination of DMBA and OA Treatment Induces Dysplasia and Oncogenic Transformation in Organoid Culture

We chose an organoid culture system supported by epidermal growth factor (EGF), Noggin and R-spondin 1 (ENR) medium to investigate whether DMBA or/and OA could induce oncogenic transformation. As previously described [27], small intestine or colon organoids predominantly exhibited a well-organized, stereotyped epithelial single-layer organization at 7 days of culture, and maintained the similar morphology over a 50-day period of culture (Figure 1). At day 7, DMBA did not affect colony (organoid)-forming efficiency. OA induced a slight increase in both colony-forming efficiency, while a combination of DMBA and OA induced a large and significant increase in colony-forming efficiency (Figure 1A). At day 50, DMBA did not affect organoid morphology, OA induced mild enlargement in part of the epithelial layer with crowded nuclei, while a combination of DMBA and OA induced a very large malformation involving the entire epithelium with a confluent sheet of nuclear pleomorphism (Figure 1B), similar to that observed only when combining *Apc*, *p53*, *Kras*^{G12D}, and *Smad4* mutations in differentiated villus cells [27]. Histological analysis revealed that organoids treated with DMBA alone had a single-layer epithelium, similar to the control. OA-treated organoids showed multi-cell-layer-changes in only a small part of the epithelium, while those treated with DMBA in combination with OA showed multi-cell-layer-changes with loss of the cell border in nearly the entire epithelium (Figure 1C), similar to the histology achieved only by quadruple mutants, *Apc/Kras*^{G12D}*/p53/Smad4* [27]. A dysplasia index quantification of proliferation, nuclear atypia, invasion, and cellular stratification in organoids indicated that DMBA did not induce dysplasia compared to the control, OA induced a marginal increase in dysplasia, while the combination of DMBA and OA induced a large and significant increase in dysplasia (Figure 1D). Furthermore, the combination of DMBA and OA, but not DMBA or OA alone, endowed organoids with robust in vivo tumorigenicity, forming alpha-smooth muscle actin (SMA)$^+$ and CK20$^+$ intestinal tumor after subcutaneous transplantation (Figure 1E–G).

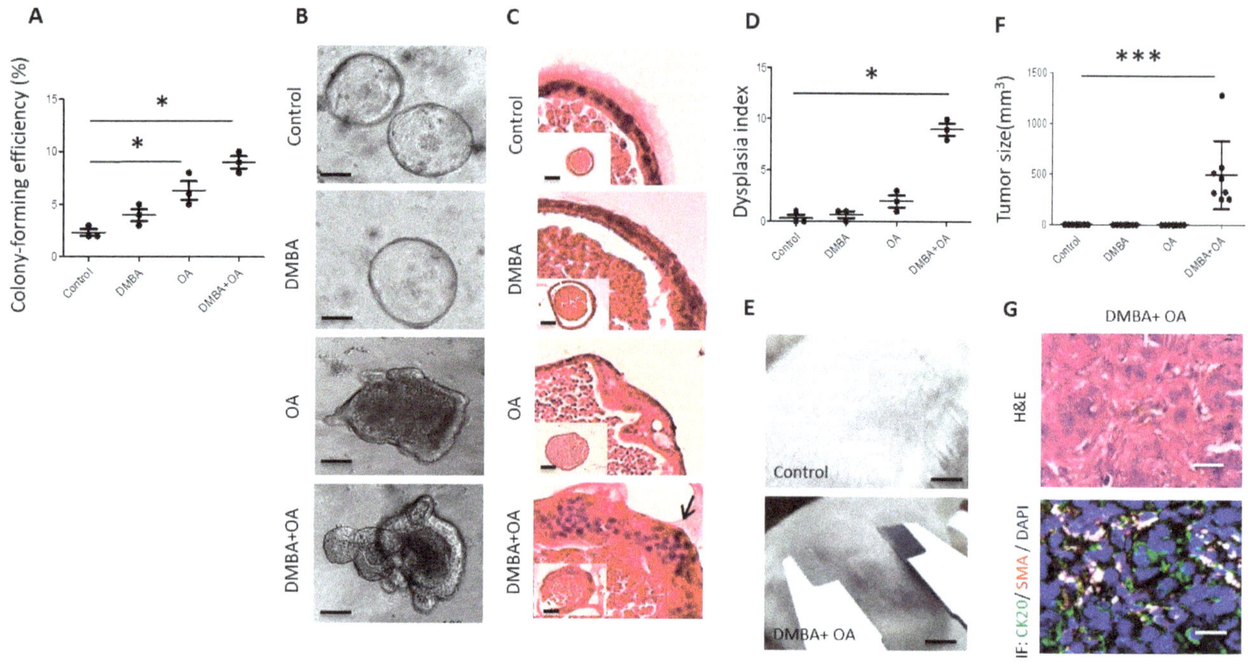

Figure 1. Combination of DMBA and okadaic acid (OA) induces dysplasia and oncogenic transformation in wild-type intestinal organoid culture. In vitro culture of wild type intestinal organoids without (control) or with DMBA or/and tamoxifen (TAM) in the presence of EGF, Noggin, and R-spondin 1 (500 single cells/well). (**A**) Colony (organoid)-forming efficiency was calculated at day 7. Experiment has been carried out in triplicate and each time 100 organoids were counted in each group. (**B**) Bright-field of organoid culture at day 50. Scale bar, 100 μm. (**C**) H&E staining and histologic characterization of cystic stratified epithelium with nuclear pleomorphism (arrow). Scale bar, 50 μm. (**D**) Dysplasia index at day 50 (experiment were repeated twice with $n = 3$ microscopic fields containing viable organoids). (**E**) Dissociated cells in Matrigel (500,000 cells/100 μL) were injected s.c. into NOD-SCID mice. In vivo tumor formation 45 days later (For those without tumor formation, observation was extended for up to 3 months, experiment were repeated twice with $n = 3$). (**F**) Tumor size 45 days after s.c. implantation ($n = 8$). (**G**) H&E staining and immunofluorescence of CK20 and SMA for tumor sections. Scale bar, 100 μm. *, $p < 0.05$; ***, $p < 0.0001$ as determined with one-way ANOVA.

3.2. Combination of DMBA Treatment and PP2A Deficiency Induces Dysplasia and Oncogenic Transformation in Organoid Culture

We further characterized the transformation effect of DMBA or/and adenovirus carrying recombinase (Ad-Cre-GFP)-mediated-PP2A deficiency in *Ppp2r1a*$^{flox/flox}$ mice-derived organoids. Similarly, DMBA had no significant effects, Ad-Cre-GFP mediated-PP2A deficiency had mild effects, and the combination of DMBA and Ad-Cre-GFP mediated-PP2A deficiency had large effects on early organoid forming efficiency, late organoid morphology, histological changes, and dysplasia (Figure 2A–D). Furthermore, only the combination of DMBA and Ad-Cre-GFP-mediated-PP2A deficiency fully endowed organoids with robust in vivo tumorigenicity (Figure 2E–G).

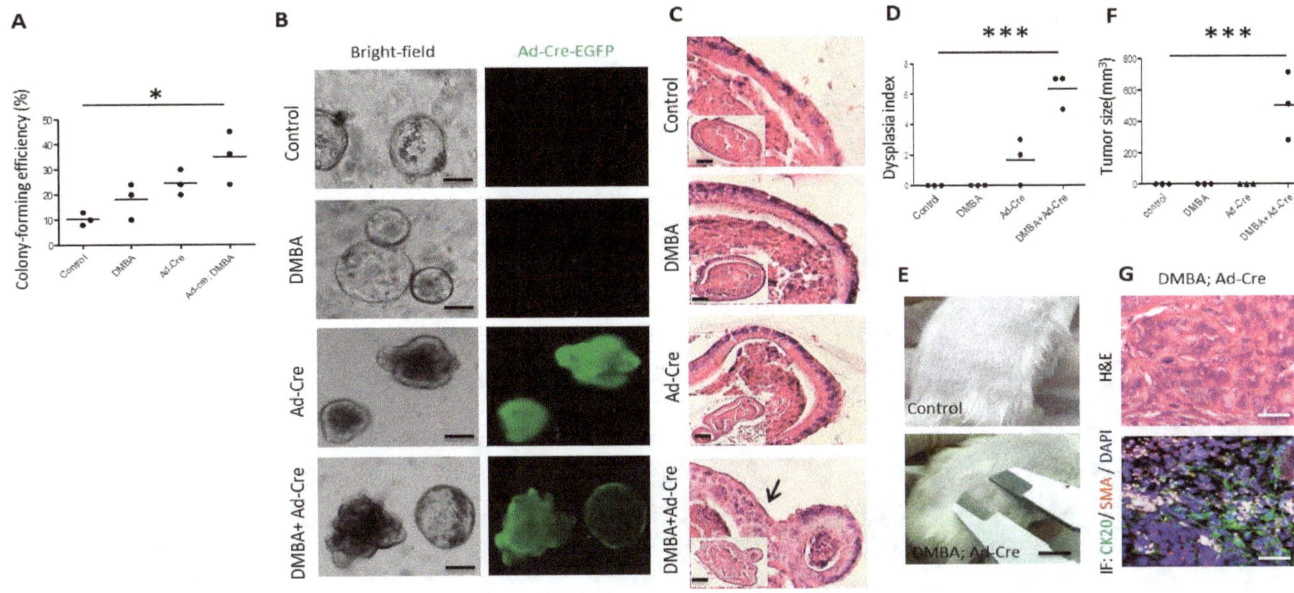

Figure 2. Combination of DMBA and Ad-Cre induces dysplasia and oncogenic transformation in Ppp2r1aflox/flox intestinal organoid culture. In vitro culture of Ppp2r1aflox/flox intestinal organoids without (control) or with DMBA or/and Ad-Cre-GFP (Ad-Cre) infection in the presence of EGF, Noggin and R-spondin 1 (500 single cells/well). (**A**) Colony (organoid)-forming efficiency was calculated at day 7. At least 100 organoids were counted in each group. (**B**) Bright-field and fluorescence images of organoid culture at day 50. Scale bar, 100 μm. (**C**) H&E staining and histologic characterization of cystic stratified epithelium with nuclear pleomorphism (arrow indicated). Scale bar, 50 μm. (**D**) Dysplasia index at day 50 (experiments were repeated twice with $n = 3$ microscopic fields containing viable organoids). (**E**) Dissociated cells in Matrigel (500,000 cells/100 μL) were injected s.c. into NOD-SCID mice. In vivo tumor formation 45 days later (for those without tumor formation, observation was extended for up to 3 months, experiments were repeated twice with $n = 3$). (**F**) Tumor size 45 days after s.c. implantation ($n = 3$). (**G**) H&E staining and immunofluorescence of CK20 and SMA for tumor sections. Scale bar, 100 μm. *, $p < 0.05$; ***, $p < 0.0001$ as determined with one-way ANOVA.

3.3. Combination of DMBA Treatment and PP2A Deficiency in Lgr5$^+$ Rather than in Differentiated Villus Cells Induces Dysplasia and Oncogenic Transformation in Organoid Culture

To investigate whether Lgr5$^+$ crypt stem cells or differentiated villus cells serve as the cell of origin of tumors, we treated organoids from *Lgr5-EGFP-CreERT2; Ppp2r1a$^{flox/flox}$* mice with DMBA or/and tamoxifen. Similarly, DMBA did not have significant effects, tamoxifen mediated-PP2A deficiency had mild effects, and the combination of DMBA and tamoxifen-mediated-PP2A deficiency had great effects on early organoid forming efficiency, late organoid morphology, histological changes, and dysplasia (Figure 3A–E). Furthermore, only the combination of DMBA and tamoxifen-mediated-PP2A deficiency endowed organoids with in vivo tumorigenicity (Figure 3F,G). Interestingly, the combination of DMBA and PP2A deficiency in organoids derived from *Villin-Cre; Ppp2r1a$^{flox/flox}$* mice did not affect early organoid forming efficiency, late organoid morphology, histological changes, and dysplasia (Figure S2A–E), and failed to induce in vivo tumorigenicity (Figure S2F). Of note, the recombinase activity in Villin-Cre mice is gradually reduced from villus to crypt [33], nevertheless, Ppp2r1a protein is only deleted in sorted villus cells but not in sorted Lgr5$^+$ cells (Figure S2A). These data suggest that Lgr5$^+$ crypt stem cells but not differentiated villus cells serve as the cell of origin of intestinal tumor.

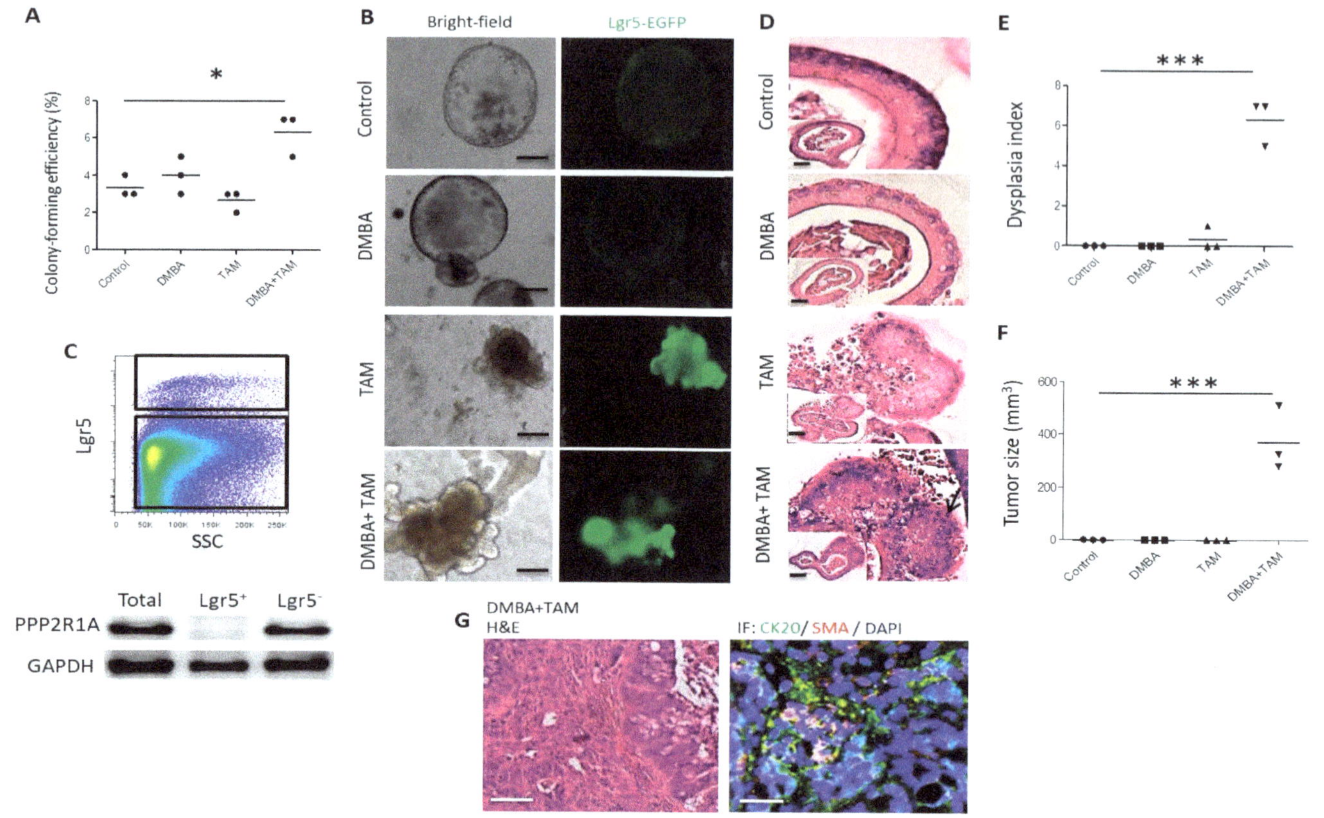

Figure 3. Combination of DMBA and TAM induces dysplasia and oncogenic transformation in *Lgr5-EGFP-CreERT2; Ppp2r1a^flox/flox^* intestinal organoid culture. In vitro culture of *Lgr5-EGFP-CreERT2; Ppp2r1a^flox/flox^* intestinal organoids without (control) or with DMBA or/and tamoxifen (TAM) in the presence of EGF, Noggin and R-spondin 1 (500 single cells/well). (**A**) Colony (organoid)-forming efficiency was calculated at day 7. At least 100 organoids were counted in each group. (**B**) Bright-field and fluorescence images of organoid culture at day 50. Scale bar, 100 μm. (**C**) Fluorescence-activated cell sorting (FACS) isolation of $Lgr5^+$ and $Lgr5^-$ populations. After FACS, PPP2R1A protein levels were detected by western blot. (**D**) H&E staining and histologic characterization of cystic stratified epithelium with nuclear pleomorphism (arrow). Scale bar, 50 μm. (**E**) Dysplasia index at day 50 (experiments were repeated twice with $n = 3$ microscopic fields containing viable organoids). (**F**) Dissociated cells in Matrigel (500,000 cells/100 μL) were injected s.c. into NOD-SCID mice. In vivo tumor formed 45 days later (for those without tumor formation, observation was extended for up to 3 months, experiments were repeated twice with $n = 3$). (**G**) H&E staining and immunofluorescence of CK20 and SMA for tumor sections. Scale bar, 100 μm. *, $p < 0.05$; ***, $p < 0.0001$ as determined with one-way ANOVA.

3.4. Combination of DMBA Treatment and PP2A Deficiency in $Lgr5^+$ Cells Induces Upregulation in Stem Cell and EMT Markers, Downregulation in Differentiated Markers, and Tumorigenicity in Organoid Culture

Flow cytometric analysis (Figure 4A) revealed that DMBA did not increase the $Lgr5^+$ cell ratio (as assayed by Lgr5-EGFP). PP2A deficiency induced a marginal increase, while the combination of DMBA and PP2A deficiency induced a significant increase in the $Lgr5^+$ cell ratio. Comparative gene expression analysis of RNA samples isolated from organoid culture of *Lgr5-EGFP-CreERT2; Ppp2r1a^flox/flox^* mice 50 days after DMBA and tamoxifen administration revealed marked upregulation of stem cell genes, such as *Lgr5*, *CD44*, *Ephb3*, *Egr2*, *Notch1*, and *Sox4*; as well as EMT markers, such as *Snail1*, *Snail2*, *Twist1*, *fibronectin*, and *vimentin*; and a marked downregulation of genes associated with differentiated cells, such as Paneth, enterocyte, goblet, and secretory cells compared to other treatment groups (Figure 4B). Furthermore, a small number (10^3) of $Lgr5^+$ but not $Lgr5^-$ cells isolated from organoid culture treated with DMBA and tamoxifen possessed in vivo tumorigenicity (Figure 4C).

Collectively, these data suggest that the combination of DMBA and PP2A deficiency converted Lgr5$^+$ crypt stem cells into CSCs.

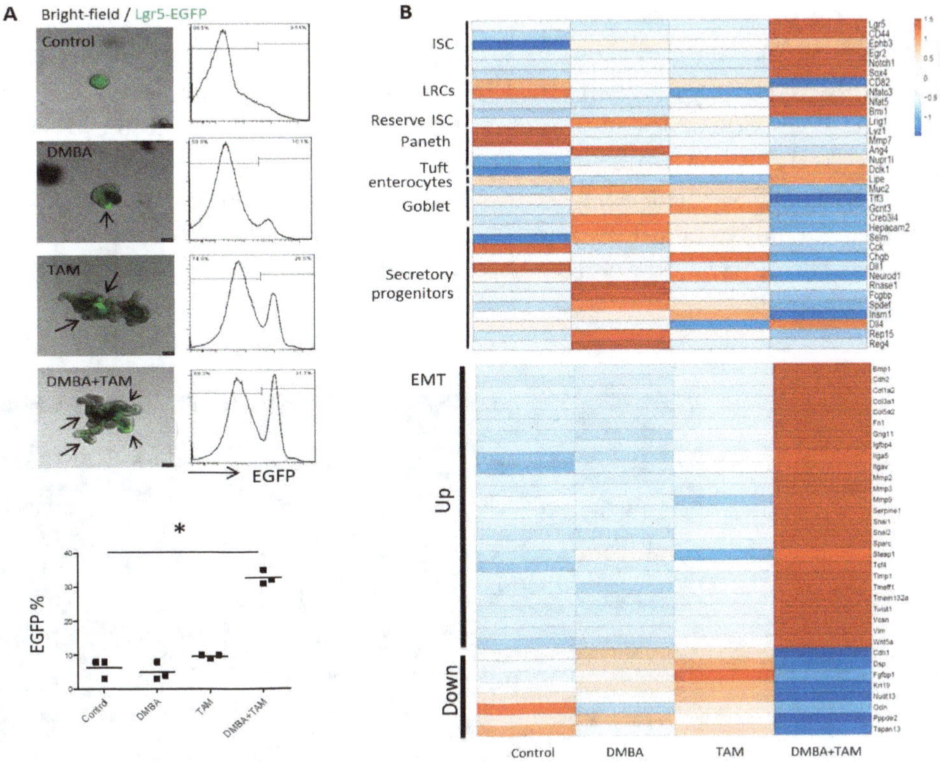

C

	2W	4W	6W	8W
Lgr5$^-$				
10^5	0/4	0/4	0/4	0/4
10^4	0/4	0/4	0/4	0/4
10^3	0/4	0/4	0/4	0/4
Lgr5$^+$				
10^5	3/4	3/4	4/4	4/4
10^4	2/4	2/4	2/4	3/4
10^3	0/4	1/4	2/4	3/4

Figure 4. Combination of DMBA and PP2A deficient induces upregulation in stem cell and EMT markers and downregulation in differentiated markers in intestinal organoid. (**A**) In vitro culture of *Lgr5-EGFP-CreERT2; Ppp2r1a$^{flox/flox}$* intestinal organoids. Merged bright-field and fluorescence images of organoid culture treated without (control) or with DMBA or/and tamoxifen (TAM) for 50 days in the presence of EGF, Noggin and R-spondin 1. *Lgr5-EGFP* were denoted as arrows. Scale bar, 100 μm. Flow cytometric analysis and quantification of GFP expression (bottom panel). No fluorescence organoid culture was serve as negative control to decide the threshold. *, $p < 0.05$ as determined with one-way ANOVA. (**B**) RNA-*sequencing* (seq) analysis of transcriptomes for 50-day organoids. The upper heat map shows clustering to previously reported RNA-seq data of sorted ISC (Intestinal Stem Cell); reserve ISC; paneth cell; LRCs (label retaining cells); tuft; enterocytes; goblet and secretory progenitor cells. The lower heat map shows clustering to EMT (epithelial–mesenchymal transition). (**C**) Tumor incidence in limiting dilution assay. Tumorigenic potential characterization of indicated numbers of Lgr5$^+$ and Lgr5$^-$ cells from the 50-day organoid culture. Dissociated cells in Matrigel (100 μL) were injected s.c. into NOD-SCID mice. Incidence of tumor formation was calculated more than 2 months.

3.5. Combination of DMBA Treatment and PP2A Deficiency Generates CSCs through PI3K, ERK, and Wnt Activation

To understand the molecular characteristics of the genes and pathways involved in organoid culture of *Lgr5-EGFP-CreERT2; Ppp2r1a$^{flox/flox}$* mice after DMBA and tamoxifen administration, RNA-seq and gene set enrichment analysis (GSEA) were performed. Similar to previous findings that intestinal tumor begins with specific molecular alterations in Wnt-beta-catenin pathway [13,23], Wnt signaling was upregulated upon oncogenic transformation of organoid culture (Figure 5A). Western blotting of nuclear proteins (Figure 5B) and immunofluorescence (Figure 5C) revealed that nuclear accumulation of beta-catenin was predominantly observed in organoids with both of DMBA treatment and PP2A deficiency. Screening with a serine/threonine phosphorylation protein array (Figure S3) followed by confirmation with western blotting further revealed that PI3K/AKT/GSK-3beta and Raf/ERK were activated in organoids with both DMBA treatment and PP2A depletion compared to other treatment groups (Figure 5D,E). Interestingly, treatment with the PI3K inhibitor LY294002, MEK inhibitor PD98059, and Wnt inhibitor DKK1 reduced the formation of malformed organoids, inhibited dysplasia (Figure 5F,G), and completely blocked in vivo tumorigenicity (Figure 5H). Notably, not only the WNT signaling was the most, but also the PI3K and RAS-MAPK (ERK) signalings were common altered pathways in human intestinal tumor, as revealed by The Cancer Genome Atlas (TCGA) Project [14]. These data suggest that CSC generation by DMBA treatment and PP2A deficiency depends on the activation of PI3K, ERK, and Wnt signals.

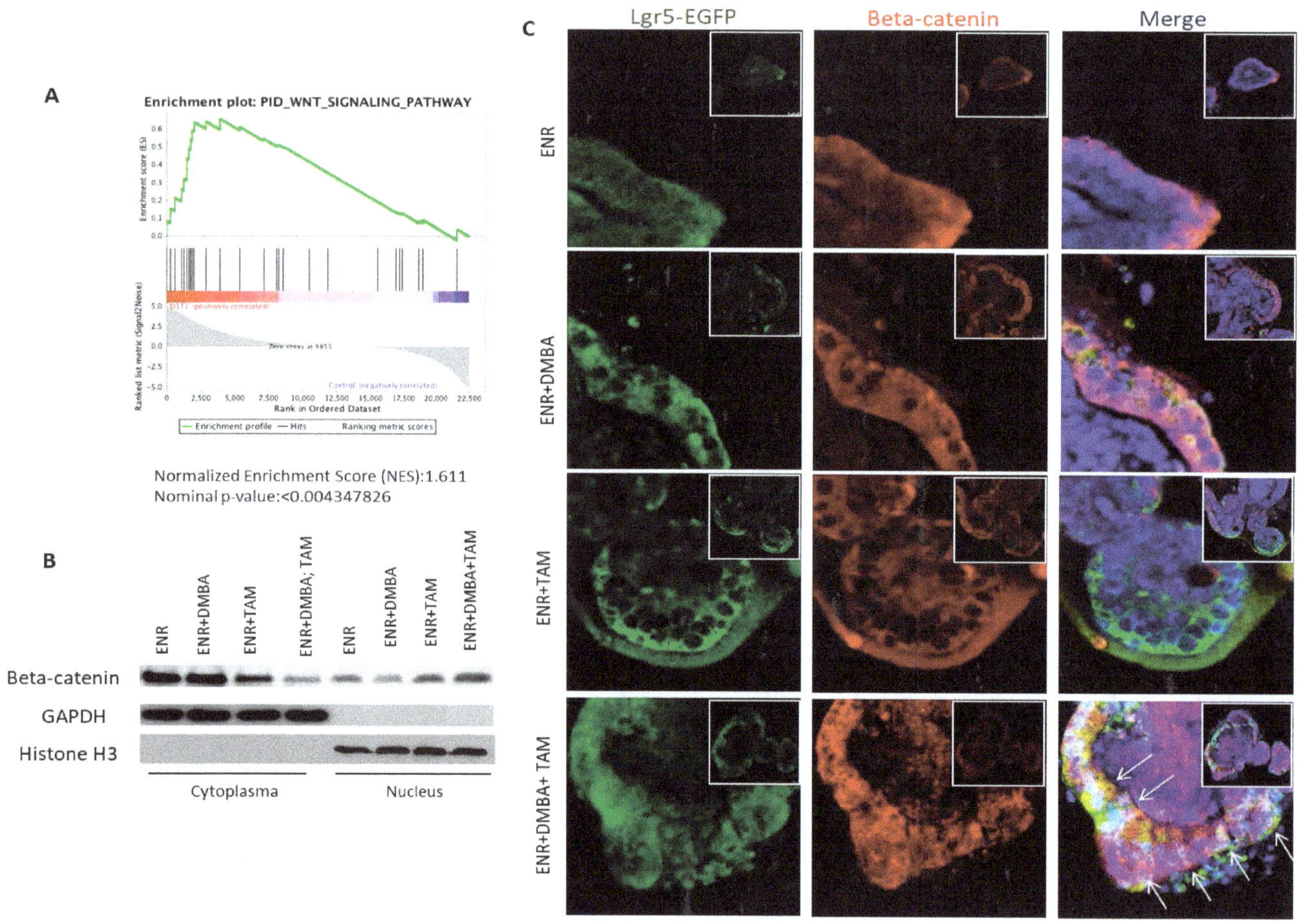

Figure 5. *Cont.*

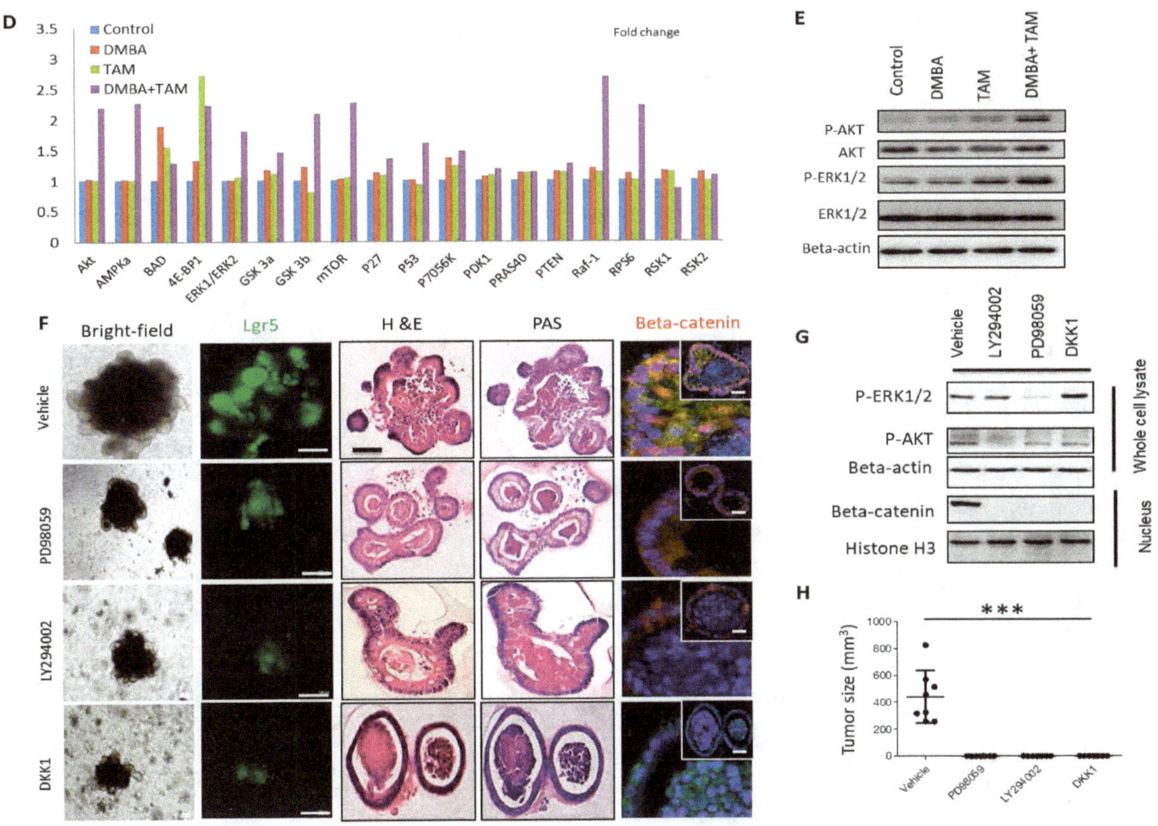

Figure 5. Beta-catenin activation caused by PI3K and ERK mediates dysplasia and oncogenic transformation in organoid culture. *Lgr5-EGFP-CreERT2; Ppp2r1a*$^{flox/flox}$ intestinal organoids were treated without (control) or with DMBA or/and tamoxifen (TAM) for 50 days in the presence of EGF, Noggin and R-spondin 1. (**A**) Gene set enrichment analysis (GSEA) shows "WNT SIGNALING PATHWAY" for the organoids treated with DMBA and TAM group versus control group, $p < 0.05$. (**B**) Western blotting of nuclear and cytoplasmic fractions. GAPDH and Histone H3 were used as protein loading controls for the cytoplasmic and nuclear fractions, respectively. (**C**) Immunofluorescence of Lgr5-EGFP and beta-catenin expression in organoid (denoted by arrows). Scale bar, 100 μm. (**D**) Graphs of mouse AKT pathway phosphorylation protein expression array (original data in Extended Data Figure 3) and densitometric analyses, $n = 1$. (**E**) Western blotting. (**F**) Bright-field, fluorescence images, H&E, PAS staining and beta-catenin immunofluorescence images. (**G**) Western blot analysis of whole cell lysate and nuclear fraction from organoid cultures treated with EGF, Noggin, R-spondin 1 (ENR), DMBA, and TAM for 50 days in the absence (Vehicle) or presence of indicated inhibitor treatment. (**H**) Dissociated cells in Matrigel (500,000 cells/100 μL) were injected s.c. into NOD-SCID mice. In vivo tumor formed 45 days later (For those without tumor formation, observation was extended for up to 3 months, $n = 8$). ***, $p < 0.0001$ as determined with one-way ANOVA.

3.6. Lgr5$^+$ CSCs Are R-Spondin 1-Dependent

CSCs isolated from human intestinal tumor specimens express the Lgr5 crypt marker [34]. R-spondin is expressed by the intestinal stroma and is differentially upregulated during *Citrobacter rodentium*- and dextran sulfate sodium (DSS)-induced colitis in mice, which reflects human ulcerative colitis, a precancerous stage [35]. These data suggest a role for R-spondin 1 and its receptor Lgr5 in the maintenance of undifferentiated status and tumorigenesis of human colorectal CSCs. In contrast to tumorigenesis initiated by dedifferentiation (tumor generated from intestinal epithelial cell of *villin-creER*^{T2}*/APC* $^{lox/lox}$*/K-ras*$^{G12D/+}$ mice) [4], where tumor cells were Lgr5$^-$ and generated independently of R-spondin 1, tumor cells in the current study were Lgr5$^+$ and generated dependently on R-spondin 1 (Figure 6), suggesting that tumor cells generated by DMBA treatment and PP2A deficiency in mouse Lgr5$^+$ cells were indeed intestinal CSCs and could serve as surrogates of human colorectal CSCs [34]. Previous reports using tumorigenesis models generated by genetic manipulation in differentiated villus cells, even with low efficiency in tumorigenesis [27], also demonstrated important roles for Lgr5$^+$ cells in tumorigenesis/metastasis [7,8].

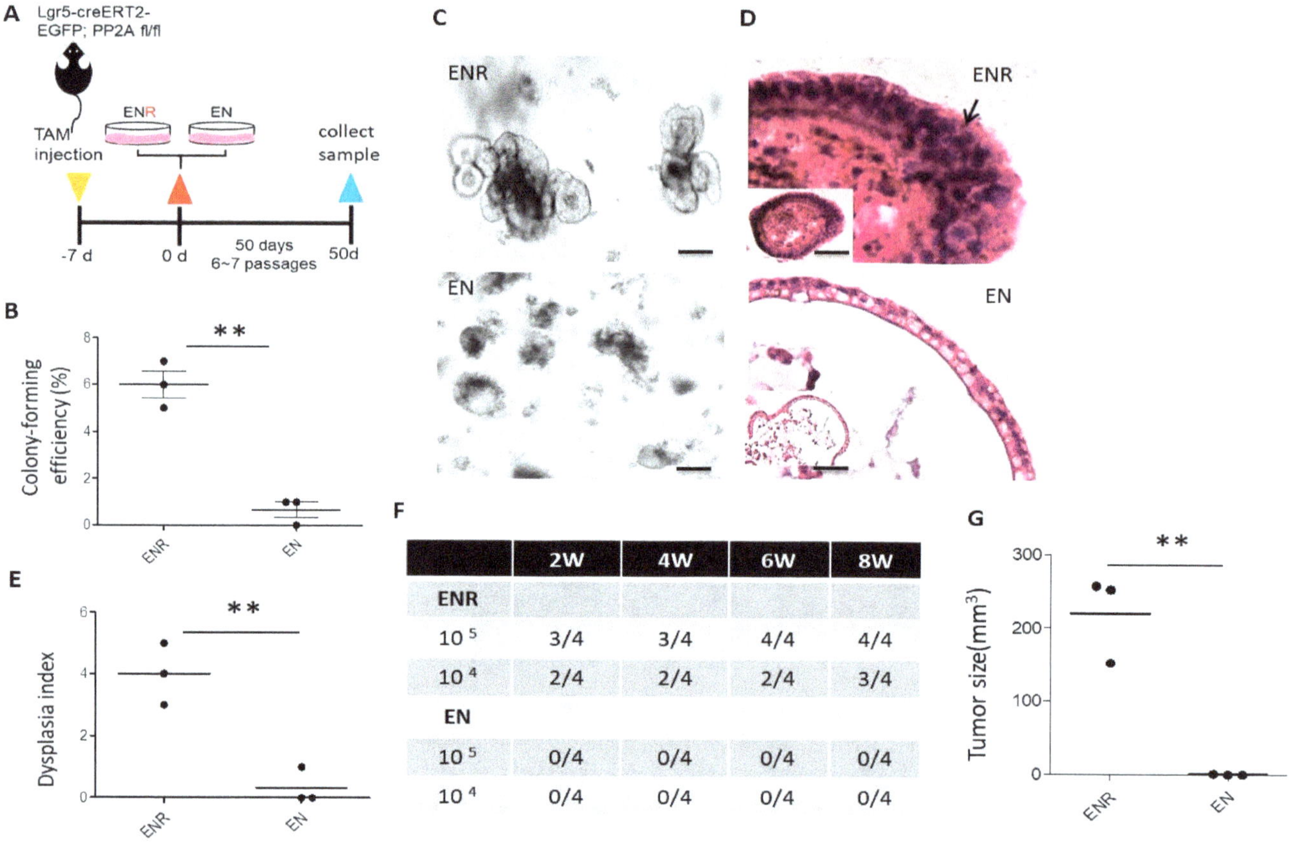

	2W	4W	6W	8W
ENR				
10 5	3/4	3/4	4/4	4/4
10 4	2/4	2/4	2/4	3/4
EN				
10 5	0/4	0/4	0/4	0/4
10 4	0/4	0/4	0/4	0/4

Figure 6. R-spondin 1 drives Wnt-dependent dysplasia and oncogenic transformation in the intestinal organoid culture. In vitro culture of *Lgr5-EGFP-CreERT2; Ppp2r1a*$^{flox/flox}$ intestinal organoids (500 single cells/well) treated with DMBA and tamoxifen (TAM) in the presence of EGF, Noggin, and with (ENR) or without R-spondin-1 (EN). (**A**) Schematic illustration of the experimental design. (**B**) Colony (organoid)-forming efficiency at day 7. At least 100 organoids were counted in each group. (**C**) Bright-field images at day 50. Scale bar, 100 μm. (**D**) H&E staining and histologic characterization of cystic stratified epithelium with nuclear pleomorphism (arrow). Scale bar, 50 μm. (**E**) Dysplasia index at day 50 (experiment was repeated twice with $n = 3$ microscopic fields containing viable organoids). (**F**) Incidence of tumor formation at indicated time periods, and (**G**) tumor size in NOD-SCID mice after s.c. injection 45 days later of indicated cell numbers from 50-day organoid. **, $p < 0.01$ as determined with Student's *t*-test.

3.7. Combination of DMBA Treatment with PP2A Deficiency in Lgr5$^+$ Also Induces Tumor Formation In Vivo

We investigated whether tumorigenesis generated by DMBA treatment and PP2A deficiency in mouse Lgr5$^+$ cells reflects tumor formation in vivo. *Lgr5-EGFP-CreERT2; Ppp2r1a*$^{flox/flox}$ mice treated with DMBA and tamoxifen formed adenocarcinoma in the small intestine and colorectal regions 36 days later. Histological analysis revealed that mice receiving DMBA and tamoxifen increased the incidence of multiple foci adenoma compared to other groups (Figure 7A). IHC analysis revealed that beta-catenin accumulated in the nuclei of tumor cells mainly in the crypt area (Figure 7B) and quantitative evaluation of nucleus beta-catenin accumulation [29] further showed that nucleus beta-catenin H-score was significantly greater in mice receiving DMBA and tamoxifen compared to other groups (Figure 7C,D). The efficiency and rapidity were much greater than tumorigenesis generated by *Apc* deficiency in either Lgr5$^+$ crypt stem cell- [3] or differentiated villus cell-based models [4].

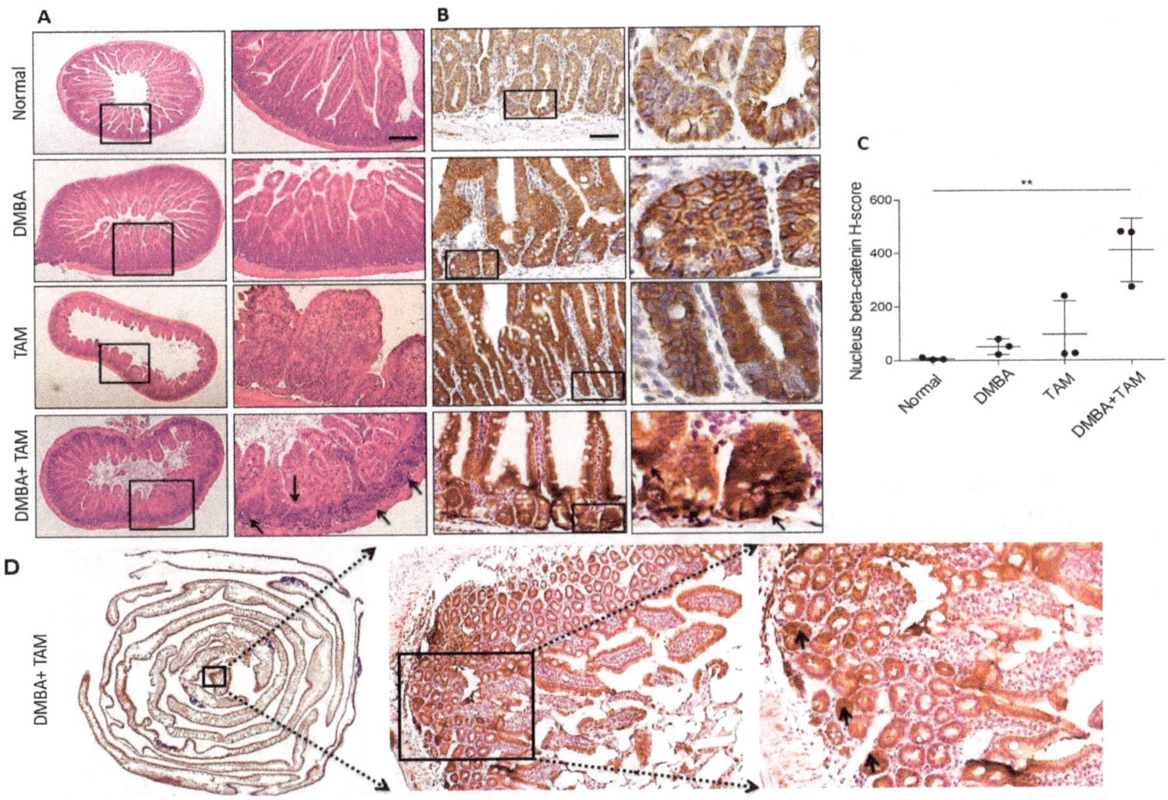

Figure 7. Combination of DMBA treatment with PP2A deficient in Lgr5$^+$ drives intestinal neoplasia in both the small intestine and colon. (**A**) H&E staining and (**B**) Beta-catenin IHC were performed on the serial sections. Multiple beta-catenin^{high} adenomas were observed throughout the colon 36 days after induction. (**C**) Quantitative analysis of nucleus beta-catenin H-score from (**B**). (**D**) High level expression of beta-catenin was apparent in the transformed stem cell compartment (Peyer's patches are stained blue). Multiple beta-catenin^{high} transformed cells were observed throughout the intestinal 36 day after induction (magnified at right panel, denoted by arrows). Original magnifications: (**A**) left, 4×; (**A**) right, 10×; (**B**) left, 20×; (**B**) right, 40×; (**D**) center, 20×; (**D**) right, 40×; Scale bar, 100 μm. **, $p < 0.01$ as determined with one-way ANOVA.

3.8. Not Only DMBA but Also Other Carcinogens Induce Tumors from Lgr5$^+$ Intestinal Stem Cells of PP2A Deficient Mice

To demonstrate carcinogen-induced tumor from Lgr5$^+$ intestinal stem cells of PP2A deficient mice was not limited to DMBA, *N*-methyl-*N*-nitrosourea (MNU) and 2-amino-1-methyl-6-phenylimidazo-[4,5-b]pyridine (PhIP) were also administered in combination with tamoxifen treatment in organoid cultures derived from *Lgr5-EGFP-CreERT2; Ppp2r1a$^{flox/flox}$* mice and from *Villin-Cre; Ppp2r1a$^{flox/flox}$* mice. Interestingly, increased early organoid forming efficiency, late organoid with irregular nuclei and prominent nucleoli morphology and dysplasia, and in vivo tumorigenicity were only observed in organoid cultures derived from *Lgr5-EGFP-CreERT2; Ppp2r1a$^{flox/flox}$* mice but not from *Villin-Cre; Ppp2r1a$^{flox/flox}$* mice (Figure S4A,B). More importantly, we found that tumorigenicity induced by combination of MUN or PhIP with PP2A deficiency also depended on the activation of PI3K, ERK, and Wnt signals (Figure S4C–F). Similarly, combination of MUN or PhIP treatment with PP2A deficiency in Lgr5$^+$ cells also induced tumor formation in vivo (Figure S4G).

3.9. Mutational Landscapes of Intestinal Tumors Derived from Lgr5$^+$ Cells Treated with Carcinogen and PP2A Deficiency

To gain insight of the genetic alterations that drove these pathways in the models, we performed whole-exome sequencing (WES) (Figure S5A). Somatic variants in each chemical induced tumor organoid sample were identified with a tumor-control paired strategy by removing the variants in their paired control samples and the variants affecting protein coding sequence were further filtrated (Figure S5B). The numbers of somatic mutations, including synonymous and nonsynonymous mutations, for organoid cultures with combined DMBA, MUN, or PhIP treatment with PP2A deficiency in Lgr5$^+$ cells were shown in Figure S5C. IPA analysis of 270 mutated genes shared by DMBA, MUN, and PhIP treatment with PP2A deficiency in Lgr5$^+$ cells (Figure S5D) revealed 10 top significantly enriched pathways (Figure S5E), including the intestinal tumor metastasis signaling (p-value $= 2.12 \times 10^{-57}$) and Wnt/beta-catenin signaling (p-value $= 1.40 \times 10^{-55}$). Furthermore, beta-catenin (*CTNNB1*) seemed to be a common downstream hub as identified by the Path Explorer function in IPA (Figure S5F). There were five core pathways related with human intestinal tumor found by TCGA, including p53, RAS-MAPK, PI3K, TGF-beta, and WNT pathways (Figure 8) [14]. We identified several recurrent mutations with FDR < 0.1, including *Braf, erbb2, kras, pten, Smad2, Smad4, Apc, DKK2, Wnt4, Wnt5a*, and *Wnt5b* (Figure 8). Based on these analyses, the mouse intestinal tumor organoid models based on combined use of chemical carcinogen and genetic modification can recapitulate the human colorectal cancer development process in response to complex interactions between the genotype and environmental factor.

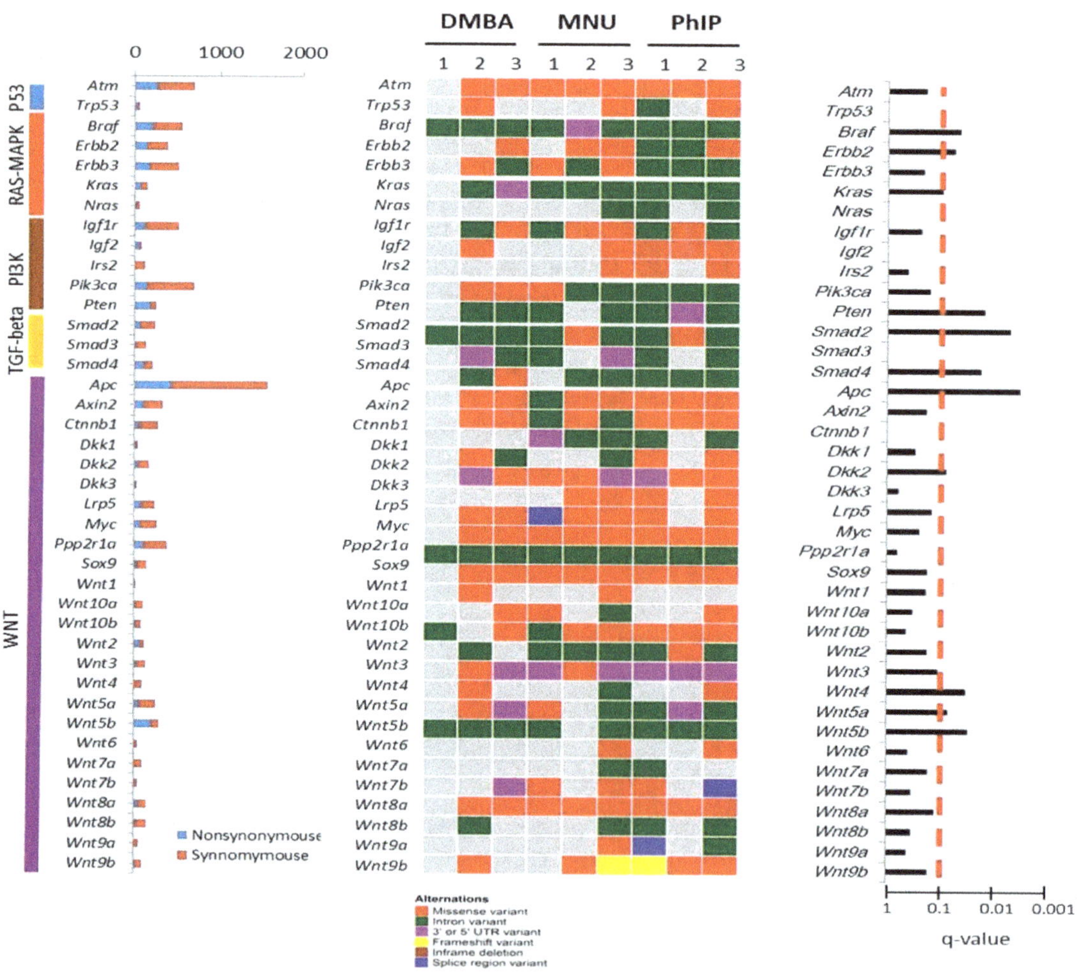

Figure 8. Landscape of somatic mutant genes in three carcinogen treated organoid cultures. Data matrix shows number of somatic mutant genes in each carcinogen treated organoid cultures and were classified according to their pathway. Somatic mutations are presented according to the type of mutation (missense variant, intron variant, 3′ or 5′ prime UTR variant, frameshift variant, inframe deletion, or splice region variant) On the left, the total number of mutations of each gene within all three groups is shown with a bar plot, while the q-value of each significantly mutated gene is shown on the right.

4. Discussion

To improve the efficiency of tumorigenesis in GEMMs, multiple gene mutations are necessary. Given that the frequency of point mutations varies from less than 0.1 to greater than 50 mutations per megabase [36], GEMMs are a greatly oversimplified view of the numbers and types of mutations found in human cancers [37]. The use of GEMMs in combination with carcinogenesis increased the tumor spectrum or speeded the tumor formation observed in some GEMMs, such as the $p53^{-/-}$ mouse model in combination with exposure to carcinogens or radiation [21,38]. In the current study, $Ppp2r1a^{-/-}$ mouse model in combination with exposure to different carcinogens induced tumorigenesis through similar signaling pathways, while GEMM or carcinogen exposure alone did not induce tumorigenesis. These data suggest that the combination of GEMMs with exposure to carcinogen or other environmental agents is a logical approach in studying tumorigenesis and cancer progression.

PP2A is a tumor suppressor that regulates many oncogenic pathways. In fact, decreased PP2A activity has been reported as a common event in colorectal cancer [39]. DMBA and MNU are important environmental carcinogens. PhIP, a food-borne carcinogen produced while cooking meat and fish, models human colon cancer in rodents [40]. Here, we demonstrated that DMBA, MNU, or PhIP each induced intestinal organoid transformation when combined with PP2A deficiency in $Lgr5^+$ intestinal stem cells but not in differentiated villus cells, suggesting that PP2A plays a role in suppressing colorectal tumorigenesis induced by chemical carcinogen exposure. Our results provide further experimental evidence to demonstrate that the cell of origin of intestinal tumor is crypt stem cells instead of differentiated villus cells. We did not focus on the molecular mechanism underlying the differential tumorigenicity pathways between stem cells and differentiated cells, which requires further investigation.

A recent study shows that unless the R-spondin and Wnt ligands are both present, the default fate of $Lgr5^+$ crypt stem cells is differentiation [41]. However, gain-of-function studies reveal that R-spondin and Wnt ligands have qualitatively distinct, non-interchangeable roles in crypt stem cells. Wnt proteins confer a basal competency unto the $Lgr5^+$ crypt stem cell by inducing R-spondin receptor expression that enables R-spondin-driven $Lgr5^+$ crypt stem cell self-renewal. In the current study, using a combination of carcinogen treatment and PP2A deficiency, intestinal organoids derived from Lgr5-cre; Ppp2r1aflox/flox, but not Villin-Cre; Ppp2r1aflox/flox, underwent oncogenic transformation and exhibited CSC phenotypes that were dependent on the presence of R-spondin 1 in the culture media. These studies together suggest the important roles of R-spondin signaling in stem cell self-renewal and preventing differentiation. The discrepancy of R-spondin 1 dependence between tumorigenicity induced by the current method and previous GEMM relied on *Apc* depletion or *CTNNB1* activation [4] is supported by the mutual exclusion of *R-spondin* fusion and *Apc* or *CTNNB1* mutation identified in human intestinal tumors [20]. The Lgr5/R-spondin 1 complex degrades Rnf43 and Znrf3, two transmembrane E3 ligases that remove Wnt receptors from the stem cell surface. Consistently, simultaneous deficient of these two E3 ligases in the intestinal epithelium induced the formation of unusual adenomas consisting entirely of $Lgr5^+$ stem cells and their niche [42].

Simultaneous carcinogen treatment and conditional deletion of PP2A in villus cells did not induce transformation or increased proliferation or dysplasia of intestine organoids. The current study did not aim to study the underlying mechanism that differentiates $Lgr5^+$ crypt stem cells from differentiated villus cells in terms of vulnerability to carcinogenesis. Based on our current results, both Wnt and Rspondin/Lgr5 signaling pathways play important roles in nuclear-beta catenin localization, which is important for stem cell self-renewal and may also initiate tumorigenesis once dysregulated.

5. Conclusions

In summary, we demonstrated that carcinogen-induced cancer arises from $Lgr5^+$ crypt stem cells in $Ppp2r1a^{-/-}$ mice. In addition, combing carcinogenesis with GEMM recapitulated the developmental process of environmentally induced human tumor, while increasing the rate and percentage of tumorigenesis. Interestingly, exposure to different carcinogens, such as DMBA, MNU, or PhIP, when combined with PP2A deficiency in Lgr5^{+} intestinal stem cells induced tumorigenesis that was dependent on the activation of pathways including Wnt, PI3K, and RAS-MAPK signalings, the common altered pathways revealed by TCGA human intestinal tumor project [14]. Together, these data suggest that PP2A functions as a tumor suppressor in intestine carcinogenesis. This organoid platform provides experimental evidence as to its usefulness in detection of key oncogenes and suppressor genes as early molecular epidemiological biomarkers of carcinogenesis, and is useful in human cancer prevention practice as well.

Supplementary Materials

Figure S1: Screening strategy for phosphatase hub gene in TCGA-COAD data set, Figure S2: DMBA and PP2A deficiency did not induce dysplasia and oncogenic transformation in *Villin-Cre; Ppp2r1a$^{flox/flox}$* intestinal organoid culture, Figure S3: Serine/threonine phosphorylation protein array screening in individual organoid groups, Figure S4: Combination of MNU or PhIP and TAM induces dysplasia and oncogenic transformation in *Lgr5-EGFP-CreERT2; Ppp2r1a$^{flox/flox}$* intestinal organoid culture, Figure S5: Ingenuity pathway analysis (IPA) to identify significant canonical pathways and interactome in three carcinogens induced oncogenic transformation in *Lgr5-EGFP-CreERT2; Ppp2r1a$^{flox/flox}$* intestinal organoids.

Author Contributions: Y.-T.Y. designed and performed experiments, analyzed the data and wrote the paper. M.C. contributed to Lgr5-cre and Vilin-cre PP2A knockout mice analysis and helped tissue processing. Y.-C.L. performed the RNA-seq analysis. D.-P.C. performed the WES analysis. C.-M.C. reviewed and edited the paper. M.-C.H. reviewed and edited the paper. S.-C.H. wrote the proposal, designed and supervised the study, analyzed the data and wrote the paper. All authors have read and agreed to the published version of the manuscript.

Acknowledgments: We thank Bin Tean Teh (Cancer Science Institute of Singapore, National University of Singapore, Centre for Life Sciences, Singapore) for discussion of our data.

References

1. Poirier, M.C. Chemical-induced DNA damage and human cancer risk. *Nat. Rev. Cancer* **2004**, *4*, 630–637. [CrossRef]
2. Visvader, J.E. Cells of origin in cancer. *Nature* **2011**, *469*, 314–322. [CrossRef] [PubMed]
3. Barker, N.; Ridgway, R.A.; van Es, J.H.; van de Wetering, M.; Begthel, H.; van den Born, M.; Danenberg, E.; Clarke, A.R.; Sansom, O.J.; Clevers, H. Crypt stem cells as the cells-of-origin of intestinal cancer. *Nature* **2009**, *457*, 608–611. [CrossRef] [PubMed]
4. Schwitalla, S.; Fingerle, A.A.; Cammareri, P.; Nebelsiek, T.; Goktuna, S.I.; Ziegler, P.K.; Canli, O.; Heijmans, J.; Huels, D.J.; Moreaux, G.; et al. Intestinal tumorigenesis initiated by dedifferentiation and acquisition of stem-cell-like properties. *Cell* **2013**, *152*, 25–38. [CrossRef] [PubMed]
5. Tetteh, P.W.; Basak, O.; Farin, H.F.; Wiebrands, K.; Kretzschmar, K.; Begthel, H.; van den Born, M.; Korving, J.; de Sauvage, F.; van Es, J.H.; et al. Replacement of lost Lgr5−positive stem cells through plasticity of their enterocyte-lineage daughters. *Cell Stem Cell* **2016**, *18*, 203–213. [CrossRef]
6. Westphalen, C.B.; Asfaha, S.; Hayakawa, Y.; Takemoto, Y.; Lukin, D.J.; Nuber, A.H.; Brandtner, A.; Setlik, W.; Remotti, H.; Muley, A.; et al. Long-lived intestinal tuft cells serve as colon cancer-initiating cells. *J. Clin. Investig.* **2014**, *124*, 1283–1295. [CrossRef]
7. De Sousa e Melo, F.; Kurtova, A.V.; Harnoss, J.M.; Kljavin, N.; Hoeck, J.D.; Hung, J.; Anderson, J.E.; Storm, E.E.; Modrusan, Z.; Koeppen, H.; et al. A distinct role for Lgr5^{+} stem cells in primary and metastatic colon cancer. *Nature* **2017**, *543*, 676–680. [CrossRef]
8. Shimokawa, M.; Ohta, Y.; Nishikori, S.; Matano, M.; Takano, A.; Fujii, M.; Date, S.; Sugimoto, S.; Kanai, T.; Sato, T. Visualization and targeting of LGR5(+) human colon cancer stem cells. *Nature* **2017**, *545*, 187–192. [CrossRef]
9. Ruvolo, P.P. The broken "Off" switch in cancer signaling: PP2A as a regulator of tumorigenesis, drug resistance, and immune surveillance. *BBA Clin.* **2016**, *6*, 87–99. [CrossRef]
10. Bardelli, A.; Parsons, D.W.; Silliman, N.; Ptak, J.; Szabo, S.; Saha, S.; Markowitz, S.; Willson, J.K.; Parmigiani, G.; Kinzler, K.W.; et al. Mutational analysis of the tyrosine kinome in colorectal cancers. *Science* **2003**, *300*, 949. [CrossRef]

11. Wang, Z.; Shen, D.; Parsons, D.W.; Bardelli, A.; Sager, J.; Szabo, S.; Ptak, J.; Silliman, N.; Peters, B.A.; van der Heijden, M.S.; et al. Mutational analysis of the tyrosine phosphatome in colorectal cancers. *Science* **2004**, *304*, 1164–1166. [CrossRef]
12. Samuels, Y.; Wang, Z.; Bardelli, A.; Silliman, N.; Ptak, J.; Szabo, S.; Yan, H.; Gazdar, A.; Powell, S.M.; Riggins, G.J.; et al. High frequency of mutations of the PIK3CA gene in human cancers. *Science* **2004**, *304*, 554. [CrossRef] [PubMed]
13. Parsons, D.W.; Wang, T.L.; Samuels, Y.; Bardelli, A.; Cummins, J.M.; DeLong, L.; Silliman, N.; Ptak, J.; Szabo, S.; Willson, J.K.; et al. Colorectal cancer: Mutations in a signalling pathway. *Nature* **2005**, *436*, 792. [CrossRef] [PubMed]
14. Cancer Genome Atlas, N. Comprehensive molecular characterization of human colon and rectal cancer. *Nature* **2012**, *487*, 330–337. [CrossRef] [PubMed]
15. Westermarck, J.; Hahn, W.C. Multiple pathways regulated by the tumor suppressor PP2A in transformation. *Trends Mol. Med.* **2008**, *14*, 152–160. [CrossRef] [PubMed]
16. Lin, S.P.; Lee, Y.T.; Wang, J.Y.; Miller, S.A.; Chiou, S.H.; Hung, M.C.; Hung, S.C. Survival of cancer stem cells under hypoxia and serum depletion via decrease in PP2A activity and activation of p38-MAPKAPK2-Hsp27. *PLoS ONE* **2012**, *7*, e49605. [CrossRef] [PubMed]
17. Lin, S.P.; Lee, Y.T.; Yang, S.H.; Miller, S.A.; Chiou, S.H.; Hung, M.C.; Hung, S.C. Colon cancer stem cells resist antiangiogenesis therapy-induced apoptosis. *Cancer Lett.* **2013**, *328*, 226–234. [CrossRef] [PubMed]
18. Liu, C.C.; Lin, S.P.; Hsu, H.S.; Yang, S.H.; Lin, C.H.; Yang, M.H.; Hung, M.C.; Hung, S.C. Suspension survival mediated by PP2A-STAT3-Col XVII determines tumour initiation and metastasis in cancer stem cells. *Nat. Commun.* **2016**, *7*, 11798. [CrossRef] [PubMed]
19. Cheng, Y.; Ren, X.; Yuan, Y.; Shan, Y.; Li, L.; Chen, X.; Zhang, L.; Takahashi, Y.; Yang, J.W.; Han, B.; et al. eEF-2 kinase is a critical regulator of Warburg effect through controlling PP2A-A synthesis. *Oncogene* **2016**, *35*, 6293–6308. [CrossRef]
20. Seshagiri, S.; Stawiski, E.W.; Durinck, S.; Modrusan, Z.; Storm, E.E.; Conboy, C.B.; Chaudhuri, S.; Guan, Y.; Janakiraman, V.; Jaiswal, B.S.; et al. Recurrent R-spondin fusions in colon cancer. *Nature* **2012**, *488*, 660–664. [CrossRef]
21. Donehower, L.A.; Harvey, M.; Slagle, B.L.; McArthur, M.J.; Montgomery, C.A., Jr.; Butel, J.S.; Bradley, A. Mice deficient for p53 are developmentally normal but susceptible to spontaneous tumours. *Nature* **1992**, *356*, 215–221. [CrossRef] [PubMed]
22. Cheng, J.L.; Futakuchi, M.; Ogawa, K.; Iwata, T.; Kasai, M.; Tokudome, S.; Hirose, M.; Shirai, T. Dose response study of conjugated fatty acid derived from safflower oil on mammary and colon carcinogenesis pretreated with 7,12-dimethylbenz[a]anthracene (DMBA) and 1,2-dimethylhydrazine (DMH) in female Sprague-Dawley rats. *Cancer Lett.* **2003**, *196*, 161–168. [CrossRef]
23. Brown, K.; Buchmann, A.; Balmain, A. Carcinogen-induced mutations in the mouse c-Ha-ras gene provide evidence of multiple pathways for tumor progression. *Proc. Natl. Acad. Sci. USA* **1990**, *87*, 538–542. [CrossRef] [PubMed]
24. Kimura, K.; Satoh, K.; Kanno, A.; Hamada, S.; Hirota, M.; Endoh, M.; Masamune, A.; Shimosegawa, T. Activation of Notch signaling in tumorigenesis of experimental pancreatic cancer induced by dimethylbenzanthracene in mice. *Cancer Sci.* **2007**, *98*, 155–162. [CrossRef] [PubMed]
25. Reddy, B.S.; Narisawa, T.; Maronpot, R.; Weisburger, J.H.; Wynder, E.L. Animal models for the study of dietary factors and cancer of the large bowel. *Cancer Res.* **1975**, *35*, 3421–3426. [PubMed]
26. Wakabayashi, K.; Nagao, M.; Esumi, H.; Sugimura, T. Food-derived mutagens and carcinogens. *Cancer Res.* **1992**, *52*, 2092s–2098s.
27. Li, X.; Nadauld, L.; Ootani, A.; Corney, D.C.; Pai, R.K.; Gevaert, O.; Cantrell, M.A.; Rack, P.G.; Neal, J.T.; Chan, C.W.; et al. Oncogenic transformation of diverse gastrointestinal tissues in primary organoid culture. *Nat. Med.* **2014**, *20*, 769–777. [CrossRef]

28. Sato, T.; Vries, R.G.; Snippert, H.J.; van de Wetering, M.; Barker, N.; Stange, D.E.; van Es, J.H.; Abo, A.; Kujala, P.; Peters, P.J.; et al. Single Lgr5 stem cells build crypt-villus structures in vitro without a mesenchymal niche. *Nature* **2009**, *459*, 262–265. [CrossRef]
29. Bilalovic, N.; Sandstad, B.; Golouh, R.; Nesland, J.M.; Selak, I.; Torlakovic, E.E. CD10 protein expression in tumor and stromal cells of malignant melanoma is associated with tumor progression. *Mod. Pathol.* **2004**, *17*, 1251–1258. [CrossRef]
30. Mahe, M.M.; Aihara, E.; Schumacher, M.A.; Zavros, Y.; Montrose, M.H.; Helmrath, M.A.; Sato, T.; Shroyer, N.F. Establishment of gastrointestinal epithelial organoids. *Curr. Protoc. Mouse Biol.* **2013**, *3*, 217–240. [CrossRef]
31. Metsalu, T.; Vilo, J. ClustVis: A web tool for visualizing clustering of multivariate data using Principal Component Analysis and heatmap. *Nucleic Acids Res.* **2015**, *43*, W566–W570. [CrossRef] [PubMed]
32. Subramanian, A.; Tamayo, P.; Mootha, V.K.; Mukherjee, S.; Ebert, B.L.; Gillette, M.A.; Paulovich, A.; Pomeroy, S.L.; Golub, T.R.; Lander, E.S.; et al. Gene set enrichment analysis: A knowledge-based approach for interpreting genome-wide expression profiles. *Proc. Natl. Acad. Sci. USA* **2005**, *102*, 15545–15550. [CrossRef] [PubMed]
33. Madison, B.B.; Dunbar, L.; Qiao, X.T.; Braunstein, K.; Braunstein, E.; Gumucio, D.L. Cis elements of the villin gene control expression in restricted domains of the vertical (crypt) and horizontal (duodenum, cecum) axes of the intestine. *J. Biol. Chem.* **2002**, *277*, 33275–33283. [CrossRef] [PubMed]
34. Kemper, K.; Prasetyanti, P.R.; De Lau, W.; Rodermond, H.; Clevers, H.; Medema, J.P. Monoclonal antibodies against Lgr5 identify human colorectal cancer stem cells. *Stem Cells* **2012**, *30*, 2378–2386. [CrossRef]
35. Kang, E.; Yousefi, M.; Gruenheid, S. R-spondins are expressed by the intestinal stroma and are differentially regulated during citrobacter rodentium- and DSS-induced colitis in mice. *PLoS ONE* **2016**, *11*, e0152859. [CrossRef]
36. Vogelstein, B.; Papadopoulos, N.; Velculescu, V.E.; Zhou, S.; Diaz, L.A., Jr.; Kinzler, K.W. Cancer genome landscapes. *Science* **2013**, *339*, 1546–1558. [CrossRef]
37. Westcott, P.M.; Halliwill, K.D.; To, M.D.; Rashid, M.; Rust, A.G.; Keane, T.M.; Delrosario, R.; Jen, K.Y.; Gurley, K.E.; Kemp, C.J.; et al. The mutational landscapes of genetic and chemical models of Kras-driven lung cancer. *Nature* **2015**, *517*, 489–492. [CrossRef]
38. Lee, C.L.; Mowery, Y.M.; Daniel, A.R.; Zhang, D.; Sibley, A.B.; Delaney, J.R.; Wisdom, A.J.; Qin, X.; Wang, X.; Caraballo, I.; et al. Mutational landscape in genetically engineered, carcinogen-induced, and radiation-induced mouse sarcoma. *JCI Insight* **2019**, *4*. [CrossRef]
39. Cristobal, I.; Manso, R.; Rincon, R.; Carames, C.; Senin, C.; Borrero, A.; Martinez-Useros, J.; Rodriguez, M.; Zazo, S.; Aguilera, O.; et al. PP2A inhibition is a common event in colorectal cancer and its restoration using FTY720 shows promising therapeutic potential. *Mol. Cancer Ther.* **2014**, *13*, 938–947. [CrossRef]
40. Kakiuchi, H.; Watanabe, M.; Ushijima, T.; Toyota, M.; Imai, K.; Weisburger, J.H.; Sugimura, T.; Nagao, M. Specific 5′-GGGA-3′–>5′-GGA-3′ mutation of the Apc gene in rat colon tumors induced by 2-amino-1-methyl-6-phenylimidazo [4,5-b] pyridine. *Proc. Natl. Acad. Sci. USA* **1995**, *92*, 910–914. [CrossRef]
41. Yan, K.S.; Janda, C.Y.; Chang, J.; Zheng, G.X.Y.; Larkin, K.A.; Luca, V.C.; Chia, L.A.; Mah, A.T.; Han, A.; Terry, J.M.; et al. Non-equivalence of Wnt and R-spondin ligands during Lgr5(+) intestinal stem-cell self-renewal. *Nature* **2017**, *545*, 238–242. [CrossRef] [PubMed]
42. Koo, B.K.; Spit, M.; Jordens, I.; Low, T.Y.; Stange, D.E.; van de Wetering, M.; van Es, J.H.; Mohammed, S.; Heck, A.J.; Maurice, M.M.; et al. Tumour suppressor RNF43 is a stem-cell E3 ligase that induces endocytosis of Wnt receptors. *Nature* **2012**, *488*, 665–669. [CrossRef] [PubMed]

6

Traditional and Advanced Cell Cultures in Hematopoietic Stem Cell Studies

Antonio Carlos Ribeiro-Filho [1], Débora Levy [2], Jorge Luis Maria Ruiz [3], Marluce da Cunha Mantovani [1] and Sérgio Paulo Bydlowski [1,2,4,*]

1. Organoid Development Team, Center of Innovation and Translational Medicine (CIMTRA), University of São Paulo School of Medicine, Sao Paulo 05360-130, Brazil; antonio_biomed@hotmail.com (A.C.R.-F.); marlucem@gmail.com (M.d.C.M.)
2. Lipids, Oxidation and Cell Biology Team, Laboratory of Immunology (LIM19), Heart Institute (InCor), University of São Paulo School of Medicine, Sao Paulo 05403-900, Brazil; d.levy@hc.fm.usp.br
3. Life and Nature Science Institute, Federal University of Latin American Integration-UNILA, Foz de Iguaçú, PR 858570-901, Brazil; jorge.ruiz@unila.edu.br
4. National Institute of Science and Technology in Regenerative Medicine (INCT-Regenera), CNPq, Rio de Janeiro 21941-902, Brazil

* Correspondence: spbydlow@usp.br

Abstract: Hematopoiesis is the main function of bone marrow. Human hematopoietic stem and progenitor cells reside in the bone marrow microenvironment, making it a hotspot for the development of hematopoietic diseases. Numerous alterations that correspond to disease progression have been identified in the bone marrow stem cell niche. Complex interactions between the bone marrow microenvironment and hematopoietic stem cells determine the balance between the proliferation, differentiation and homeostasis of the stem cell compartment. Changes in this tightly regulated network can provoke malignant transformation. However, our understanding of human hematopoiesis and the associated niche biology remains limited due to accessibility to human material and the limits of in vitro culture models. Traditional culture systems for human hematopoietic studies lack microenvironment niches, spatial marrow gradients, and dense cellularity, rendering them incapable of effectively translating marrow physiology ex vivo. This review will discuss the importance of 2D and 3D culture as a physiologically relevant system for understanding normal and abnormal hematopoiesis.

Keywords: hematopoiesis; hematopoietic stem cells; stem cell culture; 2D culture; 3D culture

---

1. Introduction

Blood is a connective tissue made up of approximately 34% cells and 66% plasma, transporting nutrients, gases and molecules in general to the whole body. Hematopoiesis is the process by which blood cells are formed, replenishing the blood system over the life of an individual. The hematopoietic process is a highly hierarchical phenomenon, in which hematopoietic stem cell (HSCs) differentiation and proliferation are of vital importance. Each cell within the hematopoietic hierarchy can be distinguished based on specific surface markers, which contain epitopes that are recognized by antibodies [1–3]. Figure 1 shows the main markers in human hematopoietic hierarchy. The hematopoietic process in humans starts in the yolk sac (mesoblastic phase). Then, it is transferred to the liver and spleen. Finally, the bone marrow becomes the main organ responsible for hematopoiesis. In the bone marrow, HSCs have the capacity of unlimited self-renewal, producing progeny that is the same as the original cell. They are generally in the G0 phase of the cell cycle and have the capacity to differentiate into specialized cells.

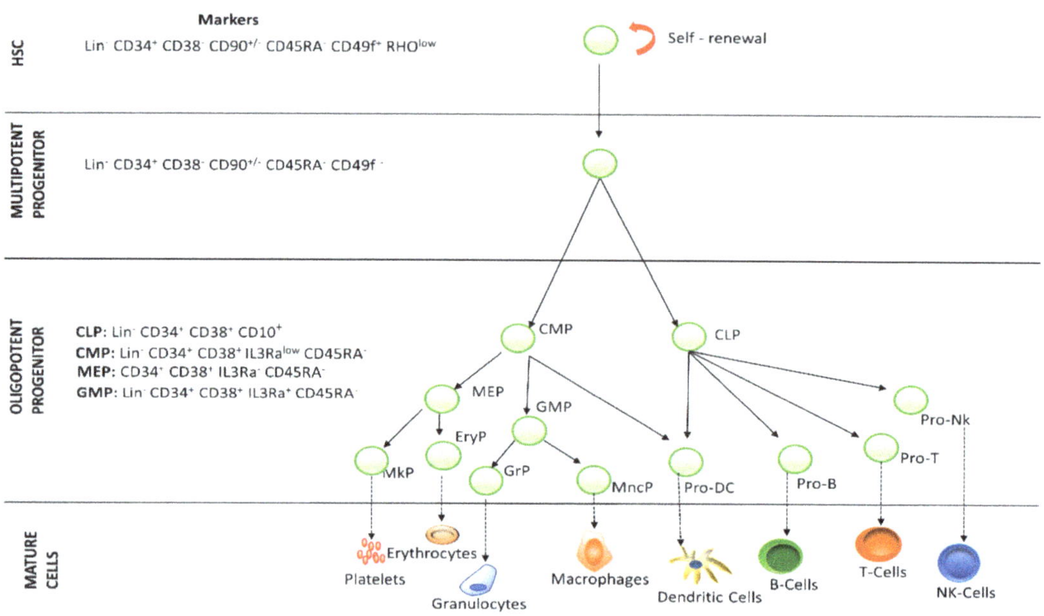

Figure 1. Hierarchy of human hematopoiesis. LT-HSC: Long Term-Hematopoietic Stem Cell; ST-HSC: Short Term-Hematopoietic Stem Cell; MPP: Multipotent Progenitor; OPP: Oligopotent Progenitor; LRP: Lineage Restricted Progenitor; MEC: Mature Effector Cell. The markers of the most important lineages are listed: Common Lymphoid Progenitor (CLP); Common Myeloid Progenitor (CMP); Megakaryocyte-Erythrocyte Progenitor (MEP); Granulocyte-Macrophage Progenitor (GMP). Restricted lineage progenitor cells: Megakaryocyte Progenitor (MkP); Erythrocytic Progenitor (EryP); Granulocytic Progenitor (GrP); Monocyte Progenitor (MncP); Dendritic Progenitor Cell (Pro DC); Progenitor Cell-T (Pro-T); Progenitor Cell-B (Pro-B); Progenitor Cell-Nk (Pro-Nk).

2. Hematopoietic Stem Cells

The medullary microenvironment participates in the quiescence, self-renewal, proliferation, maturation and apoptosis of HSCs and contains several cells (i.e., mesenchymal stem cells, endothelial cells, fibroblasts, osteoblasts, reticular cells, adipocytes). These cells are sources of cytokines, growth factors, glycoproteins and glycosaminoglycans, among other regulators. Different combinations of these molecules lead to the formation of specific microenvironments within the medullary cavity, known as niches [4]. Histologically defined microenvironments are subdivided into four regions: endosteal, subendosteal, central, and perisinusoidal. Granulocytes and monocytes are found in all regions of the bone marrow, whereas erythroblasts proliferate preferentially in the central region [5]. Concerning the dynamics of the lymphoid lineage, B lymphocyte precursors are found in the subendosteal region, gradually decreasing toward the central region, whereas mature B cells are found throughout the bone marrow [6]. HSCs are located in the endosteal region, also known as the osteoblastic niche, but studies suggest that HSCs may migrate to the perisinusoidal region or vascular niche and remain quiescent or differentiate depending on the needs of the organism [2,5,7–9]. In fact, studies with new markers for HSCs and niche cells, new image techniques, including labeling protocols, have shown that most HSCs reside adjacent to sinusoidal vessels, leading to the proposed existence of a perivascular niche for HSCs [10]. It is assumed that in the bone marrow, there are at least two different niches: the endosteal niche, which would harbor quiescent HSCs, and the perivascular niche, which would harbor cycling HSCs [11]. Although most studies have been done on non-humans, researchers suppose that the data reflect what happens in humans. It has previously been proposed that HSCs are maintained in the endosteal (osteoblastic) niche; however, the available evidence does not seem to support this model. Nevertheless, the endosteal niche seems to support the maintenance of a subset of lymphoid progenitors [10]. Approximately 80% of dividing and non-dividing HSCs have been described to be associated with sinusoidal vessels, with another 10% of HSCs being adjacent to arterioles, and almost another 10% in transition zone vessels [10]. A small percentage of HSCs are located in the endosteum. In fact, a quantitative model of cellular components that could define these microenvironments, and the preferential location of HSCs in the bone marrow are still lacking [12]. Obstacles to recognizing the HSCs in the bone marrow include the low frequencies at which HSCs are found in the bone marrow and the cellular complexity of the bone marrow microenvironment.

Cell signaling in the HSC niche is a complex process and passes through an extensive signaling network balancing self-renewal and differentiation [13]. This signaling involves several substances, such as growth factors and cytokines that are secreted by both the medullary stroma cells and hematopoietic stem cells and are important signaling factors for hematopoiesis, proliferation, and differentiation [14].

The Jak/STAT, Ras/Erk, and PI3K/Akt signaling pathways have been described as important inducers of erythropoiesis transduction by activation of the erythropoietin receptor, and these intracellular pathways are responsible for the survival, proliferation, and differentiation of normal erythropoietic progenitors [15].

Other intracellular signaling pathways that are important for the control of hematopoiesis have been described, such as the Notch, Wingless-type (Wnt), and Hedgehog pathways [16], which have been associated with self-renewal and maintenance of HSCs. Notch proteins are highly conserved receptors on the surface of the cell membrane that regulate the development of stem cells, and mutations in this receptor may cause leukemia [17] and breast cancer [18]. Activation of the Notch pathway is necessary to keep HSCs undifferentiated. This pathway is more active in HSCs and less active in differentiated cells. Inhibition of the Notch pathway potentiates the differentiation of HSCs and loss of the bone marrow reconstitution capacity of sub-lethally irradiated animals; thus, Notch has been used as a marker of undifferentiated HSCs [19]. Wnt protein regulates several phenomena during fetal development, and this protein has been related to the self-renewal of stem cells [20]. Hedgehog (Hh) protein has been described as regulating embryonic and adult stem cell activity. In mammals, three genes are known to be responsible for this protein—Sonic Hedgehog (SHh), Indian Hedgehog (IHh), and Desert Hedgehog (DHh) [21,22].

Soluble factors are also closely associated with the maintenance and regulation of the undifferentiated state of HSCs in the bone marrow of adults, in addition to regulating the proliferation and differentiation of this population. Stromal-cell-derived factor-1 (SDF-1/CXCL12) and its CXCR4 linker are activated to recruit endothelial progenitor cells (EPCs) and regulate HSCs [2,13,23]. Other soluble factors act to promote the maintenance of HSCs in their niche; for example, the stem cell factor (SCF/Kit-ligand) and its c-Kit receptor (CD117) are both required by HSCs for their maintenance. SCF is an important soluble cytokine for hematopoiesis, and THE c-Kit receptor is expressed on the HSCs surface; altered forms of this receptor have been associated with several types of cancer [24,25]. Thrombopoietin (TPO) and its MPL ligand are also important soluble factors necessary for the maintenance of HSCs in their niche. TPO is a primary physiological regulator responsible for the stimulation of platelet production, a primary dominant factor and megakaryocytopoiesis stimulator. In addition, recent in vitro studies have shown that TPO alone or in combination with growth factors, such as a c-Kit ligand, IL-3, or even Flt-3, stimulates the proliferation of hematopoietic progenitor cells [26,27]. Many other factors also modulate the function of HSCs but are not necessarily required, such as angiogenin, angiopoietin-1, G-CSF, IL-6, and TGFβ, among others [25].

3. Stem Cell Culture Methods

The complex microenvironment and cellular interactions are difficult to reproduce in vitro. Some improved techniques can help researchers mimic the bone marrow architecture for hematopoiesis studies and cell production in regenerative medicine applications.

Cell culture is the growth of cells from animal or plant in a favorable, artificial, and controlled environment. Today, cell culture is the basis of biology techniques and essential for regenerative medicine procedures [28]. These cells can originate directly from the tissue or after enzymatic or mechanically dissociated tissue or can be derived from an already established cell line.

Traditional cell culture, also known as 2D cell culture, is a very well-established method and easy-to-use culture model (Figure 2A,C). This technique depends on adherence to a flat surface, typically a culture flask, dish or polystyrene glass, to provide mechanical support for its growth in monolayers and access to nutrients and growth factors present in the culture media. Cells can also grow in suspension, as those derived from blood or bone marrow [29]. This method may present some advantages, such as low cost and performance of the functional assays, but this type of culture may also present some disadvantages, such as not mimicking the natural structures of the tissue and not being able to efficiently mimic cell–cell and cell–environment extracellular interactions [29].

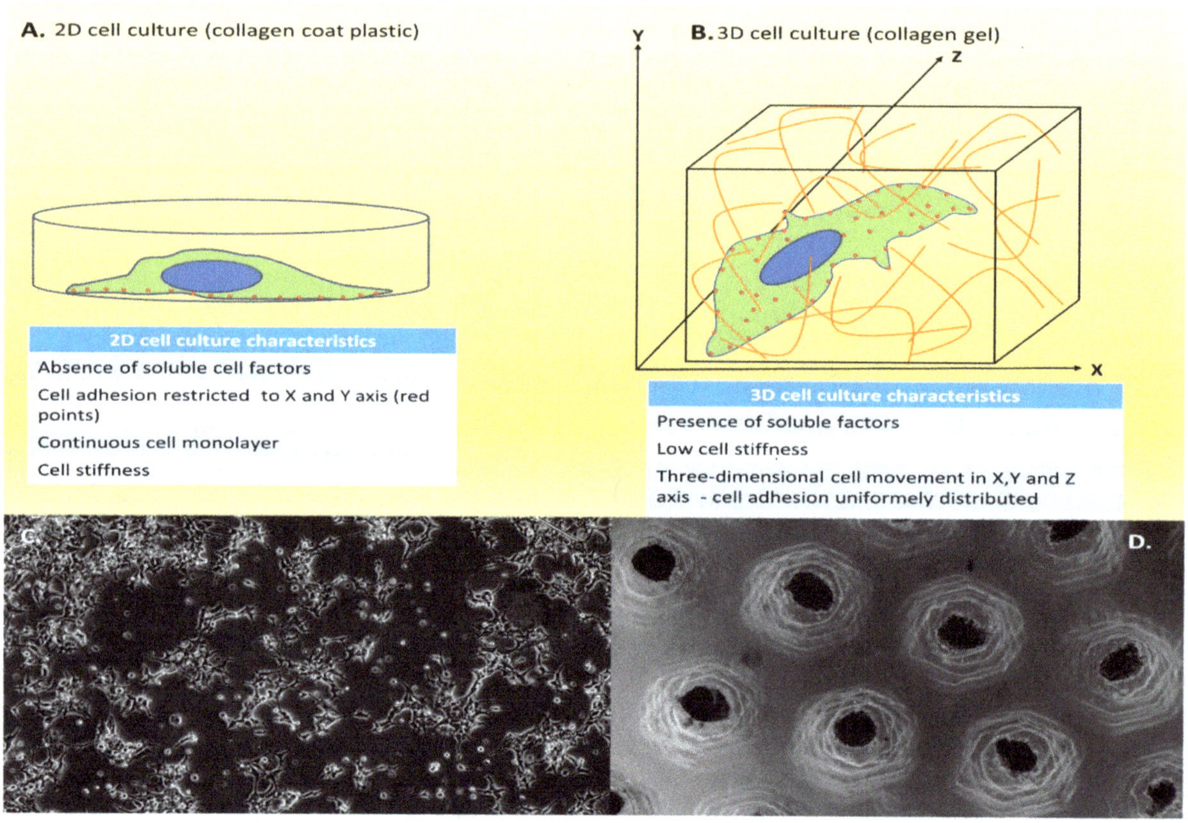

Figure 2. Two-dimensional (2D) and three-dimensional (3D) cell culture. (**A**) Schematic model of 2D cell culture. Standard model of 2D cell culture. Cells are cultured as a single layer in a culture flask. (**B**) Schematic model of 3D cell culture. 3D model of cell culture giving the notion of height, width and depth; cells are surrounded by the medium. (**C**) Example of 2D culture. HEK 293 cells in 2D cell culture; the adherence to a flat surface provides mechanical support for growth in monolayers. Scale bar 400 μm. (**D**) Example of 3D culture. HEK 293 cells 3D cell culture in agarose allowing the cells to grow or interact with their surroundings in all three dimensions. Scale bar 1000 μm.

Primary cell culture has been one of the greatest tools of cellular biology for evaluating cellular aspects such as chemoresistance, karyotypes, cellular parameters, metabolism and in vitro modeling of physiological and pathological models [30]. The isolated cells resemble their tissue of origin. Primary cell culture is complex and requires specific care, including storage, thawing-freezing procedures, or choice of enzymatic treatment, but the main problem is how to keep the primary cell in culture long enough to be established for experimental tests [31–33]. Primary culture can be representative of the cell types of the tissue from which they were isolated; although these cells are difficult to maintain, they best mimic a pathological condition or physiological function.

Immortalized cells or established cell lineages are cells that can grow in vitro indefinitely due to natural or induced transformation (e.g., embryonic stem cells or viral transformation) [34]. Immortalized cells have several advantages over primary cells, such as profitability, easy manipulation, unlimited supply of material, and lack of ethical concerns associated with the use of animal and human tissues [35]. Immortalized cells also provide a uniform and homogeneous population of cells [36]. Immortalized cell lines emerged as a solution to some problems that appeared with primary cell cultures, such as misidentification of the cell line used and genotypic and phenotypic instability. Care should also be taken to avoid the use of aged cell lines that do not maintain the original physiological characteristics. The immortalized strains constitute a simple and representative model system for functional studies and therapeutic targets. As each cell line may have unique characteristics, specific studies should take these characteristics into account [37].

The co-culture technique is used for numerous applications, including the study of natural or synthetic interactions between distinct populations of cells. Co-culture methods are of great importance in research, as they are used to observe cell–cell interactions, how cells are organized, how they participate in the development of diseases such as cancers, in which different types of cells are involved, including the microenvironment [38].

Both self-renewal and differentiation are required abilities in any cell of the hematopoietic lineage [39]. However, the expansion of these cells in vitro has been challenging for the scientific community, as regulation of these cells depends on several mechanisms of intercellular communication resulting from the secretion of local- and systemic-acting factors [40]. HSCs, which can also be obtained and expanded from umbilical cord blood, should respond to and integrate events in their microenvironment to ensure sustained production of all hematopoietic lineages [41].

Although HSCs have been extensively analyzed and characterized, their ex vivo expansion remains a problem [42–44]. It has been shown that the cell culture of HSCs is viable for at least 169 days but very much depends on the quality of the HSCs. There are currently several approaches to ex vivo expansion. Most first require the isolation of $CD34^+$ or $CD133^+$ cells from frozen or fresh hematopoietic tissue and incubation in culture medium supplemented with cytokines, granulocyte colony stimulating factor (G-CSF), stem cell factor (SCF), and thrombopoietin (TPO) [45].

Therefore, the cell culture of HSCs must take into account the microenvironment in which these cells are inserted so that it can reproduce the whole framework of the hematopoietic structure. The hematopoietic niche contains several types of cells, and mimicking this microenvironment in vitro requires a stromal layer, which controls multiple cellular parameters, including quiescence, self-renewal, differentiation, apoptosis and migration. Under the artificial culture conditions, HSCs undergo differentiation and apoptosis [46]. A support layer is required so HSCs can survive and proliferate.

Co-culture of HSCs is one of the most frequently used models for understanding how the highly specific bone marrow niche interacts ex vivo with hematopoietic cells, promoting their differentiation and expansion. The technique can focus on the importance of non-contact culture systems on the successful maintenance of hematopoietic cells [47] and on the use of the stroma as a cell layer to provide support for HSC culturing, as mentioned above. Several studies have shown that contact between HSCs and stromal cells is important for maintaining HSC function [48,49]. The stromal cells from bone marrow include osteoblasts, macrophages, endothelial cells, and mesenchymal cells [50,51]. Mesenchymal cells are multipotent progenitor cells that can differentiate into mesenchymal cells, such as osteoblasts, adipocytes, and chondrocytes, and also have the potential for differentiation into cells such as neurons and lung cells [52]. These multipotent cells have interactions with hematopoietic cells forming the framework of the HSC niche, supporting the development of hematopoiesis and acting as immunological regulators [53,54]. In addition, recent studies have indicated that other molecules, such as N6-methyladenosine (M6A), by modulating the expression of a group of YTHDF2 genes at the mRNA level, are important regulators of HSC self-renewal. Some authors have suggested that although extensive efforts have led to multiple methods for in vivo expansion of HSCs, it is not possible for some single molecules or pathways to be manipulated simultaneously due to a large number of essential targets for self-renewal of stem cells [55].

In this regard, there are still important challenges to overcome, including the development of more efficient methods for the maintenance of HSCs in vitro, and methods of ex vivo expansion for therapeutic development in regenerative medicine; then, it will be possible to have a better understanding of the hematopoietic niche and its intrinsic and extrinsic regulators from a physiological and pathophysiologic point of view. The ability to genetically reprogram HSCs for clinical therapeutic use [56] still needs to be improved.

4. 3D Hematopoietic Stem Cell Culture

Cell culture in 2D format is simple and provides excellent material for studying homogeneous populations [57]. However, it does not consider several other important parameters of cell physiology, such as cell–cell communication or communication between the cell and microenvironment or adjacent molecules [58]. Therefore, the major problem with 2D cell culture is its limits [59]. One of the main criticisms of this study format is that animal physiology cannot be mimicked using only one layer of cells, which certainly does not correspond to the original tissue considering the complexity of the cellular microenvironment in the tissue of origin [60].

As an alternative to the technical limitations of 2D cell culture, 3D cell culture allows a better simulation of the in vivo structural complexity, replicating several characteristics present in tissues (Figure 2B,D), not only the interaction of cells with their microenvironment, but also morphology, differentiation, polarity, proliferation rate, gene expression profiles, and cell heterogeneity [61–64]. In fact, 3D culture has proven to be a realer model for translating research results into in vivo applications.

Current 3D cell culture models and methods include spheroids, organoids, microcarrier cultures, organs-on-chips, and 3D bioprinting [65,66]. Organotypic explant culture methods are used mostly when a technical/specific requirement of the tissue is to be studied and mimicked [67]. However, although the complexity of the 3D system is evident, some criteria must be considered and cannot be disregarded, such as the choice of material for the scaffold and hydrogels and cell type and culture methods, which vary considerably according to the tissue studied [66].

Organoid culture is most commonly used to describe constructs derived from pluripotent stem cells (embryonic or induced cells) or adult stem cells from various organs, including the hematopoietic tissue [68]. Organoid culture is classified as either tissue organoids (i.e., organotypic) or stem cells depending on how the cell layers are formed [69]. Organotypic tissue refers to free stromal cells without parenchyma, and its application is mainly with epithelial cells because of their high intrinsic ability to self-organize. With this method, it is possible to study HSCs, and Christopher et al. [70] were able to produce mature T cells from stem cells and hematopoietic progenitors in a thymus organoid culture.

Spheroid culture models simulate the microenvironment conditions of a living cell. Compared to the classic 2D model, the spheroid culture model emphasizes the interactions between cells and their relationship to the extracellular matrix (ECM) [71,72]. Tissues are not composed of a homogeneous population of cells; they are complex structures formed by several different components, with an intricate relationship such as vessels, nerves, and stroma, which should be considered in any kind of tissue engineering. This complex interaction between cells and the extracellular compartment requires a 3D environment to best represent these interactions [73,74].

Spheroid cell culture can be used in scaffold and scaffold-free models. The scaffold methods comprise hydrogel support, decellularized extracellular matrix, and resistant polymeric material support. Hydrogel consists of polymeric groups with a water-swollen hydrophilic structure [75]. Depending on their nature, these compounds are classified into different categories, including ECM protein–containing hydrogels (ECMPs), natural hydrogels, or synthetic hydrogels, each one with its own properties [76,77]. The decellularization technique consists of removing cells from the native tissue by chemical treatment, preserving the ECM. The tissue can then be replenished with cultured cells and grafted in vivo or used for ex vivo cell differentiation. In contrast to artificial scaffolds, decellularized scaffolds preserve the vascular structure, tridimensional niches, and chemical composition of bone marrow. Extracellular matrix components are directly related to several important factors for cell survival support, including cell behavior, signaling for survival, and proliferation, among others. Resistant polymeric material supports are structures similar to fibers or sponges. Cells maintained in this type of support exhibit a physiological behavior close to that of native tissue (such as those of cartilaginous tissue). Regarding the polymeric support, the most commonly used material is polystyrene, but biodegradable supports such as polycaprolactone have also been employed.

The development of a bioprinting system for the bone marrow microenvironment is important since it cannot be mimicked by methods such as organoids. Advanced 3D bone marrow models could serve for several different studies, including hematopoietic regulation, BM, MSCs, HSC interaction and expansion, interactions of hematological cancer cells, and evolution of several hematological diseases. In addition, these models could be used as a platform for expanding HSCs for transplantation. However, it is difficult to obtain a reliable BM model because of several critical technical challenges, from currently available 3D printing techniques to the possibility of precisely mimicking the different BM niches [78].

Relatively few studies have been done using hematopoietic stem cells in these methods. Decellularization of cartilaginous tissues has been tested for ex vivo culturing of hematopoietic cells [79]. A co-culture model of hematopoietic stem/progenitor cell (HSPC) spheroids using polydimethylsiloxane (PDMS) has shown that the effectiveness of three-dimensional culture for HSPC expansion for clinical use is still a strategy that needs further improvement [80].

A three-dimensional collagen-based culture using HSPC, bone-marrow-derived MSCs or the umbilical cord (UC) to mimic the main compartments of the bone marrow hematopoietic niche has been proposed [81]. Data analysis generated the following compartments: (I) HSPC in suspension above collagen and (II) migratory HSPC in collagen fiber matrix. The different sites were representative of the distinct microenvironments that make up the bone marrow and have a significant impact on the fate of HSPC. The authors suggested that this 3D culture system using collagen and BM-MSC allowed HSPC expansion and provided a potential platform for advanced study of niches and hematopoiesis and their regulatory mechanisms.

A bone marrow composed of two compartments, solid and liquid, that act harmoniously, has been proposed [82]. A bone marrow structure was created using a macroporous PEG hydrogel that resembled the macroporous 3D architecture of the trabecular bone, site of the red bone marrow, and, therefore, where the HSPC niches are located. This bone marrow analog was found to be suitable for HSPC culture and for enhancing HSPC expansion compared to conventional 2D cell culture. The developed model of a perfused 3D bone marrow analogue mimicked the HSCs niche under steady-state or activated cell conditions that favor the maintenance or differentiation, respectively, of HSCs and allowed drug testing [82] It was concluded that the system reflected the behavior of HSPC in the niche under physiological conditions.

Some studies have shown that the interaction of bone marrow stromal cells with leukemic cells increases the resistance of these leukemic cells. Numerous scaffolds have been created to provide a minimal structure based on the 3D leukemic microenvironment [73]. As 3D culture could result in resistance to drugs, it can be a good screening tool for drug evaluation prior to the administration of chemotherapic drugs [74,83]. A classic example of scaffolds used to recreate the leukemic niche is polycaprolactone, which is an aliphatic and biodegradable polyester with good biocompatibility [84,85].

Recently, a scaffold of degradable zwitterionic hydrogel was tested for human HSPC expansion [86]. A 73-fold increase of long-term hematopoietic stem cells was observed (LT-HSC). The viscoelasticity and smoothness of the highly hydrated zwitterionic hydrogels seems to be important for the creation of cell niches, by their unique mechanical and antifouling properties.

The organs-on-chips technique was developed to study the mechanical and physiological response of a tissue, combining concepts of tissue engineering and microfluidics [87]. A bioelectronic device based on a conductive polymer scaffold was integrated with an electrochemical transistor configuration that allows 3D cell growth and the real-time monitoring of cell adhesion and growth.

This technique consists of manipulating small amounts (10^{-9} to 10^{-18} L) of fluids using small openings with micrometer dimensions. This methodology has been widely reported because it offers several advantages, such as using small amounts of samples and reagents, performing separations and detections with high resolution, and a short time for analysis [61]. Such techniques allow the precise control of fluids and particles in a given cell culture. This culture method enables the control of such nanoliter-scale fluids as described above and further enables and facilitates the simultaneous manipulation of cultured cells from a single cell [62]. Through precise manipulation of the components of the microfluidic culture medium, it is possible to transport nutrients, hormones, and oxygen growth factors to facilitate homeostasis and recreate mechanical signals that are absent in traditional culture [88].

This cell culture system offers several advantages for basic research applications, including the precision of micromanufacturing, which allows the presentation of a controllable and reproducible microenvironment [64]. Another advantage of this model is that it provides complete control over the conditions of the cell culture, including dynamic cell control, nutrient addition, removal of metabolites, stimulation with drugs and proteins, and simultaneous image and chip format [62,89,90].

The organs-on-chips technique was described in cancer studies (cancer-on-chip), where it can replicate the microenvironment to achieve robust and reliable results [91,92]. However, up to now, it has been difficult to culture HSCs in an organs-on-chips model. A device with a central cavity that successfully mimics the bone and bone marrow has been described. This device is made from polydimethylsiloxane (PDMS) with bone inducers inserted. The central chamber of this device is composed of porous PDMS membranes, and cytokines are added in the microfluids [93]. Although several methods have been used, results using HSCs are generally poor [94].

5. Final Remarks

Bone marrow HSCs are the stem cell most used in medical practice. Finding effective ways to mimic the bone marrow HSCs and microenvironment, the hematopoietic niche, in vitro is a challenge. In addition to HSCs, several other bone marrow cell types, including megakaryocytes, macrophages, monocytes and endothelial cells, directly or indirectly regulate HSCs and niche function. Two-dimensional cell cultures have been widely used due to their low maintenance cost and easy learning. However, 2D culture has several limitations. As a result, 3D cell culture techniques have been developed. Although the current cost is relatively high, the fact that a 3D culture has characteristics much closer to the tissue being studied is a great advantage. There are several methods currently employed or in development, such as spheroids, organoids, 3D bioprinting and organs-on-chip, and these are facing the challenge of culturing HSCs while maintaining their properties. This achievement will allow ex vivo HSC expansion, bringing new medical perspectives for HSCs transplantation, drug testing, and personalized treatment.

Author Contributions: Conceptualization, A.C.R.-F.; D.L. and S.P.B.; writing—original draft preparation, A.C.R.-F.; D.L.; J.L.M.R. and M.d.C.M.; writing—review and editing, A.C.R.-F.; D.L.; J.L.M.R.; M.d.C.M. and S.P.B.; supervision, S.P.B.; funding acquisition, S.P.B.

References

1. Ng, A.P.; Alexander, W.S. Haematopoietic stem cells: Past, present and future. *Cell Death Discov.* **2017**, *3*, 17002. [CrossRef]
2. Chotinantakul, K.; Leeanansaksiri, W. Hematopoietic stem cell development, niches, and signaling pathways. *Bone Marrow Res.* **2012**, *2012*, 270425. [CrossRef]
3. Wasnik, S.; Tiwari, A.; Kirkland, M.A.; Pande, G. Osteohematopoietic stem cell niches in bone marrow. *Int. Rev. Cell Mol. Biol.* **2012**, *298*, 95–133.
4. Bianco, P.; Robey, P.G. Stem cells in tissue engineering. *Nature* **2001**, *414*, 118–121. [CrossRef]
5. Calvi, L.M.; Link, D.C. Cellular complexity of the bone marrow hematopoietic stem cell niche. *Calcif. Tissue Int.* **2014**, *94*, 112–124. [CrossRef]
6. Nilsson, S.K.; Johnston, H.M.; Coverdale, J.A. Spatial localization of transplanted hemopoietic stem cells: Inferences for the localization of stem cell niches. *Blood* **2001**, *97*, 2293–2299. [CrossRef]
7. Kiel, M.J.; Morrison, S.J. Maintaining hematopoietic stem cells in the vascular niche. *Immunity* **2006**, *25*, 862–864. [CrossRef]
8. Orkin, S.H.; Zon, L.I. Hematopoiesis: An evolving paradigm for stem cell biology. *Cell* **2008**, *132*, 631–644. [CrossRef]
9. Calvi, L.M.; Adams, G.B.; Weibrecht, K.W.; Weber, J.M.; Olson, D.P.; Knight, M.C.; Martin, R.P.; Schipani, E.; Divieti, P.; Bringhurst, F.R.; et al. Osteoblastic cells regulate the haematopoietic stem cell niche. *Nature* **2003**, *425*, 841–846. [CrossRef]
10. Oh, I.H.; Kwon, K.R. Concise review: Multiple niches for hematopoietic stem cell regulations. *Stem Cells* **2010**, *28*, 1243–1249.
11. Crane, G.M.; Jeffery, E.; Morrison, S.J. Adult haematopoietic stem cell niches. *Nat. Rev. Immunol.* **2017**, *17*, 573–590. [CrossRef]
12. Nies, C.; Gottwald, E. Advances in Tissue Engineering and Regenerative Medicine Artificial Hematopoietic Stem Cell Niches-Dimensionality Matters. *Adv. Tissue Eng. Regen. Med.* **2017**, *2*, 42.
13. Gomariz, A.; Isringhausen, S.; Helbling, P.M.; Nombela-Arrieta, C. Imaging and spatial analysis of hematopoietic stem cell niches. *Ann. N. Y. Acad. Sci.* **2019**. [CrossRef]
14. Blank, U.; Karlsson, G.; Karlsson, S. Signaling pathways governing stem-cell fate. *Blood* **2008**, *111*, 492–503. [CrossRef]
15. Majka, M.; Janowska-Wieczorek, A.; Ratajczak, J.; Ehrenman, K.; Pietrzkowski, Z.; Kowalska, M.A.; Gewirtz, A.M.; Emerson, S.G.; Ratajczak, M.Z. Numerous growth factors, cytokines, and chemokines are secreted by human CD34(+) cells, myeloblasts, erythroblasts, and megakaryoblasts and regulate normal hematopoiesis in an autocrine/paracrine manner. *Blood* **2001**, *97*, 3075–3085. [CrossRef]
16. Jafari, M.; Ghadami, E.; Dadkhah, T.; Akhavan-Niaki, H. PI3k/AKT signaling pathway: Erythropoiesis and beyond. *J. Cell. Physiol.* **2019**, *234*, 2373–2385. [CrossRef]
17. Cerdan, C.; Bhatia, M. Novel roles for Notch, Wnt and Hedgehog in hematopoesis derived from human pluripotent stem cells. *Int. J. Dev. Biol.* **2010**, *54*, 955–963. [CrossRef]
18. Aster, J.C.; Pear, W.S.; Blacklow, S.C. Notch signaling in leukemia. *Annu. Rev. Pathol.* **2008**, *3*, 587–613. [CrossRef]
19. Tekmal, R.R.; Keshava, N. Role of MMTV integration locus cellular genes in breast cancer. *Front. Biosci.* **1997**, *2*, 519–526.
20. Duncan, A.W.; Rattis, F.M.; DiMascio, L.N.; Congdon, K.L.; Pazianos, G.; Zhao, C.; Pazianos, G.; Zhao, C.; Yoon, K.; Cook, J.M.; et al. Integration of Notch and Wnt signaling in hematopoietic stem cell maintenance. *Nat Immunol.* **2005**, *6*, 314–322. [CrossRef]
21. Gordon, M.D.; Nusse, R. Wnt signaling: Multiple pathways, multiple receptors, and multiple transcription factors. *J. Biol. Chem.* **2006**, *281*, 22429–22433. [CrossRef] [PubMed]
22. Dyer, M.A.; Farrington, S.M.; Mohn, D.; Munday, J.R.; Baron, M.H. Indian hedgehog activates hematopoiesis and vasculogenesis and can respecify prospective neurectodermal cell fate in the mouse embryo. *Development* **2001**, *128*, 1717–1730.
23. Shi, X.; Wei, S.; Simms, K.J.; Cumpston, D.N.; Ewing, T.J.; Zhang, P. Sonic Hedgehog Signaling Regulates Hematopoietic Stem/Progenitor Cell Activation during the Granulopoietic Response to Systemic Bacterial Infection. *Front. Immunol.* **2018**, *9*, 349. [CrossRef] [PubMed]

24. Mendez-Ferrer, S.; Michurina, T.V.; Ferraro, F.; Mazloom, A.R.; Macarthur, B.D.; Lira, S.A.; Scadden, D.T.; Ma'ayan, A.; Enikolopov, G.N.; Frenette, P.S. Mesenchymal and haematopoietic stem cells form a unique bone marrow niche. *Nature* **2010**, *466*, 829–834. [CrossRef]
25. McNiece, I.K.; Briddell, R.A. Stem cell factor. *J. Leukoc. Biol.* **1995**, *58*, 14–22. [CrossRef]
26. Edling, C.E.; Hallberg, B. c-Kit–a hematopoietic cell essential receptor tyrosine kinase. *Int. J. Biochem. Cell Biol.* **2007**, *39*, 1995–1998. [CrossRef]
27. Nakamura-Ishizu, A.; Takizawa, H.; Suda, T. The analysis, roles and regulation of quiescence in hematopoietic stem cells. *Development* **2014**, *141*, 4656–4666. [CrossRef]
28. Solar, G.P.; Kerr, W.G.; Zeigler, F.C.; Hess, D.; Donahue, C.; de Sauvage, F.J.; Eaton, D.L. Role of c-mpl in early hematopoiesis. *Blood* **1998**, *92*, 4–10. [CrossRef]
29. Mao, A.S.; Mooney, D.J. Regenerative medicine: Current therapies and future directions. *Proc. Natl. Acad. Sci. USA* **2015**, *112*, 14452–14459. [CrossRef]
30. Duval, K.; Grover, H.; Han, L.H.; Mou, Y.; Pegoraro, A.F.; Fredberg, J.; Chen, Z. Modeling Physiological Events in 2D vs. 3D Cell Culture. *Physiology (Bethesda)* **2017**, *32*, 266–277.
31. Lerescu, L.; Tucureanu, C.; Caras, I.; Neagu, S.; Melinceanu, L.; Salageanu, A. Primary cell culture of human adenocarcinomas–practical considerations. *Roum. Arch. Microbiol. Immunol.* **2008**, *67*, 55–66.
32. Lopes, A.A.; Peranovich, T.M.; Maeda, N.Y.; Bydlowski, S.P. Differential effects of enzymatic treatments on the storage and secretion of von Willebrand factor by human endothelial cells. *Thromb. Res.* **2001**, *101*, 291–297. [CrossRef]
33. Janz Fde, L.; Debes Ade, A.; Cavaglieri Rde, C.; Duarte, S.A.; Romao, C.M.; Moron, A.F.; Zugaib, M.; Bydlowski, S.P. Evaluation of distinct freezing methods and cryoprotectants for human amniotic fluid stem cells cryopreservation. *J. Biomed. Biotechnol.* **2012**, *2012*, 649353.
34. Ruiz, J.L.; Fernandes, L.R.; Levy, D.; Bydlowski, S.P. Interrelationship between ATP-binding cassette transporters and oxysterols. *Biochem. Pharmacol.* **2013**, *86*, 80–88. [CrossRef] [PubMed]
35. Kirchhoff, C.; Araki, Y.; Huhtaniemi, I.; Matusik, R.J.; Osterhoff, C.; Poutanen, M.; Samalecos, A.; Sipilä, P.; Suzuki, K.; Orgebin-Criste, M.C. Immortalization by large T-antigen of the adult epididymal duct epithelium. *Mol. Cell. Endocrinol.* **2004**, *216*, 83–94. [CrossRef] [PubMed]
36. Kaur, G.; Dufour, J.M. Cell lines: Valuable tools or useless artifacts. *Spermatogenesis* **2012**, *2*, 1–5. [CrossRef]
37. Rosa Fernandes, L.; Stern, A.C.; Cavaglieri, R.C.; Nogueira, F.C.; Domont, G.; Palmisano, G.; Bydlowskia, S.P. 7-Ketocholesterol overcomes drug resistance in chronic myeloid leukemia cell lines beyond MDR1 mechanism. *J. Proteomics* **2017**, *151*, 12–23. [CrossRef] [PubMed]
38. Romano, P.; Manniello, A.; Aresu, O.; Armento, M.; Cesaro, M.; Parodi, B. Cell Line Data Base: Structure and recent improvements towards molecular authentication of human cell lines. *Nucleic Acids Res.* **2009**, *37*, D925–D932. [CrossRef]
39. Miki, Y.; Ono, K.; Hata, S.; Suzuki, T.; Kumamoto, H.; Sasano, H. The advantages of co-culture over mono cell culture in simulating in vivo environment. *J. Steroid Biochem. Mol. Biol.* **2012**, *131*, 68–75. [CrossRef]
40. Eridani, S.; Sgaramella, V.; Cova, L. Stem cells: From embryology to cellular therapy? An appraisal of the present state of art. *Cytotechnology* **2004**, *44*, 125–141. [CrossRef]
41. Csaszar, E.; Kirouac, D.C.; Yu, M.; Wang, W.; Qiao, W.; Cooke, M.P.; Boitano, A.E.; Ito, C.; Zandstra, P. Rapid expansion of human hematopoietic stem cells by automated control of inhibitory feedback signaling. *Cell Stem Cell* **2012**, *10*, 218–229. [CrossRef] [PubMed]
42. Le Blanc, K.; Frassoni, F.; Ball, L.; Locatelli, F.; Roelofs, H.; Lewis, I.; Lanino, E.; Sundberg, B.; Bernardo, B.M.; Remberger, M.; et al. Mesenchymal stem cells for treatment of steroid-resistant, severe, acute graft-versus-host disease: A phase II study. *Lancet* **2008**, *371*, 1579–1586. [CrossRef]
43. Zon, L.I. Intrinsic and extrinsic control of haematopoietic stem-cell self-renewal. *Nature* **2008**, *453*, 306–313. [CrossRef] [PubMed]
44. Heike, T.; Nakahata, T. Ex vivo expansion of hematopoietic stem cells by cytokines. *Biochim. Biophys. Acta* **2002**, *1592*, 313–321. [CrossRef]
45. Ando, K.; Yahata, T.; Sato, T.; Miyatake, H.; Matsuzawa, H.; Oki, M.; Miyoshi, H.; Tsuji, T.; Kato, S.; Hotta, T. Direct evidence for ex vivo expansion of human hematopoietic stem cells. *Blood* **2006**, *107*, 3371–3377. [CrossRef] [PubMed]

46. McNiece, I.K.; Stoney, G.B.; Kern, B.P.; Briddell, R.A. CD34+ cell selection from frozen cord blood products using the Isolex 300i and CliniMACS CD34 selection devices. *J. Hematother.* **1998**, *7*, 457–461. [CrossRef] [PubMed]
47. Xie, J.; Zhang, C. Ex vivo expansion of hematopoietic stem cells. *Sci. China Life Sci.* **2015**, *58*, 839–853. [CrossRef] [PubMed]
48. Vaidya, A.; Kale, V. Hematopoietic Stem Cells, Their Niche, and the Concept of Co-Culture Systems: A Critical Review. *J. Stem Cells* **2015**, *10*, 13–31.
49. Breems, D.A.; Blokland, E.A.; Siebel, K.E.; Mayen, A.E.; Engels, L.J.; Ploemacher, R.E. Stroma-contact prevents loss of hematopoietic stem cell quality during ex vivo expansion of CD34+ mobilized peripheral blood stem cells. *Blood* **1998**, *91*, 111–117. [CrossRef]
50. Walenda, T.; Bork, S.; Horn, P.; Wein, F.; Saffrich, R.; Diehlmann, A.; Eckstein, V.; Ho, A.D.; Wagner, W. Co-culture with mesenchymal stromal cells increases proliferation and maintenance of haematopoietic progenitor cells. *J. Cell Mol. Med.* **2010**, *14*, 337–350.
51. Crisan, M.; Dzierzak, E. The many faces of hematopoietic stem cell heterogeneity. *Development* **2016**, *143*, 4571–4581. [CrossRef] [PubMed]
52. Boulais, P.E.; Frenette, P.S. Making sense of hematopoietic stem cell niches. *Blood.* **2015**, *125*, 2621–2629. [CrossRef] [PubMed]
53. Takano, T.; Li, Y.J.; Kukita, A.; Yamaza, T.; Ayukawa, Y.; Moriyama, K.; Uehara, N.; Nomiyama, H.; Koyano, K.; Kukita, T. Mesenchymal stem cells markedly suppress inflammatory bone destruction in rats with adjuvant-induced arthritis. *Lab. Invest.* **2014**, *94*, 286–296. [CrossRef] [PubMed]
54. Haynesworth, S.E.; Baber, M.A.; Caplan, A.I. Cytokine expression by human marrow-derived mesenchymal progenitor cells in vitro: Effects of dexamethasone and IL-1 alpha. *J. Cell Physiol.* **1996**, *166*, 585–592. [CrossRef]
55. Devine, S.M.; Hoffman, R. Role of mesenchymal stem cells in hematopoietic stem cell transplantation. *Curr. Opin. Hematol.* **2000**, *7*, 358–363. [CrossRef] [PubMed]
56. Li, Z.; Qian, P.; Shao, W.; Shi, H.; He, X.C.; Gogol, M.; Yu, Z.; Wang, Y.; Qi, M.; Zhu, Y.; et al. Suppression of m(6)A reader Ythdf2 promotes hematopoietic stem cell expansion. *Cell Res.* **2018**, *28*, 904–917. [CrossRef]
57. Ravi, M.; Paramesh, V.; Kaviya, S.R.; Anuradha, E.; Solomon, F.D. 3D cell culture systems: Advantages and applications. *J. Cell Physiol.* **2015**, *230*, 16–26. [CrossRef]
58. Murray-Dunning, C.; McArthur, S.L.; Sun, T.; McKean, R.; Ryan, A.J.; Haycock, J.W. Three-dimensional alignment of schwann cells using hydrolysable microfiber scaffolds: Strategies for peripheral nerve repair. *Methods Mol. Biol.* **2011**, *695*, 155–166.
59. Tee, D.E.H. Culture of Animal Cells: A Manual of Basic Technique. *J. R. Soc. Med.* **1984**, *77*, 902–903.
60. Melissinaki, V.; Gill, A.A.; Ortega, I.; Vamvakaki, M.; Ranella, A.; Haycock, J.W.; Fotakis, C.; Farsari, M.; Claeyssens, F. Direct laser writing of 3D scaffolds for neural tissue engineering applications. *Biofabrication* **2011**, *3*, 045005. [CrossRef] [PubMed]
61. Whitesides, G.M. The origins and the future of microfluidics. *Nature* **2006**, *442*, 368–373. [CrossRef] [PubMed]
62. Mehling, M.; Tay, S. Microfluidic cell culture. *Curr. Opin. Biotechnol.* **2014**, *25*, 95–102. [CrossRef] [PubMed]
63. El-Ali, J.; Sorger, P.K.; Jensen, K.F. Cells on chips. *Nature* **2006**, *442*, 403–411. [CrossRef] [PubMed]
64. Dittrich, P.S.; Manz, A. Lab-on-a-chip: Microfluidics in drug discovery. *Nat. Rev. Drug Discov.* **2006**, *5*, 210–218. [CrossRef] [PubMed]
65. Pampaloni, F.; Reynaud, E.G.; Stelzer, E.H. The third dimension bridges the gap between cell culture and live tissue. *Nat. Rev. Mol. Cell Biol.* **2007**, *8*, 839–845. [CrossRef] [PubMed]
66. Haycock, J.W. 3D cell culture: A review of current approaches and techniques. *Methods Mol. Biol.* **2011**, *695*, 1–15.
67. Toda, S.; Watanabe, K.; Yokoi, F.; Matsumura, S.; Suzuki, K.; Ootani, A.; Aoki, S.; Koike, N.; Sugihara, H. A new organotypic culture of thyroid tissue maintains three-dimensional follicles with C cells for a long term. *Biochem. Biophys. Res. Commun.* **2002**, *294*, 906–911. [CrossRef]
68. Lancaster, M.A.; Huch, M. Disease modelling in human organoids. *Dis. Model Mech.* **2019**, *12*. [CrossRef]
69. Lancaster, M.A.; Knoblich, J.A. Organogenesis in a dish: Modeling development and disease using organoid technologies. *Science* **2014**, *345*, 1247125. [CrossRef]

70. Seet, C.S.; He, C.; Bethune, M.T.; Li, S.; Chick, B.; Gschweng, E.H.; Zhu, Y.; Kim, K.; Kohn, D.B.; Baltimore, D.; et al. Generation of mature T cells from human hematopoietic stem and progenitor cells in artificial thymic organoids. *Nat. Methods* **2017**, *14*, 521–530. [CrossRef]
71. Yuhas, J.M.; Li, A.P.; Martinez, A.O.; Ladman, A.J. A simplified method for production and growth of multicellular tumor spheroids. *Cancer Res.* **1977**, *37*, 3639–3643.
72. Fennema, E.; Rivron, N.; Rouwkema, J.; van Blitterswijk, C.; de Boer, J. Spheroid culture as a tool for creating 3D complex tissues. *Trends Biotechnol.* **2013**, *31*, 108–115. [CrossRef] [PubMed]
73. Tanner, K.; Gottesman, M.M. Beyond 3D culture models of cancer. *Sci. Transl. Med.* **2015**, *7*, 283ps9. [CrossRef] [PubMed]
74. Gillet, J.P.; Calcagno, A.M.; Varma, S.; Marino, M.; Green, L.J.; Vora, M.I.; Patel, C.; Orina, J.N.; Eliseeva, T.A.; Singal, V.; et al. Redefining the relevance of established cancer cell lines to the study of mechanisms of clinical anti-cancer drug resistance. *Proc. Natl. Acad. Sci. USA* **2011**, *108*, 18708–18713. [CrossRef] [PubMed]
75. Li, Y.; Huang, G.; Li, M.; Wang, L.; Elson, E.L.; Jian Lu, T.; Genin, G.M.; Xu, F. An approach to quantifying 3D responses of cells to extreme strain. *Sci. Rep.* **2016**, *6*, 19550. [CrossRef] [PubMed]
76. Polymeric Scaffolds in Tissue Engineering Application: A Review. *Int. J. Polymer Sci.* **2011**, *2011*. [CrossRef]
77. Kleinman, H.K.; Philp, D.; Hoffman, M.P. Role of the extracellular matrix in morphogenesis. *Curr. Opin. Biotechnol.* **2003**, *14*, 526–532. [CrossRef]
78. Raic, A.; Naolou, T.; Mohra, A.; Chatterjee, C.; Lee-Thedieck, C. 3D models of the bone marrow in health and disease: Yesterday, today and tomorrow. *MRS Commun.* **2019**, *9*, 37–52. [CrossRef]
79. Llames, S.; Garcia, E.; Otero Hernandez, J.; Meana, A. Tissue bioengineering and artificial organs. *Adv. Exp. Med. Biol.* **2012**, *741*, 314–336.
80. Futrega, K.; Atkinson, K.; Lott, W.B.; Doran, M.R. Spheroid Coculture of Hematopoietic Stem/Progenitor Cells and Monolayer Expanded Mesenchymal Stem/Stromal Cells in Polydimethylsiloxane Microwells Modestly Improves In Vitro Hematopoietic Stem/Progenitor Cell Expansion. *Tissue Eng. Part C Methods.* **2017**, *23*, 200–218. [CrossRef]
81. Leisten, I.; Kramann, R.; Ventura Ferreira, M.S.; Bovi, M.; Neuss, S.; Ziegler, P.; Wagner, W.; Knüchel, R.; Schneider, R.K. 3D co-culture of hematopoietic stem and progenitor cells and mesenchymal stem cells in collagen scaffolds as a model of the hematopoietic niche. *Biomaterials* **2012**, *33*, 1736–1747. [CrossRef] [PubMed]
82. Rodling, L.; Schwedhelm, I.; Kraus, S.; Bieback, K.; Hansmann, J.; Lee-Thedieck, C. 3D models of the hematopoietic stem cell niche under steady-state and active conditions. *Sci. Rep.* **2017**, *7*, 4625. [CrossRef] [PubMed]
83. Reticker-Flynn, N.E.; Malta, D.F.; Winslow, M.M.; Lamar, J.M.; Xu, M.J.; Underhill, G.H.; Hynes, R.O.; Jacks, T.E.; Bhatia, S.N. A combinatorial extracellular matrix platform identifies cell-extracellular matrix interactions that correlate with metastasis. *Nat. Commun.* **2012**, *3*, 1122. [CrossRef] [PubMed]
84. Chen, Y.; Gao, D.; Wang, Y.; Lin, S.; Jiang, Y. A novel 3D breast-cancer-on-chip platform for therapeutic evaluation of drug delivery systems. *Anal. Chim. Acta* **2018**, *1036*, 97–106. [CrossRef] [PubMed]
85. Song, K.; Wang, Z.; Liu, R.; Chen, G.; Liu, L. Microfabrication-Based Three-Dimensional (3-D) Extracellular Matrix Microenvironments for Cancer and Other Diseases. *Int. J. Mol. Sci.* **2018**, *19*, 935. [CrossRef] [PubMed]
86. Wu, J.; Xiao, Y.; Chen, A.; He, H.; He, C.; Shuai, X.; Li, X.; Chen, S.; Ren, B.; Zheng, J.; et al. Sulfated zwitterionic poly(sulfobetaine methacrylate) hydrogels promote complete skin regeneration. *Acta Biomater.* **2018**, *71*, 293–305. [CrossRef] [PubMed]
87. Yi, H.G.; Lee, H.; Cho, D.W. 3D Printing of Organs-On-Chips. *Bioengineering* **2017**, *4*, 10. [CrossRef]
88. Ozbolat, I.T. Bioprinting scale-up tissue and organ constructs for transplantation. *Trends Biotechnol.* **2015**, *33*, 395–400. [CrossRef]
89. Novo, P.; Volpetti, F.; Chu, V.; Conde, J.P. Control of sequential fluid delivery in a fully autonomous capillary microfluidic device. *Lab. Chip.* **2013**, *13*, 641–645. [CrossRef]
90. Ong, S.M.; Zhang, C.; Toh, Y.C.; Kim, S.H.; Foo, H.L.; Tan, C.H.; van Noort, D.; Park, S.; Yu, H. A gel-free 3D microfluidic cell culture system. *Biomaterials* **2008**, *29*, 3237–3244. [CrossRef]
91. Zhang, Y.S.; Duchamp, M.; Oklu, R.; Ellisen, l.W.; Langer, R.; Khademhosseini, A. Bioprinting the Cancer Microenvironment. *ACS Biomater Sci. Eng.* **2016**, *2*, 1710–1721. [CrossRef] [PubMed]

92. Shafiee, A.; Atala, A. Printing Technologies for Medical Applications. *Trends Mol. Med.* **2016**, *22*, 254–265. [CrossRef] [PubMed]
93. Torisawa, Y.S.; Spina, C.S.; Mammoto, T.; Mammoto, A.; Weaver, J.C.; Tat, T.; Collins, J.J.; Ingber, D.E. Bone marrow-on-a-chip replicates hematopoietic niche physiology in vitro. *Nat. Method.* **2014**, *11*, 663–669. [CrossRef] [PubMed]
94. Di Maggio, N.; Piccinini, E.; Jaworski, M.; Trumpp, A.; Wendt, D.J.; Martin, I. Toward modeling the bone marrow niche using scaffold-based 3D culture systems. *Biomaterials* **2011**, *32*, 321–329. [CrossRef]

7

Self-Assembling Scaffolds Supported Long-Term Growth of Human Primed Embryonic Stem Cells and Upregulated Core and Naïve Pluripotent Markers

Christina McKee [1,2], Christina Brown [1,2] and G. Rasul Chaudhry [1,2,*]

1 Department of Biological Sciences, Oakland University, Rochester, MI 48309, USA; cmmckee@oakland.edu (C.M.); brown3@oakland.edu (C.B.)
2 OU-WB Institute for Stem Cell and Regenerative Medicine, Rochester, MI 48309, USA
* Correspondence: chaudhry@oakland.edu

Abstract: The maintenance and expansion of human embryonic stem cells (ESCs) in two-dimensional (2-D) culture is technically challenging, requiring routine manipulation and passaging. We developed three-dimensional (3-D) scaffolds to mimic the in vivo microenvironment for stem cell proliferation. The scaffolds were made of two 8-arm polyethylene glycol (PEG) polymers functionalized with thiol (PEG-8-SH) and acrylate (PEG-8-Acr) end groups, which self-assembled via a Michael addition reaction. When primed ESCs (H9 cells) were mixed with PEG polymers, they were encapsulated and grew for an extended period, while maintaining their viability, self-renewal, and differentiation potential both in vitro and in vivo. Three-dimensional (3-D) self-assembling scaffold-grown cells displayed an upregulation of core pluripotency genes, *OCT4*, *NANOG*, and *SOX2*. In addition, the expression of primed markers decreased, while the expression of naïve markers substantially increased. Interestingly, the expression of mechanosensitive genes, *YAP* and *TAZ*, was also upregulated. YAP inhibition by Verteporfin abrogated the increased expression of *YAP/TAZ* as well as core and naïve pluripotent markers. Evidently, the 3-D culture conditions induced the upregulation of makers associated with a naïve state of pluripotency in the primed cells. Overall, our 3-D culture system supported the expansion of a homogenous population of ESCs and should be helpful in advancing their use for cell therapy and regenerative medicine.

Keywords: embryonic stem cells; three-dimensional; self-assembling scaffold; pluripotency; culture conditions; expansion; growth; niche

---

1. Introduction

Pluripotency is defined by the potential of a cell to differentiate into germline cells as well as the cells of all three germ layers [1,2]. During development, pluripotency ranges from the formation of the epiblast until gastrulation and lineage commitment, resulting in a population of cells representing a range of pluripotent states [3,4]. Pluripotency in vitro is determined by the developmental stage from which the cell line is derived and by culture conditions [5–8] with distinct naïve and primed states of pluripotency corresponding to in vivo pre-implantation and post-implantation epiblast cells, respectively [4].

Embryonic stem cells (ESCs) display unlimited self-renewal and differentiation potential in vitro [9], making them an ideal source for the development of cell therapies and regenerative medicine applications. However, the clinical use of ESCs requires a high quality and quantity of cells, which is limited by currently used culturing techniques [10]. ESCs are typically grown as a monolayer in two-dimensional (2-D) plastic culture plates coated with extracellular matrix (ECM) components (such as gelatin from porcine skin [11], matrigel [12], laminin [13], fibronectin [14], vitronectin [15], or collagen [16]) or a mouse embryonic fibroblast (MEF) feeder layer to aid in attachment [17]. Monolayer culture also necessitates routine passaging and the removal of spontaneously differentiated colonies to maintain the self-renewal and potency of cells [18], which pose a major impediment to the large-scale expansion of cells. Generally, 2-D culturing methods often lead to heterogeneous cell populations as well as batch-to-batch variation.

Moreover, 2-D culture conditions lack the intricacies necessary to mimic the ESC niche, dynamic, and specialized three-dimensional (3-D) microenvironments, which are critical for regulating cell fate in vivo. Furthermore, native 3-D niches allow for complex spatial interactions between cells ECM components as well as gradients of nutrients, oxygen, and metabolic waste [19–24]. The microenvironment is important for the self-renewal of ESCs, since cell fate and function is affected by the composition and organization of the ECM [25–27] as well as mechanical forces generated between cells and attachment substrates [25,28–31]. Furthermore, mechanical forces generated by the expansion of the blastocoel have been shown to play an important role in blastocyst lineage formation, stimulating the generation of pluripotent cells [32]. These early morphogenic events in the mammalian embryo indicate a significant interaction between mechanical forces, cell polarity, and the regulation of gene expression in cell fate determination [33]. We hypothesized that 3-D culture would better mimic the in vivo microenvironment, promoting the proliferation and maintenance of human ESCs.

We have previously demonstrated that 3-D self-assembling scaffolds composed of thiolated dextran and 4-arm polyethylene glycol (PEG) functionalized with acrylate groups (Dex-SH/PEG-4-Acr) supported the growth and maintenance of naïve mouse ESCs [34]. However, these 3-D culture conditions failed to support the growth of human primed ESCs. Unlike mouse ESCs, human cells display poor viability and clonogenicity following single cell dissociation [9,10].

In this study, we describe 3-D hydrogel scaffolds that support the long-term growth and maintenance of human primed ESCs (H9). Chemically cross-linked hydrogels were formed by a Michael-type addition reaction by combining two 8-arm PEG polymers functionalized with either thiol (PEG-8-SH) or acrylate end groups (PEG-8-Acr). The PEG-8-SH/PEG-8-Acr scaffold provided a microenvironment that maintained self-renewal and pluripotency for an extended time period. Interestingly, H9 cells displayed an upregulation of core pluripotency markers during 3-D culture. However, the expression of core markers reverted to normal levels when 3-D grown cells were subcultured under 2-D culture conditions. Interestingly, 3-D cultured H9 cells also exhibited a significantly higher expression of naïve pluripotent markers when compared to human naïve ESCs (Elf1) cultured under 2-D conditions. Our results suggest the importance of the 3-D scaffold microenvironment in maintaining the stemness of ESCs.

2. Materials and Methods

2.1. *Maintenance of Human ESCs in 2-D culture*

H9 cells, derived from a human blastocyst [35], obtained from WiCell (Madison, WI, USA) were maintained in the culture medium containing Knockout/F12 Dulbecco's Modified Eagle Media (DMEM; Life Technologies, Carlsbad, CA, USA) with 20% KnockOut serum replacement (Life Technologies), 0.1 mM 2-mercaptoehtanol (Life Technologies), 1% GlutaMax (Life Technologies), and 1% non-essential amino acid solution (Life Technologies), supplemented with 20 ng/mL of basic fibroblast growth factor (FGF2, Prospec, Ness Ziona, Israel) and 10 μM ROCK inhibitor, Y-27632 (Cayman Chemical, Ann Arbor, MI, USA) and subcultured by manual passaging. For dissociation into single cells, H9 were treated with Accutase (Thermo Fisher Scientific, Waltham, MA, USA) for 5 min, and then cells were centrifuged at 200 g for 5 min. The supernatant was aspirated, and the cell pellet was resuspended in H9 culture media for plating.

Elf1 cells, isolated from a cryopreserved 8-cell human embryo [36], were maintained in the culture medium containing KnockOut/F12 DMEM with GlutaMax (Life Technologies) with 20% KnockOut serum replacement (Life Technologies), 1 mM sodium pyruvate (Life Technologies), 0.1mM

2-mercaptoehtanol (Life Technologies), 1% non-essential amino acids (Life Technologies), and 0.2% penicillin–streptomycin solution (Life Technologies), supplemented with 12 ng/mL of FGF2 (Prospec), 1.5 μM CHIR99021 (Caymen Chemical), 0.4 μM PD03296501 (Caymen Chemical), and 0.01 μg/mL human LIF (Prospec) and cultured according to the published protocol [36].

2.2. Selection and Composition of Scaffolds for 3-D Culture of Human ESCs

Several polymers end functionalized with thiol and acrylate end groups were tested for their ability to self-assemble and form hydrogels scaffolds via a thiol–Michael addition reaction, allowing for the formation of covalent bonds by the addition of a nucleophile to a nucleophile acceptor containing an α,β-unsaturated carbonyl compound [37]. Then, the scaffolds were tested in various compositions to determine the optimal concentration, molar ratio, and degree of modification that would best support the growth of human ESCs. Our preliminary results showed that scaffolds made using 8-arm PEG polymers yielded optimal growth. These scaffolds were prepared using 8-arm PEG-thiol (PEG-8-SH, 20 kDa) and 8-arm PEG-acrylate (PEG-8-Acr, 20 kDa) purchased from JenKem Technology USA (Plano, TX, USA). Functionalized PEG polymers were stored at −20 °C and protected from light.

The preparation of scaffolds for the 3-D culture of ESCs is depicted in Figure 1. Briefly, PEG-8-SH and PEG-8-Acr polymers were dissolved at a concentration of 2.5 *w/v* % (dry weight of polymer per volume of culture medium), combined at a 1:1 molar ratio and mixed with cells. Then, the resulting mixture was transferred to a 1 cc syringe mold for polymerization. After self-assembly, scaffolds were placed in a 24-well culture plate (Fisher Scientific, Pittsburgh, PA, USA), supplemented with culture medium, and maintained in a 5% CO_2 incubator at 37 °C. The medium was changed daily or as needed. Cell growth in the scaffolds was monitored by phase-contrast microscopy.

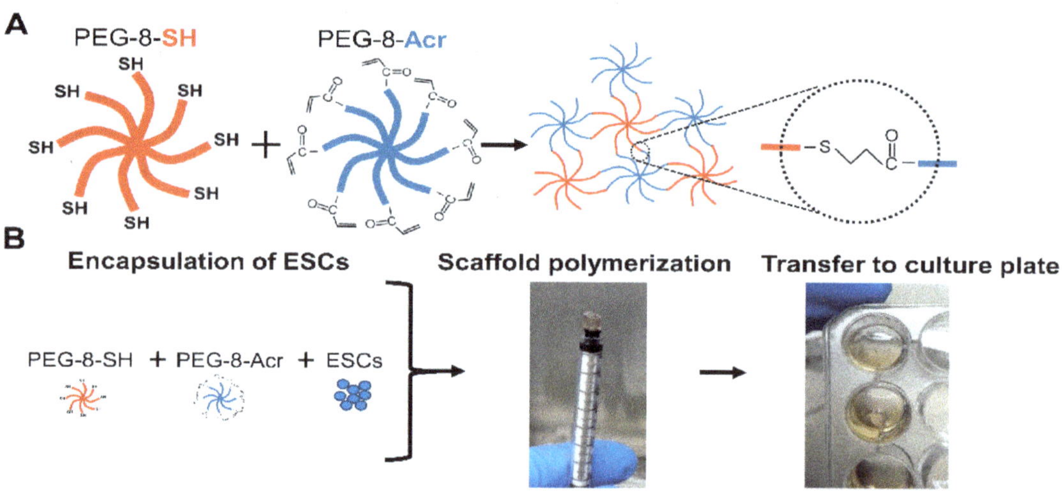

Figure 1. Schematic of self-assembling scaffolds. (**A**) Self-assembly of functionalized polymers, 8-arm polyethylene glycol functionalized with thiol (PEG-8-SH) and acrylate (PEG-8-Acr) via a thiol–Michael addition reaction. (**B**) The encapsulation of H9 cells human embryonic stem cells (ESCs), was achieved upon mixing with the self-assembling polymers in a syringe mold. Following polymerization, the scaffolds were then incubated in culture plates containing medium.

2.3. Cell Proliferation and Viability Assays

The growth rate of cells grown under 2-D and 3-D culture conditions were analyzed at various time intervals using a proliferation assay. Briefly, triplicate samples were treated with 5 mg/mL 3-(4,5-dimethylthiazol-2-yl)-2,5-diphenyltetrazolium bromide (MTT) reagent (Sigma, St. Louis, MO, USA), protected from light, and incubated at 37 °C for 4 h to obtain insoluble formazan, which was then solubilized using 15:1 isopropanol/hydrochloride. Then, the absorbance of the solubilized formazan was measured at 570 nm using an Epoch microplate reader (BioTek, Winooski, VT), and the background absorbance of the cells was subtracted from all measured values. The viability of encapsulated cells was determined by direct microscopic counts and trypan blue exclusion assay. Briefly, cells were counted using a hemocytometer and cells stained blue were considered non-viable.

2.4. *Differentiation of Human ESCs*

Germ layer differentiation was achieved by the spontaneous formation of embryoid bodies (EBs). ESCs were allowed to spontaneously aggregate for 3 days in non-adherent flat-bottomed 96-well plates in their respective ESC culture medium containing growth factors. Then, the resultant EBs were transferred to 0.1% gelatin-coated wells for adherent growth and grown in high-glucose DMEM supplemented with 10% fetal bovine serum (FBS). Spontaneous differentiation into all three germ layers was assessed by germ layer marker expression by quantitative real time-polymerase chain reaction (qRT-PCR) and immunocytochemistry.

2.5. *Teratoma Assay*

For teratoma formation, ESCs were harvested following accutase treatment, washed and resuspended in PBS, and mixed with an equal volume of matrigel (BD Biosciences, San Jose, CA, USA). Cells (1×10^6) were subcutaneously injected (20 μL) using a Hamilton syringe into 4-week-old male immune-compromised SCID (severe combined immunodeficient) Beige mice (Fox Chase SCID Beige, Charles River, Wilmington, MA, USA). Animals were monitored daily and humanely euthanized by CO_2 overdose after teratoma formation at 10–12 weeks. Teratomas were explanted, and teratoma tissue was either fixed for histological analysis or flash frozen in liquid nitrogen for RNA isolation. Teratoma assays were performed in triplicate. All the procedures involving animals were reviewed and approved by the Institutional Animal Care and Use Committee of Oakland University (IACUC protocol number: 17031).

2.6. *Gene Expression Analysis*

Transcriptional analysis was performed by qRT-PCR. Briefly, cells, scaffolds, and teratoma tissue (100–250 mg) were harvested and total cellular mRNA was isolated following the manufacturer's instructions using the GeneJET RNA purification kit (Thermo Fisher Scientific) and RNeasy Midi kit (Qiagen, Germantown, MD, USA), respectively. cDNA was synthesized with the iScript kit (BioRad, Hercules, CA, USA). qRT-PCR was performed using SsoAdvanced SYBR Green Supermix (Bio-Rad) and the CFX90 Real-Time PCR system. The primers (IDT Technologies, Coralville, IA, USA) used in this study are in Table 1. All reactions were prepared in triplicate and normalized to reference genes, *HMBS*, *GAPDH*, and *β-ACTIN*.

2.7. *Immunocytochemical Analysis*

Protein expression was determined by immunocytochemical staining using selected antibodies. H9 cells grown under 2-D culture conditions on coverslips, H9 cells encapsulated in the self-assembling scaffolds, and harvested teratoma tissue were fixed in 4% paraformaldehyde for 10 min, 30 min, and overnight, respectively. Subsequently, teratoma tissue and scaffolds were embedded and frozen in optimal cutting temperature (O.C.T) compound and cryosectioned into 10 μm sections. For immunochemical analysis, fixed cells and cryosections were permeabilized with 0.5% Triton X-100 (Sigma) for 10 min and blocked with 2% bovine serum albumin (BSA; Sigma) for 1 h at room temperature. Next, samples were incubated with primary antibodies (1:100 dilution) overnight at 4 °C. Primary antibody-treated samples were washed three times with phosphate buffer saline (PBS), stained with secondary antibody at 1:200 dilutions for 2 h at 37 °C, and counterstained with 1 mg/mL 4′,6-diamidino-2-phenylindole (DAPI; Life Technologies). The stained samples were visualized by using confocal microscopy. The antibodies used are listed in Table 2.

Table 1. List of human primer sequences used in qRT-PCR.

Gene	Primer Sequence		
	Forward (5′-3′)	**Reverse (5′-3′)**	**Product Length**
ACTIN	AATCTGGCACCACACCTTCTAC	ATAGCACAGCCTGGATAGCAAC	170
BRACHYURY	TGCTTCCCTGAGACCCAGTT	GATCACTTCTTTCCTTTGCATCAAG	121
DNMT3L	CTGCTCCATCTGCTGCTCC	ATCCACACACTCGAAGCAGT	85
DPPA3	AGACCAACAAACAAGGAGCCT	CCCATCCATTAGACACGCAGA	88
ESRRB	GACATTGCCTCTGGCTACCA	CTCCGTTTGGTGATCTCGCA	131
FGF4	CGTGGTGAGCATCTTCGGC	GTAGGACTCGTAGGCGTTGT	145
FOXA2	GGGAGCGGTGAAGATGGA	TCATGTTGCTCACGGAGGAGTA	89
FOXF1	AAGCCGCCCTATTCCTACATC	GCGCTTGGTGGGTGAACT	63
GAPDH	ACAACTTTGGTATCGTGGAAGG	GCCATCACGCCACAGTTTC	101
GATA4	TCCCTCTTCCCTCCTCAAAT	TCAGCGTGTAAAGGCATCTG	194
GATA6	CCCACAACACAACCTACAGC	GCGAGACTGACGCCTATGTA	131
GDF3	GTCTCCCGAGACTTATGCTACG	AGTAGAGGAGCTTCTGCAGGCA	136
GP130	GGAGTGAAGAAGCAAGTGGGA	AGGCAATGTCTTCCACACGA	128
HMBS	AGGAGTTCAGTGCCATCATCCT	CACAGCATACATGCATTCCTCA	104
ITGA2	TTGCGTGTGGACATCAGTCT	GCTGGTATTTGTCGGACATCT	158
ITGA5	GCCGATTCACATCGCTCTCAA	GTCTTCTCCACAGTCCAGCAA	139
ITGA6	CGAAACCAAGGTTCTGAGCCC	CTTGGATCTCCACTGAGGCAG	151
ITGAV	AGGAGAAGGTGCCTACGAAGC	GCACAGGAAAGTCTTGCTAAGG	105
ITGB1	GGATTCTCCAGAAGGTGGTTT	TGCCACCAAGTTTCCCATCT	143
ITGB3	CATGGATTCCAGCAATGTCCTC	TTGAGGCAGGTGGCATTGAAG	126
ITGB5	GCCTTTCTGTGAGTGCGACAA	CCGATGTAACCTGCATGGCAC	111
KLF17	TCAGGAAGGGACTGGTAGAA	GTACCCGCATATGTCGTCTAAG	206
KLF2	CCAAGAGTTCGCATCTGAAGG	CCGTGTGCTTTCGGTAGTG	132
KLF4	CGAACCCACACAGGTGAGAA	TACGGTAGTGCCTGGTCAGTTC	75
KLF5	ACCCTGGTTGCACAAAAGTT	CAGCCTTCCCAGGTACACTT	100
LATS1	CTCTGCACTGGCTTCAGATG	TCCGCTCTAATGGCTTCAGT	145
LATS2	ACATTCACTGGTGGGGACTC	GTGGGAGTAGGTGCCAAAAA	147
LIFR	CACCTTCCAAAATAGCGAGTATGG	ATGGTTCCGACCGAGACGAGTT	159
MIXL1	CCGAGTCCAGGATCCAGGTA	CTCTGACGCCGAGACTTGG	58
NANOG	AAAGAATCTTCACCTATGCC	GAAGGAAGAGGAGAGACAGT	110
N-CADHERIN	TGTTTGGCCTGGCGTTCTTT	AGGAGACAGAAACGAAGCCA	156
NCAM	AGGAGACAGAAACGAAGCCA	GGTGTTGGAAATGCTCTGGT	161
OCT4	CCCCTGGTGCCGTGAA	GCAAATTGCTCGAGTTCTTTCTG	97
PAX6	CTTTGCTTGGGAAATCCGAG	AGCCAGGTTGCGAAGAACTC	103
PRDM14	CCTTGTGTGGTATGGAGACTGC	CTTTCACATCTGTAGCCTTCTGC	126
RAC1	ATGTCCGTGCAAAGTGGTATC	CTCGGATCGCTTCGTCAAACA	249
RHOA	CATCCGGAAGAAACTGGT	TCCCACAAAGCCAACTC	168
ROCK1	GGTGGTCGGTTGGGGTATTTT	CGCCCTAACCTCACTTCCC	196
SALL2	GGCTTGCCTTATGGTATGTCCG	TGGCACTGAGTGCTGTTGTGGA	115
SOCS3	ATTCGGGACCAGCCCCC	AAACTTGCTGTGGGTGACCA	121
SOX11	CGACGACCTAATGTTCGACC	GACAGGGATAGGTTCCCCG	105
SOX17	CGCACGGAATTTGAACAGTA	GGATCAGGGACCTGTCACAC	182
SOX2	TTGCTGCCTCTTTAAGACTAGGA	CTGGGGCTCAAACTTCTCTC	75
SOX7	ACGCCGAGCTCAGCAAGAT	TCCACGTACGGCCTCTTCTG	73
STAT3	CTTTGAGACCGAGGTGTATCAC	GGTCAGCATGTTGTACCACAG	133
TAZ	GAGGGTGTATGGTGGAGATAAA	CCAACTGTAGCAAACAGGATTAG	86
TBX3	GGACACTGGAAATGGCCGAAG	GCTGCTTGTTCACTGGAGGAC	123
TBX3	CGGACATACTTGTTCCCCGA	GCAGGGTGAGCTGTTTTCTTTT	154
TEAD4	CCAAGCTCTGGATGTTGGAGTTC	GATGTCCACGGCTTCGAGGTA	161
TFCP2L1	TTTGTGGGACCCTGCGAAG	TGCTTAAACGTGTCAATCTGGA	129
TUJ1	GGCCAAGTTCTGGGAAGTCA	CGAGTCGCCCACGTAGTTG	70
YAP	GCTGCCACCAAGCTAGATAA	GTGCATGTGTCTCCTTAGATCC	101
ZFP42	CGCAATCGCTTGTCCTCAGA	GCTCTCAACGAACGCTTTCC	130
ZIC2	CGCTCCGAGAACCTCAAGAT	CCCTCAAACTCACACTGGAA	71

Table 2. List of primary and secondary antibodies used in immunocytochemical staining.

Antibody	Primary	Secondary
BRACHYURY	Rabbit Polyclonal	Anti-Rabbit Alexa Fluor 488
GATA4	Mouse Polyclonal	Anti-Mouse-Cy3
NANOG	Rabbit Polyclonal	Anti-Rabbit Alexa Fluor 488
OCT4	Rabbit Polyclonal	Anti-RabbitAlexa Fluor 488
SOX2	Rabbit polyclonal	Anti-Mouse Alexa Fluor 488
TUJ1	Mouse Polyclonal	Anti-Mouse-Cy3
YAP	Rabbit Polyclonal	Anti-Rabbit Alexa Fluor 488

2.8. *Effect of YAP Inhibitor on Human ESCs Grown in 3-D Self-Assembling Scaffolds*

For analysis of yes-associated protein (YAP) signaling in 3-D culture, human ESCs were grown in self-assembling scaffolds for 7 days, and then treated with 2 μM YAP inhibitor (YAPi), Verteporfin (VP, R&D Systems, Minneapolis, MN), for an additional 7 days. Cell growth in the scaffolds was monitored by light microscopy. ESCs grown in 3-D self-assembling scaffolds were harvested after 14 days of culture with and without YAPi treatment as a control. Cells were harvested for RNA and immunocytochemical analysis.

2.9. *Statistical analysis*

Data are presented as mean ± standard error of the mean (SE). One-way ANOVA analysis was performed and analyzed for unequal variances using post hoc tests for multiple comparisons. Results with a p-value less than 0.05 were considered to be significant (* $p < 0.05$ and ** $p < 0.01$). All analyses were performed using SPSS version 26 (SPSS Inc., Chicago, IL., USA).

3. Results

3.1. *Growth and Characterization of H9 Cells Grown under 3-D Culture Conditions*

H9 cells encapsulated in self-assembling scaffolds composed of PEG-8-SH and PEG-8-Acr polymers grew for extended periods without requiring routine passaging or manipulation. The optimal growth of ESCs was achieved by using a concentration of 2.5 *w/v* % (dry weight of polymer per volume of culture medium) at a 1:1 molar ratio of PEG-8-SH and PEG-8-Acr. The results depicted in Figure 2 show steady cell growth up to 21 days as observed by light microscopy (Figure 2A–D). Cells grew clonally in a time-dependent manner. Visual observations were consistent with the results obtained by MTT assay. A significant and continuous increase in the proliferation of cells was observed during day 7 to 21. However, cells grew more rapidly between day 14 and 21 than between day 1 and 14 (Figure 2E), suggesting that the growth of ESCs required a longer acclimation period in 3-D self-assembling scaffolds compared to 2-D culture conditions. The viability of 3-D cultured ESCs was further validated by direct cell counts, as depicted in Figure 2F. The results showed that while cell proliferation significantly and consistently increased from day 1 to day 21, the number of dead cells remained low.

To confirm whether H9 cells remained pluripotent during 3-D culture, the transcriptional and translational analysis of selected ESC-specific markers was performed using qRT-PCR and immunocytochemical analysis, respectively. The results indicated a successive increase in the expression of *OCT4*, *NANOG*, and *SOX2* throughout the duration of the 3-D culture, which increased 2.85, 2.23, and 3.69-fold, respectively in ESCs grown in 3-D scaffolds for 21 days as compared to cells grown under 2-D culture conditions (Figure 2G). The protein expression of these markers was also increased in cells cultured in the 3-D scaffolds (Figure 2H), which was consistent with the transcriptional upregulation.

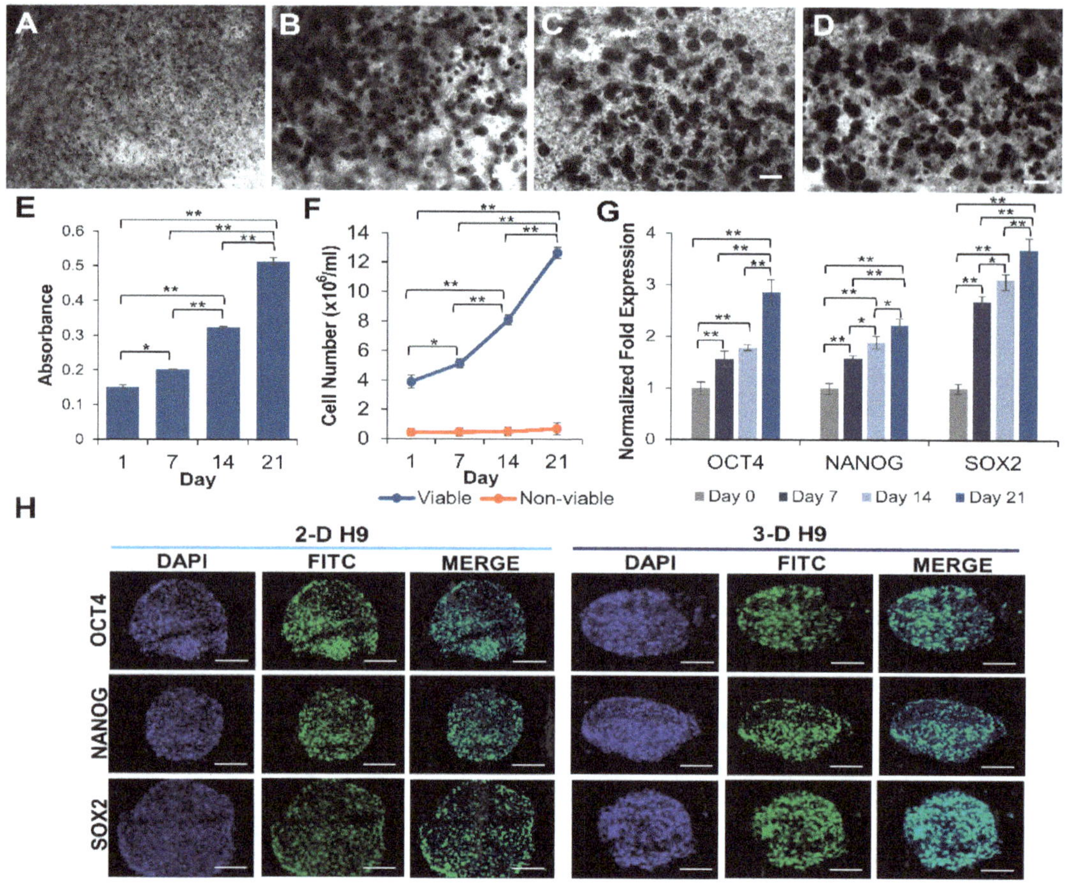

Figure 2. Growth of pluripotent human ESCs in 3-D self-assembling scaffolds. (**A**–**D**) Clonal growth of ESCs (H9 cells) encapsulated in PEG-8-SH/PEG-8-Acr scaffolds and incubated in culture medium was observed by light microscopy at 0, 7, 14, and 21 days. (**E**) Quantitative determination of cell proliferation by 3-(4,5-dimethylthiazol-2-yl)-2,5-diphenyltetrazolium bromide (MTT) assay using microplate reader. Results were expressed as the absorbance ± standard error (SE) with a significant increase in cell number. (**F**) Growth of ESCs encapsulated in 3-D scaffolds was assayed by direct counts using a hemocytometer, and cell viability was determined by trypan blue exclusion assay at various time intervals. Data presented as cell number ($\times 10^6$ cells/mL) ± SE. (**G**) Expression of selected pluripotency markers, *OCT4*, *NANOG*, and *SOX2*, in ESCs cultured in self-assembling scaffolds for 0, 7, 14, and 21 days as determined by qRT-PCR. The expression of genes at day 0 was set to 1 and results were expressed as the fold expression ± SE normalized to reference genes *HMBS*, *GAPDH*, and *β-ACTIN* (* $p < 0.05$ and ** $p < 0.01$). (**H**) Confocal images (20X) of 2-D and 3-D grown ESCs displaying the expression of pluripotent proteins, OCT4, NANOG, and SOX2. All scale bars represent 100 μm.

3.2. Maintenance of Pluripotency in H9 Cells Grown under 3-D Culture Conditions

To investigate if the 3-D grown H9 cells maintained their pluripotency after 21 days in 3-D culture, cells were subcultured under 2-D culture conditions and analyzed for cell morphology and the expression of pluripotent markers. The results showed that there were no morphological differences between cells that were passaged from 3-D to 2-D culture conditions and the initial ESCs used for encapsulation in the 3-D self-assembling scaffolds (Figure 3A–C). When H9 cells grown in 3-D self-assembling scaffolds for 21 days were subsequently subcultured under 2-D culture conditions, the expression of core pluripotent markers, *OCT4* and *NANOG*, was statistically similar to that of initial 2-D grown cells, while the expression of *SOX2* (1.55 fold) was slightly upregulated (Figure 3D). Apparently, repeated the subculturing of 3-D grown cells in 2-D culture conditions reverted the expression of these markers back to the normal level of expression. Despite the high expression of key pluripotent markers, it was prudent to further investigate the 3-D grown cells for the maintenance of their differentiation potential both in in vitro and in vivo.

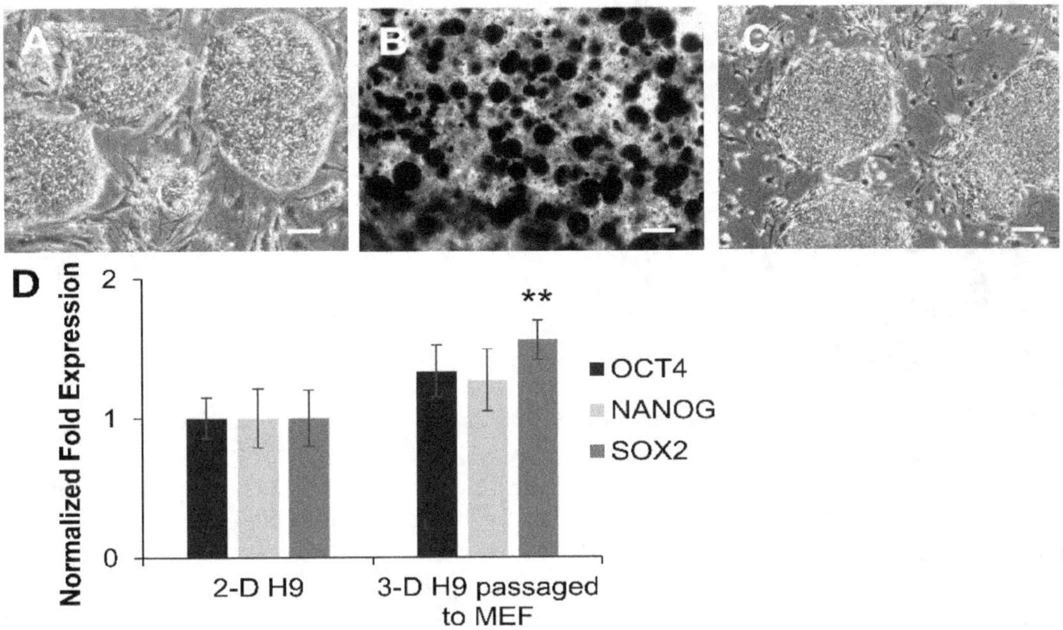

Figure 3. Self-renewal and pluripotency of ESCs were maintained following extended culture in 3-D scaffolds. (**A–C**) Cell morphology of ESCs (H9 cells) grown in 2-D cultures prior to encapsulation, in self-assembling scaffolds for 3-D culture for 21 days, and then subsequently subcultured back to 2-D culture conditions, respectively, as determined by light microscopy. All scale bars represent 100 μm. (**D**) Comparison of expression of *OCT4*, *NANOG*, and *SOX2* in ESCs grown in 2-D conditions and first in 3-D self-assembling scaffolds, and then passaged in 2-D culture conditions as determined by qRT-PCR. Results were expressed as the fold expression ± SE normalized to reference genes *HMBS*, *GAPDH*, and *β-ACTIN* (* $p < 0.05$ and ** $p < 0.01$).

3.3. Differentiation of H9 Cells Grown in 3-D Self-Assembling Scaffolds

The maintenance of the pluripotency of 3-D grown H9 cells was further investigated by the induction of differentiation into three germ layers following EB formation. The results presented in Figure 4 show that EBs from 3-D grown cells spontaneously differentiated into endoderm, mesoderm, and ectoderm germ layers, as evident by the protein expression of specific markers, GATA4, BRACHYURY, and TUJ1, respectively (Figure 4A). Transcriptional analysis also confirmed that differentiated derivatives of 3-D grown cells expressed markers of all three germ layers, including *SOX7* and *GATA6*, *BRACHYURY* and *MIXL1*, and *PAX6* and *NCAM*, representing the endoderm, mesoderm, and ectoderm, respectively (Figure 4B).

Next, the pluripotency of 3-D grown H9 cells was validated in vivo by an analysis of teratoma formation. The results showed that the 3-D grown cells injected into SCID Beige mice formed teratomas (Figure 5). An analysis of teratoma tissue showed the expression of GATA4, BRACHYURY, and TUJ1 proteins, indicating that teratoma tissue had cells representing the endoderm, mesoderm, and ectoderm, respectively (Figure 5C). The transcriptional analysis of teratoma tissues also showed the expression of genes of all three germ layers: the endoderm (*SOX7*, and *SOX17*), mesoderm (*BRACHYURY*, and *MIXL1*), and ectoderm (*PAX6*, and *TUJ1*) (Figure 5D). Taken together, these results indicated that H9 cells grown in 3-D self-assembling scaffolds maintained their pluripotency and differentiation potential.

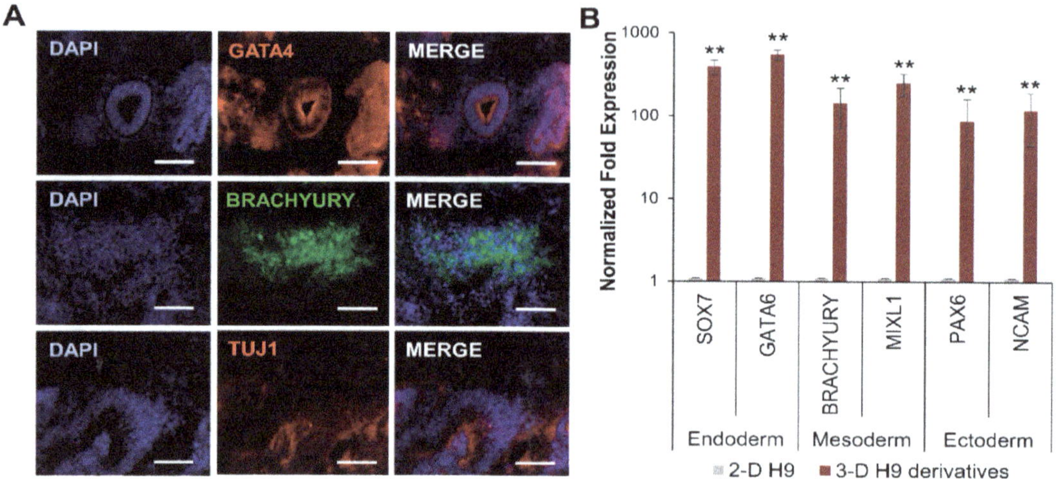

Figure 4. Differentiation potential of human ESCs grown in 3-D self-assembling scaffolds. Embryoid bodies (EBs) derived from 3-D grown ESCs (H9 cells) spontaneously differentiated into three germ layers and expressed specific proteins analyzed by immunocytochemistry. (**A**) Differentiated derivatives of 3-D grown ESCs expressed GATA4, BRACHYURY, and TUJ1 proteins representing the endoderm, mesoderm, and ectoderm, respectively, as shown by confocal images (20X). All scale bars represent 100 μm. (**B**) Differentiated derivatives of 3-D grown ESCs expressed germ layer-specific genes *SOX7* and *GATA6* (endoderm), *BRACHYURY* and *MIXL1* (mesoderm), and *PAX6* and *NCAM* (ectoderm) as determined by qRT-PCR. Results are expressed as the fold expression ± SE normalized to reference genes *HMBS*, *GAPDH*, and *β-ACTIN* (* $p < 0.05$ and ** $p < 0.01$).

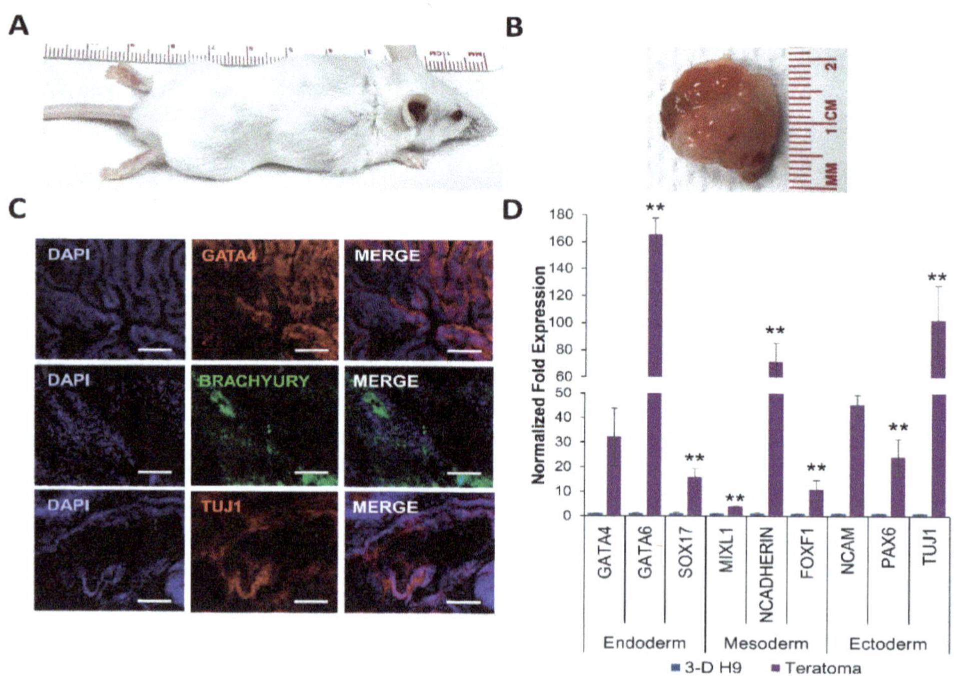

Figure 5. Teratoma formation by 3-D grown human ESCs in severe combined immunodeficient (SCID) Beige mice. (**A**) Tumor growth was observed in all mice ($n = 3$) injected with 3-D grown ESCs (H9 cells). (**B**) Explanted tumor at 4 weeks showed encapsulated, lobular, and well-circumscribed gross morphology consistent with teratoma growth. (**C**) Expression of GATA4, BRACHYURY, and TUJ1 proteins representing the endoderm, mesoderm, and ectoderm, respectively in excised teratomas, as shown by confocal images (20X). All scale bars represent 100 μm. (**D**) Expression of germ layer-specific genes, *SOX7* and *GATA6* (endoderm), *BRACHYURY* and *MIXL1* (mesoderm), and *PAX6* and *NCAM* (ectoderm) in excised teratomas, as determined by qRT-PCR. Results are expressed as the fold expression ± SE normalized to reference genes *HMBS*, *GAPDH*, and *β-ACTIN* (* $p < 0.05$ and ** $p < 0.01$).

3.4. Expression of Naïve Pluripotent Markers in H9 Cells Grown under 3-D Culture Conditions and the Effect of YAP Inhibition

Since a significant upregulation of core pluripotent markers, OCT4, NANOG, and SOX2, was observed in ESCs cultured in the 3-D scaffolds, we also assessed the expression of both primed and naïve pluripotent markers. A comparative transcriptional analysis of H9 cells grown in 3-D scaffolds and 2-D cultured H9 and Elf1 cells (representing primed and naïve ESC lines, respectively) is depicted in Figure 6. Expected primed pluripotent markers (*FOXA2, ZIC2, SALL2,* and *SOX11*) were expressed at higher levels in H9 than Elf1 cells cultured under 2-D conditions. Interestingly, the expression of primed pluripotent markers was significantly decreased in H9 cells grown in 3-D scaffolds to levels comparable to Elf1 cells grown under 2-D conditions. More strikingly, 3-D grown H9 cells expressed significantly higher levels of naïve markers (*KLF2, ESRRB, DNMT3L, KLF17, STAT3, DPPA3, TBX3, PRDM14, KLF5, ZFP42, TFCP2L1, FGF4,* and *GDF3*) in comparison to 2-D cultured Elf1 cells, suggesting that the 3-D scaffold microenvironment modulated gene expression.

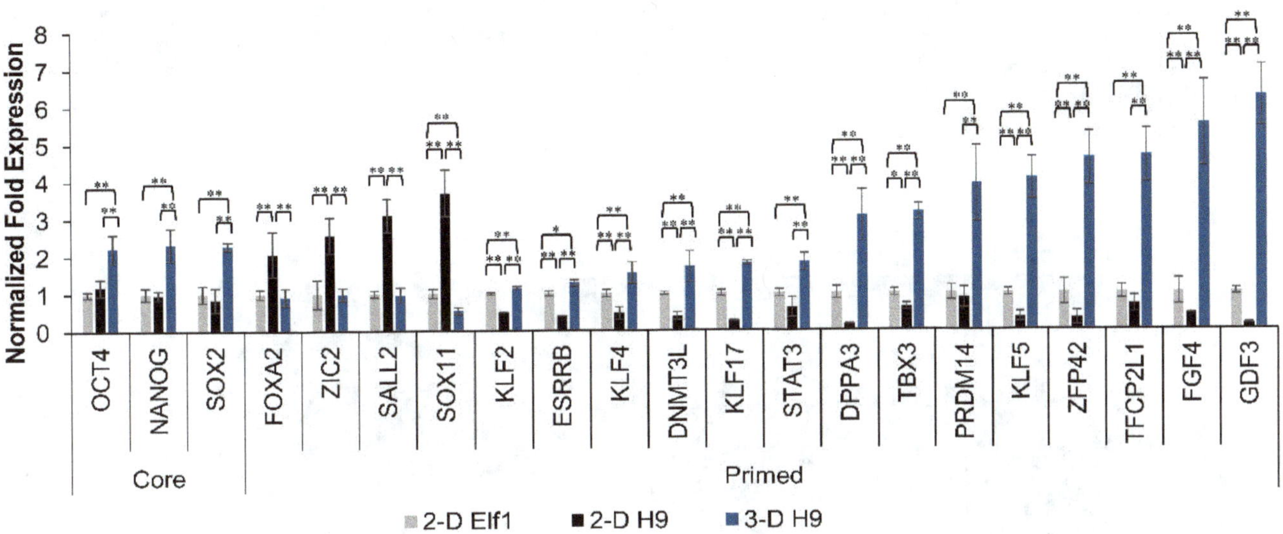

Figure 6. Effect of 3-D culture on the expression of naïve pluripotent markers in primed ESCs. (**A**) Expression of select core (*OCT4, NANOG,* and *SOX2*), primed (*FOXA2, ZIC2, SALL2,* and *SOX11*), and naïve (*KLF2, ESRRB, DNMT3L, KLF17, STAT3, DPPA3, TBX3, PRDM14, KLF5, ZFP42, TFCP2L1, FGF4,* and *GDF3*) pluripotent markers in ESCs cultured in 3-D scaffolds for 21 days and 2-D grown primed ESCs (H9 cells) and naïve ESCs (Elf1 cells, set to control) was analyzed by qRT-PCR. Results were expressed as the fold expression ± SE normalized to reference genes *HMBS, GAPDH,* and *β-ACTIN* (* $p < 0.05$ and ** $p < 0.01$).

Many reports have stated that 3-D scaffolds induce differential gene expression due to mechanical and biological stimuli [10,38,39]. Since the overexpression of YAP has been shown to induce the naïve state of pluripotency in primed ESCs [40], we investigated the effect of YAP in 3-D cultured H9 cells using VP, which is a YAP inhibitor (YAPi). Our results in Figure 7A showed that mechanosensitive genes, *YAP* and *TAZ,* were upregulated in H9 cells grown in 3-D scaffolds in comparison to 2-D cultured cells. In contrast, *YAP* and *TAZ* expression significantly decreased when 3-D grown H9 cells were subjected to YAPi. A similar trend was observed at a translational level, where it appears that the immunofluorescence signal for YAP in 3-D grown H9 cells was brighter than in 2-D cultured cells. Moreover, an increased signal for YAP was observed in the cytoplasm of 3-D grown cells treated with YAPi (Figure 7B).

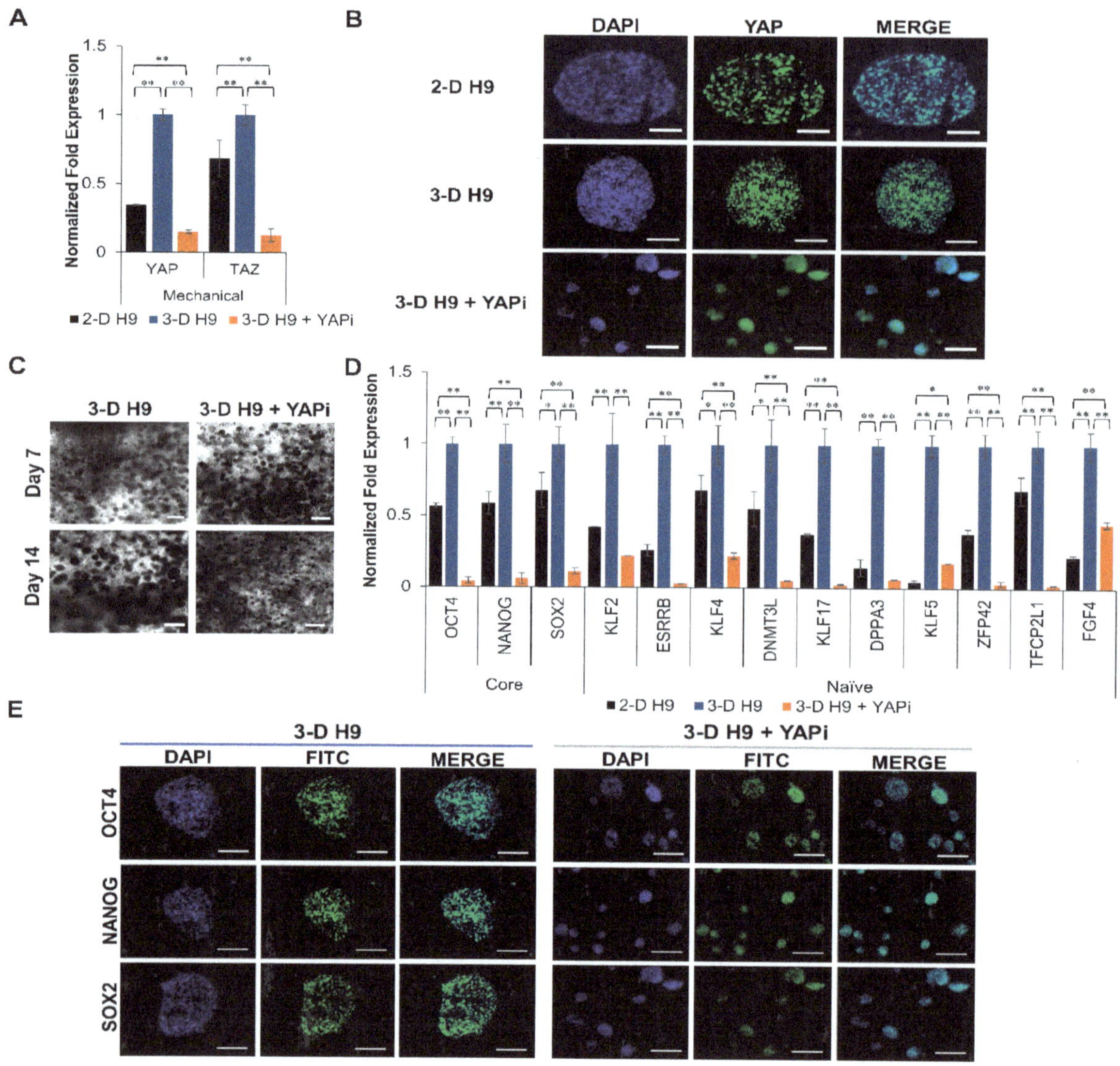

Figure 7. Effect of Yes-associated protein (YAP) inhibition (YAPi) on cell growth and expression of pluripotent markers in human ESCs cultured under 3-D conditions. ESCs (H9 cells) were grown under 3-D culture conditions for 14 days and incubated in the absence (control) or presence of a YAPi. (**A**) The effect of YAPi on the expression of mechanosensitive markers, *YAP* and *TAZ*, in H9 cells grown in 2-D and 3-D culture conditions as determined by qRT-PCR. (**B**) Merged confocal images (20X) showing YAP protein expression in 2-D and 3-D grown H9 cells. (**C**) Light micrographs depicting the cell morphology and colony size of 3-D grown H9 cells incubated in the presence or absence of YAPi. (**D**) Comparison of relative gene expression of core and naïve pluripotent markers in H9 cells grown under 2-D, 3-D, and 3-D + YAPi conditions as determined by qRT-PCR. Results were expressed as the fold expression ± SE normalized to reference genes *HMBS*, *GAPDH*, and *β-ACTIN* (* $p < 0.05$ and ** $p < 0.01$). (**E**) Merged confocal images (20X) of 3-D H9 cells with and without YAPi treatment for the protein expression of OCT4, NANOG, and SOX2. All scale bars represent 100 μm.

The incubation of YAP also affected the clonal growth of H9 cells encapsulated in 3-D scaffolds, with YAPi-treated H9 cells exhibiting a significantly smaller colony size in comparison to untreated cells (Figure 7C). Interestingly, YAPi treatment also abrogated the upregulation of core (*OCT4*, *NANOG*, and *SOX2*), and naïve (*ESRRB*, *KLF4*, *DNMT3L*, *KLF17*, *DPPA3*, *KLF5*, *ZFP42*, *TFCP2L1*, and *FGF4*) pluripotent markers in 3-D cultured cells (Figure 7D). In addition, a concurrent decrease in the expression of core pluripotent proteins, OCT4, NANOG, and SOX2 was observed when compared with the untreated 3-D grown cells (Figure 7E).

3.5. Mechanism of Regulation of Pluripotent Genes in H9 Cells Grown under 3-D Culture Conditions

To determine the basis for the upregulation of pluripotent markers observed in H9 cells grown in 3-D self-assembling scaffolds, we screened multiple signaling pathways with YAP-associated mechanotransduction using transcriptional analysis (Figure 8). The results depicted in Figure 8A show that 3-D cultured cells expressed higher levels of genes encoding integrin subunits, *ITGA5, ITGA6, ITGAV, ITGB1,* and *ITGB3* as well as G-coupled protein receptors (GCPRs), *LPAR1, LPAR2, S1PR1,* and *S1PR3*. In addition, these cells displayed the upregulation of *RHOA* and *RAC1* (Rho signaling), *YAP, TAZ,* and *TEAD4* (Hippo signaling), *LIFR, GP130, SOCS3,* and *TBX3* (LIF signaling). However, the expression of *ITGA2, ROCK1, LATS1,* and *LATS2,* which is associated with integrin, Rho, and Hippo signaling, respectively, decreased significantly, while *ITGB5* remained unchanged in 3-D cultured cells.

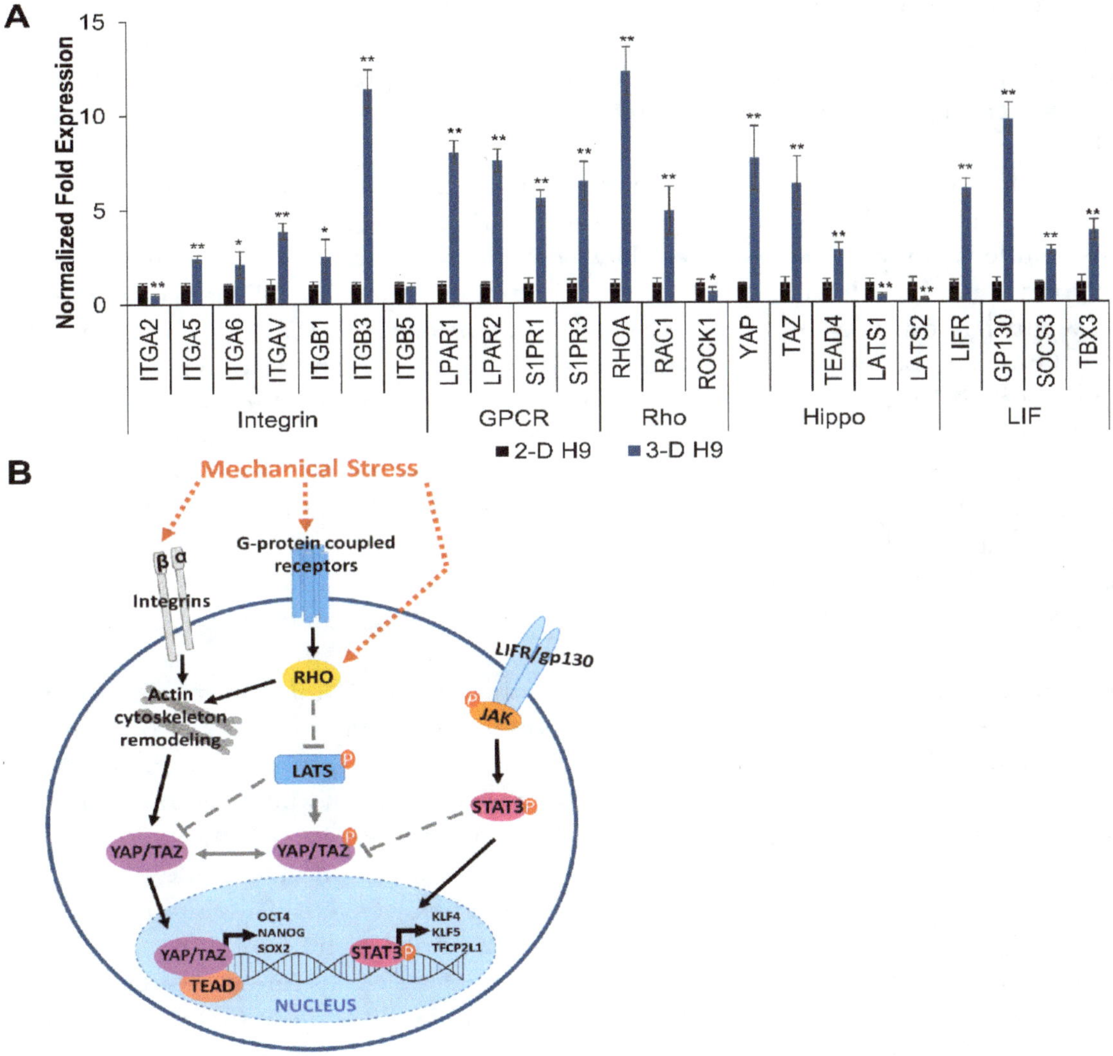

Figure 8. Molecular mechanism of upregulation of pluripotent markers in human ESCs grown in 3-D scaffolds. (**A**) The effect of 3-D culture of ESCs (H9 cells) on the expression of markers associated with integrin (*ITGA2, ITGA5, ITGA6, ITGAV, ITGB1, ITGB3,* and *ITGB5*) and G-coupled protein receptors (GCPRs; *LPAR1, LPAR2, S1PR1,* and *S1PR3*), Hippo (*YAP, TAZ, TEAD4, LATS1,* and *LATS2*), Rho (*RHOA, RAC1,* and *ROCK1*), and LIF (*LIFR, GP130, SOCS3,* and *TBX3*) signaling pathways as determined by qRT-PCR. Results were expressed as the fold expression ± SE normalized to reference genes *HMBS, GAPDH,* and *β-ACTIN* (* $p < 0.05$ and ** $p < 0.01$). (**B**) Proposed pathway involved in the induction of naïve pluripotency in primed human ESCs (H9 cells) encapsulated in 3-D self-assembling scaffolds.

These results further indicate that 3-D self-assembling scaffolds stimulated mechanosensitive signaling resulting in the upregulation of integrin receptors and GPCRs, thus promoting the activation of Rho signaling, which is associated with actin cytoskeleton remodeling; in turn, this led to the upregulation of mechanosensitive YAP/TAZ signaling. Activated YAP/TAZ act as transcription factors in the nucleus assisted by transcription co-factor, TEAD4, which binds to the DNA, to stimulate the expression of pluripotent genes. The simultaneous upregulation of LIF signaling also contributed to the upregulation of naïve pluripotent markers in H9 cells grown in 3-D self-assembling scaffolds, as proposed in Figure 8B.

4. Discussion

The expansion of human ESCs using traditional 2-D culture techniques is technically challenging and requires routine manipulation and passaging by dissecting colonies via enzymatic digestion or non-enzymatic methods [41–43]. These manipulations can result in poor viability, large batch-to-batch variation, and spontaneous differentiation. To address these problems, we hypothesized that 3-D culture may better mimic the in vivo environment from which ESCs are derived, which would improve both the long-term growth and maintenance of these cells.

Our study investigated the effect of the microenvironment in stemness by developing 3-D scaffolds made of two functionalized PEG polymers that self-assembled via a Michael addition reaction. When ESCs were included in the polymer mixtures, they were encapsulated upon self-assembly of the scaffolds. It has been previously reported that fully hydrated hydrogels mimic the 3-D native microenvironment, which allow nutrient diffusion and promote the growth of cells [44]. We have previously shown that soft 3-D scaffolds composed of Dex-SH and PEG-4-Acr self-assembling polymers were capable of supporting mouse ESC pluripotency for over 6 weeks [34]. Several other studies have also reported that mouse ESC self-renewal could be maintained on soft and low stiffness substrates in 2-D culture [45–47]. In addition, mechanically stiffer prefabricated scaffolds have been shown to promote the differentiation of mouse ESCs [48], while softer 3-D scaffolds supported the differentiation of human adult stem cells [49]. The culture of H9 cells in soft scaffolds did not support the viability of encapsulated cells, and growth was severely curtailed. This is consistent with previous studies, which showed that stiffer substrates promoted the maintenance of human ESCs [22,50,51]. Taken together, this led us to develop stiffer scaffolds made of functionalized PEG polymers.

To optimize scaffold polymerization as well as the encapsulation and growth of H9 cells, we tested several self-assembling polymers at various polymer concentrations in our preliminary studies (unpublished data). These studies showed that scaffolds made of PEG-8-SH and PEG-8-Acr prepared at 2.5% *w/v* improved the clonal growth of H9 cells in comparison to scaffolds with lower cross-linking densities and higher swelling ratios (i.e., Dex-SH/PEG-4-Acr and PEG-4-SH/PEG-4-Acr). These results were in line with other the reported studies performed using scaffolds made of multi-arm PEG functionalized with vinyl sulfone (VS) [52]. In these studies, H9 cells grew upon encapsulation in scaffolds made of VS functionalized 4-arm and 8-arm but not 3-arm PEG hydrogels [52]. However, this study was performed using clumps of H9 cells and not single cells for encapsulation. Whereas, in our study, PEG-8-SH/PEG-8-Acr scaffolds supported cell viability, allowing for even dispersal and clonal growth of H9 cells encapsulated as single cells. This has important implications because the growth of single cell inoculations and the generation of homologous populations of pluripotent cells is necessary for cell-based therapeutics [53]. Furthermore, the maintenance and growth of H9 cells was achieved using PEG-8-SH/PEG-8-Acr self-assembling scaffolds for long-term 3-D culture without passaging or manipulation. The pluripotency of the 3-D grown H9 cells was further demonstrated by their ability to differentiate into three germ layers and teratoma formation in vitro and in vivo, respectively. Additionally, core pluripotent markers, *OCT4*, *NANOG*, and *SOX2*, were upregulated during growth in the self-assembling scaffolds, showing significantly higher expression on day 21 of 3-D culture as compared to cells grown in 2-D culture. When 3-D grown H9 cells were passaged back to 2-D culture conditions, they exhibited undifferentiated morphology, and the expression of pluripotent markers decreased to levels similar to the 2-D cultured cells, suggesting that the encapsulated cells cultured under 3-D conditions were not altered.

Changes in the expression of core and naïve pluripotent markers in H9 cells cultured in 3-D self-assembling scaffolds can be attributed to multiple factors, including matrix dimensionality, stiffness, and/or bioinductive signaling [10]. While the incorporation of natural biomaterials has been shown to increase biological signaling, synthetic biomaterials lack biological activities minimizing batch-to-batch variation, but still allow for biophysical modifications, including pore size and mechanical stiffness [54]. Scaffolds composed of natural polymers including hyaluronic acid [55], chitosan, and alginate [56] have been shown to support human ESC self-renewal without a significant change in pluripotent marker expression. Whereas thermoresponsive synthetic hydrogels composed of PEG functionalized with poly-N-isopropylacrylamide allowed for the continuous 3-D culture of cells for 60 passages but only yielded cells 95% positive for *OCT4* [57]. In contrast, we observed the upregulation of not only *OCT4* but also *NANOG* and *SOX2* during the maintenance of human ESCs in 3-D scaffolds.

Previous studies with mouse ESCs reported that the differential upregulation of pluripotent markers in 3-D culture was dependent on scaffold composition and stiffness [34,58]. A recent study using single cell inoculation and the expansion of human ESCs in large-scale bioreactors resulted in the maintenance of *OCT4* levels but the upregulation of *SOX2* in 3-D grown ESC aggregates [59]. In another study, a 3-D culture of H9 clumps in VS functionalized PEG scaffolds resulted in upregulated gene expression of *SOX2* and *KLF4* but not *OCT4* and *NANOG* when compared to 2-D cultured cells [52]. Interestingly, we observed that primed ESCs, H9 cells, encapsulated in the 3-D self-assembling scaffolds exhibited a decreased expression of primed pluripotency markers, *FOXA2*, *ZIC2*, *SALL2*, and *SOX11*, which is characteristic of post-implantation epiblast cells [60,61], to levels comparable to naïve Elf1 cells. Furthermore, a 3-D culture of H9 cells resulted in an increase in the expression of naïve pluripotent markers, (*KLF2*, *ESRRB*, *DNMT3L*, *KLF17*, *STAT3*, *DPPA3*, *TBX3*, *PRDM14*, *KLF5*, *ZFP42*, *TFCP2L1*, *FGF4*, and *GDF3*), which was associated with naïve pre-implantation epiblast cells [7,60–66]. Recent studies have shown that traditionally derived primed human ESC lines can be reprogrammed to naïve pluripotency using the ectopic expression of select genes and specific media conditions containing inhibitor cocktails [67,68]. In one study, a transcriptomic comparison of primed and reprogrammed naïve human demonstrated the differential expression of certain genes such as *KLF2*, *KLF4*, *GDF3*, *SOCS3*, *STAT3*, and *TBX3* expressed at higher levels in naïve than primed cells [63]. However, the expression of core pluripotent markers *OCT4* and *NANOG* remained unchanged, while *SOX2* levels decreased upon reversion to naïve pluripotency [63]. Here, we observed an increase in core and naïve pluripotent markers during 3-D culture, which was presumably influenced by the 3-D scaffold microenvironment.

It has been well established that the transduction of biophysical signals, including substrate stiffness, cell–cell interactions, and mechanical forces can influence human ESC fate and self-renewal in 2-D culture conditions [39]. The interplay between external and internal mechanical stresses of cells and their ECM play important roles in tensional homeostasis of tissues in vivo [69]. However, the effect of 3-D culture on the regulation of the pluripotency of ESCs has not been fully explored.

Physical interactions with cells or the ECM can be transduced into biological signals and influence actin dynamics via mechanosensitive receptors, such as integrin receptors and GPCRs [70]. Evidence has shown that integrin heterodimers, $\alpha5\beta1$, $\alpha6\beta1$, and $\alpha v\beta3$, mediate interactions between ESCs and various ECM substrates [13,71–73]; they also play an important role in the maintenance of pluripotency [74], and matrix stiffness regulates integrin binding [75]. Whereas GPCRs, including lysophosphatidic acid receptors (LPARs) and sphingosine-1-phosphate receptors (S1PRs), play a role in the YAP signaling axis. In response to LPA and S1P ligand binding, the dephosphorylation of YAP and TAZ allows them to enter the nucleus to activate transcription [76]. Treatment with exogenous LPA has been shown to aid in the reversion of primed pluripotency to naïve pluripotency in ESCs due to the role that both YAP and LPA serve in suppressing differentiation associated with GSK3 inhibition [40]. However, the transduction of mechanical signals has also been shown to activate G-protein signaling independent of ligand binding [77]. We observed the upregulation of numerous integrin subunits as well as LPARs and S1PRs in 3-D grown H9 cells. This suggests that the 3-D scaffolds promoted mechanical signaling pathways in H9 cells; however, future studies are required to determine the mechanism by which these receptors are activated.

Mechanical signals such as substrate stiffness have been shown to lead to the activation of Rho signaling, which regulates actin cytoskeleton organization, leading to increased F-actin and nuclear localization of YAP and TAZ [78]. YAP is a mechanosensitive transcription factor in the Hippo signaling pathway, which plays a crucial role in cancer, regeneration, and the regulation of organ size [79]. In addition, Hippo signaling regulates the first cell fate specification to the trophoectoderm and the inner cell mass (ICM) of early blastocysts via mechanically sensitive pathways [80]. YAP is mostly retained in the cytoplasm in the ICM of early blastocysts, but becomes active during epiblast specification with a strong nuclear signal [64]. YAP and its transcription cofactor, TAZ, act as major downstream effectors in the Rho signaling pathway and have been shown to control the pluripotent state, allowing for the long-term survival and expansion of human ESCs in vitro [81].

Furthermore, cell culture on stiffer substrates has also been shown to increase the nuclear function of YAP/TAZ in human ESCs [82]. TAZ itself is required for the maintenance of self-renewal marker expression in ESCs, and the loss of TAZ leads to the inhibition of transforming growth factor beta (TGFβ) signaling and differentiation into a neuroectoderm lineage [83]. Likewise, a knockdown of YAP results in a loss of pluripotency in mouse ESCs, whereas the ectopic expression of YAP prevents differentiation [84].

In the nucleus, YAP mediates transcriptional enhanced associate domain (TEAD) transcription, and YAP/TAZ complexes associate with TEAD to regulate pluripotency by activating *OCT4* expression [85]. Pluripotency is determined by an autoregulatory core transcriptional circuitry comprised of *OCT4*, *NANOG*, and *SOX2*, which inhibits lineage specification genes [86]. The phosphorylation of YAP by LATS kinases prevents interaction with TEAD and results in cytoplasmic retention and the inactivation of YAP [87]. In contrast, YAP overexpression has been shown to promote the reversion of primed pluripotency into naïve pluripotency [40].

When 3-D grown H9 cells were cultured with a YAPi (VP), the upregulation of core and naïve pluripotent markers was abrogated. VP acts to inhibit the interaction of YAP and TEAD, disrupting downstream transcriptional activation, and sequestering YAP in the cytoplasm for inactivation [88]. Therefore, we postulate that mechanical forces in 3-D scaffolds stimulated the upregulation of mechanosensitive receptors, including integrins and GPCRs, leading to enhanced Rho signaling and higher levels of YAP/TAZ. In addition, the mechanical forces generated by the expansion of cells may also have contributed to the inhibition of phosphorylated LATS kinases [87], which in turn allowed YAP/TAZ to enter the nucleus, activating the transcription of pluripotent genes. This was concurrent with the observed increase in LIF signaling receptors and downstream naïve pluripotent target genes including *KLF4*, *KLF5*, and *TFCP2L1* [89]. While LIF/STAT3 signaling fails to maintain the self-renewal of primed human ESCs [90], naïve ESCs are dependent on LIF signaling [63]. Although these results are interesting and suggest a role of mechanical signaling in the regulation of cell fate in vitro, further analysis is needed to confirm the mechanism of upregulation of pluripotency markers and the genetic stability of ESCs grown in 3-D culture conditions.

Overall, the 3-D scaffolds made of PEG-8-SH/PEG-8-Acr support the clonal growth of primed ESCs as well as the enhanced expression of both core and naïve pluripotency markers, suggesting that the scaffold provided a permissive microenvironment for the induction of a naïve-like state of pluripotency. Our 3-D culture method is robust, simple, and less labor-intensive for the long-term amplification of homogenous populations of ESCs, which could promote their use in basic science and therapeutic applications.

Author Contributions: Conceptualization, G.R.C.; methodology, C.M.; validation, C.M., and C.B.; data curation, C.M., and C.B.; formal analysis, G.R.C.; writing—original draft preparation, C.M.; writing—review and editing, C.B, and G.R.C.; supervision, G.R.C.

Acknowledgments: The study was supported by the OU-WB Institute for Stem Cell and Regenerative Medicine (ISCRM), Oakland University, and Michigan Head and Spine Institute. C. McKee received the Provost Graduate Research Award from Oakland University for this project. We are thankful for C. Govind and S. Dinda for reviewing the manuscript.

References

1. Reubinoff, B.E.; Pera, M.F.; Fong, C.Y.; Trounson, A.; Bongso, A. Embryonic stem cell lines from human blastocysts: Somatic differentiation in vitro. *Nat. Biotech.* **2000**, *18*, 399–404. [CrossRef]
2. Takahashi, K.; Yamanaka, S. Induction of pluripotent stem cells from mouse embryonic and adult fibroblast cultures by defined factors. *Cell* **2006**, *126*, 663–676. [CrossRef]
3. Nakamura, T.; Okamoto, I.; Sasaki, K.; Yabuta, Y.; Iwatani, C.; Tsuchiya, H.; Seita, Y.; Nakamura, S.; Yamamoto, T.; Saitou, M. A developmental coordinate of pluripotency among mice, monkeys and humans. *Nature* **2016**, *537*, 57–62. [CrossRef]
4. Nichols, J.; Smith, A. Naive and primed pluripotent states. *Cell Stem Cell* **2009**, *4*, 487–492. [CrossRef]
5. Gafni, O.; Weinberger, L.; Mansour, A.A.; Manor, Y.S.; Chomsky, E.; Ben-Yosef, D.; Kalma, Y.; Viukov, S.; Maza, I.; Zviran, A.; et al. Derivation of novel human ground state naive pluripotent stem cells. *Nature* **2013**, *504*, 282–286. [CrossRef]
6. Morgani, S.; Nichols, J.; Hadjantonakis, A.-K. The many faces of pluripotency: In vitro adaptations of a continuum of in vivo states. *BMC Dev. Biol.* **2017**, *17*, 7. [CrossRef] [PubMed]
7. Theunissen, T.W.; Powell, B.E.; Wang, H.; Mitalipova, M.; Faddah, D.A.; Reddy, J.; Fan, Z.P.; Maetzel, D.; Ganz, K.; Shi, L.; et al. Systematic identification of culture conditions for induction and maintenance of naive human pluripotency. *Cell Stem Cell* **2014**, *15*, 471–487. [CrossRef] [PubMed]
8. Ying, Q.L.; Wray, J.; Nichols, J.; Batlle-Morera, L.; Doble, B.; Woodgett, J.; Cohen, P.; Smith, A. The ground state of embryonic stem cell self-renewal. *Nature* **2008**, *453*, 519–523. [CrossRef] [PubMed]
9. Wu, J.; Yamauchi, T.; Izpisua Belmonte, J.C. An overview of mammalian pluripotency. *Development* **2016**, *143*, 1644–1648. [CrossRef] [PubMed]
10. McKee, C.; Chaudhry, G.R. Advances and challenges in stem cell culture. *Colloids Surf. B Biointerfaces* **2017**, *159*, 62–77. [CrossRef] [PubMed]
11. Li, Y.; Lin, C.; Wang, L.; Liu, Y.; Mu, X.; Ma, Y.; Li, L. Maintenance of human embryonic stem cells on gelatin. *Chin. Sci. Bull.* **2009**, *54*, 4214. [CrossRef]
12. Xu, C.; Inokuma, M.S.; Denham, J.; Golds, K.; Kundu, P.; Gold, J.D.; Carpenter, M.K. Feeder-free growth of undifferentiated human embryonic stem cells. *Nat. Biotechnol.* **2001**, *19*, 971–974. [CrossRef] [PubMed]
13. Rodin, S.; Domogatskaya, A.; Strom, S.; Hansson, E.M.; Chien, K.R.; Inzunza, J.; Hovatta, O.; Tryggvason, K. Long-term self-renewal of human pluripotent stem cells on human recombinant laminin-511. *Nat. Biotechnol.* **2010**, *28*, 611–615. [CrossRef] [PubMed]
14. Kalaskar, D.M.; Downes, J.E.; Murray, P.; Edgar, D.H.; Williams, R.L. Characterization of the interface between adsorbed fibronectin and human embryonic stem cells. *J. R. Soc. Interface* **2013**, *10*, 20130139. [CrossRef]
15. Braam, S.R.; Zeinstra, L.; Litjens, S.; Ward-van Oostwaard, D.; van den Brink, S.; van Laake, L.; Lebrin, F.; Kats, P.; Hochstenbach, R.; Passier, R.; et al. Recombinant vitronectin is a functionally defined substrate that supports human embryonic stem cell self-renewal via alphavbeta5 integrin. *Stem Cells* **2008**, *26*, 2257–2265. [CrossRef]
16. Richards, M.; Fong, C.Y.; Chan, W.K.; Wong, P.C.; Bongso, A. Human feeders support prolonged undifferentiated growth of human inner cell masses and embryonic stem cells. *Nat. Biotechnol.* **2002**, *20*, 933–936. [CrossRef]
17. Jozefczuk, J.; Drews, K.; Adjaye, J. Preparation of mouse embryonic fibroblast cells suitable for culturing human embryonic and induced pluripotent stem cells. *J. Vis. Exp.* **2012**. [CrossRef]
18. Kent, L. Culture and maintenance of human embryonic stem cells. *J. Vis. Exp.* **2009**. [CrossRef]
19. Akhmanova, M.; Osidak, E.; Domogatsky, S.; Rodin, S.; Domogatskaya, A. Physical, spatial, and molecular aspects of extracellular matrix of in vivo niches and artificial scaffolds relevant to stem cells research. *Stem Cells Int.* **2015**, *2015*, 35. [CrossRef]

20. Cosson, S.; Lutolf, M.P. Chapter 7—Microfluidic patterning of protein gradients on biomimetic hydrogel substrates. In *Methods in Cell Biology*; Piel, M., Théry, M., Eds.; Academic Press: Cambridge, MA, USA, 2014; Volume 121, pp. 91–102.
21. Infantes, E.C.; Prados, A.B.H.; Contreras, I.D.; Cahuana, G.M.; Hmadcha, A.; Bermudo, F.M.; Soria, B.; Huamán, J.R.T.; Bergua, F.J.B. Nitric oxide and hypoxia response in pluripotent stem cells. *Redox Biol.* **2015**, *5*, 417–418. [CrossRef]
22. Musah, S.; Morin, S.A.; Wrighton, P.J.; Zwick, D.B.; Jin, S.; Kiessling, L.L. Glycosaminoglycan-binding hydrogels enable mechanical control of human pluripotent stem cell self-renewal. *ACS Nano* **2012**, *6*, 10168–10177. [CrossRef] [PubMed]
23. Peerani, R.; Rao, B.M.; Bauwens, C.; Yin, T.; Wood, G.A.; Nagy, A.; Kumacheva, E.; Zandstra, P.W. Niche-mediated control of human embryonic stem cell self-renewal and differentiation. *EMBO J.* **2007**, *26*, 4744–4755. [CrossRef] [PubMed]
24. Szablowska-Gadomska, I.; Zayat, V.; Buzanska, L. Influence of low oxygen tensions on expression of pluripotency genes in stem cells. *Acta Neurobiol. Exp.* **2011**, *71*, 86–93.
25. Burdick, J.A.; Vunjak-Novakovic, G. Engineered microenvironments for controlled stem cell differentiation. *Tissue Eng. Part A* **2009**, *15*, 205–219. [CrossRef]
26. Janmey, P.A.; Miller, R.T. Mechanisms of mechanical signaling in development and disease. *J. Cell Sci.* **2011**, *124*, 9–18. [CrossRef]
27. Provenzano, P.P.; Keely, P.J. Mechanical signaling through the cytoskeleton regulates cell proliferation by coordinated focal adhesion and rho gtpase signaling. *J. Cell Sci.* **2011**, *124*, 1195–1205. [CrossRef]
28. Baker, B.M.; Chen, C.S. Deconstructing the third dimension: How 3d culture microenvironments alter cellular cues. *J. Cell Sci.* **2012**, *125*, 3015–3024. [CrossRef]
29. Lund, A.W.; Yener, B.; Stegemann, J.P.; Plopper, G.E. The natural and engineered 3d microenvironment as a regulatory cue during stem cell fate determination. *Tissue Eng. Part B Rev.* **2009**, *15*, 371–380. [CrossRef]
30. Lutolf, M.P.; Gilbert, P.M.; Blau, H.M. Designing materials to direct stem-cell fate. *Nature* **2009**, *462*, 433–441. [CrossRef]
31. Saha, K.; Pollock, J.F.; Schaffer, D.V.; Healy, K.E. Designing synthetic materials to control stem cell phenotype. *Curr. Opin. Chem. Biol.* **2007**, *11*, 381–387. [CrossRef]
32. Mammoto, A.; Mammoto, T.; Ingber, D.E. Mechanosensitive mechanisms in transcriptional regulation. *J. Cell Sci.* **2012**, *125*, 3061–3073. [CrossRef] [PubMed]
33. Krupinski, P.; Chickarmane, V.; Peterson, C. Simulating the mammalian blastocyst—Molecular and mechanical interactions pattern the embryo. *PLOS Comput. Biol.* **2011**, *7*, e1001128. [CrossRef] [PubMed]
34. McKee, C.; Perez-Cruet, M.; Chavez, F.; Chaudhry, G.R. Simplified three-dimensional culture system for long-term expansion of embryonic stem cells. *World J. Stem Cells* **2015**, *7*, 1064–1077. [PubMed]
35. Thomson, J.A.; Itskovitz-Eldor, J.; Shapiro, S.S.; Waknitz, M.A.; Swiergiel, J.J.; Marshall, V.S.; Jones, J.M. Embryonic stem cell lines derived from human blastocysts. *Science* **1998**, *282*, 1145–1147. [CrossRef]
36. Ware, C.B.; Nelson, A.M.; Mecham, B.; Hesson, J.; Zhou, W.; Jonlin, E.C.; Jimenez-Caliani, A.J.; Deng, X.; Cavanaugh, C.; Cook, S.; et al. Derivation of naive human embryonic stem cells. *Proc. Natl. Acad. Sci. USA* **2014**, *111*, 4484–4489. [CrossRef]
37. Nair, D.P.; Podgórski, M.; Chatani, S.; Gong, T.; Xi, W.; Fenoli, C.R.; Bowman, C.N. The thiol-michael addition click reaction: A powerful and widely used tool in materials chemistry. *Chem. Mater.* **2014**, *26*, 724–744. [CrossRef]
38. Antoni, D.; Burckel, H.; Josset, E.; Noel, G. Three-dimensional cell culture: A breakthrough in vivo. *Int. J. Mol. Sci.* **2015**, *16*, 5517–5527. [CrossRef]
39. Kraehenbuehl, T.P.; Langer, R.; Ferreira, L.S. Three-dimensional biomaterials for the study of human pluripotent stem cells. *Nat. Methods* **2011**, *8*, 731–736. [CrossRef]
40. Qin, H.; Hejna, M.; Liu, Y.; Percharde, M.; Wossidlo, M.; Blouin, L.; Durruthy-Durruthy, J.; Wong, P.; Qi, Z.; Yu, J.; et al. Yap induces human naive pluripotency. *Cell Rep.* **2016**, *14*, 2301–2312. [CrossRef]
41. Beers, J.; Gulbranson, D.R.; George, N.; Siniscalchi, L.I.; Jones, J.; Thomson, J.A.; Chen, G. Passaging and colony expansion of human pluripotent stem cells by enzyme-free dissociation in chemically defined culture conditions. *Nat. Protoc.* **2012**, *7*, 2029–2040. [CrossRef]

42. Chen, G.; Gulbranson, D.R.; Hou, Z.; Bolin, J.M.; Ruotti, V.; Probasco, M.D.; Smuga-Otto, K.; Howden, S.E.; Diol, N.R.; Propson, N.E.; et al. Chemically defined conditions for human ipsc derivation and culture. *Nat. Methods* **2011**, *8*, 424–429. [CrossRef] [PubMed]
43. Nie, Y.; Walsh, P.; Clarke, D.L.; Rowley, J.A.; Fellner, T. Scalable passaging of adherent human pluripotent stem cells. *PLoS ONE* **2014**, *9*, e88012. [CrossRef] [PubMed]
44. Lee, J.; Cuddihy, M.J.; Kotov, N.A. Three-dimensional cell culture matrices: State of the art. *Tissue Eng. Part B Rev.* **2008**, *14*, 61–86. [CrossRef] [PubMed]
45. Chowdhury, F.; Li, Y.; Poh, Y.C.; Yokohama-Tamaki, T.; Wang, N.; Tanaka, T.S. Soft substrates promote homogeneous self-renewal of embryonic stem cells via downregulating cell-matrix tractions. *PLoS ONE* **2010**, *5*, e15655. [CrossRef]
46. Chowdhury, F.; Na, S.; Li, D.; Poh, Y.C.; Tanaka, T.S.; Wang, F.; Wang, N. Material properties of the cell dictate stress-induced spreading and differentiation in embryonic stem cells. *Nat. Mater.* **2010**, *9*, 82–88. [CrossRef]
47. Higuchi, S.; Watanabe, T.M.; Kawauchi, K.; Ichimura, T.; Fujita, H. Culturing of mouse and human cells on soft substrates promote the expression of stem cell markers. *J. Biosci. Bioeng.* **2014**, *117*, 749–755. [CrossRef]
48. McKee, C.; Hong, Y.; Yao, D.; Chaudhry, G.R. Compression induced chondrogenic differentiation of embryonic stem cells in three-dimensional polydimethylsiloxane scaffolds. *Tissue Eng. Part A* **2017**, *23*, 426–435. [CrossRef]
49. McKee, C.; Beeravolu, N.; Brown, C.; Perez-Cruet, M.; Chaudhry, G.R. Mesenchymal stem cells transplanted with self-assembling scaffolds differentiated to regenerate nucleus pulposus in an ex vivo model of degenerative disc disease. *Appl. Mater. Today* **2019**. [CrossRef]
50. Keung, A.J.; Asuri, P.; Kumar, S.; Schaffer, D.V. Soft microenvironments promote the early neurogenic differentiation but not self-renewal of human pluripotent stem cells. *Integr. Biol.* **2012**, *4*, 1049–1058. [CrossRef]
51. Sun, Y.; Villa-Diaz, L.G.; Lam, R.H.W.; Chen, W.; Krebsbach, P.H.; Fu, J. Mechanics regulates fate decisions of human embryonic stem cells. *PLoS ONE* **2012**, *7*, e37178. [CrossRef]
52. Jang, M.; Lee, S.T.; Kim, J.W.; Yang, J.H.; Yoon, J.K.; Park, J.C.; Ryoo, H.M.; van der Vlies, A.J.; Ahn, J.Y.; Hubbell, J.A.; et al. A feeder-free, defined three-dimensional polyethylene glycol-based extracellular matrix niche for culture of human embryonic stem cells. *Biomaterials* **2013**, *34*, 3571–3580. [CrossRef] [PubMed]
53. Chen, K.G.; Mallon, B.S.; McKay, R.D.G.; Robey, P.G. Human pluripotent stem cell culture: Considerations for maintenance, expansion, and therapeutics. *Cell Stem Cell* **2014**, *14*, 13–26. [CrossRef] [PubMed]
54. Kharkar, P.M.; Kiick, K.L.; Kloxin, A.M. Designing degradable hydrogels for orthogonal control of cell microenvironments. *Chem. Soc. Rev.* **2013**, *42*, 7335–7372. [CrossRef] [PubMed]
55. Gerecht, S.; Burdick, J.A.; Ferreira, L.S.; Townsend, S.A.; Langer, R.; Vunjak-Novakovic, G. Hyaluronic acid hydrogel for controlled self-renewal and differentiation of human embryonic stem cells. *Proc. Natl. Acad. Sci. USA* **2007**, *104*, 11298–11303. [CrossRef] [PubMed]
56. Li, Z.; Leung, M.; Hopper, R.; Ellenbogen, R.; Zhang, M. Feeder-free self-renewal of human embryonic stem cells in 3d porous natural polymer scaffolds. *Biomaterials* **2010**, *31*, 404–412. [CrossRef]
57. Lei, Y.; Schaffer, D.V. A fully defined and scalable 3d culture system for human pluripotent stem cell expansion and differentiation. *Proc. Natl. Acad. Sci. USA* **2013**, *110*, E5039–E5048. [CrossRef]
58. Wei, J.; Han, J.; Zhao, Y.; Cui, Y.; Wang, B.; Xiao, Z.; Chen, B.; Dai, J. The importance of three-dimensional scaffold structure on stemness maintenance of mouse embryonic stem cells. *Biomaterials* **2014**, *35*, 7724–7733. [CrossRef]
59. Li, X.; Ma, R.; Gu, Q.; Liang, L.; Wang, L.; Zhang, Y.; Wang, X.; Liu, X.; Li, Z.; Fang, J.; et al. A fully defined static suspension culture system for large-scale human embryonic stem cell production. *Cell Death Dis.* **2018**, *9*, 892. [CrossRef]
60. Collier, A.J.; Panula, S.P.; Schell, J.P.; Chovanec, P.; Plaza Reyes, A.; Petropoulos, S.; Corcoran, A.E.; Walker, R.; Douagi, I.; Lanner, F.; et al. Comprehensive cell surface protein profiling identifies specific markers of human naive and primed pluripotent states. *Cell Stem Cell* **2017**, *20*, 874–890. [CrossRef]
61. Rostovskaya, M.; Stirparo, G.G.; Smith, A. Capacitation of human naïve pluripotent stem cells for multi-lineage differentiation. *Development* **2019**, *146*, dev172916. [CrossRef]
62. Guo, G.; von Meyenn, F.; Santos, F.; Chen, Y.; Reik, W.; Bertone, P.; Smith, A.; Nichols, J. Naive pluripotent stem cells derived directly from isolated cells of the human inner cell mass. *Stem Cell Rep.* **2016**, *6*, 437–446. [CrossRef] [PubMed]

63. Hanna, J.; Cheng, A.W.; Saha, K.; Kim, J.; Lengner, C.J.; Soldner, F.; Cassady, J.P.; Muffat, J.; Carey, B.W.; Jaenisch, R. Human embryonic stem cells with biological and epigenetic characteristics similar to those of mouse escs. *Proc. Natl. Acad. Sci. USA* **2010**, *107*, 9222–9227. [CrossRef] [PubMed]
64. Hashimoto, M.; Sasaki, H. Epiblast formation by tead-yap-dependent expression of pluripotency factors and competitive elimination of unspecified cells. *Dev. Cell* **2019**, *50*, 139–154. [CrossRef]
65. Kilens, S.; Meistermann, D.; Moreno, D.; Chariau, C.; Gaignerie, A.; Reignier, A.; Lelièvre, Y.; Casanova, M.; Vallot, C.; Nedellec, S.; et al. Parallel derivation of isogenic human primed and naive induced pluripotent stem cells. *Nat. Commun.* **2018**, *9*, 360. [CrossRef] [PubMed]
66. Takashima, Y.; Guo, G.; Loos, R.; Nichols, J.; Ficz, G.; Krueger, F.; Oxley, D.; Santos, F.; Clarke, J.; Mansfield, W.; et al. Resetting transcription factor control circuitry toward ground-state pluripotency in human. *Cell* **2014**, *158*, 1254–1269. [CrossRef]
67. Sim, Y.J.; Kim, M.S.; Nayfeh, A.; Yun, Y.J.; Kim, S.J.; Park, K.T.; Kim, C.H.; Kim, K.S. 2i maintains a naive ground state in escs through two distinct epigenetic mechanisms. *Stem Cell Rep.* **2017**, *8*, 1312–1328. [CrossRef]
68. Van der Jeught, M.; Taelman, J.; Duggal, G.; Ghimire, S.; Lierman, S.; Chuva de Sousa Lopes, S.M.; Deforce, D.; Deroo, T.; De Sutter, P.; Heindryckx, B. Application of small molecules favoring naive pluripotency during human embryonic stem cell derivation. *Cell. Reprogramming* **2015**, *17*, 170–180. [CrossRef]
69. Zhang, Y.; Liao, K.; Li, C.; Lai, A.C.K.; Foo, J.J.; Chan, V. Progress in integrative biomaterial systems to approach three-dimensional cell mechanotransduction. *Bioengineering* **2017**, *4*, 72. [CrossRef]
70. Finch-Edmondson, M.; Sudol, M. Framework to function: Mechanosensitive regulators of gene transcription. *Cell. Mol. Biol. Lett.* **2016**, *21*, 28. [CrossRef]
71. Baxter, M.A.; Camarasa, M.V.; Bates, N.; Small, F.; Murray, P.; Edgar, D.; Kimber, S.J. Analysis of the distinct functions of growth factors and tissue culture substrates necessary for the long-term self-renewal of human embryonic stem cell lines. *Stem Cell Res.* **2009**, *3*, 28–38. [CrossRef]
72. Mei, Y.; Saha, K.; Bogatyrev, S.R.; Yang, J.; Hook, A.L.; Kalcioglu, Z.I.; Cho, S.W.; Mitalipova, M.; Pyzocha, N.; Rojas, F.; et al. Combinatorial development of biomaterials for clonal growth of human pluripotent stem cells. *Nat. Mater.* **2010**, *9*, 768–778. [CrossRef] [PubMed]
73. Vitillo, L.; Kimber, S.J. Integrin and fak regulation of human pluripotent stem cells. *Curr. Stem Cell Rep.* **2017**, *3*, 358–365. [CrossRef] [PubMed]
74. Lee, S.T.; Yun, J.I.; Jo, Y.S.; Mochizuki, M.; van der Vlies, A.J.; Kontos, S.; Ihm, J.E.; Lim, J.M.; Hubbell, J.A. Engineering integrin signaling for promoting embryonic stem cell self-renewal in a precisely defined niche. *Biomaterials* **2010**, *31*, 1219–1226. [CrossRef] [PubMed]
75. Sun, Y.; Chen, C.S.; Fu, J. Forcing stem cells to behave: A biophysical perspective of the cellular microenvironment. *Annu. Rev. Biophys.* **2012**, *41*, 519–542. [CrossRef]
76. Yu, F.X.; Zhao, B.; Panupinthu, N.; Jewell, J.L.; Lian, I.; Wang, L.H.; Zhao, J.; Yuan, H.; Tumaneng, K.; Li, H.; et al. Regulation of the hippo-yap pathway by g-protein-coupled receptor signaling. *Cell* **2012**, *150*, 780–791. [CrossRef]
77. Lam, R.M.; Chesler, A.T. Shear elegance: A novel screen uncovers a mechanosensitive gpcr. *J. Gen. Physiol.* **2018**, *150*, 907. [CrossRef]
78. Varelas, X. The hippo pathway effectors taz and yap in development, homeostasis and disease. *Development* **2014**, *141*, 1614–1626. [CrossRef]
79. Elosegui-Artola, A.; Andreu, I.; Beedle, A.E.M.; Lezamiz, A.; Uroz, M.; Kosmalska, A.J.; Oria, R.; Kechagia, J.Z.; Rico-Lastres, P.; Le Roux, A.L.; et al. Force triggers yap nuclear entry by regulating transport across nuclear pores. *Cell* **2017**, *171*, 1397–1410. [CrossRef]
80. Sasaki, H. Roles and regulations of hippo signaling during preimplantation mouse development. *Dev. Growth Differ.* **2017**, *59*, 12–20. [CrossRef]
81. Ohgushi, M.; Minaguchi, M.; Sasai, Y. Rho-signaling-directed yap/taz activity underlies the long-term survival and expansion of human embryonic stem cells. *Cell Stem Cell* **2015**, *17*, 448–461. [CrossRef]
82. Hsiao, C.; Lampe, M.; Nillasithanukroh, S.; Han, W.; Lian, X.; Palecek, S.P. Human pluripotent stem cell culture density modulates yap signaling. *Biotechnol. J.* **2016**, *11*, 662–675. [CrossRef] [PubMed]
83. Varelas, X.; Sakuma, R.; Samavarchi-Tehrani, P.; Peerani, R.; Rao, B.M.; Dembowy, J.; Yaffe, M.B.; Zandstra, P.W.; Wrana, J.L. Taz controls smad nucleocytoplasmic shuttling and regulates human embryonic stem-cell self-renewal. *Nat. Cell Biol.* **2008**, *10*, 837. [CrossRef] [PubMed]

84. Lian, I.; Kim, J.; Okazawa, H.; Zhao, J.; Zhao, B.; Yu, J.; Chinnaiyan, A.; Israel, M.A.; Goldstein, L.S.; Abujarour, R.; et al. The role of yap transcription coactivator in regulating stem cell self-renewal and differentiation. *Genes Dev.* **2010**, *24*, 1106–1118. [CrossRef] [PubMed]
85. Papaspyropoulos, A.; Bradley, L.; Thapa, A.; Leung, C.Y.; Toskas, K.; Koennig, D.; Pefani, D.E.; Raso, C.; Grou, C.; Hamilton, G.; et al. Rassf1a uncouples wnt from hippo signalling and promotes yap mediated differentiation via p73. *Nat. Commun.* **2018**, *9*, 424. [CrossRef]
86. Young, R.A. Control of the embryonic stem cell state. *Cell* **2011**, *144*, 940–954. [CrossRef]
87. Meng, Z.; Moroishi, T.; Guan, K.L. Mechanisms of hippo pathway regulation. *Genes Dev.* **2016**, *30*, 1–17. [CrossRef]
88. Wang, C.; Zhu, X.; Feng, W.; Yu, Y.; Jeong, K.; Guo, W.; Lu, Y.; Mills, G.B. Verteporfin inhibits yap function through up-regulating 14-3-3σ sequestering yap in the cytoplasm. *Am. J. Cancer Res.* **2015**, *6*, 27–37.
89. Onishi, K.; Zandstra, P.W. Lif signaling in stem cells and development. *Development* **2015**, *142*, 2230. [CrossRef]
90. Daheron, L.; Opitz, S.L.; Zaehres, H.; Lensch, M.W.; Andrews, P.W.; Itskovitz-Eldor, J.; Daley, G.Q. Lif/stat3 signaling fails to maintain self-renewal of human embryonic stem cells. *Stem Cells* **2004**, *22*, 770–778. [CrossRef]

8

Human Induced Pluripotent Stem Cell-Derived 3D-Neurospheres are Suitable for Neurotoxicity Screening

Julianna Kobolak [1], Annamaria Teglasi [1], Tamas Bellak [1], Zofia Janstova [1], Kinga Molnar [2], Melinda Zana [1], Istvan Bock [1], Lajos Laszlo [2] and Andras Dinnyes [1,3,4,*]

1 BioTalentum Ltd., H-2100 Gödöllő, Hungary; julianna.kobolak@biotalentum.hu (J.K.); annamaria.teglasi@biotalentum.hu (A.T.); tamas.bellak@biotalentum.hu (T.B.); Zofia.Janstova@biotalentum.hu (Z.J.); melinda.zana@biotalentum.hu (M.Z.); biotalentum.bock@yahoo.com (I.B.)
2 Department of Anatomy, Cell and Developmental Biology, Eötvös Loránd University, H-1117 Budapest, Hungary; kinga.molnar@ttk.elte.hu (K.M.); laszlo@elte.hu (L.L.)
3 Molecular Animal Biotechnology Laboratory, Szent István University, H-2101 Gödöllő, Hungary
4 Translational Biomedicine Institute, University of Szeged, H-6720 Szeged, Hungary
* Correspondence: manuscript.dinnyes@biotalentum.hu

Abstract: We present a hiPSC-based 3D in vitro system suitable to test neurotoxicity (NT). Human iPSCs-derived 3D neurospheres grown in 96-well plate format were characterized timewise for 6-weeks. Changes in complexity and homogeneity were followed by immunocytochemistry and transmission electron microscopy. Transcriptional activity of major developmental, structural, and cell-type-specific markers was investigated at weekly intervals to present the differentiation of neurons, astrocytes, and oligodendrocytes. Neurospheres were exposed to different well-known toxicants with or without neurotoxic effect (e.g., paraquat, acrylamide, or ibuprofen) and examined at various stages of the differentiation with an ATP-based cell viability assay optimized for 3D-tissues. Concentration responses were investigated after acute (72 h) exposure. Moreover, the compound-specific effect of rotenone was investigated by a panel of ER-stress assay, TUNEL assay, immunocytochemistry, electron microscopy, and in 3D-spheroid based neurite outgrowth assay. The acute exposure to different classes of toxicants revealed distinct susceptibility profiles in a differentiation stage-dependent manner, indicating that hiPSC-based 3D in vitro neurosphere models could be used effectively to evaluate NT, and can be developed further to detect developmental neurotoxicity (DNT) and thus replace or complement the use of animal models in various basic research and pharmaceutical applications.

Keywords: induced pluripotent stem cells; neurospheres; 3D culture; neurite outgrowth; neurotoxicity

---

1. Introduction

Environmental stressors, such as chemicals or the medical drugs could have a toxic effect on humans which may occur at any stage of their life, during fetal development, childhood, or adult life. The toxic effects of environmental agents coupled with inherited susceptibility of individuals make the toxicology prediction difficult. Conventional animal-based toxicity and safety tests have high costs, use large numbers of animals (mainly rats) and in many cases, they do not provide clearly translatable results for humans [1–3]. Consequently, in line with legislation, there is an increasing need to develop alternative testing methods which could handle thousands of drugs or chemicals with affordable time and cost and with human-relevant neurotoxicology (NT) outcome [4–7]. Development of new approach methods (NAM), would be important both for NT and developmental neurotoxicology (DNT) tests, providing data on the effect of chemicals and the potential adverse outcomes (AOs) [6,8–13].

An increasing number of studies use primary cell cultures and recently, pluripotent stem cells (PSCs) to create in vitro systems for NT and DNT screenings. It is widely accepted concept that 3D cell cultures can mimic better the original tissue environment including tissue-specific architecture, mechanical and biochemical features, cell-to-cell communication and signaling, and differentiation capability, while 2D cell culture systems are less complex and more artificial in this sense [14–17], although the existing assays provide meaningful readouts for specific neurodevelopmental processes [6,8,11]. Only one in vitro assay will not be able to cover the complexity of the in vivo development, therefore a battery of assays and fit-for-purpose applications should be used to cover the relevant processes [10,18]. Human induced pluripotent stem cell-derived (hiPSC) neurospheres, grown on 3D scaffolds or self-forming, can establish cell–cell interactions and model certain neurodevelopmental processes, therefore, might be used as an in vitro screening platform not only for NT but in DNT studies or drug development [19–21].

Neural stem cells (NSCs) are considered as a multipotent and self-renewing pool of cells in the mammalian central nervous system (CNS), occurring in vivo in the developing embryonic neural tissue as early as the neural tube formation happens. These cells have the capacity to differentiate into all neuronal cell types, as well as into glial cells, therefore they are able to emulate some fetal neurodevelopmental processes [22,23]. Moreover, NSCs can be differentiated in vitro from pluripotent stem cells (PSCs) thus providing an attractive and almost unlimited in vitro tool for toxicology studies including drug development [24,25].

Despite the developments mentioned above, still, a limited number of human iPSC derived 3D neuronal culture-based studies were published focusing on the development of NT or DNT models [21,26–31]. A limiting factor for the further development of such test methods is the lack of high-throughput screening (HTS) read-outs on 3D cell cultures [16], despite the developments using high-content image analysis (HCA) [27].

Here, we present a reliable model system where hiPSC-derived NSCs are differentiated towards subtypes of neurons, astrocytes, and oligodendrocytes, forming free-floating 3D neurospheres in 96-well plate format. Over 6 weeks the various differentiation stages are characterized, and three selected stages are exposed with different compounds to investigate their cytotoxic effect. The generated neuronal spheroids used for NT measurements at different time-points resembling various differentiation stages, therefore, provide an excellent platform for further DNT test system developments.

2. Methods

2.1. Chemicals and Plasticware

The chemicals were purchased from Sigma-Aldrich (St Louis, MO, USA) and all cell culture reagents and plasticware from Thermo Fisher Scientific Inc. (Waltham, MA, USA), unless otherwise specified.

2.2. Human iPSC Culture

Human iPSC line (Ctrl-2), derived from a healthy mid-age Caucasian female donor peripheral blood mononuclear cells (PBMCs), established and characterized earlier [32,33], was used in this study. Cells were cultured on BD Matrigel™ matrix (BD Biosciences, Franklin Lakes, NJ, USA) with mTeSR™1 medium (Stem Cell Technologies, Vancouver, Canada), using Gentle Cell Dissociation Reagent for passages, according to the manufacturer's instruction. Representative hiPSC colony morphology, pluripotency staining for POU5F1, NANOG, SSEA4, and TRA1-60 and karyotype analysis are presented in Figure S1. For mycoplasma screening, the Venor®GeM-Advance (Minerva Biolabs) Mycoplasma Detection Kit was used according to the manufacturer's protocol in every fifth passage during maintenance and before freezing. Cells were cultured at 37 °C in a humidified atmosphere containing 5% CO_2. In the current study cultures from passage 16 and 17 were used for differentiation.

2.3. *Neuronal Differentiation and Maintenance*

The Ctrl-2 hiPSC line was differentiated to neural stem cells (NSCs) by dual SMAD inhibition procedure [34], following the detailed protocol of Shi et al. [35]. Briefly, when hiPSC cultures reached 90% confluence, the culture was passaged onto poly-L-ornithine and laminin (POL/L; 0.002%/1 μg/cm^2) coated plates, using conventional iPSC passage (as detailed above). On the next day media was changed to neural induction medium (NIM; DMEM/F12: Neurobasal medium, supplemented with 1× N2, 2× B27, 2 mM glutamine, 1× non-essential amino acid (NEAA), 100 μM β-mercaptoethanol, 5 μg/mL insulin) supplemented with 10 μM SB431542, 500 ng/mL Noggin (R&D Systems, Inc., Minneapolis, MN, USA) and 5 ng/mL basic fibroblast growth factor (bFGF) to induce the neuroectodermal lineage. Neuronal induction media was applied until day 10 of the differentiation. The forming neural rosette-like structures were manually picked under a stereomicroscope (Olympus SZX2; Olympus Ltd. Tokyo, Japan) and re-plated onto POL/L (POL/L; 0.002%/1 μg/cm^2) plates. NSCs were expanded in neural maintenance medium (NMM; DMEM-F12: Neurobasal medium, supplemented with 1× N2, and 2× B27, 2 mM glutamine, 1× NEAA), supplemented with 10 ng/mL bFGF and 10 ng/mL Epidermal growth factor (EGF) and maintained in a monolayer on dishes coated with POL/L (0.002%/1 μg/cm^2). When reached 100% confluence, cells were passaged using Accutase (Sigma) and seeded as single cells (50,000 cells/cm^2) for further expansion on POL/L (0.002%/1 μg/cm^2) coated dishes. After 4 passages NSCs were frozen in 1 mL freezing medium (90% Fetal Bovine Serum, FBS; heat-inactivated, Thermo Fisher Sci.; Cat N.: 10500-064, LOT: 08F1180K; 10% DMSO; 2 million cells/vial), using Accutase passage. Before freezing the NSCs, a mycoplasma test was performed (see above), repeated every fifth passage during maintenance.

2.4. *3D Neurosphere Culture*

After thawing, NSCs (generated as detailed above) were cultured on POL/L (0.002%/1 μg/cm^2) coated plates (50,000 cells/cm^2). When reached 100% confluence NSCs were propagated with Accutase treatment and single-cell suspension was plated onto a low-adherent 96-well plate with 10,000 cells/well in NMM medium. Neurospheres were formed within 48 h after plating and NMM half of the media was changed in every 3 days until analyzing the samples. Samples were collected at Day 0 (D0), Day 2 (D2), and weekly intervals from the end of the 1st week (D7) until the end of the 6th week (D14, D21, D28, D35, and D42).

2.5. *Cryosectioning and Immunocytochemistry (ICC) Staining*

3D neurospheres (untreated, vehicle or compound-treated) were fixed with 4% paraformaldehyde (PFA) in 0.1 mol/L phosphate buffer for 1 h at RT and washed 3 times with PBS. The fixed samples were cryoprotected in 30% sucrose in PBS containing 0.01% sodium azide at 4 °C until embedding in Shandon Cryomatrix gel (Thermo Fischer Scientific). The 16 μm parallel sections were made using cryostat (Leica CM 1850 Cryostat, Leica GmbH), mounted to Superfrost™ Ultra Plus Adhesion Slides (Thermo Fisher Scientific) and stored at −20 °C until use. After 10 min air-drying, the sections were blocked for 1 h at RT with blocking solution (3% BSA in PBS), supplemented with 0.2% TritonX-100. The sections were then incubated with primary antibodies (Table S1) overnight at 4 °C. Next day, sections were washed in PBS 3 times, and isotype-specific secondary antibodies (Table S1) were diluted in blocking buffer and applied for 1 h at RT. The sections were washed 3 times with PBS and covered using Vectashield® mounting medium containing DAPI (1.5 μg/mL; Vector Laboratories), that labelled the nuclei of the cells (at least 1 h at RT). Negative controls for the secondary antibodies were performed by omitting the primary antibodies. Immunoreactive sections were analyzed using a BX-41 epifluorescent microscope (objectives: 20× 0.50 NA; 40× 0.75 NA; Olympus) equipped with a DP-74 digital camera and its CellSens software (V1.18; Olympus). For confocal imaging, Olympus Fv10i-W compact confocal microscope system (objective: 60× 1.35 NA; Olympus) with Fv10i software

(V2.1; Olympus) was applied. All images were further processed using the GNU Image Manipulation Program (GIMP 2.10.0) and NIH ImageJ analysis software (imagej.nih.gov/ij).

Quantification of the immunocytochemistry data was performed using ImageJ software according to Tieng et al. [36]. Briefly, images were taken by confocal microscopy at 120× magnification. The numbers of Ki-67-, NESTIN-, GFAP-, AQP4-, TUBB3-, NF200kD-, MBP-, VAMP2-, and MAP2-immunoreactive pixels were measured in 5 neurosphere (highest diameter middle sections, 5 randomly selected fields/slide) at every time points. Data was normalized with DAPI positive nuclei number. Data were expressed as a percentage of marker/DAPI ratio ± SEM ($p < 0.05$).

2.6. Apoptosis Assay

Embedding and cryosectioning of 3D samples were performed as above. To detect apoptotic activity, the DeadEnd™ Colorimetric TUNEL System (Promega) was used on the middle cryosections (highest diameter) of the spheroids, following the instructions of the manufacturer. In brief, apoptosis was detected by immersing the slides in PBS for 5 min (at RT), adding 20 µg/mL Proteinase K solution and incubating for 10–30 min (at RT). After 5–10 min treatment in Equilibration buffer, recombinant terminal deoxynucleotidyl transferase (rTdT) was added to the reaction mixture. Next, the sections were incubated for 60 min at 37 °C inside of a humidified chamber to allow the end-labelling reaction to occur. The reaction was terminated by immersing the slides in saline-sodium citrate for 15 min (RT). Endogenous peroxidases were blocked by immersing the slides in 0.3% hydrogen peroxide in PBS for 3–5 min (RT). Streptavidin-HRP was added to slides, incubated for 30 min (RT), stained with diaminobenzidine (DAB) solution for 5 min until a light brown background appeared. For hematoxylin–eosin (HE) staining Mayer's Hematoxylin solution was used for 3 min. Sections were rinsed with tap water and placed into distilled water for 30 s, then into 96% alcohol for 30 s. One percent Eosin solution in distilled water was used for 3 min. Stained sections were dehydrated through alcohols, clear in xylene and mount in DPX. Microphotographs were made with DP-74 digital camera (Olympus) using a light microscope (BX-41, objectives: 20× 0.50 NA; 40× 0.75 NA; Olympus) and CellSens software (V1.18; Olympus). For counting the apoptotic and total (Hematoxylin-stained) number of cells, NIH ImageJ analysis software was used. Five Ctrl and five ROT-treated spheroids were randomly selected, and middle sections were analyzed from each differentiation stage (D21, D28, and D42) samples in three experiments ($n = 3$).

2.7. Transmission Electron Microscopy (TEM)

Neurospheres (untreated, vehicle or compound-treated) were fixed at different differentiation stages in a fixative solution containing 3.2% PFA, 0.2% glutaraldehyde, 1% sucrose, 40 mM $CaCl_2$ in 0.1 M cacodylate buffer (pH 7.4) for 12 h at 4 °C. Samples for ultrastructural analysis were embedded in 1.5% agar (dissolved in dH_2O), post-fixed in 1% ferrocyanide-reduced osmium tetroxide [37], then dehydrated using graded series of ethanol, finally embedded in Spurr low viscosity epoxy resin medium. Ultrathin sections were collected from the middle region of the spheroids (highest diameter) on copper slot grids coated with formwar (Agar Sci., Essex, UK) and counterstained with uranyl acetate and Reynolds's lead citrate. Sections were examined with a JEOL JEM 1011 transmission electron microscope (JEOL Ltd., Tokyo, Japan) equipped with a Morada 11-megapixel camera using iTEM software (Olympus).

2.8. RT-qPCR Analysis

For each sample, 12 spheroids were pooled, and 3 biological replicates were performed ($n = 3$). Total RNA was isolated using the RNeasy Plus Mini Kit (Qiagen, Hilden, Germany). For the reverse transcription, 600 ng of the isolated RNA was used applying the Maxima First Strand cDNA Synthesis Kit for RT-qPCR with dsDNase (Thermo Fisher Scientific) according to the manufacturer's instructions.

Gene-specific primers were designed using the Primer3 software [38], specified with mFOLD software [39] and Primer-BLAST software [40]. Primers were optimized using two-fold serial dilution standard curves (Table S2). As a reference, gene GAPDH was used (Table S2). Each real-time PCR reaction contained 5 ng RNA-equivalent cDNA template, 400 nM of each primer and 50% SYBR Green JumpStart Taq ReadyMix (Sigma Aldrich) in a total volume of 15 μL. PCR reactions were set up using QIAgility liquid handling robot and performed on a Rotor-Gene Q cycler (Qiagen). The cycling parameters were as follows: 94 °C for 3 min initial denaturation followed by 40 cycles of 95 °C for 5 s, 60 °C for 15 s, and 72 °C for 30 s. Melting curve analysis and agarose gel electrophoresis confirmed the specificity of the primers and the absence of gDNA contamination. Data of three replicates were analyzed for each gene, using the ddCT method [41].

2.9. XBP1-Assay of Endoplasmic Reticulum Stress

Upon accumulation of unfolded proteins in the endoplasmic reticulum (ER), a 26-nucleotide fragment from the X-box binding protein 1 mRNA (XBP1(U)) is removed with a special splicing mechanism [42]. This shorter mRNA (XBP1(S)) is a frequently used marker of ER-stress. To study the expression of XBP1(S), previously described primers [43] and our own primers were used [44] (Table S2). As a positive control, cells were treated with 5 μM and 10 μM Tunicamycin to induce ER-stress. RT-PCR reactions were performed and analyzed as above, using Phusion Hot Start II High-Fidelity DNA Polymerase and 20 ng of the cDNA samples.

2.10. Toxicity Treatments and ATP Viability Assay

Eleven well-known compounds were tested in 7 different concentrations to generate concentration-response curves. The tested compounds with the used concentrations are detailed in Table S3. The highest dosage was determined based on the solubility of the compounds, and care was taken not to reach above 0.1% DMSO levels after diluting the compound in the culture media. In each assay plate, 4 technical replicates were applied from each sample and 3 biological replicate assays were run ($n = 3$). The 3D cell cultures were exposed to the toxicants for 72 h (exposure schemes are detailed in Figure 5A or Figure 6A or Figure 7A). Vehicle control was used as a regular medium supplemented with 0.1% DMSO.

ATP viability assay was performed with CellTiter-Glo® 3D Cell Viability Assay, according to the manufacturer's protocol (Promega). Neurospheres were lysed with 100 μL CellTiter-Glo® 3D Reagent for 60 min at RT. Luminescence signal was recorded with a Thermo VarioScan Flash (Thermo Fisher Scientific) plate reader.

2.11. Diametric and Total Protein Determination of the Spheroids

3D spheroids grown in a 96-well plate were captured using the 4x objective (0.1 NA) of Olympus IX71 microscope and DP21 camera (Olympus). The images were analyzed by measuring the diameters of the spheroids using the Olympus CellSens Dimension software (V1.11). Each value represents the average of 3 experiments ($n = 3$) in each 96 spheroids were measured. The 3D spheroids were lysed individually with RIPA Lysis and Extraction Buffer supplemented with Halt™ Protease and Phosphatase Inhibitor Cocktail and Pierce™ Universal Nuclease for Cell Lysis, sonicated and the total protein concentration determined using a Pierce BCA Protein Assay Kit according to the manufacturer's instructions. Due to the small spheroid size of the D2, D7, and D14 samples, three spheroids were pooled, then the individual values were calculated accordingly. In total, three experiments were performed ($n = 3$), in which 24 spheroids were analyzed for each timepoint.

2.12. Neurite Outgrowth Assay

For neurite outgrowth measurement the method of Harris et al. was modified [45]. 3D spheroids were treated with different concentrations (vehicle, 0.1 µM, 0.5 µM, and 0.75 µM) of Rotenone (ROT) at D21 differentiation stage using the 72 h exposure scheme, similarly to the previous concentration-response experiments. After the treatment, the spheroids were plated on Matrigel-coated plates in NMM medium without compound, and 24 h later fixed with 4% PFA and immunostained against TUBB3, as described in detail under the immunostaining paragraph. First, the number of neurites was counted for each sample. Then, the "edge" of the spheroids was determined by drawing an "edge line" and every neurite length was measured from this "edge line" to the tip of each neurite. Total neurite length was calculated for each spheroid using the ImageJ software's Neurite Tracer plugin. In each experiment eight spheroids were treated in each experimental group (5 groups) and three independent experiments were performed ($n = 3$).

2.13. Statistical Analysis

All results were analyzed using Prism 5 (GraphPad Software, La Jolla, CA, USA) and handled in Microsoft Office 2010 (Microsoft, Redmond, WA, USA) software. For normalization of the concentration response, the average values of positive control were used as 0% and the average of vehicle control served as 100%. Four parameter curve fitting methods were used to determine EC10 and EC50 values, where the data on the graphs represent the average of three biological replicates. The "n" value corresponds to the number of biological replicates for each tested concentration. Analysis of data is presented as the mean ± SEM. Significance of data was determined with paired T test for RT-qPCR (* $p < 0.01$) and with One-way ANOVA for concentration response (* $p < 0.05$; ** $p < 0.01$).

3. Results

3.1. Three-Dimensional Spheroid Differentiation of iPSCs-Derived NSCs Revealed Complex Neuronal Cultures

Our first aim was to characterize the timewise differentiation of 3D neuronal spheroids free-floating in suspension culture, originating from hiPSCs. The neuronal differentiation capacity of the starting neural stem cell (NSCs) population, derived by the dual-SMAD inhibition protocol, was comprehensively characterized by us previously [32,33]. These cells expressed the major NPC markers (NESTIN, SOX1 (SRY-Box 1), PAX6 (Paired Box 6) investigated on protein level with ICC (see details in Figure S2). When the NPCs were terminally differentiated in a 2D culture system over 5 weeks, cortical neurons and glial cells were differentiated and formed a neuronal network, as we published recently [32]. In the present study differentiation in a 3D culture system was investigated in the course of 6 weeks (42 days).

In suspension culture upon withdrawing the two mitogens EGF and bFGF, NSCs formed compact 3D spheroids, the so-called neurospheres within 48 h (Figure 1A). Growth properties of the structures were investigated by measuring the diameter of the individual spheres and their total protein content. Results showed a continuous growth of the spheres during the 6 weeks culture period, in terms of diametric growth and protein content (Figure 1). While the diameter increased linearly, the protein content increased more dynamically after week 4, resembling the cellular and structural changes during differentiation (Figure 1B,C).

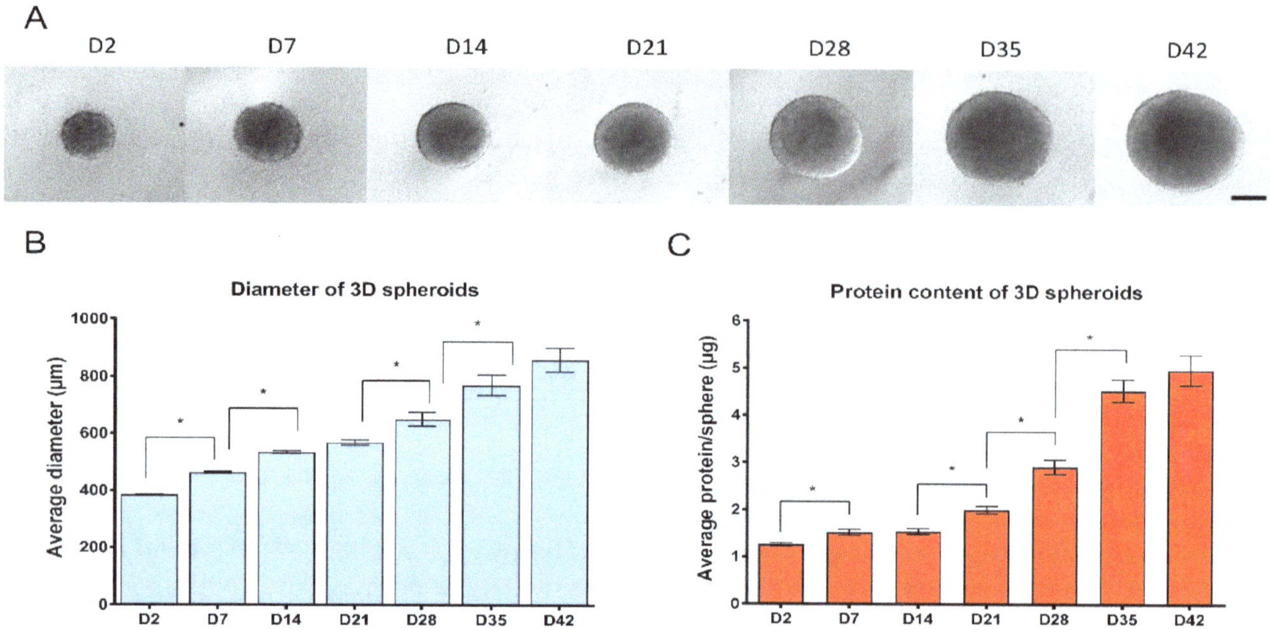

Figure 1. Growth properties of 3D neurospheres during the 6 weeks of differentiation (**A**) Representative microscopic view of 3D spheroids, investigated at weekly intervals (4×, scale bar: 200 µm). (**B**) The average diameter of 3D spheroids in µm measured by CellSens Dimension software (Olympus) ($n = 3$, in each experiment 96 spheroids were compared weekly trough 6-weeks). (**C**) Average total protein content of spheroids determined by Pierce BCA Protein Assay. Note that D2, D7 and D14 samples were measured by pooling three spheroids and individual values were calculated, while in all other timepoints spheroids were measured individually ($n = 3$, in each stage 24 spheroids were measured). ±SEM values are presented on graphs. (* $p < 0.05$).

This observation was in accordance with the expression of the cell proliferation marker Ki-67 (KI67), representing the proportion of dividing cells, monitored by ICC on protein level (Figure 2A,B) or by RT-qPCR on transcript level (Figure 3). The expression of Ki-67 and the number of Ki-67-positive cells were the highest in D2 and D7 samples, which showed a continuous decline over the differentiation period (Figure 2A,B or Figure 3). The forming 3D cell aggregates mainly expressed the early neuronal markers such as *PAX6*, *NESTIN*, and *SOX1*, which expressions decreased with maturation on a time-wise manner (RT-qPCR, Figure 3). Tubulin beta-3 chain (TUBB3) expression appeared very early in the samples, already at D2 as neuronal processes started to grow inside the spheres (ICC, Figure 2A,B). These changes were obvious when investigated at transcript level as well (RT-qPCR, *TUBB3*; Figure 3). It is important to note that no necrotic regions were detected in the center of the spheroids or in other regions, not even the highest diameter of spheroids reached 800 µm in average in D42 samples (Figure 1B or Figure 2 first line). This fact was corroborated in semithin sections where cell density proved to be high and consistent during the whole examined period. Layer organized fibers had become distinguishable on the surface from the 21st day (Figure S3a–g).

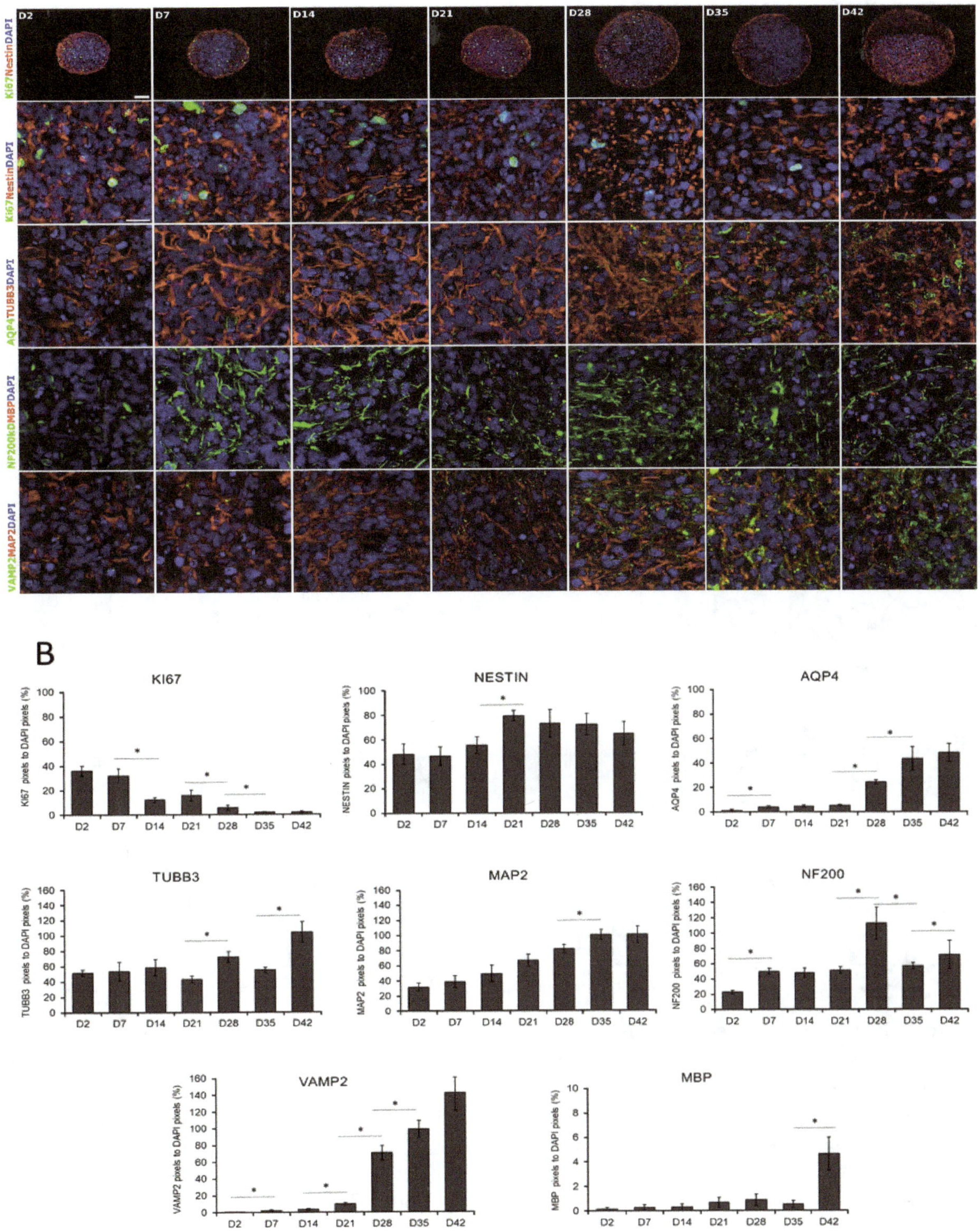

Figure 2. Immunocytochemical analysis of 3D spheroids. (**A**) Spheroids were fixed and cryosectioned then immunostained at weekly intervals from D2 until D42 stage. The first line represents the overview of the cryosectioned spheroids, while the rest of the panel shows higher magnifications. Relevant markers of proliferation (KI67), neural stem cells (NESTIN), neuronal differentiation (TUBB3 and MAP2), an intermediate filament of dendrites and axons (NF200), synaptic vesicles of neurons (VAMP2), astrocyte (AQP4), and oligodendrocyte (MBP) specific proteins were stained. Protein name IDs are indicated with colors, representing the color of the fluorophore used (e.g., green as Alexa 488; red as Alexa 594) Nuclei were counterstained with DAPI (in blue). Scale bar: 100 μm (first line only) and 25 μm.

(**B**) Quantitative analysis of the immunostainings on confocal images. The numbers of Ki-67, NESTIN, AQP4, TUBB3, NF200kD, MBP, VAMP2, and MAP2 immunoreactive pixels were measured in 5 neurospheres (middle sections, 5 randomly selected fields/slide) at every time points. Data was normalized with DAPI positive nuclei number. Data were expressed as percentage of marker/DAPI ratio ± SEM (* $p < 0.05$).

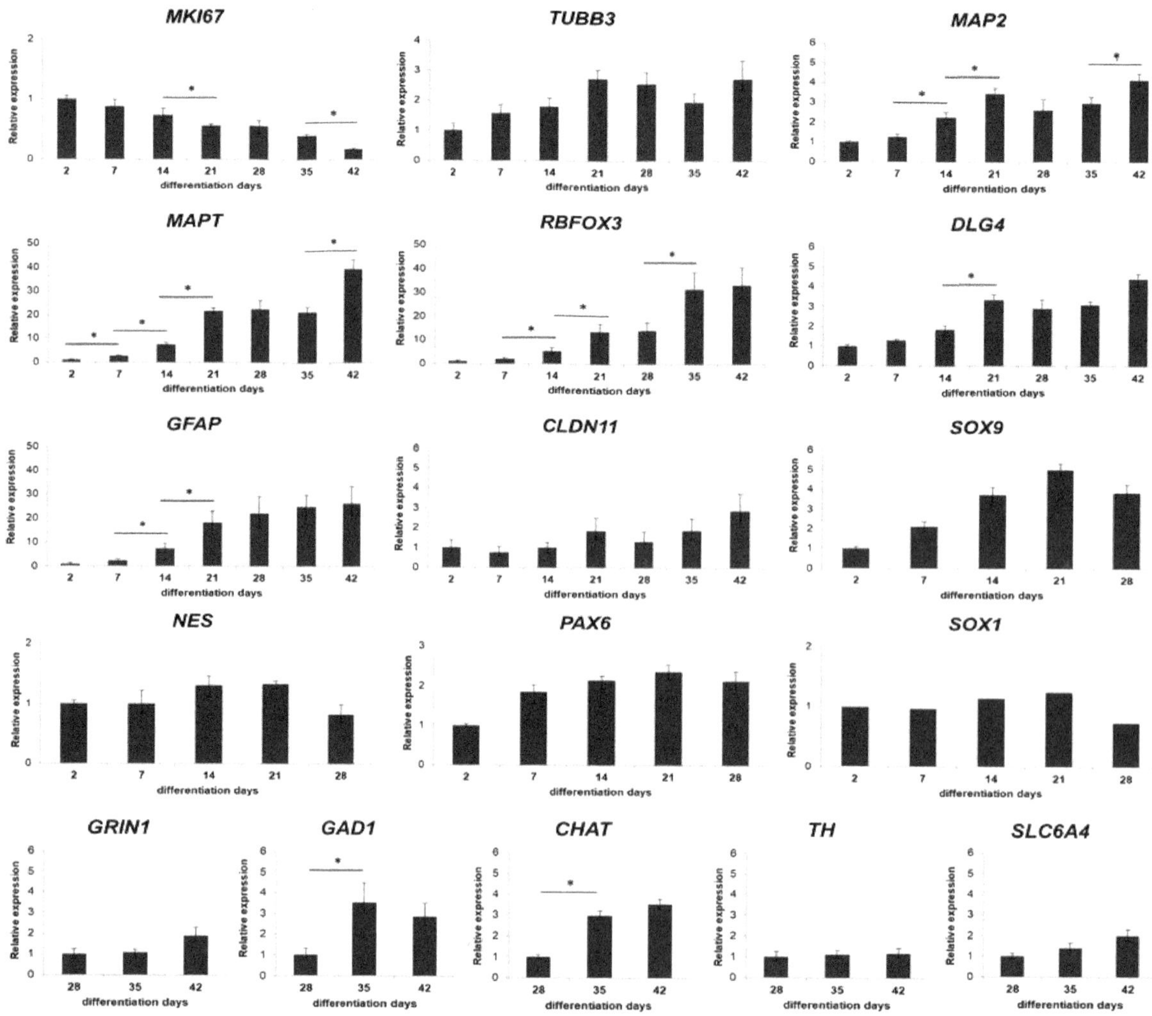

Figure 3. Real-time PCR measurements of relevant markers during the neuronal differentiation of the spheroids. Twelve 3D spheroids were pooled, lysed with RLT-buffer and used in RT-PCR analysis at each timepoints. Where the genes are already expressed in D2 samples, it is referred as 1. Day 2, 7, 14 and 21 values for *GRIN1, CHAT, TH, GAD1,* and *SLC6A4* are not indicated as these genes were not expressed at those time points. Graphs represent normalized relative expression values, analyzed for each gene using the ddCT method [41]. Mean values and ± SEM of three biological replicates ($n = 3$) are presented on graphs, significance are determined by paired T test ($* p < 0.01$). Note that y axis's scales alter according to the different relative expression values.

A weekly investigation of the gene expression revealed a continuous progress in neuronal differentiation and increase in the level of maturation markers, such as the dendritic marker Microtubule-associated protein 2 (*MAP2*) or Microtubule-associated protein tau (*MAPT*), and RNA Binding Fox-1 Homolog 3 (*RBFOX3*, also known as NeuN) (Figure 3), which were in accordance with the protein data of immunofluorescent staining of the 3D spheroids, MAP2 and Neurofilament 200 kDa (NF200); (ICC, Figure 2A,B).

Terminal differentiation resulted in the formation of neuronal networks and synapses, from the stage of D28, post-synaptic marker PSD95 (now DLG4, Discs Large MAGUK Scaffold Protein 4)

(Figures 3 and 4), and the major synaptic vesicle protein p38, Synaptophysin (SYNP) and Vesicle-associated membrane protein 2 (VAMP2) (Figures 2 and 4) expression appeared in maturing neurons. Synapse formation was assessed by transmission electron microscopy (TEM) revealing the presence of synaptic connections, presynaptic vesicles and postsynaptic density (Psd). Matured synapses were detected in D42 samples, where docked presynaptic vesicles were determined by identifiable contact point between the vesicle and the presynaptic membrane (Figure S3h,i).

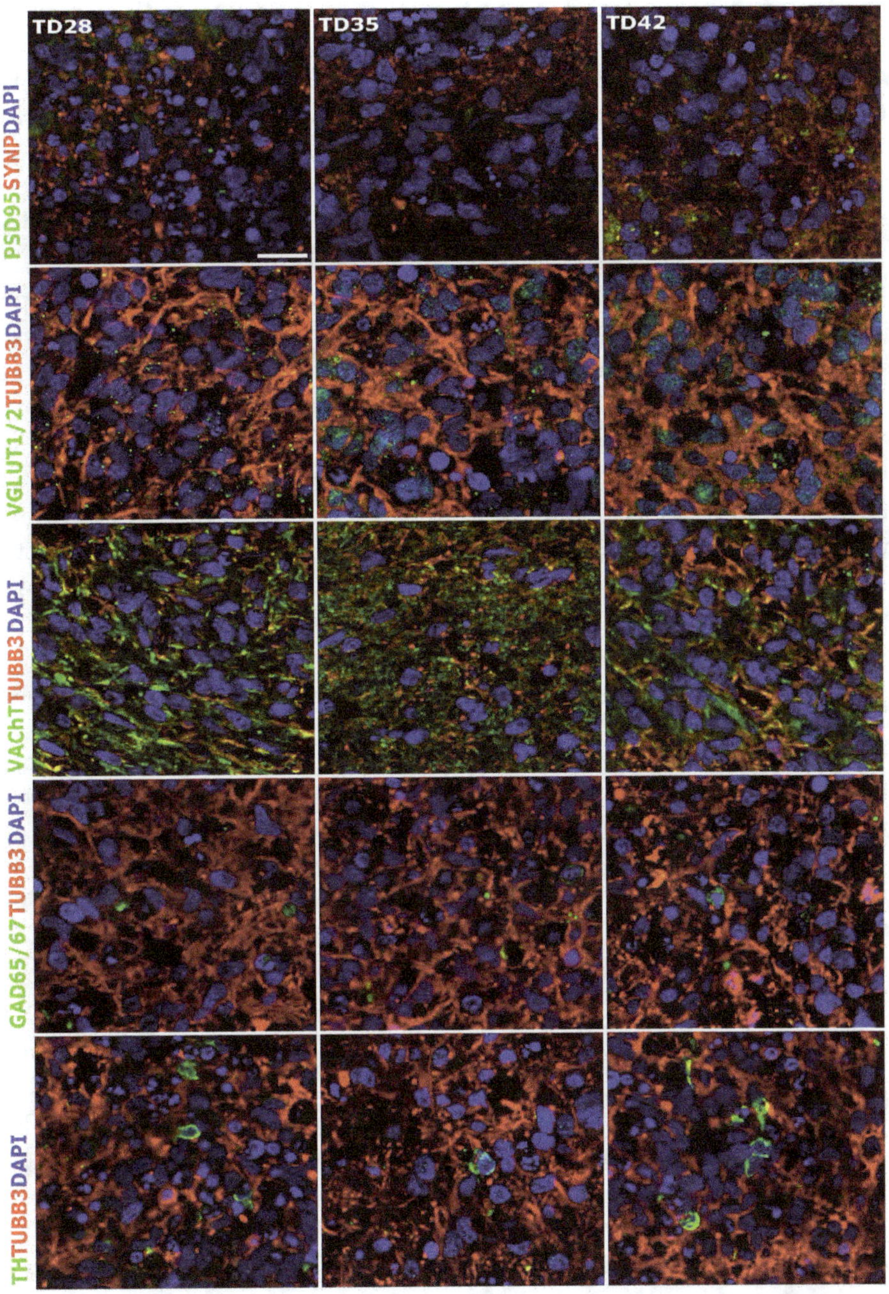

Figure 4. Immunocytochemical detection of neuronal subtypes in 3D neurospheres. Presence of synapses was determined with post-synaptic marker PSD95 and synaptic protein Synaptophysin (SYNP) double staining. Glutamatergic (VGLUT1/2), GABAergic (GAD65/67), cholinergic (VAChT) and dopaminergic (TH) neurons were detected in the developing 3D neurospheres from D28. All samples were stained with TUBB3 (in red) to label the neurites. Protein name IDs are indicated with colors, representing the color of the used fluorophore (e.g., green as Alexa 488; red as Alexa 594). Nuclei were counterstained with DAPI (in blue). Scale bar: 25 μm.

Neuronal subtypes were investigated by the expression of glutamatergic (VGLUT1/2), GABAergic (GAD65/67), cholinergic (VAChT) and dopaminergic (TH) neurons with ICC and qRT-PCR (*GRIN1*: Glutamate Ionotropic Receptor NMDA Type Subunit 1; *GAD1*: Glutamate Decarboxylase 1; *CHAT*: Choline *O*-Acetyltransferase; *TH*: Tyrosine Hydroxylase; *SLC6A4*: Solute Carrier Family 6 Member 4) (Figures 3 and 4). The expression of these markers was detectable mainly from D28 stage and gradually increased until the last investigated time point of the differentiation (D42), in line with the maturation process. It has to be remarked that the spontaneous differentiation of dopaminergic neurons was rare in the cultures, we identified only a few TH positive cells in the cultures (Figure 4).

During the neuronal differentiation, astrocytes started to emerge after week 2, based on the Aquaporin-4 (AQP4) or the Glial fibrillary acidic protein (GFAP) expression both at mRNA and protein level (Figure 2, Figure 3 and Figure S4). Oligodendrocyte differentiation followed a similar course, from D21 it was clearly detectable both at transcript (*Claudin 11*, *CLDN11*, formerly known as Oligodendrocyte marker 4; Figure 3) and protein level (Myelin basic protein, MBP; Figure 2A,B) which gradually increased from D35 with the maturation of the cells, however, still remained in a low expression level (Figure 2A,B or Figure 3). Our ICC results clearly show that the expression of mature neuronal and glial markers and neuron-neuron and neuron-glial interactions increased during the differentiation period, thus emulating the development of the human fetal neural tissue.

Overall, results at the transcript level were in accordance with those of the proteins detected by ICC, confirming the effective differentiation of hiPSC-derived NSCs towards the neural lineage. The 3D neurospheres represented a neuronal tissue-like differentiation, containing neurons, astrocytes and oligodendrocytes, what was presented by transcript-, protein-, and ultrastructural level as well. We can conclude that the 3D spheroid system provides a complex neuronal cell culture which can serve as a model of the early neuronal differentiation.

3.2. Early 3D Neurospheres as a Neurotoxicity Model

Our 3D spheroid-based model was tested in cytotoxicity assay, monitoring the viability of the cells within the neurospheres. Eleven compounds with well-known effects (drugs, pesticides, and chemicals) were selected and applied on the neurospheres in 7 different concentrations (compounds and concentrations are detailed in Table S3) at D21 by acute (72 h) exposure. Concentration-response curves were generated and evaluated. The exposure scheme is presented in Figure 5A.

According to the results, Paraquat (PQ; EC50: 1.89 log μM), Rotenone (ROT; EC50: −0.61 log μM), Mercury(II) chloride ($HgCl_2$; EC50: 1.87 log μM) and Doxorubicin (DOX; EC50: 0.67 log μM) caused maximum cell death, Hexachlorophene (HE; EC50: 1.26 log μM) and Colchicine (COL; EC50: −0.49 log μM) had a strong effect on the viability, but did not kill all the cells in the investigated concentration range. Acrylamide (ACR; EC50: 3.5 log μM) and Rifampicin (RIF; EC50: > 2 log μM) also reduced the viability, while Valproic acid (VPA; EC50: > 2.6 log μM) and Paracetamol (PAR; EC50: > 2 log μM) had minimal effect on the viability of 3D neurospheres at D21 stage. Ibuprofen (IBU; EC50: > 2 log μM) as a non-neurotoxic agent was applied as negative control which indeed, did not decrease the viability of D21 3D neurospheres (Figure 5B, Table S4). Overall, the different compounds induced different levels of cytotoxicity in a concentration-dependent manner on the 3D neurosphere cultures.

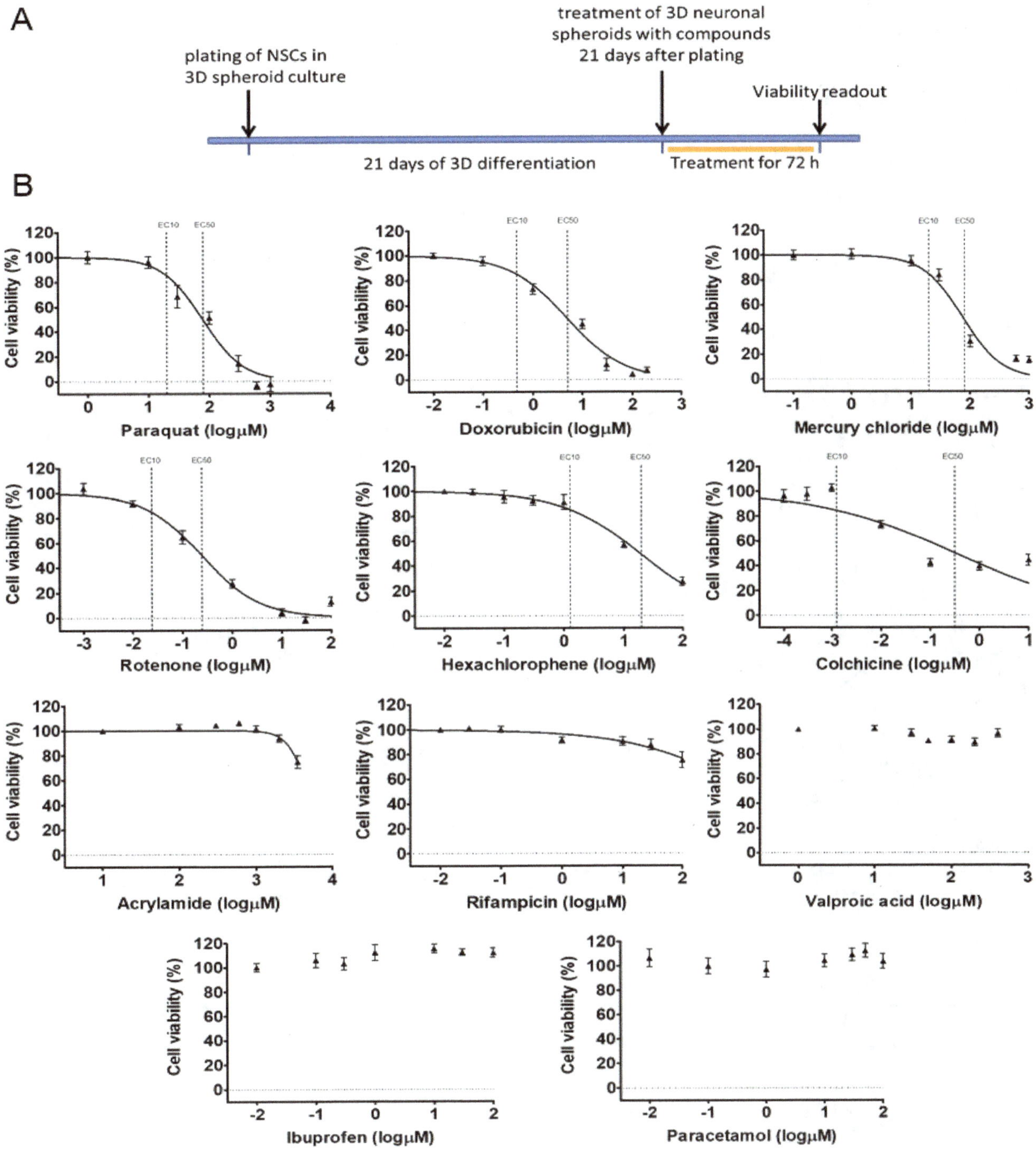

Figure 5. Cell viability measurement on D21 3D neurospheres after 72 h exposure. (**A**) Exposure scheme at D21 stage. (**B**) Concentration response curves of compounds, tested in 7 different concentrations (see concentrations listed in Table S3), representing the cell viability (%) of treated D21 3D neurospheres (n = 3). Concentration values are presented in log μM ± SEM. EC10 and EC50 values are presented on graphs where applicable.

3.3. *Different Age of the 3D Neurospheres Represent Distinct Differentiation Stages in the Cytotoxicity Model*

Next, we investigated the concentration-response of different differentiation stage-derived 3D neurospheres. D28 and D42 samples were analyzed in exposure schemes similar to that of the D21 samples (Figure 6A or Figure 7B) for all the previously tested compounds. As detailed above, in D28 samples differentiated astrocytes expressing GFAP and AQP4 are present (Figure 2 and Figure S4), axonal outgrowth is prevalent, and neuronal subtype-specific proteins start to appear (Figures 2–4). The D42 samples represent a more mature cell culture where synapsis of neurons was formed with established subtypes, mature astrocytes and a few oligodendrocytes are already present (Figure 2, Figure 4, Figures S3 and S4).

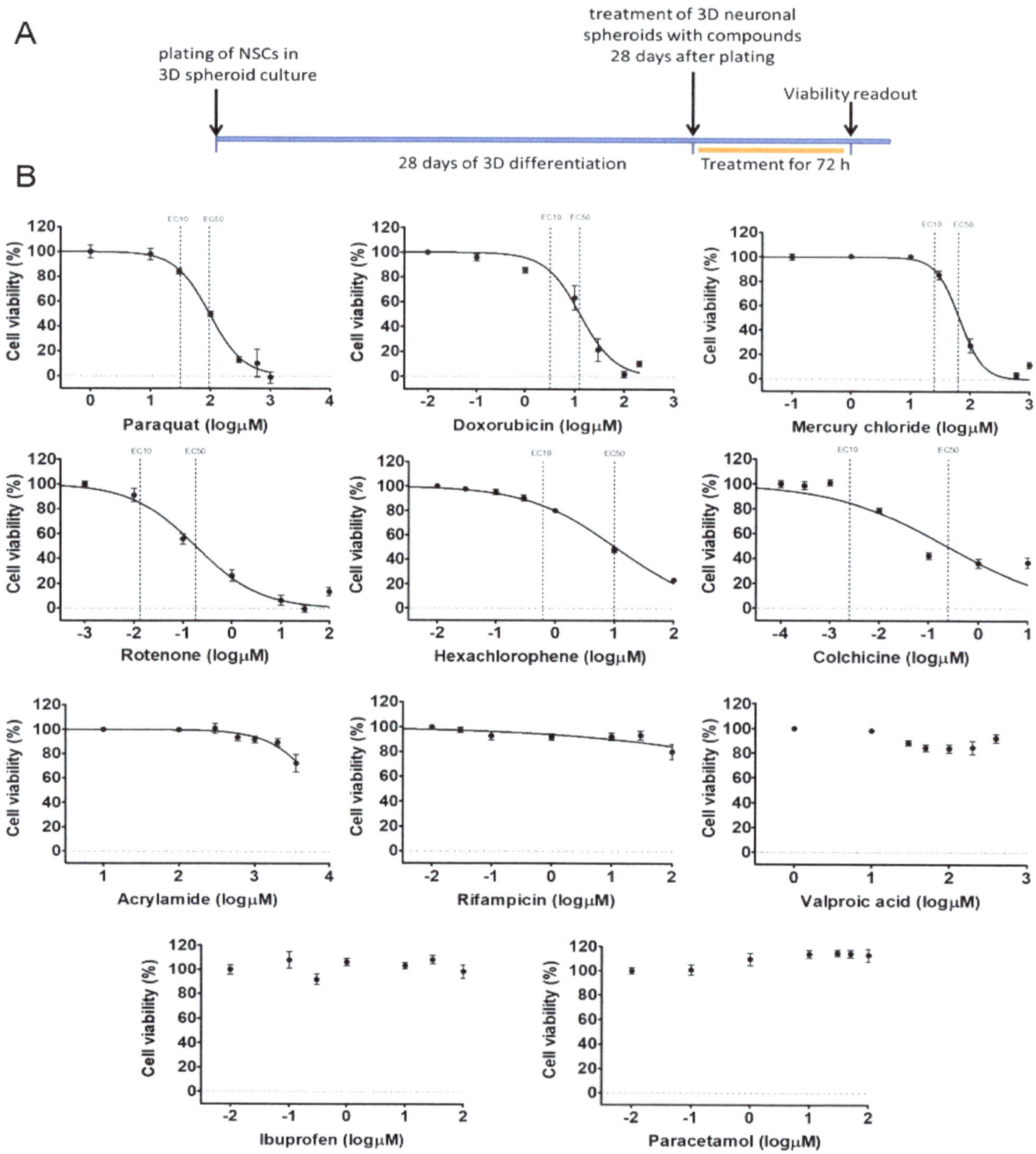

Figure 6. Cell viability measurement on D28 3D neurospheres after 72 h exposure with toxicants. (**A**) exposure scheme at D28 stage. (**B**) Concentration-response curves of compounds, tested in 7 different concentrations (see concentrations listed in Table S3), representing the cell viability (%) of treated D28 3D neurospheres (n = 3). Concentration values are presented in log μM ± SEM. EC10 and EC50 values are presented on graphs where applicable.

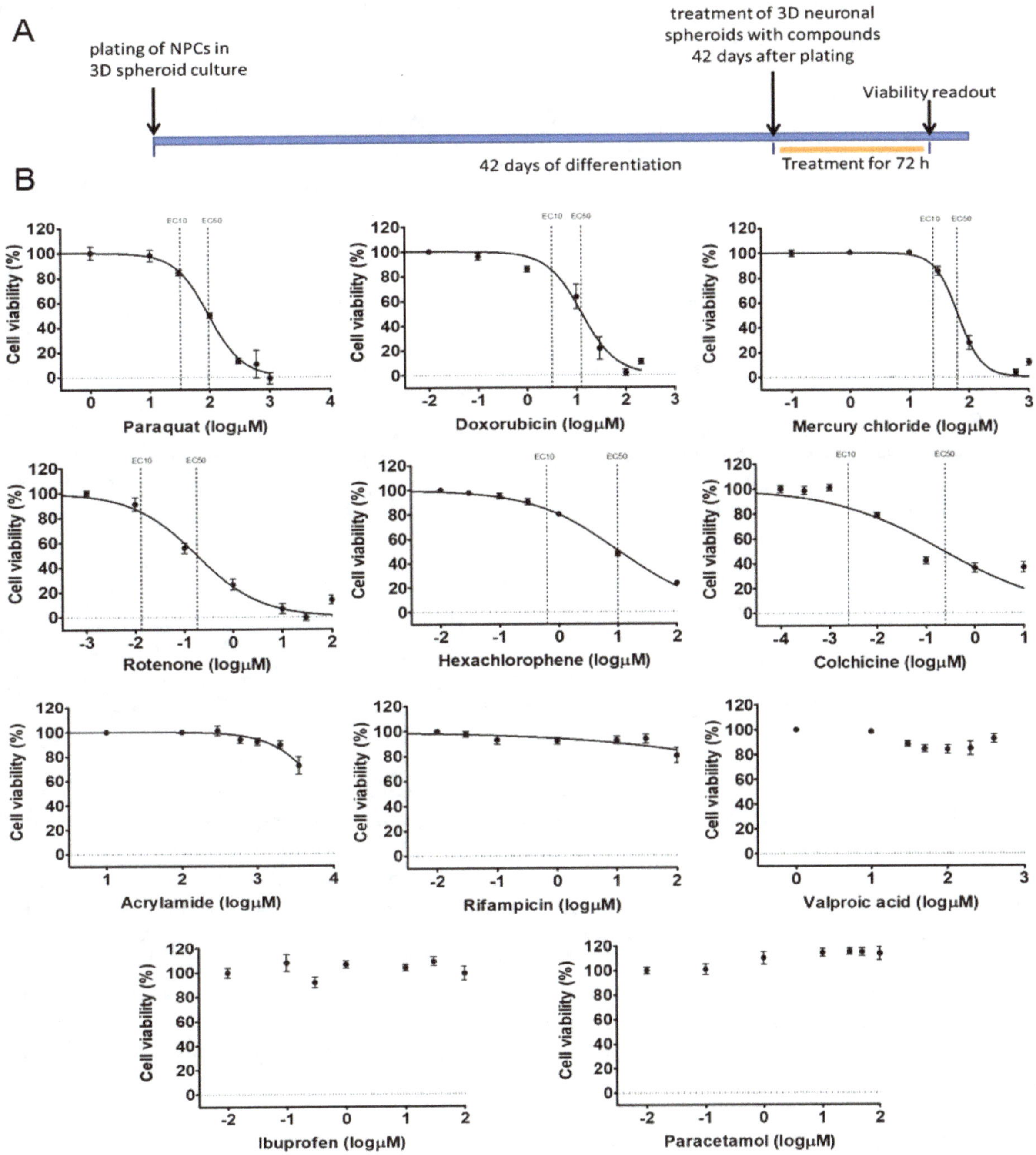

Figure 7. Cell viability measurement on D42 3D neurospheres after 72 h exposure with toxicants. (**A**) exposure scheme at D42 stage. (**B**) Concentration-response curves of compounds, tested in 7 different concentrations (see concentrations listed in Table S3), representing the cell viability (%) of treated D42 3D neurospheres ($n = 3$). Concentration values are presented in log μM ± SEM. EC10 and EC50 values are presented on graphs where applicable.

The experiments showed that both D28 and D42 stages are suitable for the viability assays without modifications in the culture system, using a 96-well plate format-based analysis. The EC50 and EC10 values for D28 and D42 samples were determined for each compound, as described above for D21 samples (Figure 6B or Figure 7B, Table S4). Some compounds (Figures 5–7, Table S4) showed less toxic effect on more matured samples (e.g., ACR: $EC50^{D21} = 4.1$ μM vs. $EC50^{D42} = 3.6$ μM) while it was the opposite action for other compounds (e.g., $HgCl_2$: $EC50^{D21} = 1.7$ μM vs. $EC50^{D42} = 1.9$ μM).

Based on the data obtained, a concentration-related ranking of the toxicant could be generated for each differentiation stage (Figure 8, Table S4).

A

Ranking of compounds based on EC50 values			
Rank	D21	D28	D42
1	Colchicine	Colchicine	Colchicine
2	Rotenone	Rotenone	Rotenone
3	Doxorubicin	Doxorubicin	Hexachlorophene
4	Hexachlorophene	Hexachlorophene	Doxorubicin
5	Paraquat	Mercury chloride	Paraquat
6	Mercury chloride	Paraquat	Mercury chloride
7	Paracetamol	Paracetamol	Paracetamol
8	Ibuprofen	Ibuprofen	Ibuprofen
9	Rifampicin	Rifampicin	Rifampicin
10	Valproic acid	Valproic acid	Valproic acid
11	Acrylamide	Acrylamide	Acrylamide

B

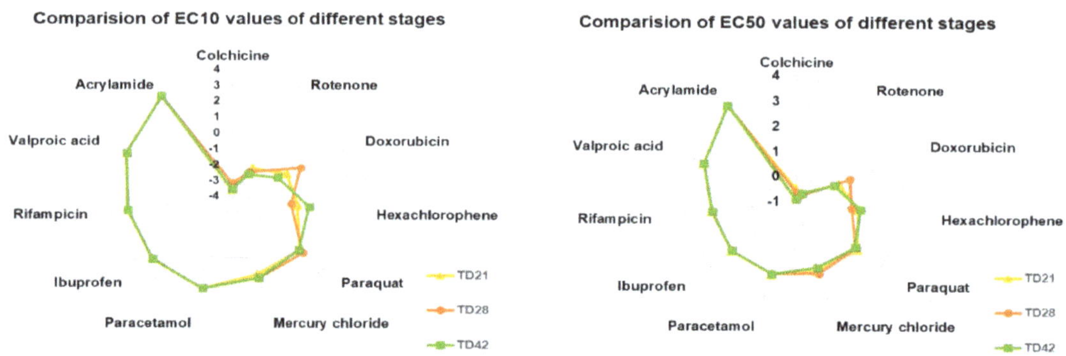

Figure 8. Comparison of the toxic effect of compounds. (**A**) Ranking of compounds at different differentiation stages based on EC50 values. Darker background color represents higher toxicity, while white color represents non-toxic compounds. (**B**) Radar chart showing the comparison of compounds based on the order of EC10 or EC50 values, at various differentiation stages (yellow line with rectangular: D21; orange line with circle: D28; green line with square: D42) (see also EC values in Table S4). Concentration values are presented in log μM.

In conclusion, we can state that a compound-specific concentration-response was detected in all stages. At the same time, a differentiation stage-relevant difference was observed, suggesting that the in vitro system can mimic the differential responses of the developing fetal human neuronal tissue, caused by distinct toxicants.

3.4. Compound-Specific Cellular Events Can be Detected in the 3D Neurospheres

ATP-based viability assay can detect cell death but is unsuitable for detecting other specific cellular events. Due to the lack of validated tests for 3D tissues, we decided to investigate the effect of a toxic compound at a subcellular level. We have chosen rotenone (ROT), a well-studied neurotoxic compound, known to interfere with the electron transport chain in mitochondria, to investigate the ultrastructure of the cells and especially mitochondria, followed by TEM. In parallel, a TUNEL assay was also applied to detect cell death. Based on the previously determined concentration-response curves and EC50 values, we treated the 3D neurospheres with 0.5 μM ROT for 72 h and analyzed the samples in the 3 differentiation stages. The data showed that ROT significantly increased the cell death in the spheroids (Figure 9A or Figure S6), compared to the vehicle-treated controls (Figure 9B), in accordance with the results of the ATP measurement (see Figure 5B or Figure 6B). Moreover, TEM revealed a change in the ultrastructure of the mitochondrial inner membrane, the cristae: Both shape and complexity were changed in the ROT treated samples, compared to the controls. Two types of mitochondria were detected, often in the same cell. Organelles with darker matrix were usually narrow and elongated, whereas lighter ones were more rounded. In control cells, cristae were straight, narrow, and long in both types. Matrix granules were observable more frequent in darker than in lighter mitochondria (Figure 9Ca–c). Effect of rotenone treatment was different on the two types of mitochondria. In darker organelles, cristae swelled and formed swirls. Several wide crista junctions became identifiable. In lighter mitochondria, cristae disintegrated and often became unrecognizable (Figure 9Cd–i). Matrix granules disappeared or showed decreased density (Figure 9Cc,i).

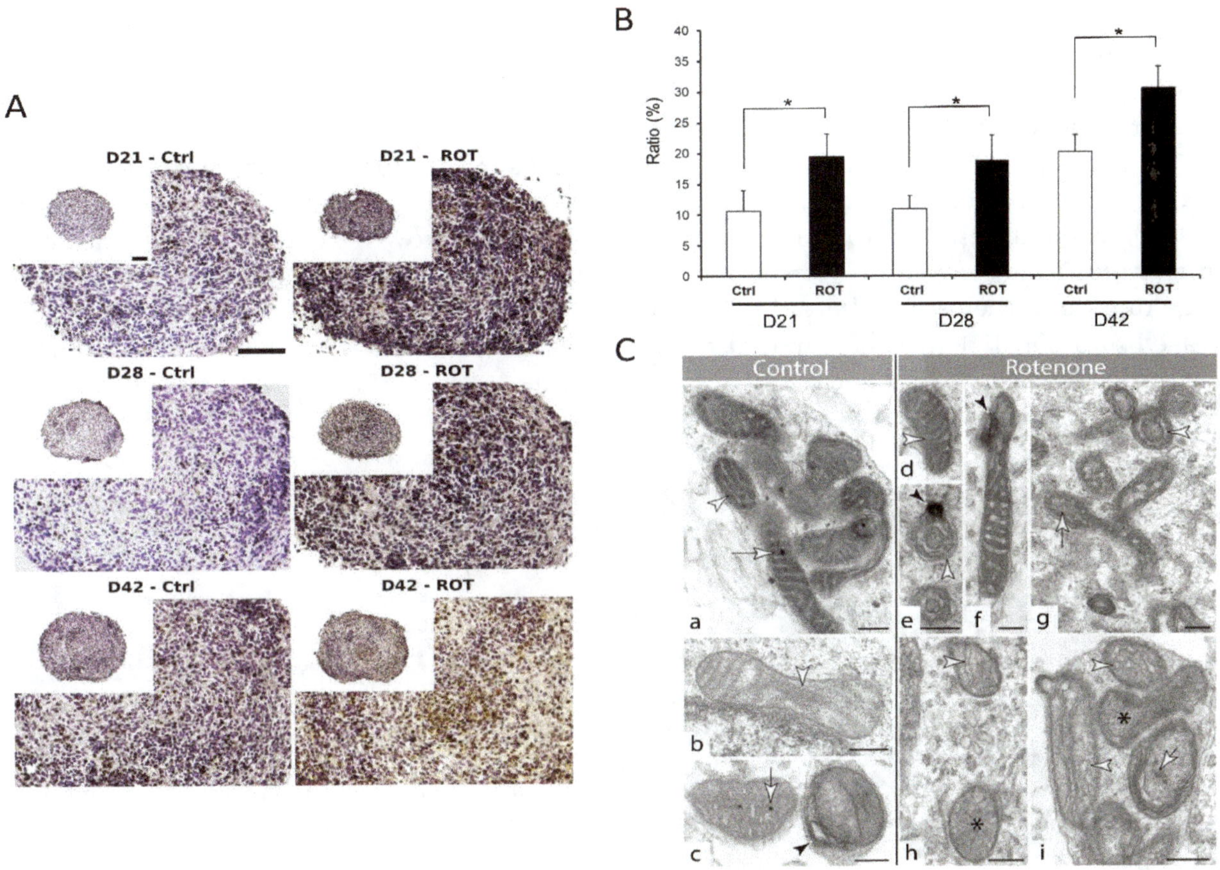

Figure 9. Effect of Rotenone (ROT) exposure on 3D neurospheres at three differentiation stages. (**A**) 3D neurospheres were treated with 0.5 μM ROT concentration for 72 h at three different maturation timepoints (D21, D28, and D42), fixed, sectioned and analyzed to detect the cellular effect of ROT by TUNEL assay (DeadEnd™ Colorimetric TUNEL System, Promega), compared to the vehicle (0.1% DMSO) treated control (scale bar: 100 μm). (**B**) ROT treatment revealed in average a 15% increase in the apoptotic cell number compared to the control in each stage (* $p < 0.05$). Average values are presented on graphs ($n = 3$). (**C**) Ultrastructure of mitochondria in control (panel a, b, c) and ROT treated (d–i) neurons in the 3D spheroids. See the alteration of the internal membrane (cristae) morphology (white arrowheads) in ROT treated cells (d, e, h, i). Black star: unidentifiable cristae morphology in lighter mitochondria (h, i); black arrowhead: membrane swirl in darker organelles (e, f); white arrows: matrix with and without matrix granules (control cells: a, c; treated cells: g, i) Note the density difference between these granules in control (c) and ROT treated mitochondria (panel i) (D21: a, d, f, h; D28: i; D42: b, c, g) (scale bar: 250 nm).

Finally, we used an ER-stress assay to detect if the observed change is compound-specific or an "overall" cell death event is detected with the ATP or TUNEL assay. The XBP1 assay was performed to identify ER-stress in the ROT treated samples, where ER stress was not expected. As a positive control, tunicamycin was used. The XBP1 assays demonstrated that ROT (using EC50 concentration) has no ER-stress inducing effect in the D21, D28 or D42 neurospheres (Figure S6B), while cell death was observed in the cryosectioned ROT-treated samples (Figure 9A), providing a strong evidence that compound-specific effects can be determined in 3D spheroids upon treatment. Immunocytochemical investigation reflected some disorganization of the treated group of spheroids compared to the controls, however other marked differences were not observed (Figure S6A).

3.5. Neurite Outgrowth Assay is Suitable to Determine the Effect of NT Compounds in the 3D Model

Functional readouts which connect the effect of a tested compound to a given cell type are essential to determine their tissue-specific effect. 3D spheroids represent a complex cell culture with a network of neurons, astrocytes, and oligodendrocytes, and this complexity makes more challenging to analyze minor differences in neurite length affected by chemical exposure due to the limitations of the current detection systems. Therefore, we investigated if neurite growth could be investigated using 3D neurospheres. We found that Matrigel-coated surface sufficiently supports the rapid and strong attachment of free-floating D21 3D neurospheres in order to generate a robust and reproducible procedure, where the effect of toxic compounds on neurite outgrowth can be investigated. The effect of different ROT concentrations were demonstrated comparing the total neurite outgrowth and the average number of neurites/spheroid ratio. ROT administration resulted significantly shorter and decreased number of neurites compared to the untreated and vehicle-treated control groups in all concentration levels in D21 spheroids (** $p < 0.01$) (Figure 10B).

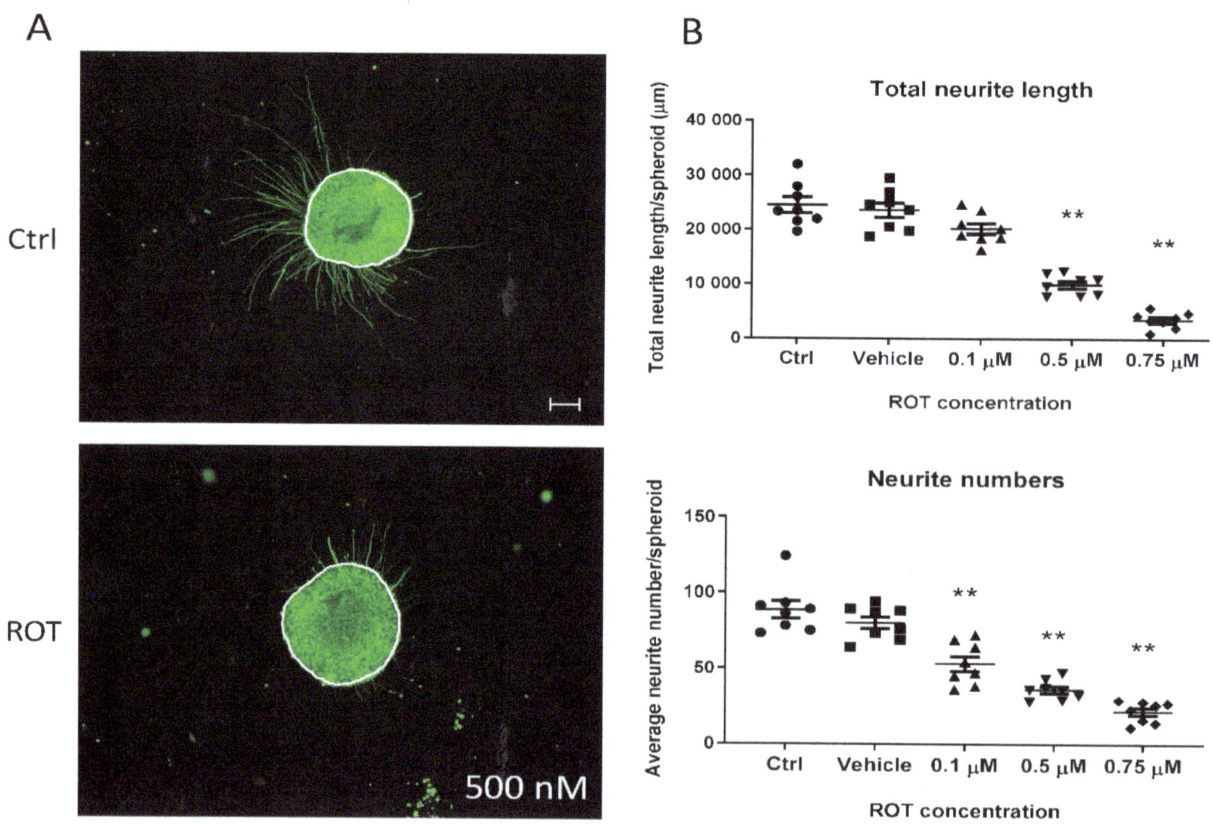

Figure 10. Neurite outgrowth measurement on D21 3D spheroids, exposed for 72 h with ROT. (**A**) Representative photograph of control (untreated) and ROT (0.5 µM) treated D21 spheroid immunolabeled with TUBB3 (in green). White lines represent the border of the spheroids where the neurite outgrowth was determined from, using ImageJ software (scale bar: 100 µm). (**B**) Total neurite length/spheroid (presented in µm ± SEM) and the average number of neurites/spheroids were determined 24 h after plating the treated spheroids. Different symbols denote treatment groups. ($n = 3$, in each experiment 8 spheroids were treated in each group) (** $p < 0.01$).

4. Discussion

The development of in vitro platforms for neurotoxicology screenings is driven by the urgent needs of the chemical, food, cosmetic, and pharma industries. Most NT studies are carried out in rodents or rodent derived primary cells, resulting in relatively high cost and lower translational value of the results due to the species differences [3,46]. Major international initiatives have started to convert the traditional animal-based neurotoxicity tests to in vitro assays using both mammalian brain cells and human cells to detect and predict chemical hazards [47,48]. However, there is only a limited number of human neuronal cell lines (e.g., carcinoma cell lines such as SH-SY5Y; BE2-M17 or immortalized cell lines like LUHMES) are available and hard to obtain primary human CNS tissue suitable for NT studies. Overall, the highly complex structure of the human brain makes in vitro modeling very difficult. Human iPSCs could fill this niche and offer the advantage that other cells and tissue types (e.g., kidney, liver, cardiac, neuronal, intestinal) sharing the same individual genetic background can be created using specific differentiation protocols in a replicable manner. This may provide a very effective in vitro tool for toxicologist for capturing the individual variability in the human population [49]. For example, in 2D neurotoxicity screening, the usage of human iPSC derived neuronal cultures, especially the commercially available QC controlled neuron and astrocyte cultures, where the differentiation of iPSCs is not required for the "users", is dynamically increasing [48].

In recent years, numerous in vitro models have been created in order to study the human CNS in a more physiological way, but the field of NT and DNT using 3D neuronal tissues did not develop as fast as disease or developmental modeling. It is due to the difficulty to find a compromise between biological complexity and technical reproducibility which are necessary for drug or toxicity screening [29].

An important approach was the application of human cell lines in NT assay development. For example, the LUHMES immortalized human fetal tissue-derived mesencephalic cell line can be efficiently differentiated towards dopamine-like neurons upon tetracycline administration [50,51] and provide a suitable system for neurotoxicology screenings [52,53] or Parkinson's disease-related drug testing [54] both in 2D or 3D cultures [45,55–57]. Despite the advantages in straightforward handling, they represent only one specific cell type of the CNS. New developments using stem cell-derived astrocytes or microglia in co-culture with LUHMES cells could reveal new potential both in NT and DNT tests [58], but also highlight that full-PSC derived systems could provide great advantages over the conventional cell lines.

To develop an efficient, highly reproducible neurotoxicology test system, an improved 3D culture of microtissues differentiated from human PSCs would be beneficial. For example, in neuronal disease modeling, a major step that opened new perspectives, was the development of cortical layer-organized 3D brain microtissues, providing a complex system forming under in vitro conditions from human PSCs [59,60]. However, for toxicological studies, the very low throughput potential of such complex systems is a significantly limiting factor at the moment.

Huang and his colleagues demonstrated that human brain organoids could be applied as an in vitro model for CNS drug screening to evaluate structural, cellular, and molecular changes. They used neurotoxic tranylcypromine in hiPSC-derived brain organoids leading to decreased proliferation activity and apoptosis induction [61]. HiPSC-derived cerebral organoids were treated with different concentrations of vincristine for 48 h, and the expansion of the treated organoids was measured, showing concentration-dependent neurotoxicity. Vincristine inhibited fibronectin, tubulin, and MMP10 expression in the cerebral organoids, which was specific for its well-known effect on microtubule dynamics [62]. Recently, iPSC-derived cortical neurons and astrocytes were co-cultured in 3D to detect calcium oscillations upon a chemical compound treatment, analyzing multiple parameters, highlighting the potential of such readouts in neurotoxicity assessment [31]. Another study has presented a novel 3D heterotypic glioblastoma-brain sphere (gBS) model applied for screening new anti-glioblastoma agents [63]. This new application highlighted the flexibility of iPSC-derived platforms to be used in disease modeling, drug testing, and toxicology.

In this study, we have analyzed the complexity of our hiPSC-derived 3D neurosphere system demonstrating intensive glial-neuron interaction with the astrocytes and oligodendrocytes present in the neurospheres. Complex characterization was performed with gene expression, protein level analysis in line with immunocytochemical investigations, including morphological evidence of neurocrine communication at the end of the examined period. Morphologically, synapses are composed of precisely opposed pre- and postsynaptic membranes decorated with electron-dense thickening, synaptic cleft filled with a fine meshwork of electron-dense material, and presynaptic vesicles [64]. One of the two populations of presynaptic vesicles belong to the active zone matrix, is the docked vesicles [65]. The membrane of them is in contact with the presynaptic membrane, therefore they form the readily releasable pool (RRP) of synaptic vesicles. [66].

Our system showed similarities in morphology and maturation properties with a previous study where 3D organoids were generated and characterized over 8 weeks [29]. In our system, there was no need to use BDNF, GDNF, or a special electrophysiology media; a basal medium was sufficient to promote the differentiation of complex spheroids within 6 weeks of culture. Although the spheroids showed a continuous growth and development/maturation, the homogeneity of the plates was not compromised, both the diameter and total protein content of the spheroids within the 96-well plate showed very low variation at a given time point, represented by the low SEM values (Figure 1). Low variability of samples and high homogeneity is crucial when the aim is the development of a reliable HTS for DNT studies.

Here, we performed a medium-throughput 96-well plate assay on 3D neurospheres to detect the cytotoxic effect of the selected compounds, based on ATP-release measurements. We tested drugs, pesticides and well-known chemicals with or without neurotoxic effect in different stages of the neuronal differentiation. Our results highlighted a compound-specific and differentiation stage-related effect of the tested chemicals, providing the possibility to determine EC50 and EC10 values for the compounds (Table S3). For example, colchicine has a strong toxic effect on neurospheres with very similar EC50 values in all differentiation stages. Its effect can be detrimental, it blocks basic cellular (protein assembly, endocytosis, exocytosis, cellular motility, etc.) and neuronal functions at the same time (assembly of tubulin filaments in the microtubules of neurites), which we supposed to happen in all differentiation stages in our neurospheres. A comparable neurotoxic effect was described on iPSC derived 2D cultures of NSCs, neurons, and astrocytes [67], and LUHMES cells [57]. By comparing EC10 or EC50 values, another example is doxorubicin, where D42 samples were the most sensitive for the treatment when intensive protein synthesis could happen both in neurons and glial cells (e.g., neurotransmitter synthesis, axonal growth, oligodendrocyte maturation), therefore the potential blocking of the transcription machinery might have a major effect on cell viability (see Table 1 for a summary).

Table 1. Known effects of the used compounds.

Compound Name (CAS Number)	Known Effects of the Compound
Acrylamide 79-06-1	- Water-soluble crystalline amide that polymerizes rapidly; - widely used in chemical industry (e.g., water treatment industry, paper industry, textile treatment industry) and cosmetics; - occurs in food upon heat treatment—both during home cooking or industrial processing of food— [68]; - neurotoxic: occupational exposures [69,70]; - could cross the placental barrier and appears in breast milk [71]; and - probable oral lethal human dose is between 50–500 mg/kg.

Table 1. *Cont.*

Compound Name (CAS Number)	Known Effects of the Compound
Colchicine 64-86-8	- Can bind to tubulin and inhibit tubulin polymerization leading to inhibition in mitosis; - interacts with the P-glycoprotein transporter (MDR1/ABCB1) and the CYP3A4 enzyme (both involved in toxin metabolism) [72,73]; - medication used to treat gout and Behçet's disease; - probable oral lethal dose in humans is less than 5 mg/kg; and - in vitro cytotoxicity limit: 0.02 μM.
Doxorubicin 25316-40-9	- Intercalating chemotherapy drug; - inhibiting the movement of topoisomerase II, which leads to the blocking of both replication and transcription; and - LD50: 21.8 mg/kg (rat, subcutaneous).
Hexachlorophene 70-30-4	- Disinfectant; - blocks the electron transport chain through acting on membrane-anchoring subunit of succinate dehydrogenase (SDHD) and Glutamate dehydrogenase 1, mitochondrial (GLUD1); - probable oral lethal dose in humans is not determined; LD50: 66 mg/kg (rat, oral); and - in vitro cytotoxicity limit: 1.86 μM.
Ibuprofen 15687-27-1	- Drug (pain killer); - effectively reduces fever, as a non-steroid anti-inflammatory drug (NSAID), acts on inhibiting cyclooxygenase (COX) enzymes (COX-1 and 2); and - overdose symptoms appear in individuals consumed more than 99 mg/kg; LD50: 636 mg/kg (rat, oral).
Mercury(II) chloride 7487-94-7	- Component of pesticides; - corrosive, toxic; - could accumulate in the kidney; - probable oral lethal dose is 5–50 mg/kg; and - in vitro cytotoxicity limit: 1.37 μM.
Paracetamol 103-90-2	- Drug (pain killer); - ioverdosing could cause liver toxicity; and - hepatic toxicity in humans occurred with acute overdoses more than 10 g; LD50: 2400 mg/kg (rat, oral).
Paraquat 7 5365-73-0	- Herbicide; - neurotoxic: occupational exposures leading to Parkinson's disease [74]; - widely investigated neurotoxic mechanism [75,76]; and - probable oral lethal dose in humans is 35 mg/kg.
Rifampicin 13292-46-1	- Antibiotic; - stops RNA synthesis in bacteria; - could cause liver toxicity; - robable lethal oral dose in humans is 14–60 gLD50: 1570 mg/kg (rat); and - in vitro cytotoxicity limit: 4.37 μM.

Table 1. *Cont.*

Compound Name (CAS Number)	Known Effects of the Compound
Rotenone 83-79-4	- Pesticide; - inhibits mitochondrial Complex I of the electron transport chain; - in rats Parkinson's disease- like symptoms were developed, →considered as an environmental risk factor for PD [77]; - probable lethal dose in human 0.3–0.5 g/kg; and - in vitro cytotoxicity limit: 0.22 μM.
Valproic acid 1069-66-5	- Drug used in epilepsy, bipolar disorders, and for the prevention of seizures; - known to block voltage-gated sodium channels and increase brain levels of gamma-aminobutyric acid (GABA); - well-known teratogen induces different congenital malformations and neural tube defects (NTDs); and - LD50: 670 mg/kg (rat, oral).

Note: probable lethal dose data is provided upon PubChem database (https://pubchem.ncbi.nlm.nih.gov/), in vitro multicellular cytotoxicity "comptox" data were collected from EPA's Chemical dashboard (https://comptox.epa.gov/dashboard).

Continuing the line with hexachlorophene that blocks the electron transport chain, by acting on GLUD1, it affects the turnover of an excitatory neurotransmitter, glutamate. This effect could explain the observed differentiation stage-related sensitivity differences of the neurospheres, still under an acute exposure scheme. Likewise, less cytotoxic effect was reported on iPSC derived 2D cultures of NSCs, early neurons (14 days old) and astrocytes when 10 or 100 μM HE was exposed for 24 h [67], while more pronounced cytotoxicity and neurite outgrowth inhibition of LUHMES cells were reported [57].

Valproate is known as a DNT positive compound. However, in our cell viability assay, it has no detectable impact on the investigated concentration range, which might be in correlation with the maturity of the treated cells, they are not in early neurodevelopment (neural tube stage), but maturing neurons and astrocytes. When LUHMES cells were exposed the EC50 value of VPA was determined as >100 μM, and significant inhibition of neurite outgrowth was detected. While the EC50 value was in a similar range with our 3D neurosphere-driven data, we did not investigate its effect on neurite outgrowth. Here, we also have to clarify that cytotoxicity read-out alone is not suitable to predict DNT, although it is suitable to predict the neurotoxic effect of compounds or detect necrosis in differentiating cultures, more relevant end-points (e.g., proliferation, migration, apoptosis, network formation, synaptogenesis, and growth of neurites) must be investigated for evaluating the DNT effect of a compound [18].

In the case of pesticide rotenone, which inhibits mitochondrial Complex I of the electron transport chain, strong cell death was detected. The effect of ROT exposure was investigated extensively. In iPSC-derived neuron-astroglia 2D cell cultures the activation of the Nrf2/ARE pathway was documented upon ROT-induced oxidative stress, which led to the activation of astrocytes and cell death of neurons [67,78,79]. Using LUHMES cells this robust cytotoxic effect was not obvious while a prominent neurite growth inhibition was reported [57]. Using "BrainSpheres" system [30], where ICC-based morphology analysis, ROS measurements and viability were compared in different developmental times and exposure schemes, dopaminergic-neuron selective toxicity and general cytotoxic concentration were determined [30]. Importantly, increased cell death and mitochondrial dysfunction were detected in our ROT exposed 3D neurospheres, similar to others' results [30]. In contrast, ibuprofen has no effect on neuronal cell viability in the investigated concentration range, as it was expected upon known in vitro and in vivo data.

Neurite outgrowth assays are well established in conventional 2D neuronal cultures, derived from cell lines, primary tissues, or iPSCs [57,80,81], however, the adaptation to 3D cultures is less tested. Determining the radial migration and neuronal density distribution within the migration area of NPCs grown as spheroids is rather described [27]. Here we adapted a simple system which provides a tissue-specific functional readout for neurospheres upon exposure, which can be automated using high content imaging (HCI) systems [45]. Based on our results, it is sensitive enough to distinguish different concentration of toxic compounds, does not require complicated read-out systems which was proved by testing the effect of ROT, previously confirmed to inhibit neurite growth of LUHMES cells [57]. Although these results agree with the reported neurite outgrowth inhibition effect, we cannot exclude the influence of cytotoxicity on the reduction of the overall neurite length we observed. Combining with LDH or lactate sampling, metabolite measurements from the media or terminal transcriptomic or proteomic assays, a complex dataset can be collected, and the effect of a given compound can be investigated systematically on the molecular level, as well to identify adverse effects. Similar to the RT-qPCR and ICC investigation that we used to characterize our 3D neurospheres these methods can be used when the effect of a given compound is investigated. Although we performed immunocytochemistry on treated spheroids using rotenone only, others successfully used such investigations to gain DNT readouts, for example detecting cell migration, neurite outgrowth, Ca^{2+} reabsorption, synaptogenesis, and PPAR pathway disruption [29,30,57,82,83].

Evaluating our concentration–response results, we faced with the lack of available in vitro human data, which makes the comparison challenging. This demonstrates that due to the lack of EC values and very different readouts, the appropriate comparison between the various test systems is often not possible. As summarized in Table 1, it is not possible to provide human in vitro NT values for every compound, the databases are incomplete. Although there are several initiatives and efforts on the field to collect and harmonize the available test methods and evaluate the data produced so far, there is still a lot to do in the field of in vitro neurotoxicity test method development [1,16].

5. Conclusions

We have established and evaluated a hiPSC-based 3D in vitro cellular model for the study of the neurotoxic effect of different compounds. Concentration–response was determined in different differentiation stages, using a well-known read-out the cell viability. For a certain compound, apoptotic activity, ER-stress assay and TEM was performed, providing possibilities to generate novel human CNS-relevant data for other compounds using hiPSC derived cells. Tissue-specific readouts such as neurite outgrowth were also investigated, however further, read-outs and DNT-relevant training compound set should be applied to evaluate the suitability of our 3D spheroids in DNT screening Nevertheless, due to the pluripotent nature of the hiPSC, this model offers an excellent tool for drug testing, gene therapy studies and toxicology studies parallelly on the same genotype using other cell or tissue types at the same time. Moreover, new advancement in gene manipulation such as CRISPR/Cas9 mediated gene targeting, makes it possible to target specific pathways and generate reporter cell lines for toxicological or other applications. Combining these new approaches with 3D cell culture-based assays could revolutionize the field of toxicology including DNT studies in the near future.

Author Contributions: Data curation, M.Z. and I.B.; Funding acquisition, A.D.; Investigation, J.K., A.T., T.B., Z.J., K.M., I.B. and L.L.; Supervision, L.L. and A.D.; Writing—original draft, J.K.; Writing—review and editing, M.Z., L.L. and A.D. All authors have read and agreed to the published version of the manuscript.

Acknowledgments: We are grateful for the help of Balázs Mihalik in study design and assay development. We thank Mónika Truszka for the technical help in the preparation of samples for TEM.

Abbreviations

3D	three dimensional;
AO	adverse outcome;
BSA	bovine serum albumin;
CAS	chemical abstracts service;
CNS	central nervous system;
Ctrl	control;
D	day;
DNT	developmental neurotoxicity;
EC	effective concentration;
ER	endoplasmic reticulum;
FBS	fetal bovine serum;
HE	Hematoxylin-Eosin;
HTS	high-throughput screening;
hiPSC	human induced pluripotent stem cell;
ICC	immunocytochemistry;
iPSC	induced pluripotent stem cell;
NA	numeric aperture;
NAM	new approach method;
NEAA	non-essential amino acids;
NIM	neural induction media;
NMM	neural maintenance media;
NPC	neural progenitor cell;
NSC	neural stem cell;
O/N	overnight;
PBMC	peripheral blood mononuclear cells;
PBS	phosphate buffer saline;
PCR	polymerase chain reaction;
PFA	Paraformaldehyde;
POL/L	Poly-L-ornithine and Laminin;
PSC	pluripotent stem cell;
RT	room temperature;
rTdT	recombinant terminal deoxynucleotidyl transferase;
RT-PCR	reverse transcription polymerase chain reaction;
TEM	transmission electron microscopy;
TUNEL	terminal deoxynucleotidyl transferase dUTP nick end labeling.

References

1. Krewski, D.; Acosta, D., Jr.; Andersen, M.; Anderson, H.; Bailar, J.C., 3rd; Boekelheide, K.; Brent, R.; Charnley, G.; Cheung, V.G.; Green, S., Jr.; et al. Toxicity testing in the 21st century: A vision and a strategy. *J. Toxicol. Environ. Health Part B Crit. Rev.* **2010**, *13*, 51–138. [CrossRef]
2. Tsuji, R.; Crofton, K.M. Developmental neurotoxicity guideline study: Issues with methodology, evaluation and regulation. *Congenit. Anom. (Kyoto)* **2012**, *52*, 122–128. [CrossRef]
3. Hartung, T. Thoughts on limitations of animal models. *Parkinsonism Relat. Disord.* **2008**, *14*, S81–S83. [CrossRef] [PubMed]
4. Crofton, K.M.; Mundy, W.R.; Shafer, T.J. Developmental neurotoxicity testing: A path forward. *Congenit. Anom. (Kyoto)* **2012**, *52*, 140–146. [CrossRef] [PubMed]
5. Bal-Price, A.; Crofton, K.M.; Leist, M.; Allen, S.; Arand, M.; Buetler, T.; Delrue, N.; FitzGerald, R.E.; Hartung, T.; Heinonen, T.; et al. International STakeholder NETwork (ISTNET): Creating a developmental neurotoxicity (DNT) testing road map for regulatory purposes. *Arch. Toxicol.* **2015**, *89*, 269–287. [CrossRef] [PubMed]
6. Bal-Price, A.; Hogberg, H.T.; Crofton, K.M.; Daneshian, M.; FitzGerald, R.E.; Fritsche, E.; Heinonen, T.; Hougaard Bennekou, S.; Klima, S.; Piersma, A.H.; et al. Recommendation on test readiness criteria for new approach methods in toxicology: Exemplified for developmental neurotoxicity. *ALTEX* **2018**, *35*, 306–352. [CrossRef]
7. Westerink, R.H. Do we really want to REACH out to in vitro? *Neurotoxicology* **2013**, *39*, 169–172. [CrossRef]
8. Bal-Price, A.K.; Hogberg, H.T.; Buzanska, L.; Lenas, P.; van Vliet, E.; Hartung, T. In vitro developmental neurotoxicity (DNT) testing: Relevant models and endpoints. *Neurotoxicology* **2010**, *31*, 545–554. [CrossRef]
9. Sachana, M.; Rolaki, A.; Bal-Price, A. Development of the Adverse Outcome Pathway (AOP): Chronic binding of antagonist to N-methyl-d-aspartate receptors (NMDARs) during brain development induces impairment of learning and memory abilities of children. *Toxicol. Appl. Pharmacol.* **2018**, *354*, 153–175. [CrossRef]
10. Fritsche, E.; Grandjean, P.; Crofton, K.M.; Aschner, M.; Goldberg, A.; Heinonen, T.; Hessel, E.V.S.; Hogberg, H.T.; Bennekou, S.H.; Lein, P.J.; et al. Consensus statement on the need for innovation, transition and implementation of developmental neurotoxicity (DNT) testing for regulatory purposes. *Toxicol. Appl. Pharmacol.* **2018**, *354*, 3–6. [CrossRef]
11. Harrill, J.A.; Freudenrich, T.; Wallace, K.; Ball, K.; Shafer, T.J.; Mundy, W.R. Testing for developmental neurotoxicity using a battery of in vitro assays for key cellular events in neurodevelopment. *Toxicol. Appl. Pharmacol.* **2018**, *354*, 24–39. [CrossRef] [PubMed]
12. Alepee, N.; Bahinski, A.; Daneshian, M.; De Wever, B.; Fritsche, E.; Goldberg, A.; Hansmann, J.; Hartung, T.; Haycock, J.; Hogberg, H.; et al. State-of-the-art of 3D cultures (organs-on-a-chip) in safety testing and pathophysiology. *ALTEX* **2014**, *31*, 441–477. [CrossRef] [PubMed]
13. Bal-Price, A.K.; Sunol, C.; Weiss, D.G.; van Vliet, E.; Westerink, R.H.; Costa, L.G. Application of in vitro neurotoxicity testing for regulatory purposes: Symposium III summary and research needs. *Neurotoxicology* **2008**, *29*, 520–531. [CrossRef] [PubMed]
14. Pampaloni, F.; Reynaud, E.G.; Stelzer, E.H. The third dimension bridges the gap between cell culture and live tissue. *Nat. Rev. Mol. Cell Biol.* **2007**, *8*, 839–845. [CrossRef]
15. F-W Greiner, J.; Kaltschmidt, B.; Kaltschmidt, C.; Widera, D. Going 3D-Cell Culture Approaches for Stem Cell Research and Therapy. *Curr. Tissue Eng.* **2013**, *2*, 8–19. [CrossRef]
16. Pamies, D.; Hartung, T. 21st Century Cell Culture for 21st Century Toxicology. *Chem. Res. Toxicol.* **2017**, *30*, 43–52. [CrossRef]
17. Edmondson, R.; Broglie, J.J.; Adcock, A.F.; Yang, L. Three-dimensional cell culture systems and their applications in drug discovery and cell-based biosensors. *Assay Drug Dev. Technol.* **2014**, *12*, 207–218. [CrossRef]
18. Bal-Price, A.; Pistollato, F.; Sachana, M.; Bopp, S.K.; Munn, S.; Worth, A. Strategies to improve the regulatory assessment of developmental neurotoxicity (DNT) using in vitro methods. *Toxicol. Appl. Pharmacol.* **2018**, *354*, 7–18. [CrossRef]
19. Smirnova, L.; Hogberg, H.T.; Leist, M.; Hartung, T. Developmental neurotoxicity—Challenges in the 21st century and in vitro opportunities. *ALTEX* **2014**, *31*, 129–156. [CrossRef]

20. Schmidt, B.Z.; Lehmann, M.; Gutbier, S.; Nembo, E.; Noel, S.; Smirnova, L.; Forsby, A.; Hescheler, J.; Avci, H.X.; Hartung, T.; et al. In vitro acute and developmental neurotoxicity screening: An overview of cellular platforms and high-throughput technical possibilities. *Arch. Toxicol.* **2017**, *91*, 1–33. [CrossRef]
21. Hofrichter, M.; Nimtz, L.; Tigges, J.; Kabiri, Y.; Schroter, F.; Royer-Pokora, B.; Hildebrandt, B.; Schmuck, M.; Epanchintsev, A.; Theiss, S.; et al. Comparative performance analysis of human iPSC-derived and primary neural progenitor cells (NPC) grown as neurospheres in vitro. *Stem Cell Res.* **2017**, *25*, 72–82. [CrossRef] [PubMed]
22. Gage, F.H. Mammalian neural stem cells. *Science* **2000**, *287*, 1433–1438. [CrossRef] [PubMed]
23. Breunig, J.J.; Haydar, T.F.; Rakic, P. Neural stem cells: Historical perspective and future prospects. *Neuron* **2011**, *70*, 614–625. [CrossRef] [PubMed]
24. Thomson, J.A.; Itskovitz-Eldor, J.; Shapiro, S.S.; Waknitz, M.A.; Swiergiel, J.J.; Marshall, V.S.; Jones, J.M. Embryonic stem cell lines derived from human blastocysts. *Science* **1998**, *282*, 1145–1147. [CrossRef]
25. Takahashi, K.; Tanabe, K.; Ohnuki, M.; Narita, M.; Ichisaka, T.; Tomoda, K.; Yamanaka, S. Induction of pluripotent stem cells from adult human fibroblasts by defined factors. *Cell* **2007**, *131*, 861–872. [CrossRef]
26. Qian, X.; Nguyen, H.N.; Song, M.M.; Hadiono, C.; Ogden, S.C.; Hammack, C.; Yao, B.; Hamersky, G.R.; Jacob, F.; Zhong, C.; et al. Brain-Region-Specific Organoids Using Mini-bioreactors for Modeling ZIKV Exposure. *Cell* **2016**, *165*, 1238–1254. [CrossRef]
27. Schmuck, M.R.; Temme, T.; Dach, K.; de Boer, D.; Barenys, M.; Bendt, F.; Mosig, A.; Fritsche, E. Omnisphero: A high-content image analysis (HCA) approach for phenotypic developmental neurotoxicity (DNT) screenings of organoid neurosphere cultures in vitro. *Arch. Toxicol.* **2017**, *91*, 2017–2028. [CrossRef]
28. Hogberg, H.T.; Bressler, J.; Christian, K.M.; Harris, G.; Makri, G.; O'Driscoll, C.; Pamies, D.; Smirnova, L.; Wen, Z.; Hartung, T. Toward a 3D model of human brain development for studying gene/environment interactions. *Stem Cell Res. Ther.* **2013**, *4*, S4. [CrossRef]
29. Pamies, D.; Barreras, P.; Block, K.; Makri, G.; Kumar, A.; Wiersma, D.; Smirnova, L.; Zang, C.; Bressler, J.; Christian, K.M.; et al. A human brain microphysiological system derived from induced pluripotent stem cells to study neurological diseases and toxicity. *ALTEX* **2017**, *34*, 362–376. [CrossRef]
30. Pamies, D.; Block, K.; Lau, P.; Gribaldo, L.; Pardo, C.A.; Barreras, P.; Smirnova, L.; Wiersma, D.; Zhao, L.; Harris, G.; et al. Rotenone exerts developmental neurotoxicity in a human brain spheroid model. *Toxicol. Appl. Pharmacol.* **2018**, *354*, 101–114. [CrossRef]
31. Sirenko, O.; Parham, F.; Dea, S.; Sodhi, N.; Biesmans, S.; Mora-Castilla, S.; Ryan, K.; Behl, M.; Chandy, G.; Crittenden, C.; et al. Functional and Mechanistic Neurotoxicity Profiling Using Human iPSC-Derived Neural 3D Cultures. *Toxicol. Sci. Off. J. Soc. Toxicol.* **2019**, *167*, 58–76. [CrossRef] [PubMed]
32. Chandrasekaran, A.; Avci, H.X.; Ochalek, A.; Rosingh, L.N.; Molnar, K.; Laszlo, L.; Bellak, T.; Teglasi, A.; Pesti, K.; Mike, A.; et al. Comparison of 2D and 3D neural induction methods for the generation of neural progenitor cells from human induced pluripotent stem cells. *Stem Cell Res.* **2017**, *25*, 139–151. [CrossRef] [PubMed]
33. Zhou, S.; Szczesna, K.; Ochalek, A.; Kobolak, J.; Varga, E.; Nemes, C.; Chandrasekaran, A.; Rasmussen, M.; Cirera, S.; Hyttel, P.; et al. Neurosphere Based Differentiation of Human iPSC Improves Astrocyte Differentiation. *Stem Cells Int.* **2016**, *2016*, 4937689. [CrossRef] [PubMed]
34. Chambers, S.M.; Fasano, C.A.; Papapetrou, E.P.; Tomishima, M.; Sadelain, M.; Studer, L. Highly efficient neural conversion of human ES and iPS cells by dual inhibition of SMAD signaling. *Nat. Biotechnol.* **2009**, *27*, 275–280. [CrossRef] [PubMed]
35. Shi, Y.; Kirwan, P.; Livesey, F.J. Directed differentiation of human pluripotent stem cells to cerebral cortex neurons and neural networks. *Nat. Protoc.* **2012**, *7*, 1836–1846. [CrossRef] [PubMed]
36. Tieng, V.; Stoppini, L.; Villy, S.; Fathi, M.; Dubois-Dauphin, M.; Krause, K.H. Engineering of midbrain organoids containing long-lived dopaminergic neurons. *Stem Cells Dev.* **2014**, *23*, 1535–1547. [CrossRef]
37. White, D.L.; Mazurkiewicz, J.E.; Barrnett, R.J. A chemical mechanism for tissue staining by osmium tetroxide-ferrocyanide mixtures. *J. Histochem. Cytochem. Off. J. Histochem. Soc.* **1979**, *27*, 1084–1091. [CrossRef]
38. Rozen, S.; Skaletsky, H. Primer3 on the WWW for general users and for biologist programmers. *Bioinform. Methods Protoc.* **1999**, *132*, 365–386. [CrossRef]
39. Zuker, M. Mfold web server for nucleic acid folding and hybridization prediction. *Nucleic Acids Res.* **2003**, *31*, 3406–3415. [CrossRef]

40. Ye, J.; Coulouris, G.; Zaretskaya, I.; Cutcutache, I.; Rozen, S.; Madden, T.L. Primer-BLAST: A tool to design target-specific primers for polymerase chain reaction. *BMC Bioinform.* **2012**, *13*, 134. [CrossRef]
41. Livak, K.J.; Schmittgen, T.D. Analysis of relative gene expression data using real-time quantitative PCR and the 2(-Delta Delta C(T)) Method. *Methods* **2001**, *25*, 402–408. [CrossRef] [PubMed]
42. Yoshida, H.; Haze, K.; Yanagi, H.; Yura, T.; Mori, K. Identification of the cis-acting endoplasmic reticulum stress response element responsible for transcriptional induction of mammalian glucose-regulated proteins. Involvement of basic leucine zipper transcription factors. *J. Biol. Chem.* **1998**, *273*, 33741–33749. [CrossRef] [PubMed]
43. Van Schadewijk, A.; van't Wout, E.F.; Stolk, J.; Hiemstra, P.S. A quantitative method for detection of spliced X-box binding protein-1 (XBP1) mRNA as a measure of endoplasmic reticulum (ER) stress. *Cell Stress Chaperones* **2012**, *17*, 275–279. [CrossRef] [PubMed]
44. Kobolak, J.; Molnar, K.; Varga, E.; Bock, I.; Jezso, B.; Teglasi, A.; Zhou, S.; Lo Giudice, M.; Hoogeveen-Westerveld, M.; Pijnappel, W.P.; et al. Modelling the neuropathology of lysosomal storage disorders through disease-specific human induced pluripotent stem cells. *Exp. Cell Res.* **2019**, *380*, 216–233. [CrossRef] [PubMed]
45. Harris, G.; Eschment, M.; Orozco, S.P.; McCaffery, J.M.; Maclennan, R.; Severin, D.; Leist, M.; Kleensang, A.; Pamies, D.; Maertens, A.; et al. Toxicity, recovery, and resilience in a 3D dopaminergic neuronal in vitro model exposed to rotenone. *Arch. Toxicol.* **2018**, *92*, 2587–2606. [CrossRef] [PubMed]
46. Baumann, J.; Gassmann, K.; Masjosthusmann, S.; DeBoer, D.; Bendt, F.; Giersiefer, S.; Fritsche, E. Comparative human and rat neurospheres reveal species differences in chemical effects on neurodevelopmental key events. *Arch. Toxicol.* **2016**, *90*, 1415–1427. [CrossRef] [PubMed]
47. Tukker, A.M.; de Groot, M.W.; Wijnolts, F.M.; Kasteel, E.E.; Hondebrink, L.; Westerink, R.H. Is the time right for in vitro neurotoxicity testing using human iPSC-derived neurons? *ALTEX* **2016**, *33*, 261–271. [CrossRef]
48. Tukker, A.M.; Wijnolts, F.M.J.; de Groot, A.; Westerink, R.H.S. Human iPSC-derived neuronal models for in vitro neurotoxicity assessment. *Neurotoxicology* **2018**, *67*, 215–225. [CrossRef]
49. Fritsche, E.; Barenys, M.; Klose, J.; Masjosthusmann, S.; Nimtz, L.; Schmuck, M.; Wuttke, S.; Tigges, J. Current Availability of Stem Cell-Based In Vitro Methods for Developmental Neurotoxicity (DNT) Testing. *Toxicol. Sci. Off. J. Soc. Toxicol.* **2018**, *165*, 21–30. [CrossRef]
50. Lotharius, J.; Falsig, J.; van Beek, J.; Payne, S.; Dringen, R.; Brundin, P.; Leist, M. Progressive degeneration of human mesencephalic neuron-derived cells triggered by dopamine-dependent oxidative stress is dependent on the mixed-lineage kinase pathway. *J. Neurosci. Off. J. Soc. Neurosci.* **2005**, *25*, 6329–6342. [CrossRef]
51. Scholz, D.; Poltl, D.; Genewsky, A.; Weng, M.; Waldmann, T.; Schildknecht, S.; Leist, M. Rapid, complete and large-scale generation of post-mitotic neurons from the human LUHMES cell line. *J. Neurochem.* **2011**, *119*, 957–971. [CrossRef] [PubMed]
52. Zuberek, M.; Stepkowski, T.M.; Kruszewski, M.; Grzelak, A. Exposure of human neurons to silver nanoparticles induces similar pattern of ABC transporters gene expression as differentiation: Study on proliferating and post-mitotic LUHMES cells. *Mech. Ageing Dev.* **2018**, *171*, 7–14. [CrossRef] [PubMed]
53. Hollerhage, M.; Moebius, C.; Melms, J.; Chiu, W.H.; Goebel, J.N.; Chakroun, T.; Koeglsperger, T.; Oertel, W.H.; Rosler, T.W.; Bickle, M.; et al. Protective efficacy of phosphodiesterase-1 inhibition against alpha-synuclein toxicity revealed by compound screening in LUHMES cells. *Sci. Rep.* **2017**, *7*, 11469. [CrossRef] [PubMed]
54. Zhang, X.M.; Yin, M.; Zhang, M.H. Cell-based assays for Parkinson's disease using differentiated human LUHMES cells. *Acta Pharmacol. Sin.* **2014**, *35*, 945–956. [CrossRef]
55. Smirnova, L.; Harris, G.; Delp, J.; Valadares, M.; Pamies, D.; Hogberg, H.T.; Waldmann, T.; Leist, M.; Hartung, T. A LUHMES 3D dopaminergic neuronal model for neurotoxicity testing allowing long-term exposure and cellular resilience analysis. *Arch. Toxicol.* **2016**, *90*, 2725–2743. [CrossRef]
56. Tong, Z.B.; Hogberg, H.; Kuo, D.; Sakamuru, S.; Xia, M.; Smirnova, L.; Hartung, T.; Gerhold, D. Characterization of three human cell line models for high-throughput neuronal cytotoxicity screening. *J. Appl. Toxicol. JAT* **2017**, *37*, 167–180. [CrossRef]
57. Delp, J.; Gutbier, S.; Klima, S.; Hoelting, L.; Pinto-Gil, K.; Hsieh, J.H.; Aichem, M.; Klein, K.; Schreiber, F.; Tice, R.R.; et al. A high-throughput approach to identify specific neurotoxicants/developmental toxicants in human neuronal cell function assays. *ALTEX* **2018**, *35*, 235–253. [CrossRef]
58. Brull, M.; Spreng, A.S.; Gutbier, S.; Loser, D.; Krebs, A.; Reich, M.; Kraushaar, U.; Britschgi, M.; Patsch, C.; Leist, M. Incorporation of stem cell-derived astrocytes into neuronal organoids to allow neuro-glial interactions in toxicological studies. *ALTEX* **2020**. [CrossRef]

59. Lancaster, M.A.; Renner, M.; Martin, C.A.; Wenzel, D.; Bicknell, L.S.; Hurles, M.E.; Homfray, T.; Penninger, J.M.; Jackson, A.P.; Knoblich, J.A. Cerebral organoids model human brain development and microcephaly. *Nature* **2013**, *501*, 373–379. [CrossRef]
60. Kadoshima, T.; Sakaguchi, H.; Nakano, T.; Soen, M.; Ando, S.; Eiraku, M.; Sasai, Y. Self-organization of axial polarity, inside-out layer pattern, and species-specific progenitor dynamics in human ES cell-derived neocortex. *Proc. Natl. Acad. Sci. USA* **2013**, *110*, 20284–20289. [CrossRef]
61. Huang, J.; Liu, F.; Tang, H.; Wu, H.; Li, L.; Wu, R.; Zhao, J.; Wu, Y.; Liu, Z.; Chen, J. Tranylcypromine Causes Neurotoxicity and Represses BHC110/LSD1 in Human-Induced Pluripotent Stem Cell-Derived Cerebral Organoids Model. *Front. Neurol.* **2017**, *8*, 626. [CrossRef] [PubMed]
62. Liu, F.; Huang, J.; Liu, Z. Vincristine Impairs Microtubules and Causes Neurotoxicity in Cerebral Organoids. *Neuroscience* **2019**, *404*, 530–540. [CrossRef] [PubMed]
63. Plummer, S.; Wallace, S.; Ball, G.; Lloyd, R.; Schiapparelli, P.; Quinones-Hinojosa, A.; Hartung, T.; Pamies, D. A Human iPSC-derived 3D platform using primary brain cancer cells to study drug development and personalized medicine. *Sci. Rep.* **2019**, *9*, 1407. [CrossRef]
64. Sudhof, T.C. The presynaptic active zone. *Neuron* **2012**, *75*, 11–25. [CrossRef]
65. Bell, M.E.; Bourne, J.N.; Chirillo, M.A.; Mendenhall, J.M.; Kuwajima, M.; Harris, K.M. Dynamics of nascent and active zone ultrastructure as synapses enlarge during long-term potentiation in mature hippocampus. *J. Comp. Neurol.* **2014**, *522*, 3861–3884. [CrossRef] [PubMed]
66. Wang, S.S.H.; Held, R.G.; Wong, M.Y.; Liu, C.; Karakhanyan, A.; Kaeser, P.S. Fusion Competent Synaptic Vesicles Persist upon Active Zone Disruption and Loss of Vesicle Docking. *Neuron* **2016**, *91*, 777–791. [CrossRef] [PubMed]
67. Pei, Y.; Peng, J.; Behl, M.; Sipes, N.S.; Shockley, K.R.; Rao, M.S.; Tice, R.R.; Zeng, X. Comparative neurotoxicity screening in human iPSC-derived neural stem cells, neurons and astrocytes. *Brain Res.* **2016**, *1638*, 57–73. [CrossRef]
68. Mottram, D.S.; Wedzicha, B.L.; Dodson, A.T. Acrylamide is formed in the Maillard reaction. *Nature* **2002**, *419*, 448–449. [CrossRef]
69. Dearfield, K.L.; Abernathy, C.O.; Ottley, M.S.; Brantner, J.H.; Hayes, P.F. Acrylamide: Its metabolism, developmental and reproductive effects, genotoxicity, and carcinogenicity. *Mutat. Res.* **1988**, *195*, 45–77. [CrossRef]
70. Kjuus, H.; Goffeng, L.O.; Heier, M.S.; Sjöholm, H.; Øvrebø, S.; Skaug, V.; Paulsson, B.; Törnqvist, M.; Brudal, S. Effects on the peripheral nervous system of tunnel workers exposed to acrylamide and N-methylolacrylamide. *Scand. J. Work Environ. Health* **2004**, 21–29. [CrossRef]
71. Sorgel, F.; Weissenbacher, R.; Kinzig-Schippers, M.; Hofmann, A.; Illauer, M.; Skott, A.; Landersdorfer, C. Acrylamide: Increased concentrations in homemade food and first evidence of its variable absorption from food, variable metabolism and placental and breast milk transfer in humans. *Chemotherapy* **2002**, *48*, 267–274. [CrossRef] [PubMed]
72. Mundy, W.R.; Tilson, H.A. Neurotoxic effects of colchicine. *Neurotoxicology* **1990**, *11*, 539–547. [PubMed]
73. Goldschmidt, R.B.; Steward, O. Comparison of the neurotoxic effects of colchicine, the vinca alkaloids, and other microtubule poisons. *Brain Res.* **1989**, *486*, 133–140. [CrossRef]
74. Tanner, C.M.; Kamel, F.; Ross, G.W.; Hoppin, J.A.; Goldman, S.M.; Korell, M.; Marras, C.; Bhudhikanok, G.S.; Kasten, M.; Chade, A.R.; et al. Rotenone, paraquat, and Parkinson's disease. *Environ. Health Perspect.* **2011**, *119*, 866–872. [CrossRef] [PubMed]
75. Rappold, P.M.; Cui, M.; Chesser, A.S.; Tibbett, J.; Grima, J.C.; Duan, L.; Sen, N.; Javitch, J.A.; Tieu, K. Paraquat neurotoxicity is mediated by the dopamine transporter and organic cation transporter-3. *Proc. Natl. Acad. Sci. USA* **2011**, *108*, 20766–20771. [CrossRef] [PubMed]
76. Kanthasamy, A.; Jin, H.; Charli, A.; Vellareddy, A.; Kanthasamy, A. Environmental neurotoxicant-induced dopaminergic neurodegeneration: A potential link to impaired neuroinflammatory mechanisms. *Pharmacol. Ther.* **2019**, *197*, 61–82. [CrossRef]
77. Betarbet, R.; Sherer, T.B.; MacKenzie, G.; Garcia-Osuna, M.; Panov, A.V.; Greenamyre, J.T. Chronic systemic pesticide exposure reproduces features of Parkinson's disease. *Nat. Neurosci.* **2000**, *3*, 1301–1306. [CrossRef]
78. Pistollato, F.; Canovas-Jorda, D.; Zagoura, D.; Bal-Price, A. Nrf2 pathway activation upon rotenone treatment in human iPSC-derived neural stem cells undergoing differentiation towards neurons and astrocytes. *Neurochem. Int.* **2017**, *108*, 457–471. [CrossRef]

79. Zagoura, D.; Canovas-Jorda, D.; Pistollato, F.; Bremer-Hoffmann, S.; Bal-Price, A. Evaluation of the rotenone-induced activation of the Nrf2 pathway in a neuronal model derived from human induced pluripotent stem cells. *Neurochem. Int.* **2017**, *106*, 62–73. [CrossRef]
80. Radio, N.M.; Mundy, W.R. Developmental neurotoxicity testing in vitro: Models for assessing chemical effects on neurite outgrowth. *Neurotoxicology* **2008**, *29*, 361–376. [CrossRef]
81. Ryan, K.R.; Sirenko, O.; Parham, F.; Hsieh, J.H.; Cromwell, E.F.; Tice, R.R.; Behl, M. Neurite outgrowth in human induced pluripotent stem cell-derived neurons as a high-throughput screen for developmental neurotoxicity or neurotoxicity. *Neurotoxicology* **2016**, *53*, 271–281. [CrossRef] [PubMed]
82. Hellwig, C.; Barenys, M.; Baumann, J.; Gassmann, K.; Casanellas, L.; Kauer, G.; Fritsche, E. Culture of human neurospheres in 3D scaffolds for developmental neurotoxicity testing. *Toxicol. Int. J. Publ. Assoc. BIBRA* **2018**, *52*, 106–115. [CrossRef] [PubMed]
83. Pamies, D.; Bal-Price, A.; Chesne, C.; Coecke, S.; Dinnyes, A.; Eskes, C.; Grillari, R.; Gstraunthaler, G.; Hartung, T.; Jennings, P.; et al. Advanced Good Cell Culture Practice for human primary, stem cell-derived and organoid models as well as microphysiological systems. *ALTEX* **2018**, *35*, 353–378. [CrossRef] [PubMed]

9

Tendon and Cytokine Marker Expression by Human Bone Marrow Mesenchymal Stem Cells in a Hyaluronate/Poly-Lactic-Co-Glycolic Acid (PLGA)/Fibrin Three-Dimensional (3D) Scaffold

Maria C. Ciardulli [1], Luigi Marino [1], Joseph Lovecchio [2], Emanuele Giordano [2], Nicholas R. Forsyth [3], Carmine Selleri [1], Nicola Maffulli [1,3,4] and Giovanna Della Porta [1,5,*]

1 Department of Medicine, Surgery and Dentistry, University of Salerno, Via S. Allende, 84081 Baronissi (SA), Italy; mciardulli@unisa.it (M.C.C.); lmarino@unisa.it (L.M.); cselleri@unisa.it (C.S.); n.maffulli@qmul.ac.uk (N.M.)

2 Department of Electrical, Electronic and Information Engineering "Guglielmo Marconi" (DEI), University of Bologna, Via dell'Università 50, 47522 Cesena (FC), Italy; joseph.lovecchio@unibo.it (J.L.); emanuele.giordano@unibo.it (E.G.)

3 Guy Hilton Research Centre, School of Pharmacy and Bioengineering, Keele University, Stoke-on-Trent, Staffordshire ST4 7QB, UK; n.r.forsyth@keele.ac.uk

4 Centre for Sport and Exercise Medicine, Queen Mary University of London, Barts and The London School of Medicine, London E1 4NL, UK

5 Department of Industrial Engineering, University of Salerno, Via Giovanni Paolo II, 84084 Fisciano (SA), Italy

* Correspondence: gdellaporta@unisa.it

Abstract: We developed a (three-dimensional) 3D scaffold, we named HY-FIB, incorporating a force-transmission band of braided hyaluronate embedded in a cell localizing fibrin hydrogel and poly-lactic-co-glycolic acid (PLGA) nanocarriers as transient components for growth factor controlled delivery. The tenogenic supporting capacity of HY-FIB on human-Bone Marrow Mesenchymal Stem Cells (hBM-MSCs) was explored under static conditions and under bioreactor-induced cyclic strain conditions. HY-FIB elasticity enabled to deliver a mean shear stress of 0.09 Pa for 4 h/day. Tendon and cytokine marker expression by hBM-MSCs were studied. Results: hBM-MSCs embedded in HY-FIB and subjected to mechanical stimulation, resulted in a typical tenogenic phenotype, as indicated by type 1 Collagen fiber immunofluorescence. RT-qPCR showed an increase of type 1 Collagen, scleraxis, and decorin gene expression (3-fold, 1600-fold, and 3-fold, respectively, at day 11) in dynamic conditions. Cells also showed pro-inflammatory (IL-6, TNF, IL-12A, IL-1β) and anti-inflammatory (IL-10, TGF-β1) cytokine gene expressions, with a significant increase of anti-inflammatory cytokines in dynamic conditions (IL-10 and TGF-β1 300-fold and 4-fold, respectively, at day 11). Mechanical signaling, conveyed by HY-FIB to hBM-MSCs, promoted tenogenic gene markers expression and a pro-repair cytokine balance. The results provide strong evidence in support of the HY-FIB system and its interaction with cells and its potential for use as a predictive in vitro model.

Keywords: hBM-MSCs; cytokines; tenogenic markers; cyclic strain; 3D microenvironment; PLGA carriers; bioreactor

---

1. Introduction

Tissue engineering strategies for tendon healing and regeneration are designed to improve existing therapies or provide new treatment possibilities. Three-dimensional (3D) bioengineered systems have the potential to promote our understanding of the physiopathology of tendinopathy and the role of stem cells in tendon regeneration. In this sense, 3D scaffold design and fabrication coupled to specific bioreactor arrangements could develop highly predictive 3D in vitro culture and differentiation systems to explore cell behaviors in response to defined external biochemical and mechanical stimuli [1]. The 3D scaffold provides a model of fidelity via its provision of a microenvironment with defined stiffness and elastic modulus as well as the necessary surfaces for cell attachment [2–4].

Detailed understanding of cell behavior when incorporated into specific biomaterials allows to develop designs with specific functionalization. These functionalization may, for instance, stimulate local stem cells, attract specific circulating nucleated blood cells, such as macrophages, and induce their polarization into M2 phenotype to accelerate tissue regeneration and healing following the biomaterials in vivo implantation [5]. For example, human Mesenchymal Stem Cells (hMSCs) are largely used in tissue engineering strategies, and their immune-modulatory activity in the development of tendon pathologies have been explored, but the precise mechanisms involved remain undetermined [6–10]. Neutrophils and macrophages infiltrate injured tendons, potentially interacting with MSCs and stimulating cytokine release at the site of repair and promoting degradation of the extracellular matrix (ECM), inflammation, apoptosis, and, in the later stages of acute tendon healing, they release anti-inflammatory cytokines to alleviate inflammation and promote tendon remodeling [11–13].

Among the biomaterials described for tendon tissue engineering [14], a promising emerging strategy is the use of a complex biomimetic matrix with a hydrogel component and extracellular matrix mimicking properties [15,16]. Tenocyte precursors can be harvested from different sources, including periosteum [17,18], bone marrow [19–21], tendon [21,22], and adipose tissue [21,23]. To overcome the intrinsic poor mechanical properties of the hydrogel, they can be merged with more force resistant biopolymers. Cells and biomaterials alone are not sufficient to achieve optimal levels of differentiation and matrix organization. Mechanical stimulation has a key role in tenogenic differentiation induction [19]. Scaffolds are therefore required to display an appropriate elastic behavior to deliver strain [24] or compression [25] inputs. Strain is a tenogenic differentiation signal [26–29], and several bioreactors have been used to impart tenogenic mechanical stimuli to cells in culture [19,30–35]. For example, Rinoldi et al. designed and fabricated 3D multilayered composite scaffolds, where an electrospun nanofibrous substrate was coated with a thin layer of GelMA-alginate composite hydrogel carrying MSCs. MSCs were subsequently differentiated by the addition of bone morphogenetic protein 12 (BMP-12) and, to mimic the natural function of tendons, the scaffolds were mechanically stimulated using a custom-built bioreactor [34]. Grier et al. described an aligned collagen-glycosaminoglycan scaffold able to enhance tenogenic differentiation of MSCs via cyclic tensile strain within a bioreactor, in the absence of growth factor supplementation [36]. Another protocol, proposed by Youngstrom et al., promoted tenogenic differentiation of MSCs cultivated on decellularized tendon scaffolds with the application of 3% cyclic strain for one hour per day for 11 days [31]. Additionally, several growth factors and other small molecules can stimulate transcriptional activation of genes involved in tenogenic differentiation [15,37,38]. Growth Differentiation Factor 5 (GDF-5), for instance, induces the expression of genes linked to the neo-tendon phenotype [39–41].

Tendinopathies associated with physical activity and age-related degeneration are a major medical issue [23], and recent healing and regeneration studies include the use of human Bone Marrow Mesenchymal Stem Cells (hBM-MSCs) [42–44]. hBM-MSCs are a multipotent population present in bone marrow that can be readily differentiated in vitro [45,46] into cells of three mesodermal lineages, namely adipocytes, chondrocytes and osteoblasts under appropriate conditions [47–50]. MSCs-based therapies include direct transplantation of MSCs populations, growth factor-loaded scaffolds for local MSCs recruitment or implantation of scaffolds containing in vitro culture-expanded MSCs populations [51,52].

We previously described an engineered multiphase three-dimensional (3D) scaffold as an in vitro model for tendon regeneration studies. The multiphase 3D construct was totally absorbable and consisted of a braided hyaluronate elastic band merged with a fibrin hydrogel containing hBM-MSCs and poly-lactic-co-glycolic acid nano-carriers (PLGA-NCs) themselves loaded with human

Growth Differentiation Factor 5 (hGDF-5) [53]. In that work, the PLGA nano-carriers were transient scaffold components to ensure sustained and controlled delivery of hGDF-5 with benefits beyond those associated with standard culture medium supplementation [53,54]. The study reported an early tenogenic commitment of hBM-MSCs after three days of cultivation under dynamic conditions.

In the present study, we describe the use of the same scaffold (named HY-FIB here) to investigate the effect of the 3D environment on hBM-MSCs for 11 days with or without mechanical stimulation and in the absence of any specific biochemical differentiation signal. HY-FIB was assembled with hBM-MSCs as previously described [53] including PLGA-NCs stratified within the 3D fibrin structure. Importantly, the PLGA-NCs carried an inactive form of human Growth Differentiation Factor 5 (ihGDF-5) enabling the overall 3D scaffold structure to be safely evaluated with or without mechanical input. Gene expression of type 1 Collagen, decorin, scleraxis, and tenomodulin were considered; type 3 collagen was also monitored, as negative control. Histology and quantitative immunofluorescence were used to monitor cell behavior and their interaction with the synthetic extracellular matrix. Moreover, to understand whether HY-FIB microenvironment configuration would stimulate any cell inflammation responses, the cells expression of cytokine markers was also monitored, including pro-inflammatory cytokines and anti-inflammatory ones. The results provide strong evidence that HY-FIB environment plus mechanical signaling, promoted tenogenic markers expression, collagen production and better pro-repair cytokine balance by hBM-MSCs.

2. Materials and Methods

2.1. hBM-MSCs Isolation and Harvesting

Human bone marrow mesenchymal stem cells (hBM-MSCs) were obtained from the bone marrow of three independent healthy donors (age 36, 38, 44 years). The donors gave written informed consent in accordance with the Declaration of Helsinki to the use of their filter residual bone marrow aspirate for research purposes, with approval from the University Hospital of San Giovanni di Dio e Ruggi d'Aragona (Salerno, IT). Review Board authorization number: (24988 achieved on April 9, 2015).

Briefly, total bone marrow aspirate was directly seeded at a concentration of 50,000 total nucleated cells/cm^2 in T75 plastic flask in Minimum Essential Medium Alpha (α-MEM) supplemented with 1% Glutagro^{TM}, 10% Fetal Bovine Serum (FBS), and 1% Pen/Strep and incubated at 37 °C in 5% CO_2 atmosphere and 95% relative humidity [55]. After 72h, non-adherent cells were removed by medium change, and the adherent cells were further fed twice a week with new medium. On day 14, colonies of adherent hBM-MSCs were detached and re-seeded at 4000 cells/cm^2 in the same culture conditions. Once the cell cultures reached 70–80% confluence, cells were detached using 0.05% trypsin-0.53mM EDTA and washed with PBS 1× (Corning Cellgro, Manassas, VA, USA), counted using Trypan Blue (Sigma-Aldrich, Milan, IT) and subcultured at a concentration of 4×10^3 cells/cm^2. Flow Cytometry analysis was performed on hBM-MSCs obtained at Passage 1 examining levels of CD90, CD105, CD73 CD14, CD34, CD45, and HLA-DR expression (Miltenyi Biotec, DE).

2.2. ihGDF-5 Effect on hBM-MSCs

These sets of experiments were performed to assure the absence of any effect of inactivated human GDF-5 (ihGDF-5, Cloud-Clone Corp., USA) on both tenogenic markers stimulation and cytokines expression by hBM-MSCs. Cells were seeded on coverslips in 12 well plates at a concentration of 4×10^3 cells/cm^2. Once the cultures reached 60% confluence, cells were treated with either 1.6 ng/mL or 100 ng/mL of ihGDF-5. Cells were fed twice a week with new medium and fresh ihGDF-5 supplementation for up to 16 days. Untreated cells for matched time-points studied were used for control purposes. Passage 3 cells were seeded in the 3D environment (~8×10^5 cells/mL) and were fed twice a week with new medium, without any growth factor added.

PLGA carriers were not tested because they cannot be supplemented to the cells planar monolayer culture, indeed, in static conditions their sedimentation on cells reduced oxygen exchange and prevented cells survival (data not shown).

2.3. *Immunofluorescence and Immunohistochemical Assays*

Cells were fixed with 3.7% formaldehyde for 30 min at room temperature (RT) followed by permeabilization with 0.1% Triton X-100 for 5 min and blocking with 1% Bovine Serum Albumine (BSA) for 1h. For type 1 and type 3 Collagen staining, cells were incubated overnight at 4 °C with a mouse monoclonal anti-type 1 Collagen antibody (1:100, Sigma-Aldrich) and a rabbit polyclonal anti-type 3 Collagen antibody (1:100, Santa Cruz Biotechnology). Following incubation with the primary antibody, cells were incubated for 1h at RT with the DyLight 649 anti-mouse IgG (1:500, BioLegend, CA) and the Alexa FluorTM 488 goat anti-rabbit IgG (1:500; Thermo Fisher Scientific, USA). Cell nuclei were stained with DAPI solution (1:1000) for 5 min. Images acquisition was at 20× magnification on a fluorescent microscope (Eclipse Ti-E Inverted Microscope; NIKON Instruments Inc., USA).

For 3D scaffold immunohistochemical analysis, slices were permeabilized with 0.1% Triton X-100 for 5 min, and non-specific staining blocked with 1% BSA for 1h at RT. For type 1 Collagen staining, slices were incubated overnight at 4 °C with a rabbit polyclonal anti-type 1 Collagen antibody (1:200, AbCam). Following incubation with the primary antibody, slices were incubated for 1h at RT with Alexa Fluor^{TM} 488 goat anti-rabbit IgG (1:400, Thermo Fisher Scientific, USA) antibody. Subsequently, cell nuclei were stained with DAPI solution (1:1000) and incubated for 5 min. Images were acquired as described above. Image quantification was performed using image analysis software (ImageJ, National Institutes of Health, USA) [56,57] by measuring the red and green areas where type 1 and type 3 collagen, respectively, are expressed. A minimum of 10 image fields was used for the image analysis at each time point. Signal intensity at each time point was normalized by the cell number (e.g., by amount of cell nuclei revealed by DAPI staining).

Sirius red staining was performed using the Picrosirius Red Stain Kit (Polysciences, Inc., USA). Sections of 15 μm of thickness were stained in hematoxylin for 8 min, then washed in water for 2 min. The sections were dipped into phosphomolybdic acid for 2 min, then washed in water for 2 m. Then they were dipped into Picrosirius Red F3BA Stain for 60 min and dipped into HCl 0.1M solution for 2 min. The sections were dehydrated in increasing ethanol gradient solutions (70–75–95–100%) and finally dipped into xylene for 5 min. Eukitt medium was used to mount the samples.

2.4. *RNA Isolation and Gene Expression Profile by Quantitative Reverse Transcription PCR (RT-qPCR)*

Total RNA was extracted from hBM-MSCs seeded into the 3D construct of each experimental group using QIAzol® Lysis Reagent (Qiagen, DE), chloroform (Sigma-Aldrich, Milan, IT) and the RNeasy Mini Kit (Qiagen, DE). For each sample, 300 ng of total RNA was reverse-transcribed using the iScriptTM cDNA synthesis kit (Bio-Rad, Milan, IT). Relative gene expression analysis was performed in a LightCycler® 480 Instrument (Roche, IT), using the SsoAdvancedTM Universal SYBR® Green Supermix (Bio-Rad) with the validated primers for COL1A1, COL3A1, DCN, IL-1β, IL-6, IL-10, IL-12A, SCX-A, TGF-β1, TNF, and TNMD (Bio-Rad), and following MIQE guidelines [58]. Amplification was performed in a 10 μL final volume, including 2 ng of complementary DNA (cDNA) as template. Specificity of the formed products was addressed via melting curve analysis. Triplicate experiments were performed for each condition explored, and data were normalized to glyceraldehyde-3-phosphate dehydrogenase (GAPDH) expression (reference gene), applying the geNorm method [59] to calculate reference gene stability between the different conditions (calculated with CFX Manager software; M <0.5). Fold changes in gene expression were determined by the $2^{-\Delta\Delta Cp}$ method, and are presented as relative levels versus hBM-MSCs just loaded within the HY-FIB system.

2.5. *PLGA-NCs Fabrication, Size, Morphology, and ihGDF-5 Release Profile*

PLGA nano-carriers (PLGA-NCs) were obtained using Supercritical Emulsion Extraction (SEE) technology enabling rapid polymer NCs production from multiple emulsions via dense gas extraction of the oily phase organic solvent utilizing a countercurrent packed tower operating in continuous mode [60]. In detail, ihGDF-5 (Cloud-Clone Corp., USA) was dissolved into 1% (*w*/*v*) human serum albumin (hSA; Sigma-Aldrich, Milan, IT) containing 0.06% polyvinyl alcohol (PVA). Human serum albumin (hSA) was included as an ihGDF-5 stabilizer. This solution was added to the oily phase formed in an Ethyl Acetate (EA, purity 99.9%) and PLGA (RG 504H, 38,000–54,000 kDa, Evonik, DE) at 5% (*w*/*w*) solution. All emulsions were processed immediately after their preparation.

SEE technology operative pressure and temperature conditions in the high-pressure column were set at 8 MPa and 38 °C, respectively, with a dense gas flow of Carbon Dioxide (CO_2) set at 1.4 kg/h with Liquid/Gas ratio of 0.1 (*w*/*w*) [61]. Carrier suspensions were collected at the bottom of the extraction column, washed, and lyophilized. Each run allowed the recovery of 98% of the loaded biopolymer. Empty and loaded NCs were produced using the same process conditions.

Carrier particle size distributions (PSDs) were measured using a laser granulometer (mod. Mastersizer S; Malvern Instruments Ltd., Worcestershire, UK), based on dynamic light scattering (DLS). Sizes are expressed as volume mean size (MS) with standard deviation (SD) in nanometers (nm). The shape and morphology of the PLGA-NCs were investigated by field emission-scanning electron microscopy (FE-SEM; mod. LEO 1525; Carl Zeiss, Oberkochen, D). Samples were placed on a double-sided adhesive carbon tape previously glued to an aluminum stub and coated with a gold film (250 A thickness) using a sputter coater (mod.108 A; Agar Scientific, Stansted, UK).

ihGDF-5 release profile was monitored in vitro from 20(±0.3) mg of carriers suspended in 1mL of α-MEM, placed in an incubator at 37 °C, and stirred continuously at 1× *g*. Every 24 h, samples were centrifuged at 160× *g* for 10 min and the supernatant completely removed and replaced with fresh media to maintain sink conditions. Released ihGDF-5 concentrations from collected samples were then measured with an Enzyme Linked Immunosorbent Assay (ELISA, Cloud-Clone Corp., USA). Release experiments were performed in triplicate (n = 3), and the curve describing the mean profile calculated as ng/g (protein released/PLGA-NCs) versus time.

2.6. *HY-FIB Preparation and Characterization*

For each sample, a mixture of 50 mg/mL fibrinogen from human plasma (Sigma-Aldrich, Milan, IT), 15,600 U/mL aprotinin (Sigma-Aldrich, Milan, IT), and α-MEM (Corning, NY, USA) supplemented with 10% FBS (referred to as growing media, GM) was added at a 1:1:1 ratio to 100 mg of PLGA-NCs (ihGDF-5 loading: 350 ng/g) and, then, to an average of 8×10^5 cells. A homogeneous cells/PLGA-NCs/fibrinogen suspension was then embedded into a mold (30 × 20 × 4.5 mm) where the braided band had been previously positioned. Free ends were left to enable HY-FIB fixing into the bioreactor. Upon addition of 100 U/mL thrombin (Sigma-Aldrich, Milan, IT), the mold was placed in a 37 °C humidified incubator for 30 min to allow fibrin polymerization. When the hydrogel was formed, the band was entrapped inside a uniformly distributed hydrogel. The construct was then transferred from the mold to either a standard polystyrene culture plate or to the bioreactor culture chamber, each containing 30 mL of the culture media, and placed in an incubator at 37 °C in a 5% CO_2 atmosphere and 95% relative humidity.

HY-FIB morphology was observed by field emission-scanning electron microscopy (FE-SEM; mod. LEO 1525; Carl Zeiss, Oberkochen, Germany). Samples were fixed in 4% PFA (4 °C, overnight) and then dehydrated by multiple passages across ethanol:water solutions (10 min each) with increasing concentrations of ethanol (10%, 20%, 30%, 50%, 70%, 90%), ending in a 100% dehydrating liquid (3 changes, 10 min each).

Samples were then lyophilized in a Critical Point Dryer (mod. K850 Emitech, Assing, Rome IT), placed on a double-sided adhesive carbon tape previously glued to an aluminum stub and coated with a gold film (250 A thickness) using a sputter coater (mod. 108 A; Agar Scientific, Stansted, United Kingdom) before observation.

HY-FIB mechanical characterization was performed according to the ASTM 1708 by a dynamometer (CMT 6000 SANS, Shenzen, China) equipped with a 1 kN load cell. The sample was conditioned in Dulbecco's Modified Essential Medium (DMEM) for 1 h, and then shaped to obtain a specimen with gauge length (Lo) of 22 mm and width (W) of 5 mm. Sample thickness (S) was measured with a thickness gauge brand at three different averaged points. Monoaxial deformation was applied to the sample at a speed of 10 mm/min, and force (F) and elongation (L) during traction were recorded. The elastic modulus and ultimate tensile strength (both expressed in MPa) were calculated from the stress/strain plot. For the immuno-histochemical analysis, at different time points, a portion of HY-FIB was fixed in 4% PFA (4 °C, overnight), cryo-protected in 30% sucrose overnight, mounted in OCT embedding compound, frozen at −20°C and then cut in slices of 10 μm of thickness using a cryostat. The remaining portion of HY-FIB was placed in QIAzol® Lysis Reagent for total RNA extraction.

2.7. *Dynamic Culture*

HY-FIB was clamped at both free ends, one motionless and one sliding (operated by a linear motor actuator) arm, into the bioreactor system culture chamber, described in detail elsewhere [23]. A maximal load, set by pre-tensioning, was relaxed to a minimum value cycling at a pre-determined frequency. In addition, continuous feedback signals provided by strain gauges located onto the fixed arm, allowed the maintenance of a defined load on the scaffold in response to physical system modifications, by automatic adjustment of the pre-tensioning position.

2.8. *Finite Element Modeling*

Finite Element Modeling (FEM) was implemented by using COMSOL Multiphysics Software (rel. 5.3a®) to assess the stress distribution into 3D constructs when a 10% mechanical cyclic strain stimulus is applied. All components were obtained using primitive geometries and Boolean operations. Linear elasticity equations were set as boundary conditions for the scaffold. A sensitivity study of the mesh obtained the most computationally efficient solution. The specific parameters used in the model are listed in Table 1. The simulation considered only the stress within the fibrin 3D environment neglecting any further contribution of the band, cells and PLGA-NCs.

Table 1. Finite Element Modeling (FEM) parameters used.

Parameters	Value	Unit
Young's modulus [1]	4.56	MPa
Poisson ratio [2]	0.25	-
Density	1050	Kg/m^3

[1] HY-FIB Young Modulus was measured experimentally [53] [2] Poisson ratio value was taken in the literature [62].

2.9. *Statistical Analysis*

Statistical analysis was performed using GraphPad Prism software (6.0 for Windows). Data obtained from multiple experiments are expressed as mean+/−SD and analyzed for statistical significance using ANOVA test, for independent groups. Differences were considered statistically significant when $p \leq 0.05$ [63].

3. Results

3.1. *hBM-MSCs Cultivation in 2D Environment*

hBM-MSCs were cultivated in a two-dimensional (2D) monolayer environment with medium supplemented with either 1.6 ng/mL or 100 ng/mL of inactive human Growth Differentiation Factor 5 (ihGDF-5) for up to 16 days. These two concentration conditions (two order of magnitude of difference) were chosen to ascertain absence of any effect of the inactive hGDF-5 form on cells expression of tenogenic markers (COL3A1, COL1A1, DCN, SCX-A and TNMD) and of cytokines (pro-inflammatory: IL-6, TNF, IL-12A, IL-1β; anti-inflammatory: IL-10, TGF-β1) by RT-qPCR (see Figure 1a–d).

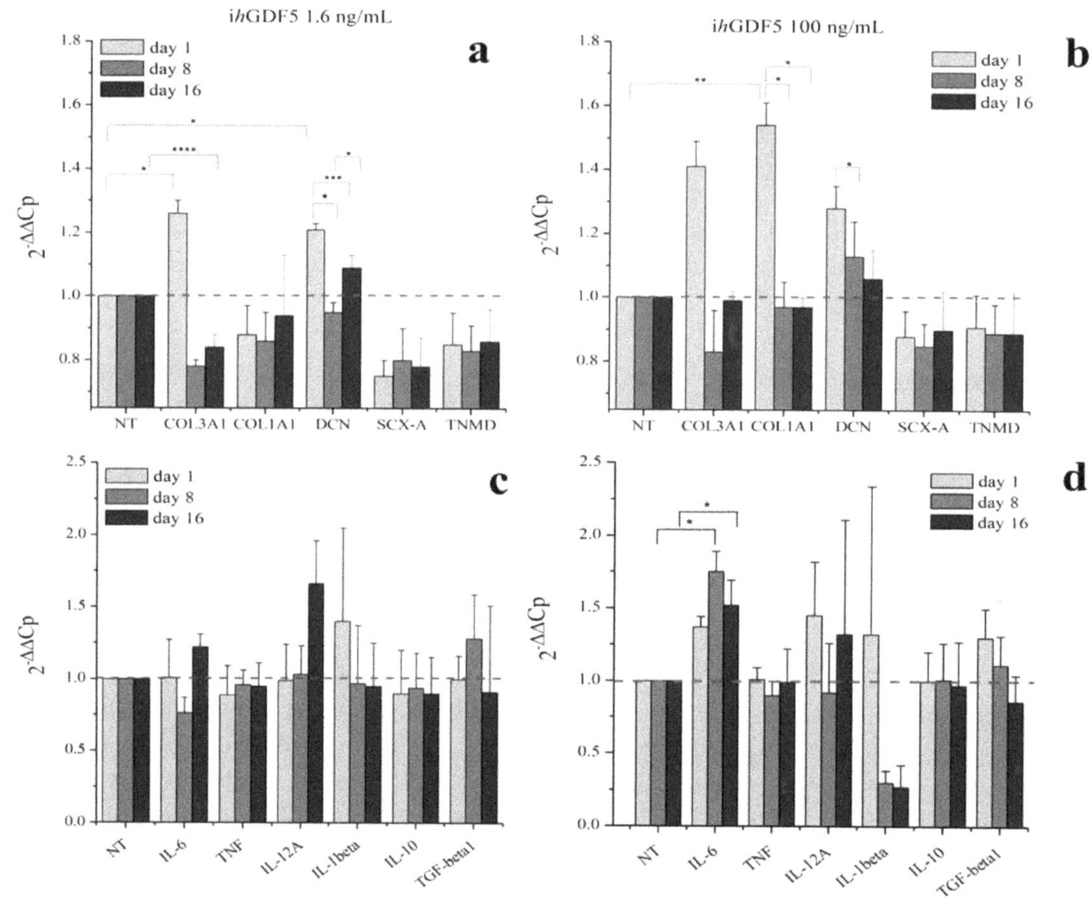

Figure 1. Gene expression profiles for tenogenic markers and pro-inflammatory and anti-inflammatory cytokines by hBM-MSCs treated with 1.6 ng/mL (**a**,**c**) and 100 ng/mL (**b**,**d**) of ihGDF-5 in monolayer 2D culture up to 16 days. mRNA levels of COL1A1, DCN, SCX-A, and TNMD were considered as tenogenic markers and COL3A1 selected as negative ones; pro-inflammatory (IL-6, TNF, IL-12A and IL-1β) and anti-inflammatory (IL-10 and TGF-β1) cytokines were monitored. ihGDF-5 not had significant impact on genes expression, especially at the lower concentration tested. Untreated cells for matched time-points were used as control. * ≤ 0.05; ** <0.01; *** <0.005; **** <0.001 N = 3 (biological replicates); n = 3 (technical replicates).

Transient and slight, though significant, upregulation of COL3A1 (0.4-fold), DCN (0.2-fold), and COL1A1 (0.5-fold) was observed at Day 1 in cultures supplemented with 100 ng/mL ihGDF-5 (Figure 1b). The reduced dose of 1.6 ng/mL induced low-level transient expression at Day 1 for only COL3A1 (0.3-fold) and DCN (0.2-fold) only (Figure 1a). No significant upregulation was noted for TNF, IL-12A, IL-1β, IL-10, or TGFβ at any time point or ihGDF-5 concentration tested (Figure 1c,d). Compared to controls, IL-6 displayed significant levels of elevation at Days 8 (0.8-fold) and 16 (0.5-fold) with 100 ng/mL ihGDF-5 (Figure 1d). Types 1 and 3 Collagen expression levels were monitored by immunofluorescence during the 16 day culture period, as illustrated in Figure 2a,b. Immunofluorescence quantitative data by image analysis were congruent with RT-qPCR outputs when 1.6 ng/mL ihGDF-5 was supplemented, in this case, both proteins signals were not significantly elevated, compared to untreated cells (see Figure 2c).

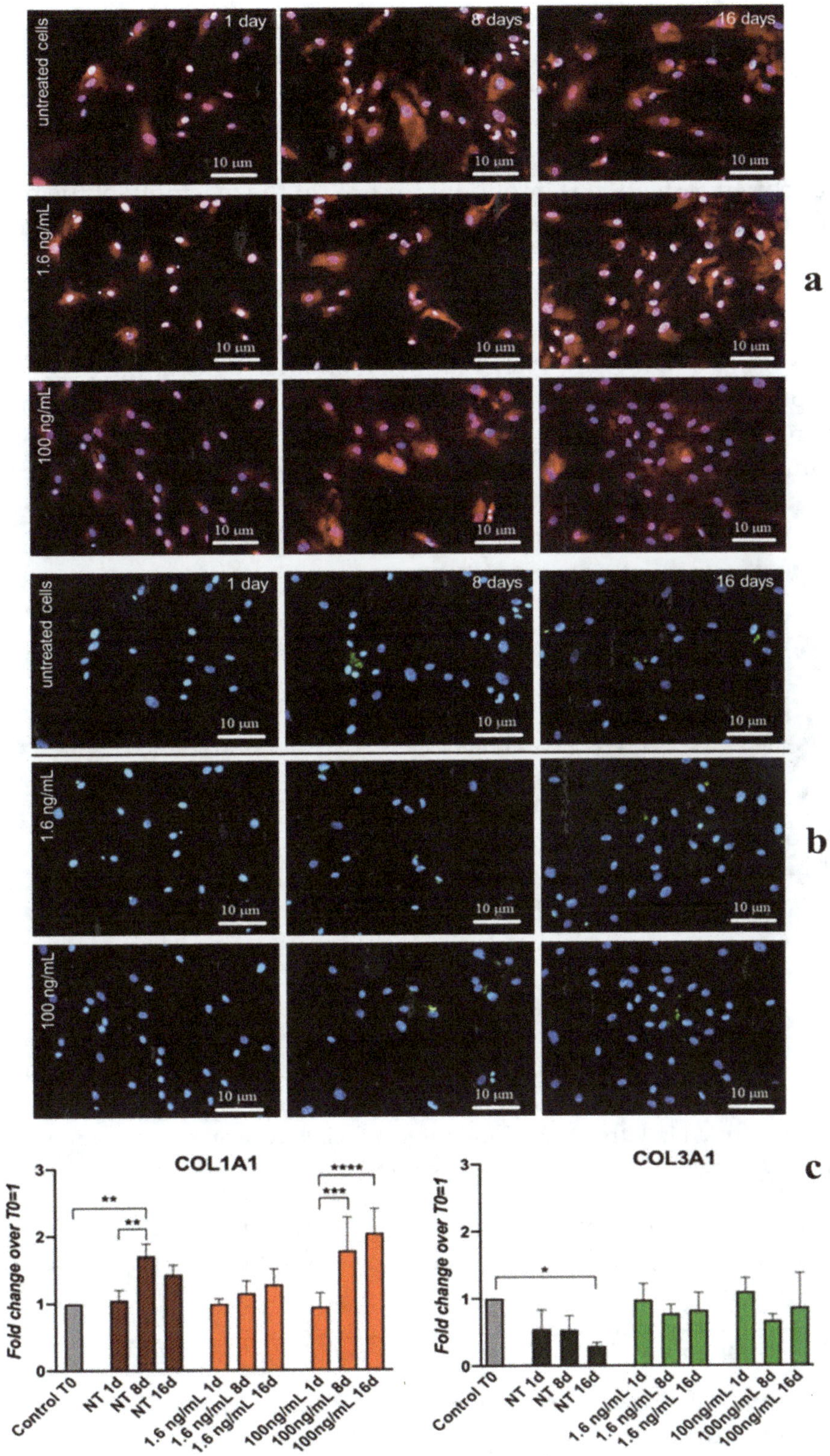

Figure 2. IF and quantitative-IF assays of type 1 Collagen (COL1A1) and type 3 Collagen (COL3A1) monitored along hBM-MSCs treatment with 1.6 ng/ and 100 ng/mL of ihGDF-5 for 16 days. Type 1 collagen was stained in red; type 3 collagen was stained in green; cell nuclei highlighted with DAPI in blue (**a**,**b**). Quantitative signal detection was performed via ImageJ software (**c**). A slight up-regulation of COL1A1 was observed when hBM-MSC were in routine culture for 16 days. Color intensity in each time point was normalized by the cell number. * ≤0.05; * *<0.01; *** <0.005; **** <0.001. n = 10 (image fields for each time point).

Quantitative image analysis displayed COL1A1 signal increase (1 fold) at Day 16 only with 100 ng/mL ihGDF-5 supplementation. This last data is in contrast with gene expression ones. COL1A1 and COL3A1 are the major components of the extracellular matrix in connective tissues, and their slight up-regulations was reported when hBM-MSC were in routine culture for 16 days [64]. However, in our case, ihGDF-5 seemed not to impact on their production, especially at the lower concentration tested. This preliminary information is important to confirm the inactivity of the biochemical input in regards to the gene expression and proteins that will be monitored in the 3D experiments.

3.2. PLGA-NCs Fabrication and ihGDF-5 Controlled Deliver

Poly-lactic-co-glycolic-acid nano-carriers (PLGA-NCs) displayed a spherical morphology with a mean size of 230 ± 80 nm (Figure 3a). PLGA-NCs had an ihGDF-5 loading of 350 ng/g and provided a daily released peptide mean concentration of 1.6 ng/mL/day (Figure 3b), when an amount of 100 mg were inserted within HY-FIB over 11 days of culture. As highlighted above, these ihGDF-5 concentration levels did not stimulate sustained impacts on hBM-MSCs gene expression in 2D monolayer culture (see Figure 1a,c).

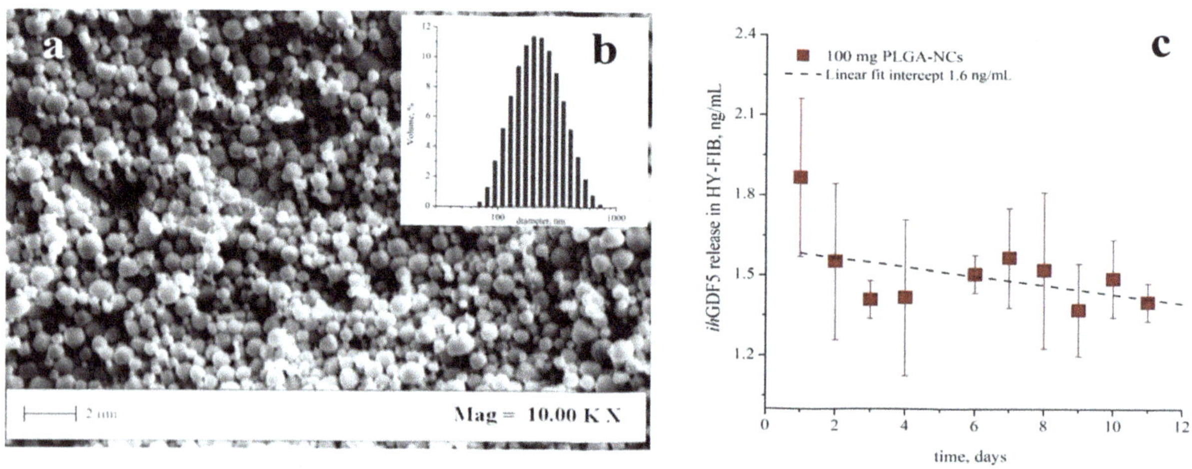

Figure 3. Poly-lactic-co-glycolic-acid (PLGA) transient carriers field emission-scanning electron microscopy (FE-SEM) image, particle size distribution, and ihGDF-5 release profile. FE-SEM images indicated spherical morphology of carriers (**a**); the size distribution set at 230 ± 80 nm their mean size (**b**). Release profiles performed in vitro at 37 °C indicated a ihGDF-5 mean concentration of 1.6 ng/mL/day released from the 100 mg of PLGA-NCs loaded within HY-FIB over 11 days of culture (**c**).

Therefore, by excluding any non-specific ihGDF-5 induction (released within the 3D scaffold by the NCs), we could now observe cell behaviors arising from the HY-FIB microenvironment in both static and dynamic conditions.

3.3. hBM-MSCs Cultivation in HY-FIB 3D Microenvironment

The HY-FIB assembly featured a braided band (3 × 10cm) joined to a fibrin hydrogel (on a band surface of 6 cm^2) containing 8 × 10^5 hBM-MSCs and 100 mg of ihGDF-5/PLGA-NCs. The picture and schematic representation of the 3D system is shown (Figure 4a,b). Field Emission Scanning Electron Microscopy (FE-SEM) images of the scaffold illustrate hyaluronate fibers, embedded within a fibrin hydrogel (Figure 4c), which provided an entrapment surface for both NCs and hBM-MSCs (Figure 4d).

HY-FIB was exposed to 10% deformation over a 1 Hz frequency for 4 h a day during the dynamic culture experiments via a cyclic strain bioreactor, illustrated in Figure 4e [24]. In greater detail, a HY-FIB braided band was held at one end by a motionless arm and at the other end by a sliding one. Motion was driven by a linear motor and transmitted through the braided band to cells embedded within the fibrin hydrogel. The motionless arm comprises a base, attached to the side wall of the culture chamber, housing the electronic components for load monitoring, and from which extended a cantilevered shelf whose deformation is measured by four strain gauges. The whole system was housed within an incubator to ensure the appropriate CO_2 gaseous environment to control the pH of the cell culture media and 37 °C operational temperature.

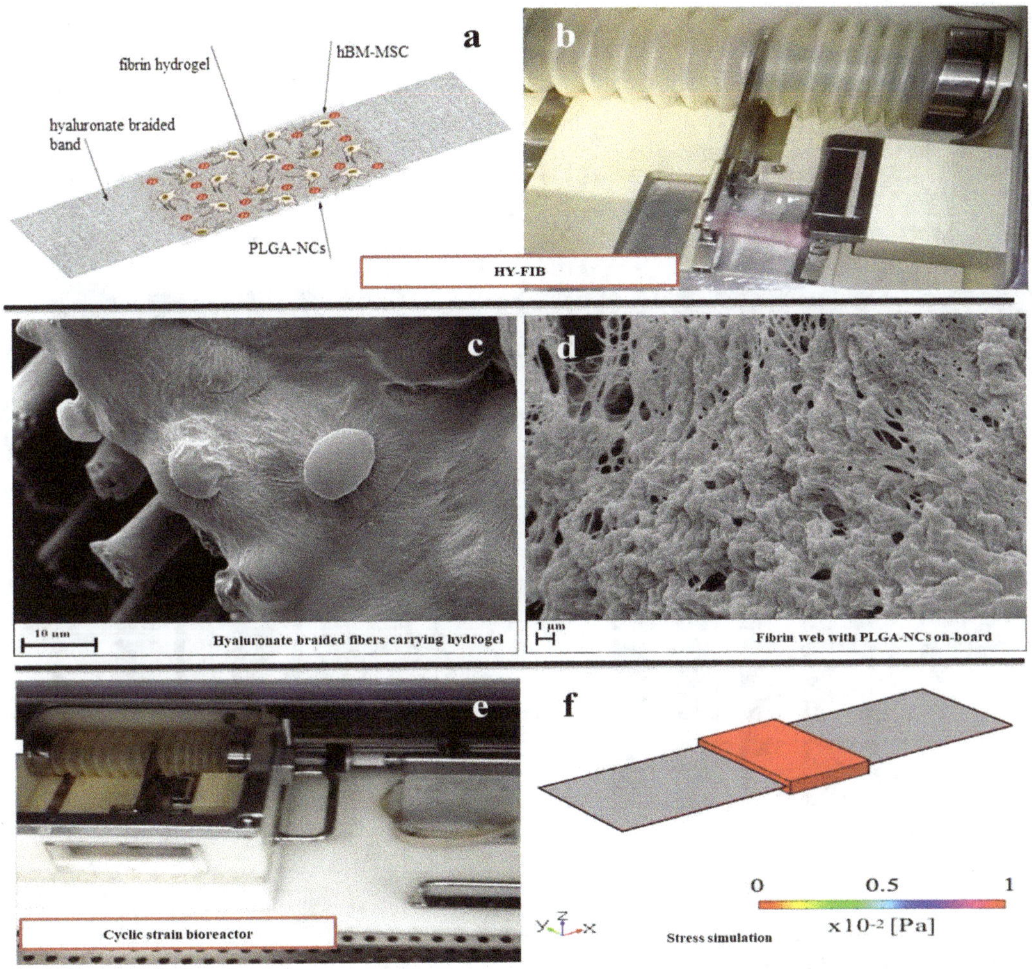

Figure 4. HY-FIB three-dimensional (3D) scaffold features and cyclic strain bioreactor. Schematic HY-FIB representation (**a**) and image of 3D scaffold (**b**). SEM images of hyaluronate braided fibers (10 μm mean diameter size) (**c**) joined to a fibrin web which entrapped both NCs and hBM-MSCs (**d**). Cyclic strain bioreactor (**e**) and in-silico study of stress distribution over HY-FIB upon mechanical strain of 10% (**f**). The simulation involved only the stress of the fibrin 3D environment, neglecting any further contribution.

The stress delivered to the cells immobilized within the system was explored via computational analysis that estimated a mean shear stress value estimated at 9×10^{-2} Pa within the fibrin 3D environment (Figure 4f). This order of magnitude of stress value was reported for tenogenic induction [30]; larger deformation for longer times were excluded to focus the study on 3D environment assembled.

HY-FIB samples were collected at Day 1, 2, 5, and 11 to monitor tenogenic and cytokine marker expression. Time points at Day 1 and Day 2 were added for 3D culture to monitor the effect of HY-FIB on cells behavior alone or in combination with cyclic strain culture. Indeed, in static conditions COL1A1 and DCN both displayed significant upregulation of 3.8 fold (COL1A1) and 2.6-fold (DCN) at Days 1 and 2 before dropping progressively to elevated but non-significant levels (Figure 5a and Figure S1 in Supplementary Materials), confirming an HY-FIB effect on this gene expression in the first days of culture. In dynamic conditions, COL1A1 levels displayed responses similar to the static culture in the first two days but progressively rising thereafter to significant levels (2.9 fold) at Day 11, probably due to strain input. DCN expression levels in response to dynamic culture were to be elevated throughout, achieving significance at Day 11 (3-fold) (Figure 5b and Figure S1 of Supplementary Materials).

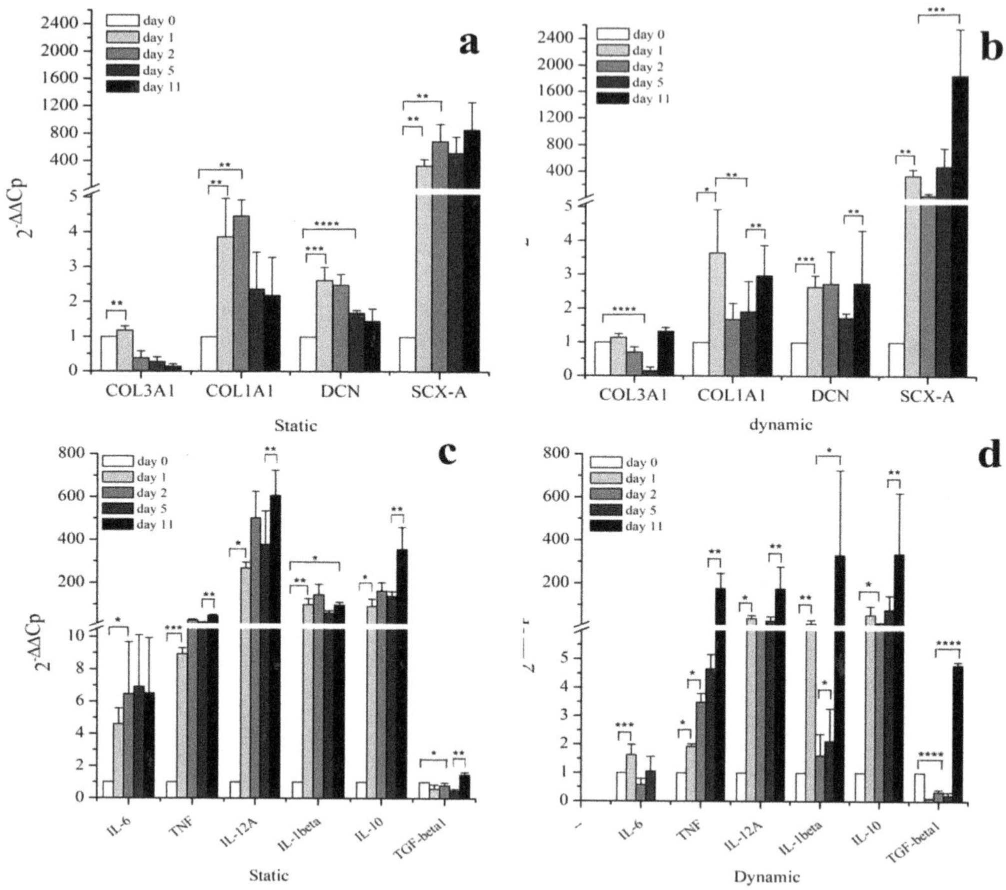

Figure 5. Gene expression profiles of tenogenic markers and pro-inflammatory and anti-inflammatory cytokines from hBM-MSCs within HY-FIB environment in static (**a**,**c**) and dynamic culture (**b**,**d**) up to 11 days. Days 1, 2, 5, and 11 were selected as time points to study the mRNA levels of positive tenogenic markers (COL1A1, DCN, SCX-A, and TNMD), negative ones (COL3A1) and pro-inflammatory (IL-6, TNF, IL-12A and IL-1β) and anti-inflammatory (IL-10 and TGF-β1) cytokines. Effect of HY-FIB environment on cells behavior was visible along the first two days of culture; a better over expression of tenogenic markers and anti-inflammatory cytokines was observed in dynamic culture at Day 11. hBM-MSCs within HY-FIB at time zero were used as control. * <0.05; ** <0.01; *** <0.005; **** <0.001. N = 3 (biological replicates); n = 3 (technical replicates).

SCX-A displayed significant upregulation (~340-fold) in both static and dynamic conditions at Day 1, suggesting an effect of HY-FIB system, on this gene expression. SCX-A levels were substantially elevated in both static and dynamic condition at all following time points studied, even if a larger and significant increase was observed in dynamic condition; an increase of 800-fold in static and of 1600-fold in dynamic culture conditions was monitored at Day 11 (Figure 5a,b and Figure S1 of Supplementary Materials). Tenomodulin gene expression was also tested by RT-qPCR, but no expression was detected, probably because it is an event occurring during late differentiation [65]. Sustained, significant, downregulation of COL3A1 was observed in either static or dynamic conditions which instead either decreased progressively (static) or decreased through to Day 5 before reestablishing Day 0 levels at Day 11 (dynamic).

The data suggested an overall effect of the 3D environment on cells behavior clearly visible along the first two days of culture; furthermore, a statistically significant COL1A1, DCN, and SCX-A over-expression was observed after 11 days when mechanical strain was provided (Figure S1 in Supplementary Materials).

Cytokine transcript expression data is illustrated in Figure 5c,d, for static and dynamic culture, respectively. HY-FIB system has an effect also on cytokines gene expression, as observed in all time points monitored with respect to Day 0, within static culture. Indeed, pro-inflammatory cytokines IL-6 (~6-fold), TNF (~10-fold), IL-12A (≤600-fold), and IL-1β (~200-fold) displayed rapid and significant upregulation that was maintained for the entirety of the experimental duration. Anti-inflammatory TGF-β1 on the other hand displayed either no change (Day 11) or significant down-regulation (other time points) while IL-10 exhibited an overall similar profile to Il-1β culminating in marked upregulation at day 11 (~300-fold) (Figure 5c).

Dynamic culture conditions had a distinct and significant effect on IL-6 with expression levels achieving a peak upregulation of 1.5-fold at Day 1 and decreasing to undetectable levels by Day 11 (Figure S2 in Supplementary Materials). TNF and IL-1β were both gradually upregulated before achieving ~200-fold and ~300-fold, respectively, upregulations at day 11 (compared to 10-fold and 100-fold in static conditions). IL-12 displayed a similar profile of upregulation in dynamic vs. static culture conditions while achieving maximal levels that were 3X less in dynamic. Anti-inflammatory IL-10 expression levels were consistent across both dynamic and static conditions. In contrast to static culture, TGF-β1 was significantly downregulated until day 5, and it underwent a 5-fold increase at Day 11 in dynamic culture conditions (Figure 5d and Figure S2 in Supplementary Materials).

Histological characterization of HY-FIB scaffold in both static and dynamic culture at Days 5 and 11 are reported in Figure 6; the overall scaffold structure was stained with Sirius Red for collagen highlighting. Despite fibrin hydrogel matrix, collected at Day 0, was only light pink stained, the same matrix was clearly stained in red at Day 5 and 11 in both samples taken from static and dynamic culture. However, a less homogeneous matrix organization and staining was observed in the samples taken from static culture. This data is in agreement with gene expression indications and confirmed that both HY-FIB alone and HY-FIB plus cyclic strain had an effect on cells phenotype commitment.

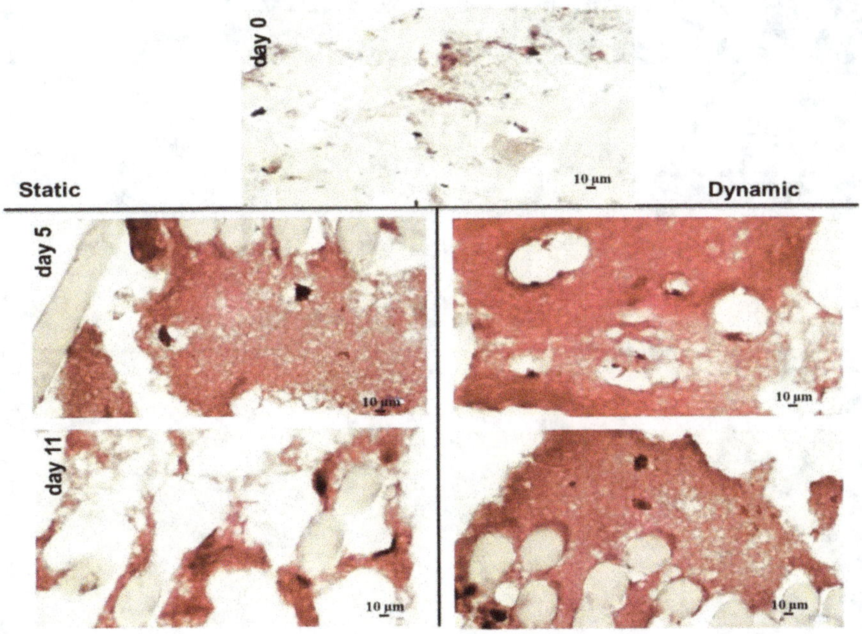

Figure 6. Histology characterization of the overall HY-FIB scaffold structure with Sirius Red staining. HY-FIB scaffolds in both static and dynamic culture at Days 5 and 11 are reported; the overall scaffold

structure was stained with Sirius Red for collagen highlighting. Fibrin hydrogel was light pink stained in the sample collected at Day 0. Fibrin matrix was clearly stained in red at Day 5 and 11 in both samples from static and dynamic culture. Less homogeneous scaffold matrix structure and staining was observed in samples taken from static culture.

The expression of type 1 Collagen, a tenogenic matrix-associated marker, was monitored by immunofluorescence over the culture time (see Figure 7). At day 1, empty areas surrounding the cells are present, probably due to absence of uniform fibrin hydrogel. These spaces were then progressively filled with the protein, presumably via secretion into the extracellular environment. The level of staining observed under static conditions decreased after Day 1 and was maintained at ~50% of original levels thereafter while levels were maintained consistent to Day 0 in dynamic culture. Moreover, in the dynamic condition a more uniform cells distribution was noted throughout the hydrogel matrix, especially at day 11.

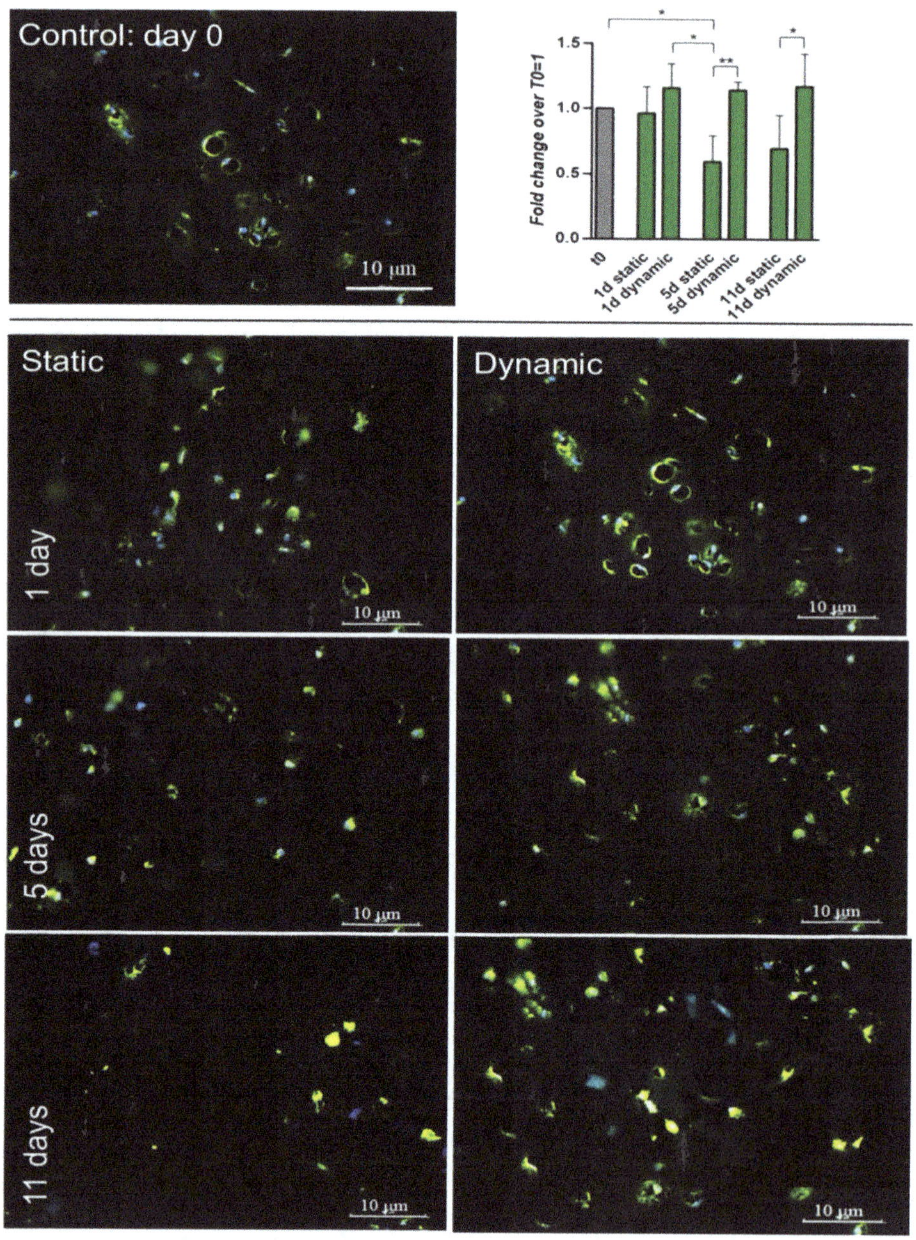

Figure 7. IF and quantitative-IF assays of type 1 Collagen (COL1A1) in 3D static and dynamic2 culture of hBM-MSCs for 11 days. At day 1, holes within the hydrogel structure are evident nearby cells

surroundings. These areas were progressively filled by COL1A1 protein (stained in green) and it happened more uniformly during the dynamic cultivation (see day 11). Fluorescence quantification by ImageJ software, reported in the plot, confirmed the presence of more abundant signal (0.2–0.3 fold changes) in dynamic condition; signal intensity in each time point was normalized by cell number (e.g., by amount of cell nuclei revealed by DAPI staining). * ≤0.05; ** <0.01. N = 3 (biological replicates); n = 3 (technical replicates).

4. Discussion

The HY-FIB system is engineered to support delivery of PLGA nanocarriers (PLGA-NCs) within the hydrogel matrix, enabling controlled delivery of specific molecules within 3D environment, e.g., drugs or other biological signals. The active form of hGDF-5 loaded into PLGA-NCs for controlled delivery within HY-FIB environment was investigated in a previous study [53]. Here, we investigated the effect of the HY-FIB 3D environment (hyaluronate band + PLGA carriers + fibrin gel) on hBM-MSCs tenogenic and cytokine marker gene expression in both static and dynamic, mechanical, input scenarios. We adopted the previous HY-FIB configuration including PLGA-NCs, but on this occasion, we delivered an inactive form of hGDF5. In this manner, the biochemical input provided by the growth factor was excluded, but the complete HY-FIB configuration was maintained, and we were thus able to investigate the impact of mechanical input alone.

HY-FIB braided fibers enabled a defined mechanical stimulation of 9×10^{-2} Pa provided to hBM-MSCs during the 4h/day dynamic culture regime. The mean shear stress was calculated by FEM modeling [62], assuming a system homogeneous behavior at a density of 1050 kg/m^3 and Young's modulus of 4.56 Mpa [53]. A Poisson ratio value of 0.25 was adopted as described elsewhere [62]. Further mechanical inputs with different intensities and durations were not investigated, not being the aim of the present work. Stress values resembling reduced physiological activity, similar that the ones used here, have been reported to direct stem cell commitment to a tenogenic phenotype [31,36,66].

COL1A1 is the major component of tendon tissue (75–85% of the dry mass of tendon), and is responsible for its mechanical strength [64]. In the static group, COL1A1 showed a ≥3-fold upregulation during the first and second day of cultivation. These data seem to suggest an overall effect of the 3D environment on cells behavior. COL1A1 expression was progressively reduced to a 2-fold upregulation at Day 11, in static environment. In dynamic conditions, its mRNA levels showed similar behavior during the first two days of cultivation (an increase up to 3-fold-changes, then reduced at Day 2). However, its expression was subsequently increased again to 2.8-fold at Day 11. Decorin (DCN), a small leucine-rich proteoglycan implicated in the regulation of fibrillogenesis, is a fundamental component of the tendon extracellular matrix (ECM) [67]. Compared to the static condition, a significant enhancement, up to 2.5-fold, of the mRNA level of DCN was shown when hBM-MSCs were cultured for 11 days with mechanical stimulation.

Scleraxis-A (SCX-A) is a neotendon marker, expressed in pro-tendon sites in the developing embryo. Specifically, SCX-A is a tendon-specific basic helix-loop-helix transcription factor responsible for the transition of MSCs into tendon progenitors [68]. We observed substantial increases in SCX-A expression, up to 800-fold in static and 1600-fold in dynamic conditions after 11 days, demonstrating a stimulatory effect via the 3D system organization and consistent with previous observations [69–71].

COL3A1 mRNA level seems to be downregulated after 2 days in the static group and after 5 days in the dynamic group. Its downregulation can be considered a positive indication of proper cell differentiation; indeed, it seems that COL3A1 is the main responsible of fibrotic and scarred tissue arrangement and has been consistently reported at the rupture site of human tendons. [64].

From both histology and immunofluorescence assays, we noted that the area surrounding the cells was progressively filled by type 1 Collagen and, at Day 11, the extracellular matrix seemed to undergo remodeling (Figures 6 and 7). Moreover, in dynamic conditions a more homogeneous cell distribution within the hydrogel matrix was observed in the IF images. These findings support the concept that 3D cultivation provides cues to the hBM-MSCs, and that dynamic signaling enables the adoption of a more uniform behavior including type 1 Collagen protein deposition in the externally available space of the fibrin hydrogel. A near total absence of type 3 Collagen was found, except for a very small fluorescence signal at day 1, in both static and dynamic conditions (data not shown). These data suggest that tenogenic commitment of hBM-MSCs cultured within HY-FIB environment may be enhanced when dynamic stretching is applied.

MSCs secrete a variety of cytokines and growth factors that promote cell recruitment, migration, proliferation, and differentiation. MSCs are also immunomodulatory, which may allow them to exert beneficial effects on the local immune cell population at the site of muscle injury [72]. To better understand the hBM-MSCs inflammatory response when cultured within HY-FIB, cytokine expression was monitored along the 11 days of culture. The balance between pro- and anti-inflammatory soluble factors in the tendon healing process exerts a major impact on successful resolution of inflammation [73]. Recent analysis of tendinopathy biopsies showed a distinct inflammatory infiltrate in the initial phase of tendinopathy with a high content of pro-inflammatory factors such as IL-6, TNF-α and IL-17 [74].

To exclude a role for ihGDF5 in cytokine expression induction we evaluated their expression in hBM-MSCs undergoing 2D planar cultivation as a negative control. Indeed, in disc degeneration models using in vitro three-dimensional cultures, human annulus cells display increased expression of pro-inflammatory cytokines, such as IL-1β and TNF-α, while exposure to TNF-α and IL-1β resulted in significant downregulation of GDF-5 [75]. Therefore, it is plausible that the GDF-5 may upregulate the expression of pro-inflammatory genes in hBM-MSCs leading to the maintenance of an autocrine feedback. However, when the ihGDF-5 was added, no statistically significant expression of pro-inflammatory cytokines was observed; therefore, ihGDF-5 did not exert any effect on cytokines expression.

The addition of PLGA-NCs enabled an informed analysis of the inductive role of the HY-FIB overall structure. Previous studies have noted that cytotoxicity of SEE-fabricated PLGA-NCs on blood mononucleate viability, monitored with MTT assay [76], was not affected after either 24 or 48 h. Here, the overall HY-FIB system (loaded with SEE fabricated NCs) does not evidence any toxic effect on hBM-MSC cultivated within it for 11 days, providing an indirect indication about SEE technology as suitable process for biomedical carriers production.

In general, we observed that pro-inflammatory gene expression was higher in static than in dynamic conditions at all-time points. On the contrary, the anti-inflammatory cytokines IL-10 was consistently upregulated in both static and dynamic conditions; while TGF-β1 was downregulated at all the time points tested except day 11, when, it showed a marked increase (4-fold) only in dynamic environment. The described behavior confirmed that MSCs respond to a variety of biophysical cues; indeed, as suggested by Qazi et al., 3D culture of MSCs on biomaterials can promote cell-cell interactions and enhance the paracrine effects of MSCs [77]. Moreover, as concluded by Ogle et al., historically, biomaterial-based therapies to promote tissue regeneration were designed to minimize the host inflammatory response. Recently, the roles that monocytes and macrophages can play in tissue repair have been highlighted. In this context, material properties and their possibility of specific biomolecule controlled delivery has been engineered to achieve a given biological response that can be tuned not only to a better integration with biological systems but also in regulating the inflammatory response [5].

The overall and statistically significant balance of pro- vs. anti-inflammatory cytokines expressed by cells provided indications regarding the importance of dynamic culture for 3D in vitro model systems. For instance, IL-6, a well-known pleiotropic cytokine delivered by tissues in response to physio-pathological changes such as physical exercise, infection, and injury, was reported to deeply alter skeletal muscle milieu, by affecting the activity and quality of cellular interactors during tissue regeneration and leading to the fibrotic response [78]. In our 3D model system, IL-6 gene expression was considerably reduced in hBM-MSCs that underwent dynamic 3D HY-FIB cultivation when compared to the same cells cultivated in static condition.

It is worth of note that there is no specific literature on cytokines response by hBM-MSCs cultivated within 3D scaffold. Almost all published studies described cell-specific differentiation toward a given phenotype, without considering how cytokines expression may be related to a 3D in vitro scaffold system. In this sense, the present investigation is the first study, which suggests cytokines expression as a further variable to monitor cell behavior and reaction when loaded into a 3D in-vitro model. Moreover, improved balance in anti-inflammatory cytokines observed for HY-FIB plus cyclic strain may be considered an indication of better cells response to the 3D in vitro system designed and proposed.

5. Conclusions

The 3D cell culture yielded evidence of type 1 Collagen expression observed by both immunohistology and gene assay. When the same 3D system was cultivated under cyclic strain stimulation, the mechanical input stimulated a statistically significant increase in tenogenic markers expression when compared to the same cells assembled into the 3D system, but cultivated in a static culture. Further studies may involve a deeper understanding of the relation between collagen type I production, cell commitment and mechanical input strain percentage or duration; in this sense, HY-FIB system can be considered a good instrument for this study. The 3D culture system activated also the expression of pro-inflammatory cytokines, and, when cyclic strain was applied, pro-inflammatory cytokine gene over-expression by hBM-MSCs was better balanced against over-expression of anti-inflammatory cytokines. It remains to be determined what the involvement and the immunomodulatory activity of hBM-MSCs are, and the role of implantable biomaterials in the stimulation of inflammatory reactions. For instance, the stimulation of local inflammation is reported as an important event in triggering repair in avascular tissues, such as cartilage and tendons [5].

On the other hand, the presence of PLGA-NCs within the fibrin hydrogel would allow the delivery of specific biomolecules that may be studied for the ability to further modulate inflammation reactions or promote regeneration/repair events. In this sense, HY-FIB provides a potential strategic approach to address a range of issues via the provision of a tightly controlled in vitro protocol. The 3D scaffold is a potential system to organize the sustained release of different biochemical signals and opens concrete perspectives for developing 3D bioengineered models to understand specific molecular and cellular composition of damaged systems.

Author Contributions: M.C.C. developed the experimental activity and optimized the protocols and methodology; she was also responsible for the paper draft preparation; L.M. isolated the stem cells and characterized them with formal analysis and validated methodology; J.L. followed the cyclic strain bioreactor protocols and provided the FEM data; E.G. supervised the bioreactor protocols and data acquisitions; N.R.F. provided contribution in data supervision and paper writing; C.S. provided the bone marrow aspirate and the methodology for hBM-MSCs cultivation; N.M. helped in the interpretation of the data, reviewed the manuscript, and was responsible for funding acquisition and management; G.D.P. was responsible for all experimental data production, curation and supervision, paper writing and editing, and research project administration. All authors have read and agreed to the published version of the manuscript.

Acknowledgments: The authors acknowledge MiUR within the framework of PON-RI 2014/2020. Action I.1–"Innovative PhDs with industrial characterization" Cicle XXXIII (D.D. n 0001377 June 5th, 2017) for the PhD fellowship entitled: "Scaffold innovation for the cure of tendon disorders: new generation of poly-hyaluronate functionalized biocomposites. The authors acknowledge Devis Galesso and R&D team at Fidia Farmaceutici S.p.A (Abano Terme, PD, Italy) for their participation into the PhD training program. We also thanks Annamaria Giordano for help in cells culture during her laboratory internship.

Patents: The SEE technology for nanocarriers fabrication was described in the US Patent US/8628802 B2 Jan 2014. Inventors: Reverchon E., Della Porta G., Continuous process for microspheres production by using expanded fluids. Applicant: University of Salerno.

References

1. Ahmed, S.; Chauhan, V.M.; Ghaemmaghami, A.M.; Aylott, J.W. New generation of bioreactors that advance extracellular matrix modelling and tissue engineering. *Biotechnol. Lett.* **2019**, *41*, 1–25. [CrossRef] [PubMed]
2. Okamoto, M. The role of Scaffolds in Tissue Engineering. In *Handbook of Tissue Engineering Scaffolds: Volume One*; Elsevier: Amsterdam, The Netherlands, 2019; pp. 23–49. ISBN 978-0-08-102563-5.
3. D'Angelo, M.; Benedetti, E.; Tupone, M.G.; Catanesi, M.; Castelli, V.; Antonosante, A.; Cimini, A. The Role of Stiffness in Cell Reprogramming: A Potential Role for Biomaterials in Inducing Tissue Regeneration. *Cells* **2019**, *8*, 1036. [CrossRef] [PubMed]
4. Cipollaro, L.; Ciardulli, M.C.; Della Porta, G.; Peretti, G.M.; Maffulli, N. Biomechanical issues of tissue-engineered constructs for articular cartilage regeneration: In vitro and in vivo approaches. *Br. Med. Bull.* **2019**, *132*, 53–80. [CrossRef] [PubMed]
5. Ogle, M.E.; Segar, C.E.; Sridhar, S.; Botchwey, E.A. Monocytes and macrophages in tissue repair: Implications for immunoregenerative biomaterial design. *Exp. Biol. Med.* **2016**, *241*, 1084–1097. [CrossRef] [PubMed]
6. Vinhas, A.; Rodrigues, M.T.; Gomes, M.E. Exploring Stem Cells and Inflammation in Tendon Repair and Regeneration. In *Cell Biology and Translational Medicine, Volume 2*; Turksen, K., Ed.; Springer International Publishing: Cham, Germany, 2018; Volume 1089, pp. 37–46. ISBN 978-3-030-04169-4.
7. Rees, J.D.; Stride, M.; Scott, A. Tendons—Time to revisit inflammation. *Br. J. Sports Med.* **2014**, *48*, 1553–1557. [CrossRef] [PubMed]
8. Dakin, S.G.; Martinez, F.O.; Yapp, C.; Wells, G.; Oppermann, U.; Dean, B.J.F.; Smith, R.D.J.; Wheway, K.; Watkins, B.; Roche, L.; et al. Inflammation activation and resolution in human tendon disease. *Sci. Transl. Med.* **2015**, *7*, 311ra173. [CrossRef] [PubMed]
9. John, T.; Lodka, D.; Kohl, B.; Ertel, W.; Jammrath, J.; Conrad, C.; Stoll, C.; Busch, C.; Schulze-Tanzil, G. Effect of pro-inflammatory and immunoregulatory cytokines on human tenocytes. *J. Orthop. Res.* **2010**, *28*, 1071–1077. [CrossRef]
10. Manning, C.N.; Havlioglu, N.; Knutsen, E.; Sakiyama-Elbert, S.E.; Silva, M.J.; Thomopoulos, S.; Gelberman, R.H. The early inflammatory response after flexor tendon healing: A gene expression and histological analysis: Early inflammatory response after flexor tendon healing. *J. Orthop. Res.* **2014**, *32*, 645–652. [CrossRef]
11. Marsolais, D.; Coté, C.H.; Frenette, J. Neutrophils and macrophages accumulate sequentially following Achilles tendon injury. *J. Orthop. Res.* **2001**, *19*, 1203–1209. [CrossRef]
12. Chisari, E.; Rehak, L.; Khan, W.S.; Maffulli, N. The role of the immune system in tendon healing: A systematic review. *Br. Med. Bull.* **2020**, *133*, 49–64. [CrossRef]
13. Giordano, L.; Della Porta, G.; Peretti, G.M.; Maffulli, N. Therapeutic potential of microRNA in tendon injuries. *Br. Med. Bull.* **2020**, *133*, 79–94. [CrossRef] [PubMed]
14. Oryan, A.; Moshiri, A.; Meimandi Parizi, A.; Maffulli, N. Implantation of a Novel Biologic and Hybridized Tissue Engineered Bioimplant in Large Tendon Defect: An In Vivo Investigation. *Tissue Eng. Part A* **2013**, *20*, 447–465. [CrossRef] [PubMed]
15. Gaspar, D.; Spanoudes, K.; Holladay, C.; Pandit, A.; Zeugolis, D. Progress in cell-based therapies for tendon repair. *Adv. Drug Deliv. Rev.* **2015**, *84*, 240–256. [CrossRef] [PubMed]
16. Lui, P.P.Y.; Wong, O.T. Tendon stem cells: Experimental and clinical perspectives in tendon and tendon-bone junction repair. *Muscles Ligaments Tendons J.* **2012**, *2*, 163–168.
17. Chang, C.-H.; Chen, C.-H.; Liu, H.-W.; Whu, S.-W.; Chen, S.-H.; Tsai, C.-L.; Hsiue, G.-H. Bioengineered Periosteal Progenitor Cell Sheets to Enhance Tendon-Bone Healing in A Bone Tunnel. *Biomedical* **2012**, *35*, 473. [CrossRef]

18. Ito, R.; Matsumiya, T.; Kon, T.; Narita, N.; Kubota, K.; Sakaki, H.; Ozaki, T.; Imaizumi, T.; Kobayashi, W.; Kimura, H. Periosteum-derived cells respond to mechanical stretch and activate Wnt andBMP signaling pathways. *Biomed. Res.* **2014**, *35*, 69–79. [CrossRef]
19. Doroski, D.M.; Levenston, M.E.; Temenoff, J.S. Cyclic Tensile Culture Promotes Fibroblastic Differentiation of Marrow Stromal Cells Encapsulated in Poly(Ethylene Glycol)-Based Hydrogels. *Tissue Eng. Part A* **2010**, *16*, 3457–3466. [CrossRef]
20. Chaudhury, S. Mesenchymal stem cell applications to tendon healing. *Muscles Ligaments Tendons J.* **2012**, *2*, 222–229.
21. Youngstrom, D.W.; LaDow, J.E.; Barrett, J.G. Tenogenesis of bone marrow-, adipose-, and tendon-derived stem cells in a dynamic bioreactor. *Connect. Tissue Res.* **2016**, *57*, 454–465. [CrossRef]
22. Bi, Y.; Ehirchiou, D.; Kilts, T.M.; Inkson, C.A.; Embree, M.C.; Sonoyama, W.; Li, L.; Leet, A.I.; Seo, B.-M.; Zhang, L.; et al. Identification of tendon stem/progenitor cells and the role of the extracellular matrix in their niche. *Nat. Med.* **2007**, *13*, 1219–1227. [CrossRef]
23. Yang, G.; Rothrauff, B.B.; Lin, H.; Gottardi, R.; Alexander, P.G.; Tuan, R.S. Enhancement of tenogenic differentiation of human adipose stem cells by tendon-derived extracellular matrix. *Biomaterials* **2013**, *34*, 9295–9306. [CrossRef] [PubMed]
24. Govoni, M.; Lotti, F.; Biagiotti, L.; Lannocca, M.; Pasquinelli, G.; Valente, S.; Muscari, C.; Bonafè, F.; Caldarera, C.M.; Guarnieri, C.; et al. An innovative stand-alone bioreactor for the highly reproducible transfer of cyclic mechanical stretch to stem cells cultured in a 3D scaffold: A novel stand-alone bioreactor to induce cell muscle phenotype. *J. Tissue Eng. Regen. Med.* **2014**, *8*, 787–793. [CrossRef] [PubMed]
25. Lovecchio, J.; Gargiulo, P.; Vargas Luna, J.L.; Giordano, E.; Sigurjonsson, O.E. A standalone bioreactor system to deliver compressive load under perfusion flow to hBMSC-seeded 3D chitosan-graphene templates. *Sci. Rep.* **2019**, *9*, 16854. [CrossRef] [PubMed]
26. Barber, J.G.; Handorf, A.M.; Allee, T.J.; Li, W.-J. Braided Nanofibrous Scaffold for Tendon and Ligament Tissue Engineering. *Tissue Eng. Part A* **2013**, *19*, 1265–1274. [CrossRef] [PubMed]
27. Wang, T.; Lin, Z.; Day, R.E.; Gardiner, B.; Landao-Bassonga, E.; Rubenson, J.; Kirk, T.B.; Smith, D.W.; Lloyd, D.G.; Hardisty, G.; et al. Programmable mechanical stimulation influences tendon homeostasis in a bioreactor system. *Biotechnol. Bioeng.* **2013**, *110*, 1495–1507. [CrossRef]
28. Delaine-Smith, R.M.; Reilly, G.C. Mesenchymal stem cell responses to mechanical stimuli. *Muscles Ligaments Tendons J.* **2016**, *2*, 169–180.
29. Chen, Y.-J.; Huang, C.-H.; Lee, I.-C.; Lee, Y.-T.; Chen, M.-H.; Young, T.-H. Effects of Cyclic Mechanical Stretching on the mRNA Expression of Tendon/Ligament-Related and Osteoblast-Specific Genes in Human Mesenchymal Stem Cells. *Connect. Tissue Res.* **2008**, *49*, 7–14. [CrossRef]
30. Govoni, M.; Muscari, C.; Lovecchio, J.; Guarnieri, C.; Giordano, E. Mechanical Actuation Systems for the Phenotype Commitment of Stem Cell-Based Tendon and Ligament Tissue Substitutes. *Stem Cell Rev. Rep.* **2016**, *12*, 189–201. [CrossRef]
31. Youngstrom, D.W.; Rajpar, I.; Kaplan, D.L.; Barrett, J.G. A bioreactor system for in vitro tendon differentiation and tendon tissue engineering. *J. Orthop. Res.* **2015**, *33*, 911–918. [CrossRef]
32. Testa, S.; Costantini, M.; Fornetti, E.; Bernardini, S.; Trombetta, M.; Seliktar, D.; Cannata, S.; Rainer, A.; Gargioli, C. Combination of biochemical and mechanical cues for tendon tissue engineering. *J. Cell. Mol. Med.* **2017**, *21*, 2711–2719. [CrossRef]
33. Woon, C.Y.L.; Kraus, A.; Raghavan, S.S.; Pridgen, B.C.; Megerle, K.; Pham, H.; Chang, J. Three-Dimensiona l-Construct Bioreactor Conditioning in Human Tendon Tissue Engineering. *Tissue Eng. Part A* **2011**, *17*, 2561–2572. [CrossRef] [PubMed]
34. Rinoldi, C.; Fallahi, A.; Yazdi, I.K.; Campos Paras, J.; Kijeńska-Gawrońska, E.; Trujillo-de Santiago, G.; Tuoheti, A.; Demarchi, D.; Annabi, N.; Khademhosseini, A.; et al. Mechanical and Biochemical Stimulation of 3D Multilayered Scaffolds for Tendon Tissue Engineering. *ACS Biomater. Sci. Eng.* **2019**, *5*, 2953–2964. [CrossRef]
35. Nam, H.Y.; Pingguan-Murphy, B.; Abbas, A.A.; Merican, A.M.; Kamarul, T. Uniaxial Cyclic Tensile Stretching at 8% Strain Exclusively Promotes Tenogenic Differentiation of Human Bone Marrow-Derived Mesenchymal Stromal Cells. *Stem Cells Int.* **2019**, *2019*, 1–16. [CrossRef] [PubMed]

36. Grier, W.G.; Moy, A.S.; Harley, B.A. Cyclic tensile strain enhances human mesenchymal stem cell Smad 2/3 activation and tenogenic differentiation in anisotropic collagen-glycosaminoglycan scaffolds. *Eur. Cell Mater.* **2017**, *33*, 227–239. [CrossRef] [PubMed]
37. Kay, A.G.; Dale, T.P.; Akram, K.M.; Mohan, P.; Hampson, K.; Maffulli, N.; Spiteri, M.A.; El Haj, A.J.; Forsyth, N.R. BMP2 repression and optimized culture conditions promote human bone marrow-derived mesenchymal stem cell isolation. *Regen. Med.* **2015**, *10*, 109–125. [CrossRef]
38. Correia, S.I.; Pereira, H.; Silva-Correia, J.; Van Dijk, C.N.; Espregueira-Mendes, J.; Oliveira, J.M.; Reis, R.L. Current concepts: Tissue engineering and regenerative medicine applications in the ankle joint. *J. R. Soc. Interface* **2013**, *11*, 20130784. [CrossRef]
39. Keller, T.C.; Hogan, M.V.; Kesturu, G.; James, R.; Balian, G.; Chhabra, A.B. Growth/differentiation factor-5 modulates the synthesis and expression of extracellular matrix and cell-adhesion-related molecules of rat Achilles tendon fibroblasts. *Connect. Tissue Res.* **2011**, *52*, 353–364. [CrossRef]
40. Hogan, M.; Girish, K.; James, R.; Balian, G.; Hurwitz, S.; Chhabra, A.B. Growth differentiation factor-5 regulation of extracellular matrix gene expression in murine tendon fibroblasts. *J. Tissue Eng. Regen. Med.* **2011**, *5*, 191–200. [CrossRef]
41. Tan, S.-L.; Ahmad, R.E.; Ahmad, T.S.; Merican, A.M.; Abbas, A.A.; Ng, W.M.; Kamarul, T. Effect of Growth Differentiation Factor 5 on the Proliferation and Tenogenic Differentiation Potential of Human Mesenchymal Stem Cells in vitro. *Cells Tissues Organs* **2012**, *196*, 325–338. [CrossRef]
42. Lui, P.P. Stem cell technology for tendon regeneration: Current status, challenges, and future research directions. *SCCAA* **2015**, *8*, 163. [CrossRef]
43. MacLean, S.; Khan, W.S.; Malik, A.A.; Snow, M.; Anand, S. Tendon Regeneration and Repair with Stem Cells. *Stem Cells Int.* **2012**, *2012*, 1–6. [CrossRef] [PubMed]
44. Liu, L.; Hindieh, J.; Leong, D.J.; Sun, H.B. Advances of stem cell based-therapeutic approaches for tendon repair. *J. Orthop. Transl.* **2017**, *9*, 69–75. [CrossRef] [PubMed]
45. Caplan, A.I. Adult mesenchymal stem cells for tissue engineering versus regenerative medicine. *J. Cell. Physiol.* **2007**, *213*, 341–347. [CrossRef] [PubMed]
46. Akram, K.M.; Samad, S.; Spiteri, M.A.; Forsyth, N.R. Mesenchymal stem cells promote alveolar epithelial cell wound repair in vitro through distinct migratory and paracrine mechanisms. *Respir. Res.* **2013**, *14*, 9. [CrossRef]
47. Pontikoglou, C.; Deschaseaux, F.; Sensebé, L.; Papadaki, H.A. Bone Marrow Mesenchymal Stem Cells: Biological Properties and Their Role in Hematopoiesis and Hematopoietic Stem Cell Transplantation. *Stem Cell Rev. Rep.* **2011**, *7*, 569–589. [CrossRef]
48. Dominici, M.; Le Blanc, K.; Mueller, I.; Slaper-Cortenbach, I.; Marini, F.C.; Krause, D.S.; Deans, R.J.; Keating, A.; Prockop, D.J.; Horwitz, E.M. Minimal criteria for defining multipotent mesenchymal stromal cells. *Cytotherapy* **2006**, *8*, 315–317. [CrossRef]
49. Chamberlain, G.; Fox, J.; Ashton, B.; Middleton, J. Concise Review: Mesenchymal Stem Cells: Their Phenotype, Differentiation Capacity, Immunological Features, and Potential for Homing. *Stem Cells* **2007**, *25*, 2739–2749. [CrossRef]
50. Mosna, F.; Sensebé, L.; Krampera, M. Human Bone Marrow and Adipose Tissue Mesenchymal Stem Cells: A User's Guide. *Stem Cells Dev.* **2010**, *19*, 1449–1470. [CrossRef]
51. Kumar, D.; Gerges, I.; Tamplenizza, M.; Lenardi, C.; Forsyth, N.R.; Liu, Y. Three-dimensional hypoxic culture of human mesenchymal stem cells encapsulated in a photocurable, biodegradable polymer hydrogel: A potential injectable cellular product for nucleus pulposus regeneration. *Acta Biomater.* **2014**, *10*, 3463–3474. [CrossRef]
52. Bullough, R.; Finnigan, T.; Kay, A.; Maffulli, N.; Forsyth, N.R. Tendon repair through stem cell intervention: Cellular and molecular approaches. *Disabil. Rehabil.* **2008**, *30*, 1746–1751. [CrossRef]
53. Govoni, M.; Berardi, A.C.; Muscari, C.; Campardelli, R.; Bonafè, F.; Guarnieri, C.; Reverchon, E.; Giordano, E.; Maffulli, N.; Della Porta, G. An Engineered Multiphase Three-Dimensional Microenvironment to Ensure the Controlled Delivery of Cyclic Strain and Human Growth Differentiation Factor 5 for the Tenogenic Commitment of Human Bone Marrow Mesenchymal Stem Cells. *Tissue Eng. Part A* **2017**, *23*, 811–822. [CrossRef] [PubMed]
54. Della Porta, G.; Ciardulli, M.C.; Maffulli, N. Microcapsule Technology for Controlled Growth Factor Release in Musculoskeletal Tissue Engineering. *Sports Med. Arthrosc. Rev.* **2018**, *26*, e2–e9. [CrossRef] [PubMed]

55. Giordano, R.; Canesi, M.; Isalberti, M.; Isaias, I.; Montemurro, T.; Viganò, M.; Montelatici, E.; Boldrin, V.; Benti, R.; Cortelezzi, A.; et al. Autologous mesenchymal stem cell therapy for progressive supranuclear palsy: Translation into a phase I controlled, randomized clinical study. *J. Transl. Med.* **2014**, *12*, 14. [CrossRef] [PubMed]
56. Jensen, E.C. Quantitative Analysis of Histological Staining and Fluorescence Using ImageJ: Histological Staining/Fluorescence Using ImageJ. *Anat. Rec.* **2013**, *296*, 378–381. [CrossRef]
57. Rinoldi, C.; Costantini, M.; Kijeńska-Gawrońska, E.; Testa, S.; Fornetti, E.; Heljak, M.; Ćwiklińska, M.; Buda, R.; Baldi, J.; Cannata, S.; et al. Tendon Tissue Engineering: Effects of Mechanical and Biochemical Stimulation on Stem Cell Alignment on Cell-Laden Hydrogel Yarns. *Adv. Healthc. Mater.* **2019**, *8*, 1801218. [CrossRef]
58. Bustin, S.A.; Benes, V.; Garson, J.A.; Hellemans, J.; Huggett, J.; Kubista, M.; Mueller, R.; Nolan, T.; Pfaffl, M.W.; Shipley, G.L.; et al. The MIQE Guidelines: Minimum Information for Publication of Quantitative Real-Time PCR Experiments. *Clin. Chem.* **2009**, *55*, 611–622. [CrossRef]
59. Hellemans, J.; Mortier, G.; De Paepe, A.; Speleman, F.; Vandesompele, J. qBase relative quantification framework and software for management and automated analysis of real-time quantitative PCR data. *Genome Biol.* **2007**, *8*, R19. [CrossRef]
60. Della Porta, G.; Falco, N.; Reverchon, E. Continuous supercritical emulsions extraction: A new technology for biopolymer microparticles production. *Biotechnol. Bioeng.* **2011**, *108*, 676–686. [CrossRef]
61. Della Porta, G.; Falco, N.; Giordano, E.; Reverchon, E. PLGA microspheres by Supercritical Emulsion Extraction: A study on insulin release in myoblast culture. *J. Biomater. Sci. Polym. Ed.* **2013**, *24*, 1831–1847. [CrossRef]
62. Duong, H.; Wu, B.; Tawil, B. Modulation of 3D Fibrin Matrix Stiffness by Intrinsic Fibrinogen–Thrombin Compositions and by Extrinsic Cellular Activity. *Tissue Eng. Part A* **2009**, *15*, 1865–1876. [CrossRef]
63. De Winter, J.C.F. Using the Student's t-test with extremely small sample sizes. *Pract. Assess Res. Eval.* **2013**, *18*, 10.
64. Pajala, A.; Melkko, J.; Leppilahti, J.; Ohtonen, P.; Soini, Y.; Risteli, J. Tenascin-C and type I and III collagen expression in total Achilles tendon rupture. An immunohistochemical study. *Histol. Histopathol.* **2009**, *24*, 1207–1211. [CrossRef] [PubMed]
65. Shukunami, C.; Yoshimoto, Y.; Takimoto, A.; Yamashita, H.; Hiraki, Y. Molecular characterization and function of tenomodulin, a marker of tendons and ligaments that integrate musculoskeletal components. *Jpn. Dent. Sci. Rev.* **2016**, *52*, 84–92. [CrossRef] [PubMed]
66. Wang, T.; Thien, C.; Wang, C.; Ni, M.; Gao, J.; Wang, A.; Jiang, Q.; Tuan, R.S.; Zheng, Q.; Zheng, M.H. 3D uniaxial mechanical stimulation induces tenogenic differentiation of tendon-derived stem cells through a PI3K/AKT signaling pathway. *FASEB J.* **2018**, *32*, 4804–4814. [CrossRef]
67. Wagenhäuser, M.U.; Pietschmann, M.F.; Sievers, B.; Docheva, D.; Schieker, M.; Jansson, V.; Müller, P.E. Collagen type I and decorin expression in tenocytes depend on the cell isolation method. *Bmc Musculoskelet. Disord.* **2012**, *13*, 140. [CrossRef]
68. Alberton, P.; Popov, C.; Prägert, M.; Kohler, J.; Shukunami, C.; Schieker, M.; Docheva, D. Conversion of Human Bone Marrow-Derived Mesenchymal Stem Cells into Tendon Progenitor Cells by Ectopic Expression of Scleraxis. *Stem Cells Dev.* **2012**, *21*, 846–858. [CrossRef]
69. Bagchi, R.A.; Czubryt, M.P. Scleraxis: A New Regulator of Extracellular Matrix Formation. In *Genes and Cardiovascular Function*; Ostadal, B., Nagano, M., Dhalla, N.S., Eds.; Springer: Boston, MA, USA, 2011; pp. 57–65. ISBN 978-1-4419-7206-4.
70. Chen, X.; Yin, Z.; Chen, J.; Shen, W.; Liu, H.; Tang, Q.; Fang, Z.; Lu, L.; Ji, J.; Ouyang, H. Force and scleraxis synergistically promote the commitment of human ES cells derived MSCs to tenocytes. *Sci. Rep.* **2012**, *2*, 977. [CrossRef]
71. Scott, A.; Danielson, P.; Abraham, T.; Fong, G.; Sampaio, A.V.; Underhill, T.M. Mechanical force modulates scleraxis expression in bioartificial tendons. *J. Musculoskelet Neuronal Interact* **2011**, *11*, 124–132.
72. Gao, F.; Chiu, S.M.; Motan, D.A.L.; Zhang, Z.; Chen, L.; Ji, H.-L.; Tse, H.-F.; Fu, Q.-L.; Lian, Q. Mesenchymal stem cells and immunomodulation: Current status and future prospects. *Cell Death Dis.* **2016**, *7*, e2062. [CrossRef]

73. Sugg, K.B.; Lubardic, J.; Gumucio, J.P.; Mendias, C.L. Changes in macrophage phenotype and induction of epithelial-to-mesenchymal transition genes following acute Achilles tenotomy and repair. *J. Orthop. Res.* **2014**, *32*, 944–951. [CrossRef]
74. Millar, N.L.; Akbar, M.; Campbell, A.L.; Reilly, J.H.; Kerr, S.C.; McLean, M.; Frleta-Gilchrist, M.; Fazzi, U.G.; Leach, W.J.; Rooney, B.P.; et al. IL-17A mediates inflammatory and tissue remodelling events in early human tendinopathy. *Sci. Rep.* **2016**, *6*. [CrossRef] [PubMed]
75. Gruber, H.E.; Hoelscher, G.L.; Ingram, J.A.; Bethea, S.; Hanley, E.N. Growth and differentiation factor-5 (GDF-5) in the human intervertebral annulus cells and its modulation by IL-1ß and TNF-α in vitro. *Exp. Mol. Pathol.* **2014**, *96*, 225–229. [CrossRef] [PubMed]
76. Govoni, M.; Lamparelli, E.P.; Ciardulli, M.C.; Santoro, A.; Oliviero, A.; Palazzo, I.; Reverchon, E.; Vivarelli, L.; Maso, A.; Storni, E.; et al. Demineralized Bone Matrix Paste Formulated with biomimetic PLGA microcarriers for the Vancomycin Hydrochloride Controlled Delivery: Release Profile, Citotoxicity and Efficacy against S. aureus. *Int. J. Pharm.* **2020**, *582*, 119322. [CrossRef] [PubMed]
77. Qazi, T.H.; Mooney, D.J.; Duda, G.N.; Geissler, S. Biomaterials that promote cell-cell interactions enhance the paracrine function of MSCs. *Biomaterials* **2017**, *140*, 103–114. [CrossRef] [PubMed]
78. Forcina, L.; Miano, C.; Scicchitano, B.; Musarò, A. Signals from the Niche: Insights into the Role of IGF-1 and IL-6 in Modulating Skeletal Muscle Fibrosis. *Cells* **2019**, *8*, 232. [CrossRef]

10

Mouse Embryonic Stem Cell-Derived Ureteric Bud Progenitors Induce Nephrogenesis

Zenglai Tan [1,*], Aleksandra Rak-Raszewska [1], Ilya Skovorodkin [1] and Seppo J. Vainio [1,2,*]

1 Center for Cell Matrix Research, Faculty of Biochemistry and Molecular Medicine, Biocenter Oulu, Laboratory of Developmental Biology, Infotech Oulu, University of Oulu, Aapistie 5A, 90220 Oulu, Finland; Aleksandra.Rak-Raszewska@oulu.fi (A.R.-R.); ilya.skovorodkin@oulu.fi (I.S.)

2 Borealis Biobank of Northern Finland, Oulu Central Hospital, 90220 Oulu, Finland

* Correspondence: zenglai.tan@oulu.fi (Z.T.); seppo.vainio@oulu.fi (S.J.V.)

Abstract: Generation of kidney organoids from pluripotent stem cells (PSCs) is regarded as a potentially powerful way to study kidney development, disease, and regeneration. Direct differentiation of PSCs towards renal lineages is well studied; however, most of the studies relate to generation of nephron progenitor population from PSCs. Until now, differentiation of PSCs into ureteric bud (UB) progenitor cells has had limited success. Here, we describe a simple, efficient, and reproducible protocol to direct differentiation of mouse embryonic stem cells (mESCs) into UB progenitor cells. The mESC-derived UB cells were able to induce nephrogenesis when co-cultured with primary metanephric mesenchyme (pMM). In generated kidney organoids, the embryonic pMM developed nephron structures, and the mESC-derived UB cells formed numerous collecting ducts connected with the nephron tubules. Altogether, our study established an uncomplicated and reproducible platform to generate ureteric bud progenitors from mouse embryonic stem cells.

Keywords: mouse embryonic stem cell; differentiation protocol; ureteric bud progenitor cells; 3D kidney organoids

---

1. Introduction

Pluripotent stem cells (PSCs) possess great potential of differentiating into multiple cell types that are widely used for studies in developmental biology and regenerative medicine [1]. Kidney organoids derived from PSCs have been shown to be able to mimic the in vivo kidney structure development and function in vitro [2–4]. Renal organoids in a four-dimensional (4D) (3D plus time) culture system self-organize into highly complex tissue-specific morphology that is sufficient to model tissue development, disease, and injury [5–8]. A combination of genome editing and stem cell technologies allows for generation of personalized kidney organoids, which provide powerful tools for kidney disease treatment, drug toxicity screening, and tissue regeneration [9,10].

Recently, multiple protocols enabling generation of renal lineages from mouse and human PSCs have been published. We and several other groups reported induction of nephron progenitors, which have the potential to develop into epithelial nephron-like structures [2–4,8,11–18]. Other groups have shown the derivation of ureteric bud (UB) progenitors [19]. However, they did not show nephron progenitor induction and also lacked the connection between collecting ducts and nephrons [2,19]. A newly published study has shown generation of UB structures from PSCs, which possessed UB-like branching morphogenesis when aggregated with the primary metanephric mesenchyme (pMM) to form chimeric kidney organoids [8]. However, the protocol is technically complex, which limits its application. Analysis of these reports suggests that we still need further studies to develop simple, reproducible, and stable protocols for UB progenitor generation.

Here, we report a simple protocol to direct differentiation of mouse embryonic stem cells (mESCs) into UB progenitor cells. The newly generated UB progenitor cells have a potential to develop into ureteric bud structures and have a capacity to induce nephrogenesis when co-cultured with dissociated pMM. In reconstructed kidney organoids, the pMM developed into nephron structures and the UB progenitor cells formed collecting ducts, which also connected with the nephron tubules. The chimeric kidney organoids also display the presence of endothelial cells forming a vascular network. In conclusion, our study established an uncomplicated and reproducible method for generation of UB progenitors from PSCs that can be used for tubulogenesis induction.

2. Materials and Methods

Animal care and experimental procedures in this study were in accordance with Finnish national legislation on the use of laboratory animals, the European Convention for the protection of vertebrate animal used for experimental and other scientific purposes (ETS 123), and the EU Directive 86/609/EEC. Animal experimentation was also authorized by the Finnish National Animal Experiment Board (ELLA) as being compliant with the EU guidelines for animal research and welfare.

2.1. Mouse ES Cell Line Generation and Maintenance

mESC line Sv129S6 was obtained from the Biocenter Oulu Transgenic core facility. The Sv129S6 mESC line was derived from Taconic's W4/129S6 inbred mouse strain and has been tested to have a normal karyotype [19]. The mESC and mESC-GFP (Green Fluorescent Protein, pcDNA3.1 transfected) line was described previously [20]. It was maintained on mitotically inactivated mouse embryonic fibroblasts (MEFs) in mESC medium: Dulbecco's Modified Eagle's Medium (DMEM; Life Technologies, Waltham, MA, USA) supplemented with 10% fetal bovine serum (FBS, Gibco, Waltham, MA, USA), 1% (*v*/*v*) nonessential amino acids (Life Technologies), 100 mM 2-mercaptoethanol (Nacalai Tesque, Kyoto, Japan), and 1000 U/mL leukemia inhibitory factor (LIF, Millipore, Espoo, Finland).

2.2. Directed Differentiation of mESCs to UB Progenitors

Mouse ESCs were cultured as described above in MEF-coated 6-well plates in mESCs medium up to 70%–90% confluency. The cells were passaged on 1% geltrex-coated 24-well plates at 30,000 cells/cm^2 in 500 µL 50% ReproFF2 (ReproCELL, Glasgow, UK)—50% CM (conditioned MEF medium) supplied with 10 ng/mL fibroblast growth factor (FGF) 2 (PeproTech Nordic, Stockholm, Sweden) and 10 ng/mL Activin A (Cell Guidance Systems, Cambridge, UK). After 2 days, the cells reached 60–80% of confluency, and the medium was changed to differentiation medium: Advanced RPMI (Thermo Fisher Scientific, Waltham, MA, USA) supplemented with 8 µM CHIR99021 (Reagents Direct, Encinitas, CA, USA) and 4–8 ng/mL Noggin (R&D systems, Minneapolis, MN, USA) for 2 days, followed by treatment of the cells with Activin A (10 ng/mL) for 2 days and then 3 days with FGF9 (40 ng/mL) (PeproTech Nordic, Stockholm, Sweeden) in Advanced RPMI medium (ThermoFisher Scientific).

2.3. Gene Expression Analysis

A RNeasy kit (Qiagen, Germantown, MD, USA) was used according to the manufacturer's recommendations to extract total RNA. cDNA synthesis (First Strand cDNA Synthesis Kit, ThermoFisher Scientific, Waltham, MA, USA) was performed using standard protocols. qPCR was run using a CFX96 BioRad thermocycler. Brilliant III SYBR® Green QPCR Master Mix (Agilent Technologies, Santa Clara, CA, USA) was used according to the manufacturer's instructions. GAPDH probe served as a control to normalize the data. The data was analyzed using the $2^{-\Delta\Delta CT}$ method. Gene expression measurements were performed in triplicates on three independent experiments. The primer sequences are given in Supplementary Table S1.

2.4. *Chimera 3D Kidney Organoids Formation*

At day 9 of mESCs differentiation, which represents the UB progenitor cells stage, cells were dissociated into single cell suspension using TrypLE select (Life Technologies). After three washes in phosphate buffer saline (PBS), cells were reconstituted in organ culture medium (DMEM supplemented with 10% fetal bovine serum, 1% penicillin–streptomycin). For generation of kidney organoids, we mixed dissected and dissociated pMM cells (from CD-1 pregnant females, at E11.5), as described before [21], with differentiated UB progenitors at a ratio of 3:1, respectively. Briefly, kidney rudiments were dissected out of E11.5 embryos and treated with trypsin/pancreatin solution for 30–40 s to separate the UB from the pMM [22]. The MM cells were dissociated using 2.0 mg/mL Collagenase IV solution (Wortington, Lakewood, NJ, USA) in 0.1% bovine serum albumin (BSA) in 1×PBS. 10 min incubation was interrupted by pipetting and continued until cells were separated. The reaction was stopped by washing three times in complete organ culture medium [22]. The pMM- and mESC-derived UB cells were mixed in a 3:1 ratio and centrifuged for 4 min at 1380× g to form a pellet (5×10^4 cells) in Lo-Binding Eppendorf tubes. Following centrifugation, we carefully transferred the differentiated UB and pMM pellets to filter into Trowel culture to aggregate as an organoid. The organ culture medium was changed every 3–4 days.

For generation of whole kidney organoids, we dissected mouse kidney rudiments at E11.5 from CD-1 pregnant females. Kidney rudiments were dissociated into single cell suspension as described previously [19]. After dissociation, the embryonic kidney cells (7×10^4) were mixed with undifferentiated mESC or differentiated mESCs-derived UB progenitors (1×10^4) to make the pellet. We then continued the procedure as described above.

2.5. *Whole-Mount Immunohistochemistry*

Kidney organoids were washed 2 times with PBS and fixed with 100% cold methanol (–20 °C) for 30 min at room temperature (RT) or with 4% paraformaldehyde in PBS (organoid with GFP or dye) for 30 min at RT in the dark. After fixation, the organoids were washed at least three times in PBS and blocked in 0.1% Triton-X100 (Sigma, Lyon, France), 1% BSA, and 10% goat serum/0.02M glycine-PBS for 1–3 h at room temperature. Incubation of the organoids with primary antibodies was performed in a blocking buffer overnight at 4 °C. The samples were washed 6 times with PBS and incubated with secondary antibodies Alexa Fluor 405, 488, 568, 546, or 647 (1:1000, Life technologies) and fluorescein anti-LTL (Lotus Tetragonolobus Lectin, 1:350, #FL-1321, Vector Laboratories, Burlingame, CA, USA) overnight at 4 °C and counter-stained with Hoechst (Thermo Fisher Scientific). The primary antibodies used in stainings were: Wt1 (1:100, #05-753, Millipore), Pax2 (1:200, #PRB-276P, Covance, Cambridge, MA, USA), Troma1 (1:200, DSHB, Iowa City, IA, USA), Gata3 (1:20, #AF2605-SP, R&D Systems), E-cad (1:300, #610181, BD Biosciences, Franklin Lakes, NJ, USA), Synaptopodin (SYNPO) (1:4, #ABIN112223, antibody on line.com, Aachen, Germany), Umod (1:25, #LS-C150268, LSBio, Seattle, WA, USA), CD31 (1:100, #550274 BD Biosciences), Laminin (1:200, #L9393, Sigma), and Cleaved Caspase-3 (1:200, #9661s, Cell Signaling Technology, Leiden, Netherlands). Stained organoids were mounted with Shandon™ Immu-Mount™ (Thermo Scientific™). A Zeiss LSM780 microscope and Zeiss Axiolab (Zeiss, Oberkochen, Germany) were used for image capture and analysis.

2.6. *Nephrotoxicity Assay*

3D kidney organoids were cultured in organ culture medium supplemented with gentamicin at 5 mg/mL (#G1264, Sigma) for 48 h, or with cisplatin at 5, 20, or 50 μM (#P4394 Sigma) for 24 h after day 8 of organ culture. Organoids were then fixed with 100% cold methanol for 30 min for whole-mount immunohistochemistry. The Notch inhibitor, *N*-*S*-phenyl-glycine-*t*-butyl ester (DAPT, #D5942, Sigma), was used (10μM) to investigate toxicity towards proximal tubule development.

2.7. *Statistical Evaluation*

All data are presented as mean ± standard deviation (SD) and represent a minimum of three independent experiments. Student's two-tailed t-test was used for statistical evaluation. p-value < 0.05 was considered significant.

3. Results

3.1. *Direct Differentiation of mESCs into UB Progenitor Cells*

During development, both the nephron and ureteric bud progenitor cells are derived from the intermediate mesoderm (IM). To establish a protocol and direct differentiation of mESCs into UB lineage, we first differentiated mESCs into IM (Figure 1A). We treated the mESCs with FGF2 and activin A to differentiate mESCs into epiblast in monolayer culture (Supplementary Figure S1A). The differentiated cells showed expression of epiblast markers such as Fgf5 and T (Brachyury) (Supplementary Figure S1B). We then used glycogen synthase kinase-3β (GSK-3β) inhibitor, CHIR99021 (CHIR), together with a low concentration of bone morphogenetic protein (BMP) signaling inhibitor, Noggin (suppresses number of cells differentiated to lateral plate mesoderm) [2], to activate differentiation of mouse epiblast cells into primitive streak (PS). The Noggin-treated cells expressed PS markers Cdx2, T, Tbx6, and Mixl1 (Supplementary Figure S1B). Activin A has been previously used for specification of the mesodermal cells towards intermediate mesoderm [19]. Therefore, we followed for two days with activin A treatment, which differentiated the cells to the IM stage and expressed Osr1, Lhx1, and Pax2 (Supplementary Figure S1B).

Previous studies have demonstrated that FGF9 is able to induce renal lineage differentiation from the IM population [2]. Therefore, we treated these cells with a moderate concentration of FGF9 for an additional three days, directing them to differentiate into UB progenitor cells with expression of UB markers. These cells expressed UB tip markers: Ret, Wnt11, and Sox9, as well as other markers of UB: Lhx1, Ecad, Hnf1b, Wnt7b, Wnt9b, Calb1, Emx2, Gata3, Hoxb7, and Tacstd2 (Figure 1B and Supplementary Figure S1C). In addition, expression of stromal cell marker Foxd1 nephron progenitor cell markers, Six2 and Eya1 (Figure 1B), or other epithelial segment markers, were observed at day nine of differentiation (Supplementary Figure S1D). Immunofluorescence staining further revealed that the use of a moderate concentration of FGF9 induced the cells to express Pax2, E-cadherin (Ecad), and Gata3 (Figure 1C–F), which may suggest that these differentiated cells represent putative UB progenitor cells.

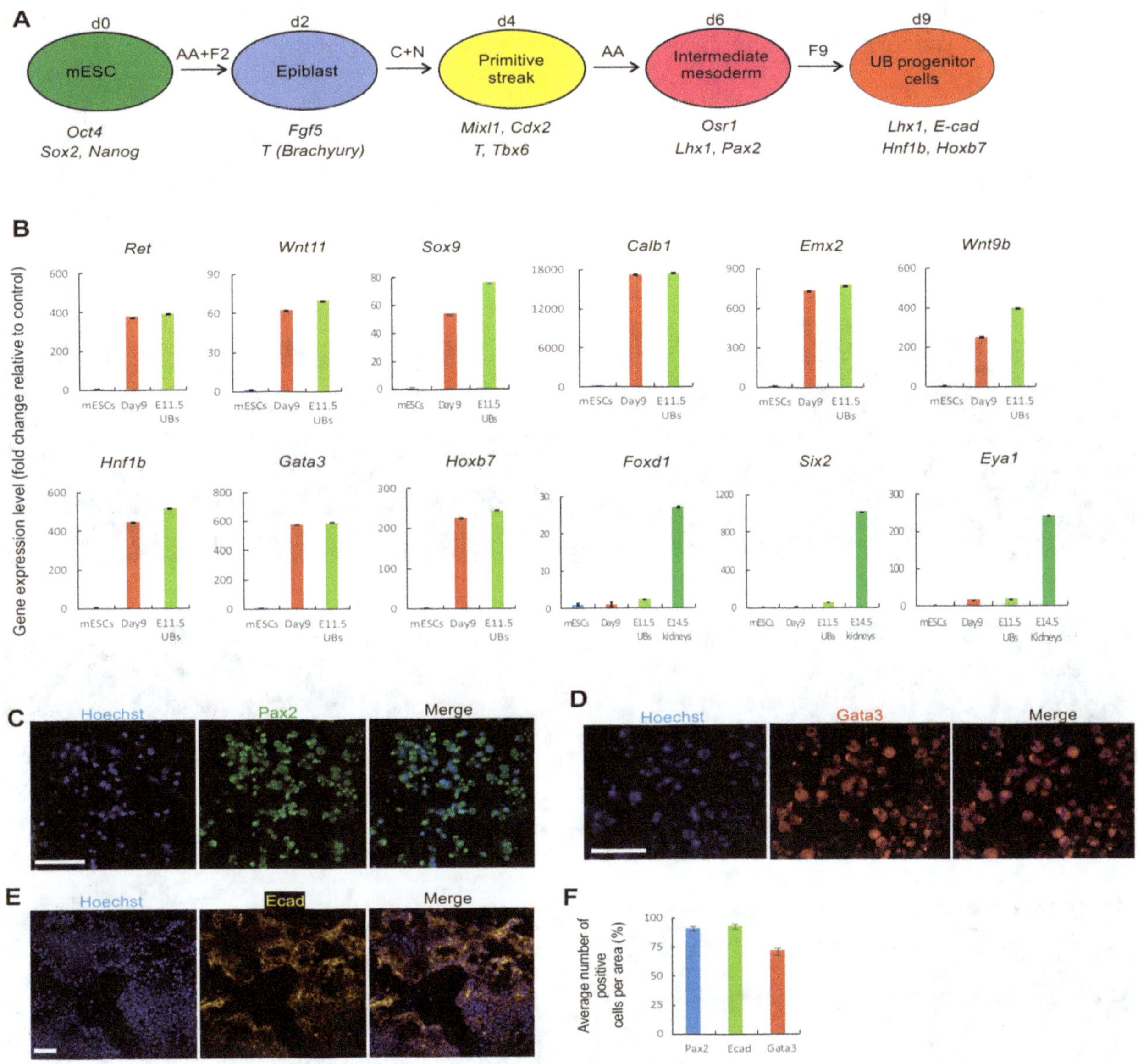

Figure 1. Differentiation of mouse embryonic stem cells (mESCs) to ureteric bud (UB) progenitor cells. (**A**) Schematic of the differentiation protocol of mESCs into UB progenitor cells. AA: Activin A; F2: FGF2; C: CHIR99021; N: Noggin; F9: FGF9. (**B**) Graphs of qPCR results showing the gene expression (fold change) of ureteric bud markers relative to mESC. No expression of stroma (Foxd1) and nephron progenitor cell markers (Six2 and Eya1) was observed at day 9 of differentiation. ($n = 3$). (**C–E**) Immunocytochemistry of Pax2, Ecad, and Gata3 in mESCs on day 9 of differentiation. Scale bars, 50 μm. (**F**) Quantification of the number of cells expressing Pax2, Ecad, and Gata3 at day 9 of differentiation. $n = 3$ samples per marker (3 randomly chosen areas in 3 independent experiments).

3.2. *Generation of Kidney Organoids by mESC-Derived UB Progenitor Cells and Dissociated Primary MM Population*

We and other groups previously reported that dissociation of mouse pMM into single cells maintains the nephron progenitor stemness. The dissociated MM population develops into nephrons when induced by the inducer such as the embryonic UB or spinal cord cells [8,21,23–27]. To establish the potential and function of the mESC-derived UB progenitor cells, we aggregated these cells with mouse E11.5-dissociated pMM cells to generate a kidney organoid. The cell aggregates were cultured for up to 11 days in a traditional Trowell organ culture system, during which they spontaneously formed kidney organoids with complex structures (Figure 2A,B). On day three, we observed Troma1+ UB structures and the formation of renal vesicles adjacent to the UB (Figure 2C–D).

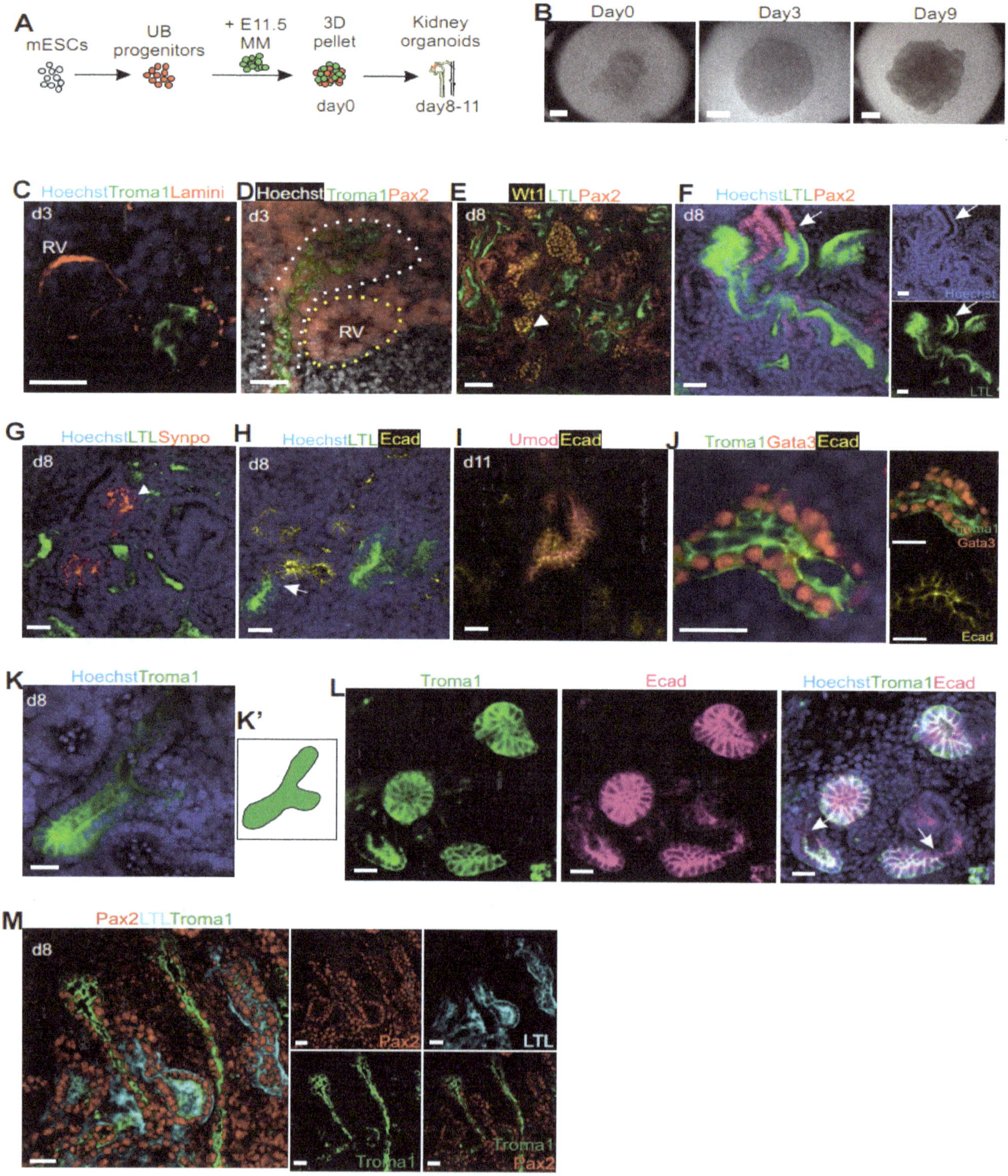

Figure 2. Generation of renal organoids by mESC-derived UB progenitors and primary metanephric mesenchyme (pMM) cells. (**A**) Schematic of generation of kidney organoids from mESC-derived UB progenitors with mouse E11.5-dissociated pMM. (**B**) Global bright field images of self-organizing kidney organoids in a Trowel organ culture system. Scale bars, 500 µm. (**C**) Immunofluorescence of

kidney organoids at day 3 show formation of a renal vesicle next to the Troma1+ structure generated by mESC-derived UB progenitor cells. (**D**) Confocal image at day 3 showing Pax2+ renal vesicle (yellow dotted line) surrounded by Troma+Pax2+ ureteric epithelium (white dotted line). (**E–M**) Immunofluorescence of kidney organoids at day 8 or 11. (**E**) Glomeruli are marked with Wt1, proximal tubules are marked with LTL, and distal tubules are marked with Pax2+LTL–. The arrowhead shows the connection location of glomeruli with proximal tubule. (**F**) Immunostaining of distal tubule marked with Pax2 and proximal tubule marked with LTL, with all nuclei stained with DAPI. The arrow shows the proximal tubules (Pax2+LTL+) connected with distal tubules (Pax2+LTL–). (**G**) Glomeruli (Synpo+) adjacent to proximal tubules (LTL+) (the arrowhead marks the place of connection). (**H**) Immunostaining of proximal tubule with LTL and distal tubule with Ecad. Confocal image shows proximal tubules (LTL+) connected with distal tubules (Ecad+LTL–) (marked with an arrow). (**I**) Confocal image of Loops of Henle marked by Umod and Ecad on day 11. (**J**) Confocal images of Troma1+Gata3+Ecad+ collecting duct structure. (**K-K′**) A "T" shaped UB structure in the kidney organoid. (**L**) mESC-derived UB progenitor cells-generated collecting ducts (Troma1+Ecad+) connecting with nephron's distal tubules (Troma1–Ecad+) (marked with an arrow). (**M**) Kidney organoids developed collecting duct trunk structures. Scale bars, (**B**) 500 μm, (**E**) 50 μm, (**C–D**, **F–M**) 20 μm.

Whole-mount immunostaining of day eight chimeric organoids showed development of nephrons with positive staining of glomerular markers Wilms tumor 1 (Wt1+) and Synaptopodin (Synpo+), proximal tubule marker Lotus Tetragonolobus Lectin (LTL+), and distal tubules markers Pax2+LTL– and Ecad+ (Figure 2E–H, Supplementary Figure S2A). Moreover, we found numerous Synpo+ and Wt1+ glomeruli adjacent to LTL+ proximal tubules, which connected with Pax2+LTL–/Ecad+ distal tubules (Figure 2E–H marked with arrowheads and arrows, Supplementary Figure S2A), indicating a proper nephron structure with glomerulus/proximal tubule/distal tubule organization. On day 11, the organoids also displayed Henle's loop with the expression of uromodulin (Umod+) and Ecad+ (Figure 2I).

We also tested whether the mESC-derived UB progenitor cells have in vivo UB capacity to form the collecting duct system. On day eight, we observed Troma1+Gata3+Ecad+Pax2+LTL– collecting duct structures in the kidney organoids by immunocytochemistry (Figure 2J–M, Supplementary Figure S2B). We also found some Troma1+ "T" shaped UB structures (Figure 2K, K′), indicating that the mESC-derived UB progenitors behave in a manner similar to UB cells in vivo. Importantly, we observed that the collecting ducts (Troma1+Ecad+) connected with the distal tubules (Ecad+Troma1–) of the nephron structure (Figure 2L, Supplementary Figure S2C), generating an interconnection between collecting ducts and nephrons, which is essential for urine drainage. The collecting ducts displayed the morphology of branches and long collecting duct trunks (Figure 2M). Altogether, the *in vitro* reconstructed organoids developed kidney structures that are similar to in vivo kidney, although it is unclear if the generated UB structures form a proper network.

We have also attempted to induce vascularization of developing glomeruli in these renal organoids. Previous studies demonstrated that rho-associated protein kinases (ROCK) are downstream effectors of vascular endothelial growth factor (VEGF), and negatively regulate the process of angiogenesis [28]. Therefore, we used the ROCK inhibitor to enhance angiogenesis in the renal organoids. However, this treatment did not increase the endothelial network area, and CD31+ cells could not be found in the developing glomerular tuft (Supplementary Figure S3).

3.3. Characterization of Kidney Organoids

To rule out the possibility that the differentiated cells (mESC-derived UB progenitors) could give rise to nephrons when induced by the embryonic UB, we aggregated day nine differentiated cells with E11.5-dissociated UB cells and grew them in organ culture. Since UB survival and development depend on the presence of metanephric mesenchyme, in the organ cultures, the purified UB failed to branch in the presence of UB-differentiated cells (Supplementary Figure S4A). Next, we wanted to verify that nephron structures in the organoids were generated via the MM population induced by the mESC-derived UB progenitors, and not by contamination of the MM with the primary UB cells. To assess this possibility, we cultured E11.5 MM tissue in isolation. The tissue already underwent apoptosis at the second day of culture and died at day three (Supplementary Figure S4B). This result also confirmed that without a suitable inducer, the MM cells will not undergo nephrogenesis and cannot survive for a long time under the organ culture conditions in vitro. These data suggest that we did differentiate mESC towards UB progenitors, and that they can further develop into collecting ducts when co-cultured with pMM.

To further confirm that the nephron structures and collecting ducts were derived from pMM- and mESC-derived UB progenitors respectively, we used pMM population isolated from mTmG mice (td-tomato+) to aggregate with differentiated mESCs—unlabeled (Figure 3A). The aggregates formed well-developed nephron structures (WT1+ glomeruli, LTL+ proximal tubules, Ecad+Troma1– distal tubules), which originated from MM cells (mTmG+) (Figure 3B) and Troma1-labeled collecting duct, which was mTmG– and therefore originated from the mESC-derived UB progenitors (Figure 3C–E). We also generated an mESC line with stable expression of the GFP. We differentiated these cells into UB progenitors and used them to induce pMM cells. Immunofluorescence analysis confirmed that the Troma1+Ecad+ collecting ducts were derived from GFP+ mESCs (Figure 3F, Supplementary Figure S4C). Altogether, these data demonstrate that the nephrogenesis occurring in generated kidney organoids is specifically derived from an interaction between the pMM and mESC-derived UB progenitors.

3.4. mESC-Derived UB Progenitor Cells Integrated into the 3D Ureteric Bud Structures

In order to identify the mESC-derived UB progenitors from wild-type UB cells in chimeric organoids, and to ensure that they will only generate UB cells, we generated an organoid with an mESC-GFP line. We dissociated a whole E11.5 embryonic kidney rudiment (MM and UB) and mixed it with undifferentiated mESC or mESC-derived UB progenitors generating chimeric kidney organoids [19,21,27] (Figure 4A).

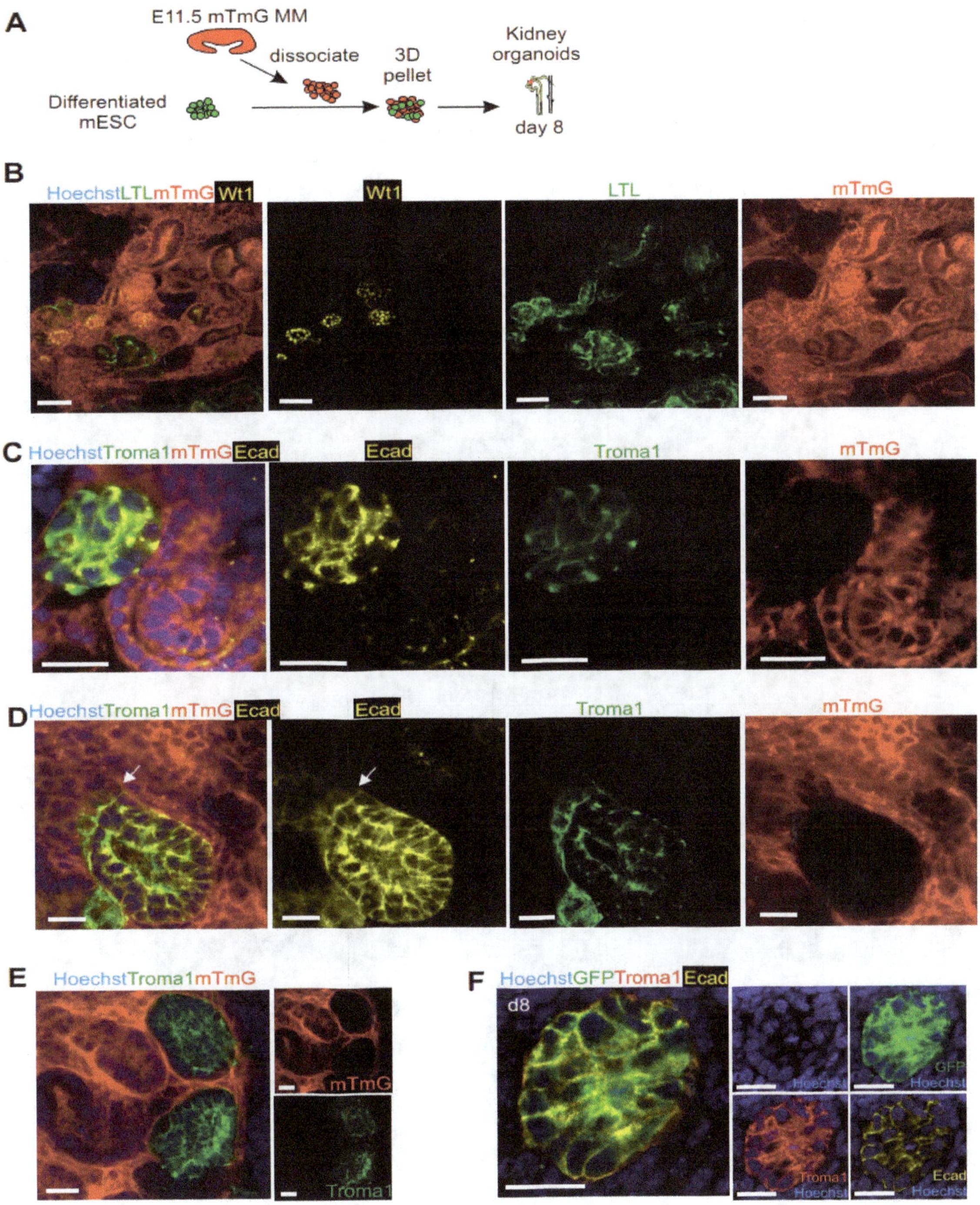

Figure 3. UB structures in kidney organoids are specifically derived from differentiated mESCs. (**A**) Schematic of generation of kidney organoids from mESC-derived UB progenitors with mouse E11.5-dissociated mTmG (td-tomato+) pMM. (**B**) Confocal images of three-dimensional (3D) kidney organoids derived from E11.5 mTmG+ pMM and mESC-derived UB progenitors. Nephron structures such as WT1+ glomeruli and LTL+ proximal tubules were derived from mTmG+ pMM. Scale bars, 50 μm. (**C**) Confocal images of kidney organoids showing ureteric bud structures (Ecad+Troma1+) being generated by mESC-derived UB progenitor cells. Scale bars, 20 μm. (**D**) Confocal images showed Troma1+Ecad+ collecting duct derived from mTmg– cells (mESC-derived UB progenitors) and connected with Troma1–Ecad+ renal tubules (connection marked with an arrow). Scare bars, 20 μm. (**E**) The Troma1+ UB structures derived from mTmG– cells (mESC-derived UB progenitors). Scale bar: 20 μm. (**F**) Troma1+Ecad+ collecting ducts are derived from GFP+ mESCs-derived UB progenitors. Scale bar: 20 μm.

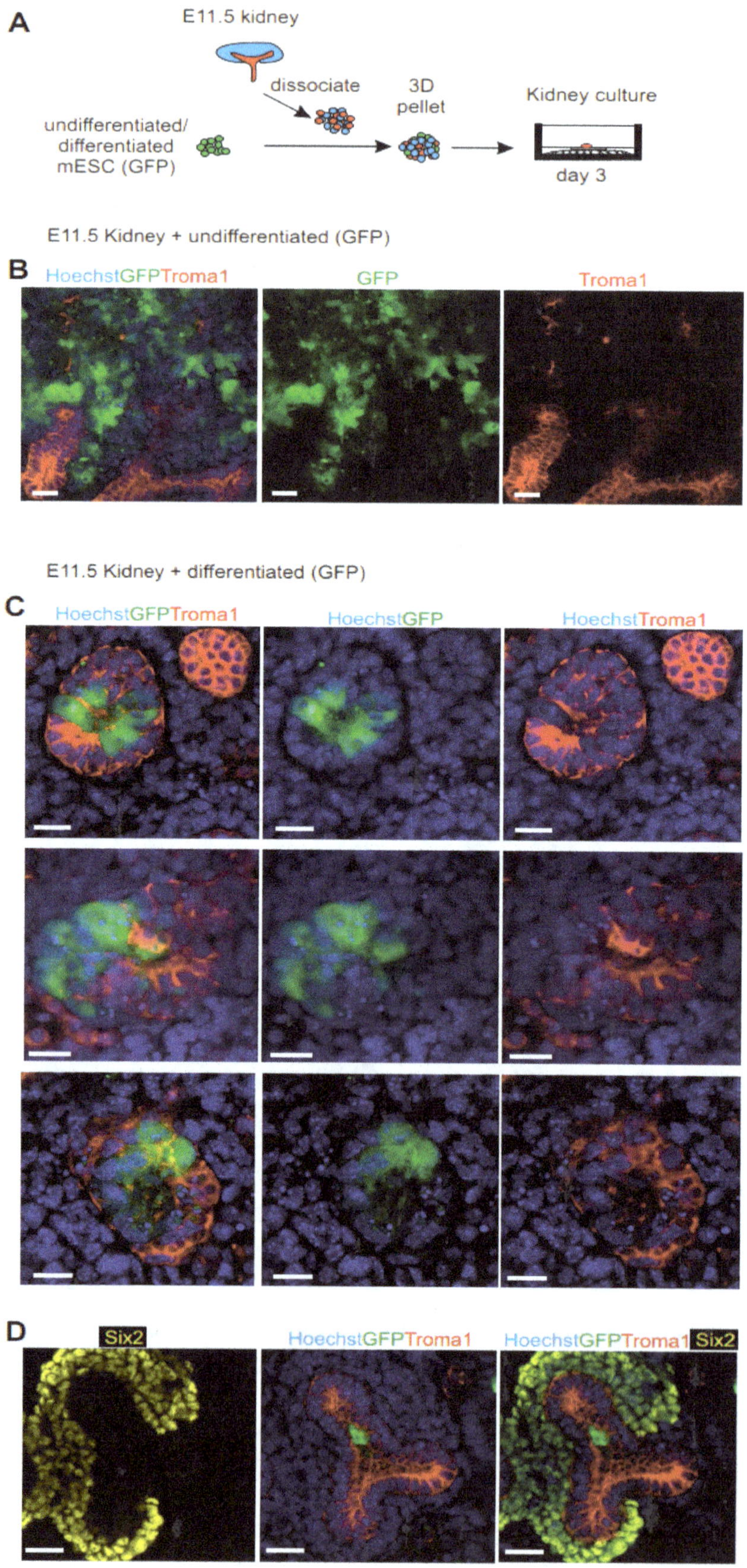

Figure 4. Mouse ESC-derived UB progenitor cells form anembryonic UB in 3D organ culture in vitro. (**A**) Schematic of kidney organoid generation from mESC-derived UB progenitors with mouse E11.5-dissociated kidney rudiments. (**B**) Immunofluorescence analysis demonstrating random

localization of undifferentiated mESCs in organ co-cultures. Scale bars, 20 μm. (**C**) Immunofluorescence analysis demonstrating mESC-derived UB progenitors integrated into the UB structures and enhanced chimeric ureteric bud formation. Scale bars, 20 μm. (**D**) Confocal image showing localization of mESC-derived UB progenitor cells (GFP+) in Troma1+ UB structures of the chimeric organoid and not in the Six2+ nephron progenitor's region. Scale bars: 20 μm.

The undifferentiated mESCs aggregated with dissociated embryonic kidney rudiments randomly localized within the organoids and had a negative effect on nephrogenesis. The undifferentiated mESC had formed groups within the renal structures; however, there were only a few structures, and very few Troma1+ ureteric buds could be observed (Figure 4B). When we aggregated the mESCs-derived UB progenitors with the dissociated embryonic kidney, we noticed much more robust tubulogenesis and numerous chimeric ureteric bud structures (Figure 4C). The UB progenitor cells did not interfere with renal development, but efficiently and specifically integrated into the mouse ureteric buds, as indicated by Troma1 and GFP+ staining (Figure 4C and D). Moreover, the immunostaining of day three organoids with nephron progenitor's marker, Six2, presented integration of mESC-derived UB progenitors (GFP+) into the UB structure (Figure 4D). Together, these data demonstrate that mESC-derived UB progenitors do integrate well into ureteric bud structures in chimeric renal organoids and are able not only to form UB de novo, but also generate chimeric structures.

3.5. Kidney Organoids as Models to Study Kidney Development and Drug Toxicity

After transfer of the 3D chimeric organoids to organ culture, they self-organized and generated a well-structured kidney organoid with proper nephron and collecting duct structures. As patterning of the nephron into its different segments begins at the renal vesicle stage during development [29,30], we postulated that developmental patterning could be changed by chemical modulation of these endogenous signals. A previous report revealed that Notch signaling is required for proximal tubule fate acquisition in a mammalian nephron [31]. We therefore treated the organoids (pMM cells mixed with mESC-derived UB progenitor cells in a ratio of 3:1, Figure 2A) with a Notch signaling inhibitor, *N*-*S*-phenyl-glycine-*t*-butyl ester (DAPT). DAPT leads to suppression of proximal tubules' formation in the human nephron organoid culture [4,31,32]. We added DAPT to the culture medium from day 3 to 11 of 3D organoid culture. Immunofluorescence analysis of the cells demonstrated that with the DAPT treatment, formation of glomerulus and ureteric bud ducts was normal, but development of proximal tubules was severely suppressed (Supplementary Figure S5A).

In order to use stem cell-derived kidney organoids for disease modelling and drug screening, they need to present functional maturation of the nephrons within the organoids. To test whether these organoids could be used to study kidney injury and toxicity in vitro, we focused on drug-induced nephrotoxicity, which has been shown as an important cause of acute kidney injury in hospitalized patients [33]. We treated the chimeric kidney organoids with gentamicin, a commonly used antibiotic with well-established proximal tubular toxicity, after nine days of organ culture for 48 h [34]. We also treated organoids with another nephrotoxicant, cisplatin, from day 10 for 24 h. Cisplatin induces caspase-mediated acute apoptosis of proximal tubular cells in the kidney [35]. Whole-mount immunostaining of control organoids with caspase 3 showed random apoptotic interstitial cells; however, both gentamicin and cisplatin induced acute apoptosis in LTL+ proximal tubules (Supplementary Figure S5B). The percentage of apoptotic proximal tubular cells induced by cisplatin increased in a dose-dependent manner: the 5 and 20 μM cisplatin doses mainly affected the proximal tubule compartment, 32% and 62% respectively, while 50 μM cisplatin led to apoptosis of almost all of the proximal tubule cells (≈96%), but was also toxic to other cell types presenting a global type of nephrotoxicity at this concentration (Supplementary Figure S5C).

To summarize, our work shows that mESC-derived UB progenitors induce nephrogenesis in pMM cells, and furthermore, chimeric renal organoids generated from these progenitors show an expected response to toxic chemicals and drugs. We also demonstrated that the kidney organoid system can be used to test nephrotoxicity of drugs and other chemicals in vitro.

4. Discussion

Kidney development starts from an interaction between two precursor tissues of the kidney, UB and MM. A major part of the MM cell population comprises the nephron progenitor cells (NPCs) which will differentiate into nephrons, and the ureteric bud will form the collecting duct tree. Protocols for differentiation of mouse and human pluripotent stem cells to renal progenitor cells, and further to self-organized kidney organoids containing nephrons, have been well established [2–4,13,36], but current methods of differentiation of pluripotent stem cells, specifically to ureteric bud progenitor cells, need further development [8,19].

Previous studies on generation of ESC-derived UB have shown derivation of a UB-like population by selective induction with metanephric mesenchyme cells [19] or through CHIR99021 treatment for nephron differentiation of both ureteric bud and nephron structures [2]. However, the ESC-derived UB-like cells did not show nephron progenitor induction [19], and therefore, an inter-nephron connection with collecting ducts was lacking [2]. Another group recently published a protocol for derivation of UB structures from PSCs, and generation of kidney organoid composed of mESCs-derived UB aggregated with pMM, or mESCs-derived UB combined with mESC-derived NPC and primary stromal progenitor cells (SPs). Although successful, their protocol requires a knock-in of markers in the PSCs, which later involves sorting of cells for the specific marker [8]. More recently, Mae and colleagues reported a protocol for generation of branching ureteric bud tissues from human pluripotent stem cells (hPSC) with a series of growth factors, but there was no evidence of nephrogenesis [37] (Supplementary Table S2). While these published protocols produce functional kidney organoids, they are technically complex. Here, we reported an establishment of a simple (directed with growth factors), efficient (>90% of Pax2+, Ecad+ and >70% Gata3+ cells), and reproducible differentiation protocol of mESC to ureteric bud progenitor cells. These mESC-derived UB progenitor cells induced pMM cells to undergo nephrogenesis leading to development of well-structured nephrons. These nephrons consisted of glomeruli, proximal tubules, loops of Henle, and distal tubules, and were connected with collecting duct structures generated by mESC-derived UB cells. Moreover, these mESC-derived UB progenitor cells formed a UB de novo when combined with pMM cells, and they generated chimeric structures when combined with kidney rudiment cells (pUB and pMM) (Supplementary Table S2).

While our culturing conditions produced well-functioning kidney organoids, further studies are needed to fine-tune the culturing conditions. Previous studies showed that ROCK kinases are VEGF downstream effectors, which negatively regulate the process of angiogenesis [28]. We added ROCK inhibitor (Y27) to culture expecting an increase in angiogenesis. However, we failed to see a difference between control and Y27-treated samples. Better results could be obtained by supplementing the medium with VEGF, which was already published by Freedman and co-workers, although they still did not observe endothelial cells entering the glomerular tuft in their system [9]. Similarly, development of a vascular glomeruli in organotypic kidney cultures and renal organoids was reported earlier by our research team [38]. It seems that in the absence of the blood flow, the endothelial cells are not able to properly interact with developing nephrons [39] and formation of glomerular vasculature does not proceed further than migration of endothelial cells into the vascular cleft region of an S-shaped stage nephron. Sufficient vascularization of the organoids may be achieved by treatment with angiogenic factors, co-culture with blood vessel organoids [40], by providing flow to the system [15], or with a combination of the aforementioned treatments.

Even though organoids generated by mESC-derived UB progenitor cells did not have a proper vascular network, the developed nephrons did respond to toxicological tests as expected. We and others [4] have shown that the use of gentamicin and cisplatin induces apoptosis in proximal tubular cells. Thus, these organoids present a functional platform to test drug-induced nephrotoxicity.

In summary, we have developed an easy and reproducible protocol for generation of UB progenitors from mESCs. This work generates a strong foundation for in vitro kidney studies, including disease modelling and drug discovery approaches, which are difficult to perform, and require animal models and/or primary cells which may not faithfully recapitulate all features of developmental or disease processes. Given the rapid progress in the field, we hope that in the near future, researchers will be able to generate fully functional nephrons in kidney organoids where the UB and MM parts are derived from PSC. Using these cells will enable generation of not only well-structured nephrons, but also the collecting duct tree. This is the first step for generating high-throughput gene discovery models and advancing tissue engineering for producing organs for transplanting. However, these organoids need to be successfully vascularized and grown to appropriate size. The studies presented here produce new insights into renal pathophysiology and open new avenues for developing new treatment options.

Supplementary Materials
Table S1: Primers (5′-3′) used in the study. Table S2: Summary of available protocols of differentiation of mouse and human PSC to UB linages. Figure S1: Differentiations of mESC into ureteric bud progenitor cells. Figure S2: Differentiated mESC induce embryonic MM to nephrogenesis. Figure S3: Vascularization of the kidney organoids. Figure S4: Characterization of kidney organoids. Figure S5: Kidney organoids model kidney development and injury.

Author Contributions: Conceptualization, Z.T., A.R.-R., I.S., and S.J.V.; Funding acquisition, S.J.V.; Investigation, Z.T.; Visualization, Z.T.; Writing—original draft, Z.T.; Writing—review and editing, A.R.-R. All authors have read and agreed to the published version of the manuscript.

Acknowledgments: We thank Paula Haipus, Hannele Härkman, Johanna Kekolahti-Liias, and Sanna Kauppinen for technical assistance; Biocenter Oulu Transgenic core facility for the mouse ESCs; Dr. Susanna Kosamo for critical reading of the manuscript and English language corrections; Dr. Jingdong Shan, Dr. Florence Naillat, MSc. Abhishek Sharma, Dr. Fariba Jian Motamedi, and Prof. Andreas Schedl for the discussion.

References

1. Robinton, D.A.; Daley, G.Q. The promise of induced pluripotent stem cells in research and therapy. *Nature* **2012**, *481*, 295–305. [CrossRef]
2. Takasato, M.; Er, P.X.; Chiu, H.S.; Maier, B.; Baillie, G.J.; Ferguson, C.; Parton, R.G.; Wolvetang, E.J.; Roost, M.S.; Chuva de Sousa Lopes, S.M.; et al. Kidney organoids from human iPS cells contain multiple lineages and model human nephrogenesis. *Nature* **2015**, *526*, 564–568. [CrossRef]
3. Freedman, B.S.; Brooks, C.R.; Lam, A.Q.; Fu, H.; Morizane, R.; Agrawal, V.; Saad, A.F.; Li, M.K.; Hughes, M.R.; Werff, R.V.; et al. Modelling kidney disease with CRISPR-mutant kidney organoids derived from human pluripotent epiblast spheroids. *Nat. Commun.* **2015**, *6*, 8715. [CrossRef]
4. Morizane, R.; Lam, A.Q.; Freedman, B.S.; Kishi, S.; Valerius, M.T.; Bonventre, J.V. Nephron organoids derived from human pluripotent stem cells model kidney development and injury. *Nat. Biotechnol.* **2015**, *33*, 1193–1200. [CrossRef]
5. Boreström, C.; Jonebring, A.; Guo, J.; Palmgren, H.; Cederblad, L.; Forslöw, A.; Svensson, A.; Söderberg, M.; Reznichenko, A.; Nyström, J.; et al. A CRISP(e)R view on kidney organoids allows generation of an induced pluripotent stem cell–derived kidney model for drug discovery. *Kidney Int.* **2018**, *94*, 1099–1110. [CrossRef] [PubMed]
6. Kim, Y.K.; Nam, S.A.; Yang, C.W. Applications of kidney organoids derived from human pluripotent stem cells. *Korean J. Intern. Med.* **2018**, *33*, 649–659. [CrossRef] [PubMed]
7. Hale, L.J.; Howden, S.E.; Phipson, B.; Lonsdale, A.; Er, P.X.; Ghobrial, I.; Hosawi, S.; Wilson, S.; Lawlor, K.T.; Khan, S.; et al. 3D organoid-derived human glomeruli for personalised podocyte disease modelling and drug screening. *Nat. Commun.* **2018**, *9*, 1–17. [CrossRef]
8. Taguchi, A.; Nishinakamura, R. Higher-Order Kidney Organogenesis from Pluripotent Stem Cells. *Cell Stem Cell* **2017**, *21*, 730–746. [CrossRef] [PubMed]

9. Czerniecki, S.M.; Cruz, N.M.; Harder, J.L.; Menon, R.; Annis, J.; Otto, E.A.; Gulieva, R.E.; Islas, L.V.; Kim, Y.K.; Tran, L.M.; et al. High-Throughput Screening Enhances Kidney Organoid Differentiation from Human Pluripotent Stem Cells and Enables Automated Multidimensional Phenotyping. *Cell Stem Cell* **2018**, *22*, 929–940. [CrossRef] [PubMed]
10. Cruz, N.M.; Song, X.; Czerniecki, S.M.; Gulieva, R.E.; Churchill, A.J.; Kim, Y.K.; Winston, K.; Tran, L.M.; Diaz, M.A.; Fu, H.; et al. Organoid cystogenesis reveals a critical role of microenvironment in human polycystic kidney disease. *Nat. Mater.* **2017**, *16*, 1112–1119. [CrossRef]
11. Araoka, T.; Mae, S.; Kurose, Y.; Uesugi, M.; Ohta, A.; Yamanaka, S.; Osafune, K. Efficient and Rapid Induction of Human iPSCs/ESCs into Nephrogenic Intermediate Mesoderm Using Small Molecule-Based Differentiation Methods. *PLoS ONE* **2014**, *9*, e84881. [CrossRef]
12. Takasato, M.; Er, P.X.; Becroft, M.; Vanslambrouck, J.M.; Stanley, E.G.; Elefanty, A.G.; Little, M.H. Directing human embryonic stem cell differentiation towards a renal lineage generates a self-organizing kidney. *Nat. Cell Biol.* **2014**, *16*, 118–126. [CrossRef] [PubMed]
13. Taguchi, A.; Kaku, Y.; Ohmori, T.; Sharmin, S.; Ogawa, M.; Sasaki, H.; Nishinakamura, R. Redefining the in vivo origin of metanephric nephron progenitors enables generation of complex kidney structures from pluripotent stem cells. *Cell Stem Cell* **2014**, *14*, 53–67. [CrossRef] [PubMed]
14. Garreta, E.; Prado, P.; Tarantino, C.; Oria, R.; Fanlo, L.; Martí, E.; Zalvidea, D.; Trepat, X.; Roca-Cusachs, P.; Gavaldà-Navarro, A.; et al. Fine tuning the extracellular environment accelerates the derivation of kidney organoids from human pluripotent stem cells. *Nat. Mater.* **2019**, *18*, 397–405. [CrossRef] [PubMed]
15. Homan, K.A.; Gupta, N.; Kroll, K.T.; Kolesky, D.B.; Skylar-Scott, M.; Miyoshi, T.; Mau, D.; Valerius, M.T.; Ferrante, T.; Bonventre, J.V.; et al. Flow-enhanced vascularization and maturation of kidney organoids in vitro. *Nat. Methods* **2019**, *16*, 255–262. [CrossRef]
16. Allison, S.J. Fluid flow enhances vascularization and maturation of kidney organoids. *Nat. Reviews. Nephrol.* **2019**, *15*, 254. [CrossRef] [PubMed]
17. Low, J.H.; Li, P.; Chew, E.G.Y.; Zhou, B.; Suzuki, K.; Zhang, T.; Lian, M.M.; Liu, M.; Aizawa, E.; Rodriguez Esteban, C.; et al. Generation of Human PSC-Derived Kidney Organoids with Patterned Nephron Segments and a De Novo Vascular Network. *Cell Stem Cell* **2019**, *25*, 373–387. [CrossRef] [PubMed]
18. Przepiorski, A.; Sander, V.; Tran, T.; Hollywood, J.A.; Sorrenson, B.; Shih, J.; Wolvetang, E.J.; McMahon, A.P.; Holm, T.M.; Davidson, A.J. A Simple Bioreactor-Based Method to Generate Kidney Organoids from Pluripotent Stem Cells. *Stem Cell Rep.* **2018**, *11*, 470–484. [CrossRef] [PubMed]
19. Xia, Y.; Nivet, E.; Sancho-Martinez, I.; Gallegos, T.; Suzuki, K.; Okamura, D.; Wu, M.; Dubova, I.; Esteban, C.R.; Montserrat, N.; et al. Directed differentiation of human pluripotent cells to ureteric bud kidney progenitor-like cells. *Nat. Cell Biol.* **2013**, *15*, 1507–1515. [CrossRef]
20. Aro, E.; Salo, A.M.; Khatri, R.; Finnilä, M.; Miinalainen, I.; Sormunen, R.; Pakkanen, O.; Holster, T.; Soininen, R.; Prein, C.; et al. Severe Extracellular Matrix Abnormalities and Chondrodysplasia in Mice Lacking Collagen Prolyl 4-Hydroxylase Isoenzyme II in Combination with a Reduced Amount of Isoenzyme I. *J. Biol. Chem.* **2015**, *290*, 16964–16978. [CrossRef]
21. Junttila, S.; Saarela, U.; Halt, K.; Manninen, A.; Pärssinen, H.; Lecca, M.R.; Brändli, A.W.; Sims-Lucas, S.; Skovorodkin, I.; Vainio, S.J. Functional genetic targeting of embryonic kidney progenitor cells ex vivo. *J. Am. Soc. Nephrol.* **2015**, *26*, 1126–1137. [CrossRef]
22. Rak-Raszewska, A. Experimental Tubulogenesis Induction Model in the Mouse. In *Kidney Organogenesis*; Methods in Molecular Biology; Vainio, S., Ed.; Humana Press: New York, NY, USA, 2019; Volume 1926, pp. 39–51.
23. Ganeva, V.; Unbekandt, M.; Davies, J.A. An improved kidney dissociation and reaggregation culture system results in nephrons arranged organotypically around a single collecting duct system. *Organogenesis* **2011**, *7*, 83–87. [CrossRef] [PubMed]
24. Hendry, C.E.; Vanslambrouck, J.M.; Ineson, J.; Suhaimi, N.; Takasato, M.; Rae, F.; Little, M.H. Direct Transcriptional Reprogramming of Adult Cells to Embryonic Nephron Progenitors. *J. Am. Soc. Nephrol. Jasn.* **2013**, *24*, 1424–1434. [CrossRef]
25. Lusis, M.; Li, J.; Ineson, J.; Christensen, M.E.; Rice, A.; Little, M.H. Isolation of clonogenic, long-term self renewing embryonic renal stem cells. *Stem Cell Res.* **2010**, *5*, 23–39. [CrossRef] [PubMed]
26. Saxén, L.; Sariola, H. Early organogenesis of the kidney. *Pediatr. Nephrol.* **1987**, *1*, 385–392. [CrossRef] [PubMed]

27. Unbekandt, M.; Davies, J.A. Dissociation of embryonic kidneys followed by reaggregation allows the formation of renal tissues. *Kidney Int.* **2010**, *77*, 407–416. [CrossRef] [PubMed]
28. Kroll, J.; Epting, D.; Kern, K.; Dietz, C.T.; Feng, Y.; Hammes, H.P.; Augustin, H.G. Inhibition of Rho-dependent kinases ROCK I/II activates VEGF-driven retinal neovascularization and sprouting angiogenesis. *Am. J. Physiol.—Heart Circ. Physiol.* **2009**, *296*, H893–H899. [CrossRef]
29. Georgas, K.; Rumballe, B.; Valerius, M.T.; Chiu, H.S.; Thiagarajan, R.D.; Lesieur, E.; Aronow, B.J.; Brunskill, E.W.; Combes, A.N.; Tang, D.; et al. Analysis of early nephron patterning reveals a role for distal RV proliferation in fusion to the ureteric tip via a cap mesenchyme-derived connecting segment. *Dev. Biol.* **2009**, *332*, 273–286. [CrossRef]
30. Little, M.; Georgas, K.; Pennisi, D.; Wilkinson, L. Kidney development: Two tales of tubulogenesis. *Curr. Top. Dev. Biol.* **2010**, *90*, 193.
31. Cheng, H.T.; Kim, M.; Valerius, M.T.; Surendran, K.; Schuster-Gossler, K.; Gossler, A.; McMahon, A.P.; Kopan, R. Notch2, but not Notch1, is required for proximal fate acquisition in the mammalian nephron. *Development* **2007**, *134*, 801–811. [CrossRef]
32. Cheng, H.; Miner, J.H.; Lin, M.; Tansey, M.G.; Roth, K.; Kopan, R. Gamma-secretase activity is dispensable for mesenchyme-to-epithelium transition but required for podocyte and proximal tubule formation in developing mouse kidney. *Development* **2003**, *130*, 5031–5042. [CrossRef]
33. Uchino, S.; Kellum, J.A.; Bellomo, R.; Doig, G.S.; Morimatsu, H.; Morgera, S.; Schetz, M.; Tan, I.; Bouman, C.; Macedo, E.; et al. Beginning and Ending Supportive Therapy for the Kidney (BEST Kidney) Investigators, for the Acute Renal Failure in Critically Ill Patients: A Multinational, Multicenter Study. *JAMA* **2005**, *294*, 813–818. [CrossRef]
34. Whiting, P.H.; Brown, P.A.J. The Relationship Between Enzymuria and Kidney Enzyme Activities in Experimental Gentamicin Nephrotoxicity. *Ren. Fail.* **1996**, *18*, 899–909. [CrossRef] [PubMed]
35. Mese, H.; Sasaki, A.; Nakayama, S.; Alcalde, R.E.; Matsumura, T. The role of caspase family protease, caspase-3 on cisplatin-induced apoptosis in cisplatin-resistant A431 cell line. *Cancer Chemother. Pharm.* **2000**, *46*, 241–245. [CrossRef] [PubMed]
36. Takasato, M.; Little, M.H. A strategy for generating kidney organoids: Recapitulating the development in human pluripotent stem cells. *Dev. Biol.* **2016**, *420*, 210–220. [CrossRef] [PubMed]
37. Mae, S.; Ryosaka, M.; Toyoda, T.; Matsuse, K.; Oshima, Y.; Tsujimoto, H.; Okumura, S.; Shibasaki, A.; Osafune, K. Generation of branching ureteric bud tissues from human pluripotent stem cells. *Biochem. Biophys. Res. Commun.* **2018**, *495*, 954–961. [CrossRef]
38. Halt, K.J.; Pärssinen, H.E.; Junttila, S.M.; Saarela, U.; Sims-Lucas, S.; Koivunen, P.; Myllyharju, J.; Quaggin, S.; Skovorodkin, I.N.; Vainio, S.J. CD146+ cells are essential for kidney vasculature development. *Kidney Int.* **2016**, *90*, 311–324. [CrossRef] [PubMed]
39. Serluca, F.C.; Drummond, I.A.; Fishman, M.C. Endothelial Signaling in Kidney Morphogenesis: A Role for Hemodynamic Forces. *Curr. Biol.* **2002**, *12*, 492–497. [CrossRef]
40. Wimmer, R.A.; Leopoldi, A.; Aichinger, M.; Wick, N.; Hantusch, B.; Novatchkova, M.; Taubenschmid, J.; Hämmerle, M.; Esk, C.; Bagley, J.A.; et al. Human blood vessel organoids as a model of diabetic vasculopathy. *Nature* **2019**, *565*, 505–510. [CrossRef]

11

Generation of Differentiating and Long-Living Intestinal Organoids Reflecting the Cellular Diversity of Canine Intestine

Nina Kramer [1,*], Barbara Pratscher [1,‡], Andre M. C. Meneses [1,2,†,‡], Waltraud Tschulenk [3], Ingrid Walter [3], Alexander Swoboda [1], Hedwig S. Kruitwagen [2], Kerstin Schneeberger [2], Louis C. Penning [2], Bart Spee [2], Matthias Kieslinger [1], Sabine Brandt [4] and Iwan A. Burgener [1]

1 Division of Small Animal Internal Medicine, Department for Small Animals and Horses, University of Veterinary Medicine, 1210 Vienna, Austria
2 Department of Clinical Sciences, Faculty of Veterinary Medicine, Utrecht University, 3584 Utrecht, The Netherlands
3 Institute of Pathology, Department for Pathobiology, University of Veterinary Medicine, 1210 Vienna, Austria
4 Research Group Oncology, Equine Surgery, Department of Small Animals and Horses, University of Veterinary Medicine, 1210 Vienna, Austria
* Correspondence: nina.kramer@vetmeduni.ac.at
† Current address: Institute of Animal Health and Production, Veterinary Hospital, Amazon Rural Federal University, Belém 66.077-830, Brazil.
‡ These authors contributed equally to this work.

Abstract: Functional intestinal disorders constitute major, potentially lethal health problems in humans. Consequently, research focuses on elucidating the underlying pathobiological mechanisms and establishing therapeutic strategies. In this context, intestinal organoids have emerged as a potent in vitro model as they faithfully recapitulate the structure and function of the intestinal segment they represent. Interestingly, human-like intestinal diseases also affect dogs, making canine intestinal organoids a promising tool for canine and comparative research. Therefore, we generated organoids from canine duodenum, jejunum and colon, and focused on simultaneous long-term expansion and cell differentiation to maximize applicability. Following their establishment, canine intestinal organoids were grown under various culture conditions and then analyzed with respect to cell viability/apoptosis and multi-lineage differentiation by transcription profiling, proliferation assay, cell staining, and transmission electron microscopy. Standard expansion medium supported long-term expansion of organoids irrespective of their origin, but inhibited cell differentiation. Conversely, transfer of organoids to differentiation medium promoted goblet cell and enteroendocrine cell development, but simultaneously induced apoptosis. Unimpeded stem cell renewal and concurrent differentiation was achieved by culturing organoids in the presence of tyrosine kinase ligands. Our findings unambiguously highlight the characteristic cellular diversity of canine duodenum, jejunum and colon as fundamental prerequisite for accurate in vitro modelling.

Keywords: intestinal organoids; canine intestine; differentiation; organoid culture

---

1. Introduction

Gastrointestinal (GI) disorders such as inflammatory bowel disease (IBD), infection-induced GI disorders and cancer have a major negative impact on human health and impose a high financial burden on healthcare systems. Infectious diarrhea is the second leading cause of death among young children [1] and GI cancers are the third most common cancer type worldwide, with one million newly diagnosed cases per year [2]. Potentially lethal GI diseases also affect livestock and companion animals, with enterotoxic bacteria and enteropathogenic viruses being frequently involved in disease onset and progression [3]. Although considerable efforts are made to establish new therapies for human and veterinary GI diseases, mortality rates remain high, because translation of biomedical research into clinical practice is hampered by the lack of epithelial models that reliably mimic the organ and recapitulate disease in patients [4–6].

Early attempts to bring the intestinal epithelium to in vitro culture were mostly short-lived, as exemplified by explanted biopsies and epithelial cells, which disintegrated after 72 h and two weeks, respectively [7–9]. To establish culture systems for long-lived homogenous cell populations tissue explants were either cultivated in collagen gels or on 3T3 feeder layers on an air-liquid interface [10,11]. However, growth factor-providing (myo)fibroblasts, which are prerequisites for prolonged cultivation in these systems, made them susceptible to inconsistencies from one experiment to another. A major step to overcome these limitations was the establishment of three-dimensional (3D) murine intestinal organoids derived from adult intestinal stem cells [12]. These organoids contained long-lived stem cells that differentiated into the main cell types of murine small intestine such as Paneth cells, enterocytes, enteroendocrine cells and goblet cells. Establishment of this innovative in vitro model was achieved by using a laminin-rich extracellular matrix (Matrigel) and growth media supplemented with epithelial growth factor (EGF), Noggin, and R-spondin as reviewed in detail by Date and Sato [13]. For the cultivation of human intestinal organoids, growth medium had to be supplemented with Wnt3a, gastrin, p38-MAPK inhibitor, nicotinamide and ALK4/5/7 inhibitor, thereby preventing the differentiation into goblet and enteroendocrine cells and retaining enterocytes in a premature state [14]. Withdrawing nicotinamide and p38-MAPK inhibitor induced de novo differentiation, thus shortening the lifespan of the intestinal stem cell pool and rendering this system into an endpoint assay. Further improvement of human 3D intestinal organoid models was achieved by microscaffolds that mimic the size and distance between crypts in transwell assays [15]. Differentiation was promoted by using a growth factor gradient system based on (i) intestinal stem cell-supporting expansion medium in lower wells, and (ii) addition of differentiation medium to upper wells, thereby inducing differentiation of migrating cells to goblet cells, enteroendocrine cells and enterocytes along the modeled crypt/villus axis. However, this technique is unsuitable for, e.g., high-throughput screenings because it is very labor-intensive, time-consuming and hampers down-stream analysis. An approach characterizing receptor tyrosine kinase signaling between the crypt base and its niche managed to omit nicotinamide and p38-MAPK inhibitor usage by a combination of two ligands, i.e., insulin-like growth factor 1 (IGF1) and basic fibroblast growth factor (FGF2). This combination sustains stem cell growth and allows simultaneous differentiation into enteroendocrine and goblet cells, mimicking corresponding natural epithelia more authentically with minimal effort.

In recent years, organoid technology has also been introduced into veterinary research, albeit to a lesser extent [16–22]. Animal organoid systems that are already available today mainly consist of stem and undifferentiated cells requiring tailor-made species-specific media to allow for differentiation [17]. Interestingly, dogs and their owners share similar environments, food and carcinogenic load, and develop similar GI diseases including GI cancer, infectious disease and IBD [23]. Consequently, canine patients represent an interesting natural model for human GI disorders, even more so since dogs show reduced genetic variation within the majority of breeds [24]. To exploit the full potential of canine intestinal organoids to mimic the corresponding organ accurately, a balance between self-renewal and differentiation of stem cells must be achieved to design a physiologically relevant system.

Herein, we report the establishment of a novel culture system for canine intestinal organoids that is based on recent findings by Fujii et al. [25] and reliably supports the sustained proliferation of duodenum-, jejunum- and colon-derived stem cells, while concomitantly allowing their differentiation into secretory lineage cells.

2. Materials and Methods

2.1. Isolation and Cultivation of Canine Intestinal Organoids

Duodenal, jejunal and colonic samples were obtained from three dogs (two males and one female) that were euthanized due to non-intestinal disease. In addition, duodenal biopsies were taken from one dog undergoing routine gastroduodenal endoscopy. Tissue sampling was approved by the institutional ethics committee in accordance with Good Scientific Practice guidelines and Austrian legislation. Based on the guidelines of the institutional ethics committee, use of tissue material collected during therapeutic excision or post mortem is included in the University's "owner's consent for treatment", which was signed by all patient owners. Intestinal crypts were isolated from tissue samples and biopsies according to established protocols [12]. In short, tissue was incubated with 5 mM EDTA (Sigma-Aldrich, St. Louis, MO, USA) in order to dissociate crypts for 30 min to one hour depending on bowel segment, followed by vigorous shaking until crypts were released. After two washing steps with PBS and Advanced Dulbecco's modified Eagle's medium/F12 (DMEM/F12, Invitrogen, Thermo Fisher Scientific, Waltham, MA, USA) 500 crypts were resuspended in 50 μL Matrigel (BD Biosciences, Franklin Lakes, NJ, USA) and seeded per well of a 24-well plate. Following Matrigel polymerization expansion medium was added. Canine intestinal growth media were prepared as indicated in Table 1. Penicillin/streptomycin, HEPES, Glutamax, B27 (with Vitamin A) and N2 supplement were obtained from Invitrogen, Thermo Fisher Scientific. N-acetylcysteine, gastrin and nicotinamide were purchased from Sigma-Aldrich. EGF was provided by Thermo Fisher Scientific. ALK5 kinase inhibitor (A83-01) was obtained from Tocris Bioscience, Bristol, UK. p38 MAPK inhibitor (SB202190) was purchased from Selleck Chemicals, Houston, TX, USA. Human hepatocyte growth factor (HGF), human Noggin, human IGF1 and human FGF2 were provided by PeproTech, Rocky Hill, NJ, USA. Cultrex R-spondin1 cells were obtained from Trevigen, Gaithersburg, MD, USA. Expansion medium was supplemented with 10 μM Rock inhibitor (Y-27632, Selleck Chemicals) for the first two days after isolation, then medium was changed to not-supplemented expansion medium. For isolations using refined medium, EGF and 10 μM Y-27632 were additionally added only for the first two days. Growth medium was changed every two to three days. For weekly passaging at 1:4 to 1:8 split ratios, organoids were harvested, mechanically disrupted using a flame-polished Pasteur pipette. Depending on the splitting ratio, the corresponding quantity of organoid fragments were then embedded in 50 μL fresh Matrigel, seeded per well of a 24-well plate and cultured with either expansion or refined growth medium. To induce differentiation, organoids cultivated in expansion medium were transferred to differentiation medium and analyzed after 4 days. For assessing derivation efficiency in expansion and refined medium, organoids were counted 10 days after initial isolation and projected area was calculated using ImageJ64 (NIH).

Table 1. List of media compositions used in this study. Basal medium consists of Advanced DMEM/F12, 2 mM GlutaMAX, 10 mM HEPES, 1x B27, 1 mM N-acetylcystein. Final concentrations are given.

Reagent	Expansion Medium	Differentiation Medium	Refined Medium
basal medium	+	+	+
N2	1x	1x	–
EGF	50 ng/mL	50 ng/mL	– [a]
Noggin	100 ng/mL	100 ng/mL	100 ng/mL
Rspo1	10% *v/v*	10% *v/v*	10% *v/v*
Wnt3a	43% *v/v*	43% *v/v*	50% *v/v*
Nicotinamide	10 mM	–	–
Gastrin	10 nM	10 nM	10 nM
A83-01	500 nM	500 nM	500 nM
SB202190	10 μM	–	–
HGF	50 ng/mL	50 ng/mL	50 ng/mL
IGF1	–	–	100 ng/mL
FGF2	–	–	50 ng/mL

[a] in refined medium 50 ng/mL EGF is only added for the first two days after isolation or medium adaption.

2.2. Gene Expression Analysis

Total RNA was purified from duodenal, jejunal and colonic organoids in expansion, differentiation and refined medium and from corresponding intestinal epithelium that was harvested during crypt isolation using ReliaPrep™ RNA Tissue Miniprep System (Promega, Madison, WI, USA) according to instructions of the manufacturer. Individual RNA concentrations were determined spectrophotometrically (Nanodrop One C, Thermo Fisher Scientific) and RNA integrity was checked using Agilent Tape Station 4200 (Agilent, Santa Clara, CA, USA). 500 ng RNA was reverse-transcribed to cDNA with oligo-dT and random hexamer primers according to recommendations of the manufacturer (GoScript™ Reverse Transcription System, Promega). Subsequently, qPCR reactions were carried out using GoTaq® qPCR Master Mix (Promega). Primer sequences are given in Table S1. Amplification conditions were as follows: 2 min of initial denaturation at 95 °C, 40 cycles of 15 s of denaturation at 95 °C, 60 s of annealing/extension at 60 °C and a read step, followed by 10 s of dissociation at 95 °C and a melting curve from 65 °C to 95 °C in 5 s per 0.5 °C increments. Data analysis was performed according to Pfaffl et al. [26] taking PCR efficiency into account. Relative quantification was achieved by normalization of values to the stably expressed canine housekeeping gene hypoxanthine phosphoribosyl transferase (HPRT). Heat maps were generated using Excel conditional formatting of log2 fold changes.

2.3. Periodic Acid-Schiff Reaction of Organoid and Tissue Sections

For the generation of formalin-fixed paraffin-embedded (FFPE) samples, organoids were fixed using 2% *v/v* paraformaldehyde (PFA, Merck, Darmstadt, Germany) for 15 min at room temperature, cast in 1.5% *w/v* agarose and dehydrated prior to embedding in paraffin (Sigma-Aldrich). 2.5 μm FFPE sections of organoids derived from duodenum, jejunum and colon and corresponding tissue were stained using PAS-Reaction staining kit and counterstained with Haematoxylin acidic after Mayer according to manufacturer's instructions (Morphisto, Frankfurt am Main, Germany). Images were acquired with a DMi8 microscope and LASX software (Leica, Wetzlar, Germany).

2.4. Transmission Electron Microscopy of Organoids

Samples were fixed in 3% *v/v* buffered glutaraldehyde (pH 7.4, Merck, Darmstadt, Germany), pre-embedded in 1.5% *w/v* agarose (Invitrogen, Thermo Fisher Scientific), washed three times in 0.1 M phosphate buffer (Soerensen, pH 7.4) afterwards and post-fixed in 1% *v/v* osmium tetroxide (Electron Microscopy Sciences, Hatfield, USA) for 2 h at room temperature. Dehydration was performed in a series of graded ethanol solutions (70%, 80%, 96% and 100%), subsequently infiltrated with propylene oxide (Sigma-Aldrich), followed by increasing ratios of epoxy resin-propylene oxide and finally pure resin (Serva, Mannheim, Germany). After an additional change, the resin was polymerized at 60 °C for 48 h. Semi-thin sections were cut at 0.8 μm and stained with toluidine blue, ultra-thin sections were cut at 70 nm, mounted on copper grids (Gröpl, Tulln, Austria) and stained with uranyl acetate (Fluka Chemie AG, Buchs, CH) and lead citrate (Merck, Darmstadt, Germany). Transmission electron micrographs were made with an EM900 (Zeiss, Oberkochen, Germany).

2.5. Proliferation Assay

In order to assess cell proliferation, the Click-iT® EdU Imaging Kit (Invitrogen, Thermo Fisher Scientific) was used. Canine intestinal organoids cultivated in expansion and refined medium were incubated with 5-ethynyl-2'deoxyuridine (EdU) at a final concentration of 10 μM for one hour at 37 °C and were then fixed with 2% *v/v* PFA for 15 min at room temperature. Staining procedure was carried out according to manufacturer's instructions. DNA was counterstained using 4',6-Diamidine-2'-phenylindole dihydrochloride (DAPI, Sigma-Aldrich). Confocal images were taken using an LSM 880 (Zeiss). For 3D reconstruction of z-stack images, the Arivis 3D plugin of Zen 2.3 lite (Zeiss) was used.

2.6. *Viability and Apoptosis Assays*

Viability and apoptosis of organoids in the three media compositions were assessed using the RealTime-Glo MT Cell Viability Assay (Promega) and RealTime-Glo Annexin V Apoptosis Assay (Promega). Organoids were trypsinized for 2 min and seeded in 24-well plates with expansion or refined medium, both supplemented with Y-27632 inhibitor. After 2 days in culture, organoids were harvested, counted and 50 organoids were seeded in six wells of white 96-well plate for each condition and assay. Detection reagents were prepared according to manufacturer and added to the different culture media before overlaying the organoids with either expansion, differentiation or refined medium. Luminescence was measured 0 h, 24 h, 48 h and 72 h after adding the substrates using a GloMax plate reader (Promega). Six wells without organoids, but with medium, served as background controls.

2.7. *Statistical Evaluation*

Data were analyzed and plotted in Prism5 (GraphPad, San Diego, CA, USA). Using a t-test for independent samples (two tailed, unpaired) p-values ≤ 0.05 were considered statistically significant. qPCR data were analyzed by two-way ANOVA (factors are growth medium and individual dogs) and Tukey multiple comparison test using R (version 3.3.1). Statistical results are shown in Tables S2–S7. *** $p < 0.001$ ** $p < 0.01$ * $p < 0.05$. $p < 0.1$.

3. Results

3.1. *Expansion Medium Does Not Support Expression of Secretory Lineage Differentiation Markers*

Canine intestinal organoids derived from duodenal, jejunal and colonic tissue were generated and cultivated in expansion medium. Initial cell isolates representing the intestinal epithelium after separation from the underlying tissue, as well as low- (≤5) and high-passage (≥10) organoids derived therefrom, were quantitatively assessed for expression of selected cell markers using RT-qPCR.

The stem cell marker *LGR5* displayed similar expression levels in low- and high-passage organoids, except high-passage duodenal organoids, which exhibited significant up-regulation of *LGR5* in comparison to their low-passage and initial cell isolate counterparts (Figure 1A). The enterocyte marker villin 1 (*VIL1*) exhibited significantly reduced expression in duodenal, jejunal and colonic organoids in low- and high-passage organoids compared to the initial cell isolates. Enteroendocrine cell marker chromogranin A (*CHGA*) [27] and goblet cell marker mucin 2 (*MUC2*) [28,29] transcripts were barely detectable in organoids when kept in expansion medium throughout passages (Figure 1A).

3.2. *Canine Intestinal Organoids Grow Efficiently in Refined Medium*

To best reflect the in vivo situation and thus allow for downstream applications including drug and toxicity assessment, microbiome studies and modelling of infectious disease, intestinal organoids should harbor stem cells, undifferentiated transit-amplifying cells and differentiated cells. Three different media were compared to identify the ideal medium composition supporting such a scenario: (i) an expansion medium for the culture of canine organoids; (ii) a differentiation medium characterized by omission of nicotinamide and p38-MAPK inhibitor [14]; and (iii) a novel refined medium based on canine expansion medium without nicotinamide, p38-MAPK inhibitor, N2 and EGF, but supplemented with IGF1 and FGF2 as suggested for human organoid culture [25]. After several passages, duodenal, jejunal and colonic canine organoids grown in expansion medium were partly transferred to refined medium for long-term cultivation. For experiments, expansion medium-grown organoids where partly transferred to differentiation medium and after four days, organoids in all three media were harvested (Figure S1A). Growth characteristics and gene expression were assessed.

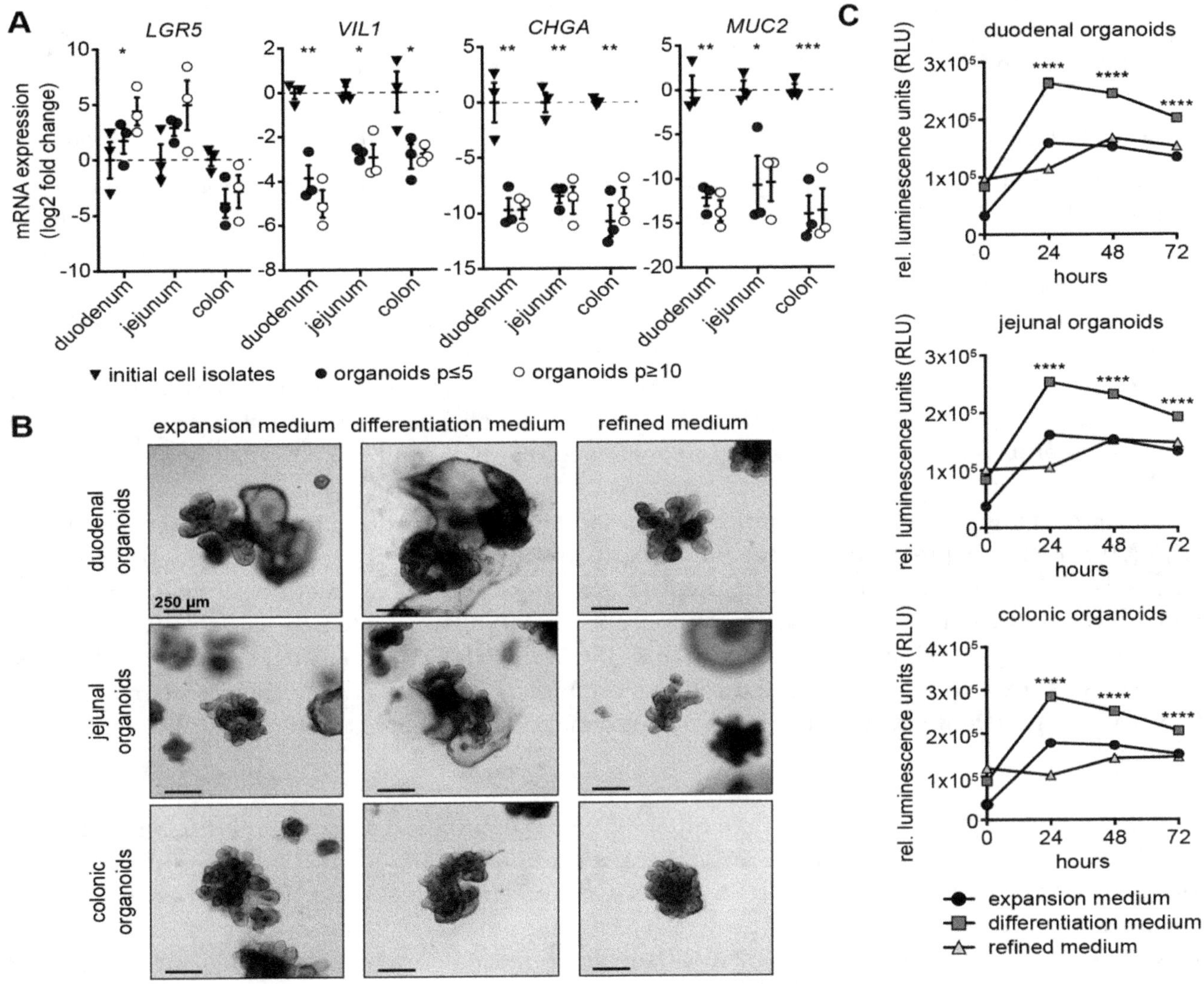

Figure 1. Establishment of new culture conditions for canine intestinal organoids. (**A**) RT-qPCR analysis of initial cell isolates (triangles), organoids with up to five passages (dark circles) and organoids with 10 or more passages (light circles) isolated from duodenum, jejunum or colon using *LGR5*, *VIL1*, *CHGA* and *MUC2* primer. log2 fold changes normalized to expression of initial cell isolates are shown with scatter dot plots; mean is shown; whiskers present SEM; * $p < 0.05$, ** $p < 0.01$ and *** $p < 0.001$, statistical analysis given in detail in Supplementary Tables S2–S4; n = 3 individual dogs. (**B**) Light microscopic images of organoids derived from duodenum, jejunum and colon cultivated in expansion, differentiation and refined medium four days after seeding; scale bars represent 250 µm. (**C**) Mean activity of Annexin V-induced luciferase is shown for 0, 24, 48 and 72 h after addition of substrate to duodenal, jejunal and colonic organoids in expansion (circle), differentiation (square) and refined medium (triangle); whiskers represent SEM; **** $p < 0.0001$; 50 organoids seeded per replicate; n = 6 replicates.

Bright field images of duodenal, jejunal and colonic organoids grown in expansion and refined medium displayed intact, budding structures (Figure 1B). In contrast, all organoid types grown in differentiation medium displayed dark and necrotic areas (Figure 1B, Figure S1B). In accordance with these observations, luciferase-based Annexin V assay revealed low and comparable apoptosis in organoids grown in expansion and refined medium, but pronounced apoptosis in organoids grown in differentiation medium (Figure 1C). Organoids in expansion and refined medium could be cultivated for prolonged period of 25 passages after adaptation of growth medium (Figure S1C). Taken together, only expansion and refined medium supported organoid growth.

3.3. Refined Medium Promotes Expression of Cell Differentiation Markers in Canine Intestinal Organoids

We next aimed at defining the cellular composition of organoids with respect to the different culture conditions. To this end, total RNA isolated from initial cell isolates and from the three different organoid types cultured in the three different media were subjected to gene expression analysis of stem cell and differentiation markers using RT-qPCR.

In general, a heat map representing log2 fold changes of organoids to respective initial cell isolates showed that transcription of stem cell markers was up-regulated in intestinal organoids grown in all three media. Conversely, transcription of differentiation markers was exclusively up-regulated in intestinal organoids grown in differentiation or refined medium (Figure 2A).

In more detail, duodenal organoids displayed significant up-regulation of stem cell marker *LGR5* [30,31] gene expression irrespective of culture conditions (Figure 2B). Culture of duodenal organoids in differentiation and refined medium enhanced mRNA expression of the differentiation markers *NEUROG3* (enteroendocrine progenitor cells) [32,33], *CHGA* (enteroendocrine cells) and *MUC2* (Goblet cells). Transcription of stem cell markers *ASCL2*, *EPHB2*, *BMI1*, *NOTCH1* and *PROM1* [30,31] did not vary significantly throughout duodenal organoids analyzed, while enterocyte marker *VIL1* mRNA was significantly down-regulated in duodenal organoids irrespective of culture conditions (Figure S2A).

Jejunal organoids grown in the three different media exhibited no significant difference in *LGR5*, *ASCL2*, *EPHB2*, *BMI1*, *NOTCH1* or *PROM1* transcription compared to corresponding initial cell isolates (Figure 2C, Figure S2B). *NEUROG3*, *CHGA* and *MUC2* gene expression was significantly down-regulated in jejunal organoids grown in expansion medium. Transcription of differentiation markers *NEUROG3*, *CHGA*, and *MUC2* mRNA was up-regulated when culturing jejunal organoids in refined medium. Expression of *VIL1* was slightly, but significantly, reduced throughout jejunal organoids (Figure S2B).

In colonic organoids, medium compositions had no significant impact on *LGR5* expression, while *NEUROG3*, *CHGA* and *MUC2* transcription was significantly down-regulated in organoids kept in expansion medium (Figure 2D). Culturing colonic organoids in differentiation and refined medium induced an increase of *CHGA* and *MUC2* expression almost to levels displayed by initial cell isolates. In contrast to duodenal and jejunal organoids, colonic organoids showed aberrant *ASCL2*, *EPHB2*, *BMI1* and *PROM1* transcription in comparison to initial cell isolates, whereas *NOTCH1* expression remained grossly unchanged (Figure S2C). *VIL1* transcription was generally slightly lower expressed in colonic organoids irrespective of culture conditions.

Marker transcription levels in organoids varied to some extent with respect to the individual dog they were derived from. These variations were less pronounced when cultivating organoids—notably jejunal and colonic organoids—in refined medium (Figure 2, Figure S2). Despite these individual variations the presented data indicate that differentiation and refined medium supported cell differentiation to secretory lineage cells within organoids.

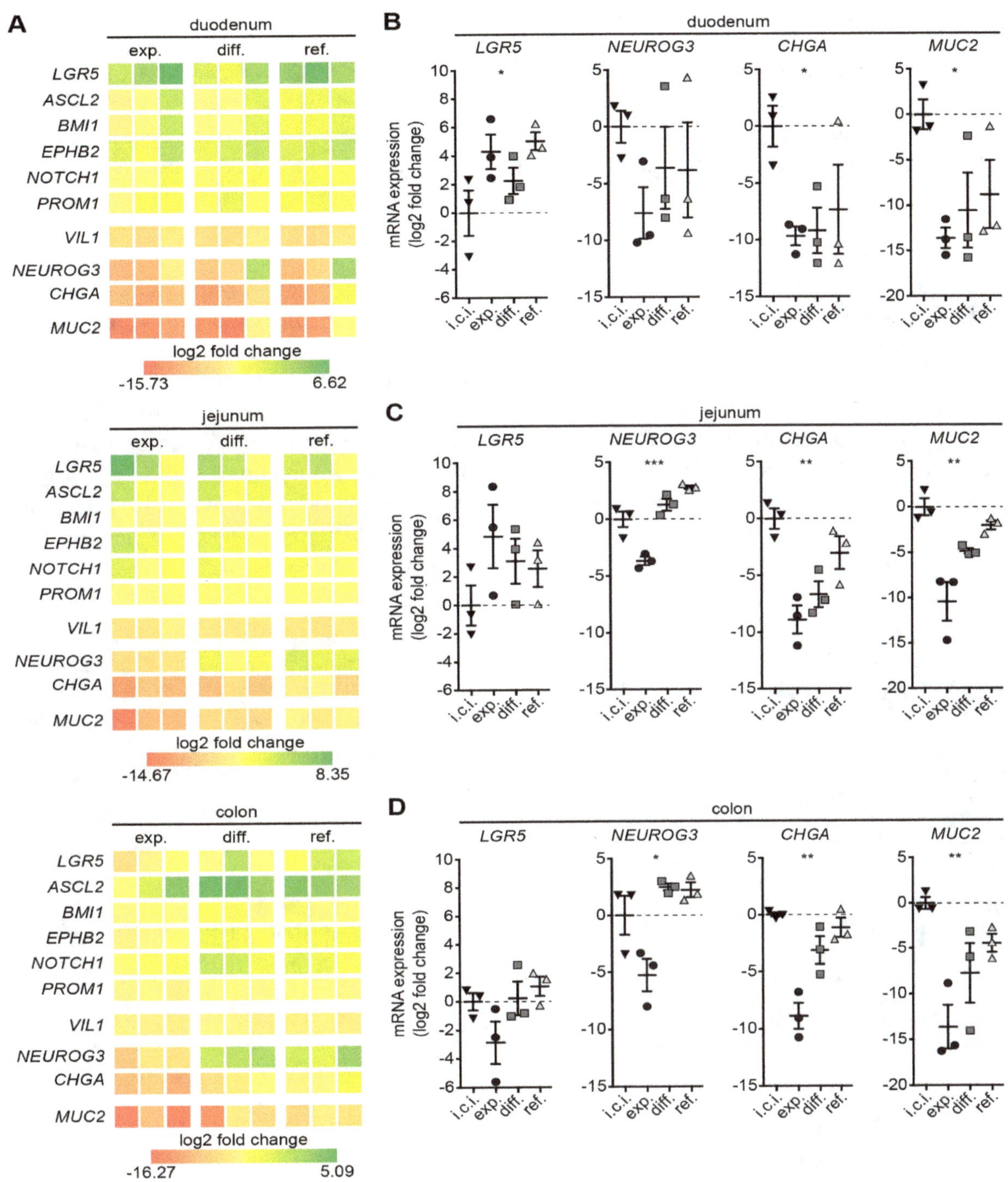

Figure 2. Refined medium and differentiation medium show elevated marker expression of enteroendocrine and goblet cells. (**A**) Heat map of log2 fold changes derived from RT-qPCR data of duodenal, jejunal and colonic organoids cultivated four days in expansion, differentiation and refined medium normalized to initial cell isolates (i.c.i. = 0); data from three individuals are shown in side-by-side columns. (**B–D**) Individual scatter dot plots of gene expression data from (**A**) shown for stem cell marker *LGR5*, secretory lineage precursor marker *NEUROG3*, enteroendocrine cell marker *CHGA* and goblet cell marker *MUC2* for organoids derived from duodenum (**B**), jejunum (**C**) and colon (**D**) four days after seeding; mean is shown, whiskers are SEM; * $p < 0.05$, ** $p < 0.01$, *** $p < 0.001$, statistical analysis given in detail in Supplementary Tables S5–S7; n = 3 dogs; i.c.i., initial cell isolates; exp., expansion medium; diff., differentiation medium; ref., refined medium.

3.4. Refined and Differentiation Medium Both Support Organoid Differentiation

The detection of markers for goblet and enteroendocrine cells by RT-qPCR prompted us to confirm and localize these cells within organoids. Therefore, periodic-acid Schiff (PAS) reaction and transmission electron microscopy (TEM) were used. PAS reaction of mucin as revealed by a purple signal demonstrated the presence of goblet cells that were evenly distributed in duodenal and jejunal tissue sections along the crypt-villus axis, and at higher frequency around the crypt base in colon tissue sections (Figure 3A). In organoid culture, expansion medium failed to induce differentiation into goblet cells, while growth in differentiation and refined medium gave rise to goblet cells irrespective of the intestinal segment. These observations were in accordance with *MUC2* transcription data (see Figure 2) and further substantiated by TEM, which revealed the presence of goblet cells with apical large mucin granules and enteroendocrine cells characterized by small dark basal granules in organoids grown in differentiation and refined medium (Figure 3B and Figure S3B). In contrast, TEM of organoids cultivated in expansion medium revealed uniformly appearing cells harboring microvilli (Figure S3A). These cells were also present in all other conditions suggesting the presence of enterocytes, which is correlating with *VIL1* expression data (see Figure 2 and Figure S2). Interestingly, cell-to-cell interactions such as tight junctions, adherens junctions and desmosomes could be visualized by organoid TEM throughout culture conditions (Figure S3C). These data demonstrate the presence of secretory lineage cells in canine organoids upon cultivation in refined and differentiation medium.

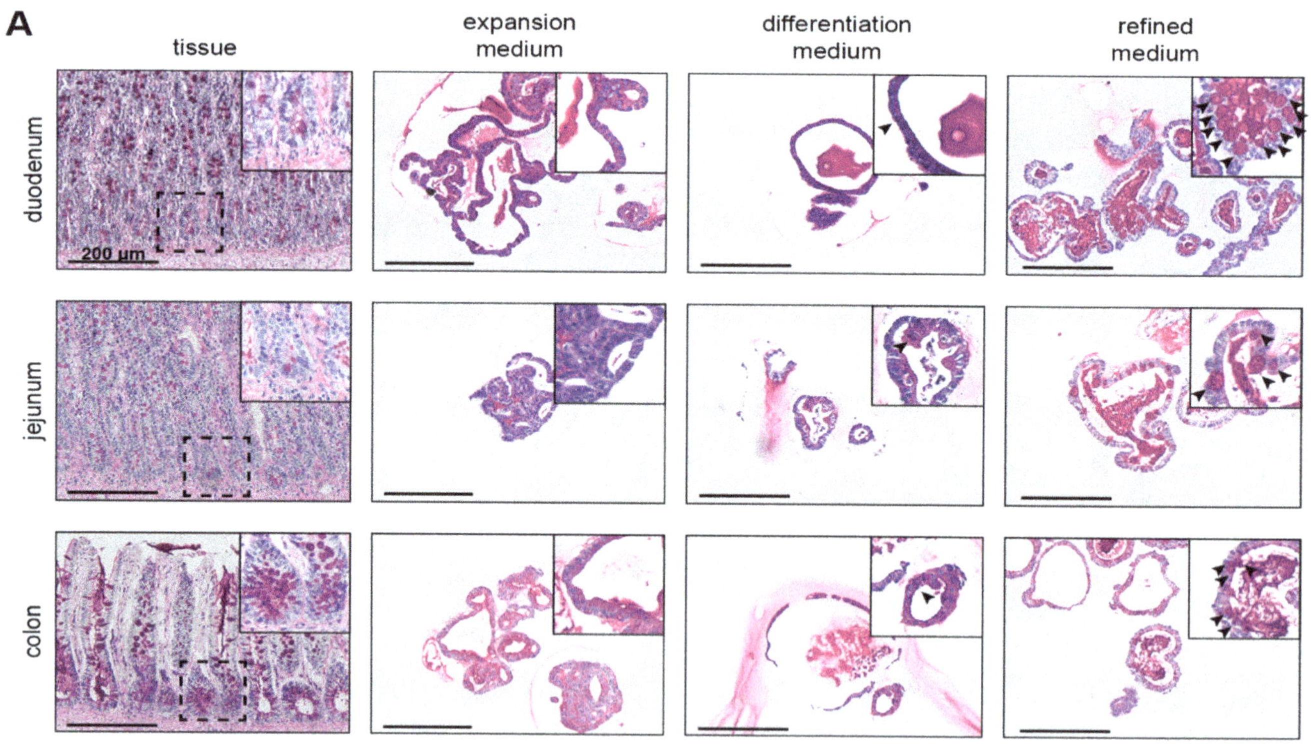

Figure 3. *Cont.*

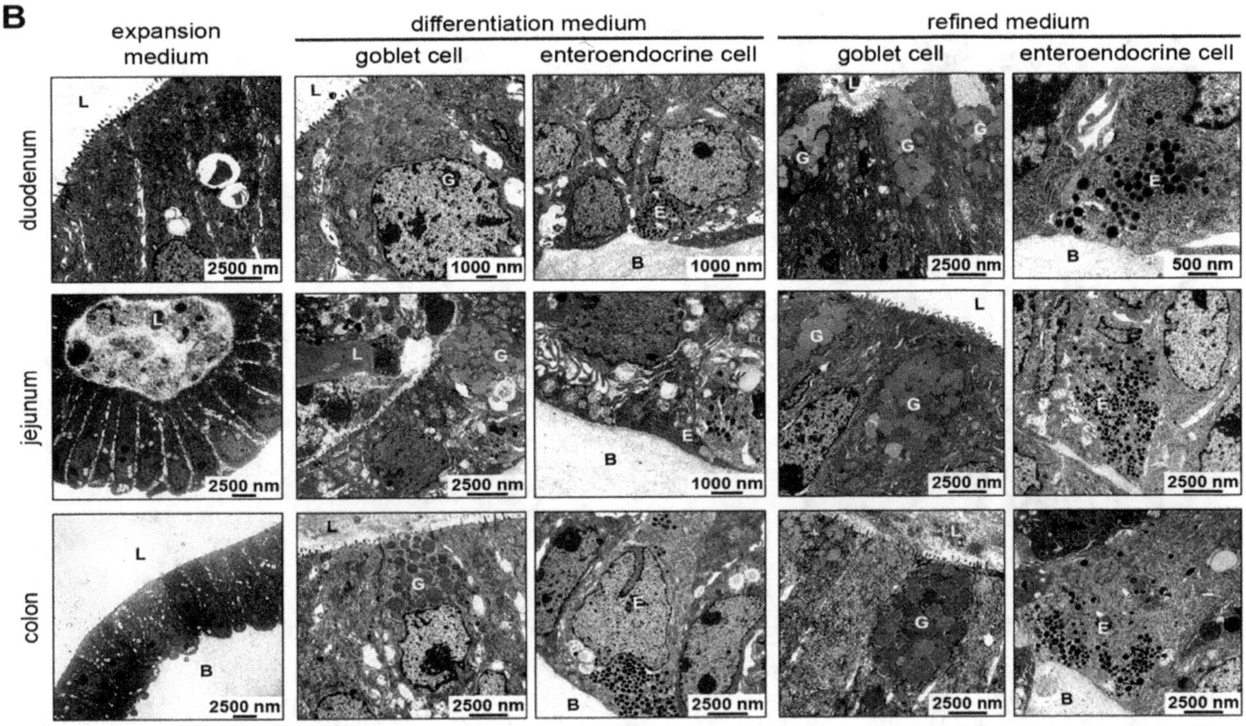

Figure 3. Goblet and enteroendocrine cells are present in organoids upon cultivation in differentiation and refined medium, but not in expansion medium. (**A**) PAS reaction in FFPE sections of canine intestinal tissue or organoids in expansion, differentiation and refined medium derived from duodenum, jejunum and colon; counterstained with Hematoxylin, images of duodenal and jejunal tissue were cropped in order to visualize crypts; scale bars 200 µm, inset depicts higher magnification; arrow heads indicate PAS positive goblet cells; one representative image is shown. (**B**) TEM images of duodenal, jejunal and colonic organoids in expansion, differentiation and refined medium; scale bars are indicated; L, lumen; B, basal lamina; G, goblet cell; E, enteroendocrine cell.

3.5. Refined Medium Supports Continuous Growth and Derivation of Organoids

So far, we have been able to verify that organoids in refined medium were differentiating similarly or even better than organoids in differentiation medium (see Figures 2 and 3) without concomitant induction of apoptosis (see Figure 1B,C). Therefore, we conclude that a refined medium composition is beneficial over the use of a separate differentiation medium. In a next step we assessed the effectiveness of refined versus expansion medium on the continuous growth of organoids.

Importantly, all organoid types showed a similar proliferation pattern, as revealed by a scattered distribution of single, EdU-positive cells within organoids when exposed to the two different medium conditions (Figure 4A). However, viability of duodenal and jejunal organoids in expansion medium was reduced compared to colonic organoids, which could be significantly enhanced by using refined medium (Figure 4B).

To assess the suitability of refined medium for the establishment of new organoid lines, duodenal biopsies were taken during routine gastroduodenoscopy. Following isolation, starting cell material was cultured either with expansion medium containing Rock inhibitor, or with EGF- and Rock inhibitor-supplemented refined medium for the first two days, then changed to unsupplemented expansion and refined medium. After 10 days in culture, an almost 4-fold higher number of organoids had grown in refined medium compared to in expansion medium (Figure 4C). Organoids exposed to refined medium also displayed a higher projected area compared to organoids formed in expansion medium, indicating that refined medium allows for easier adaption of canine intestinal cells to in vitro culture conditions.

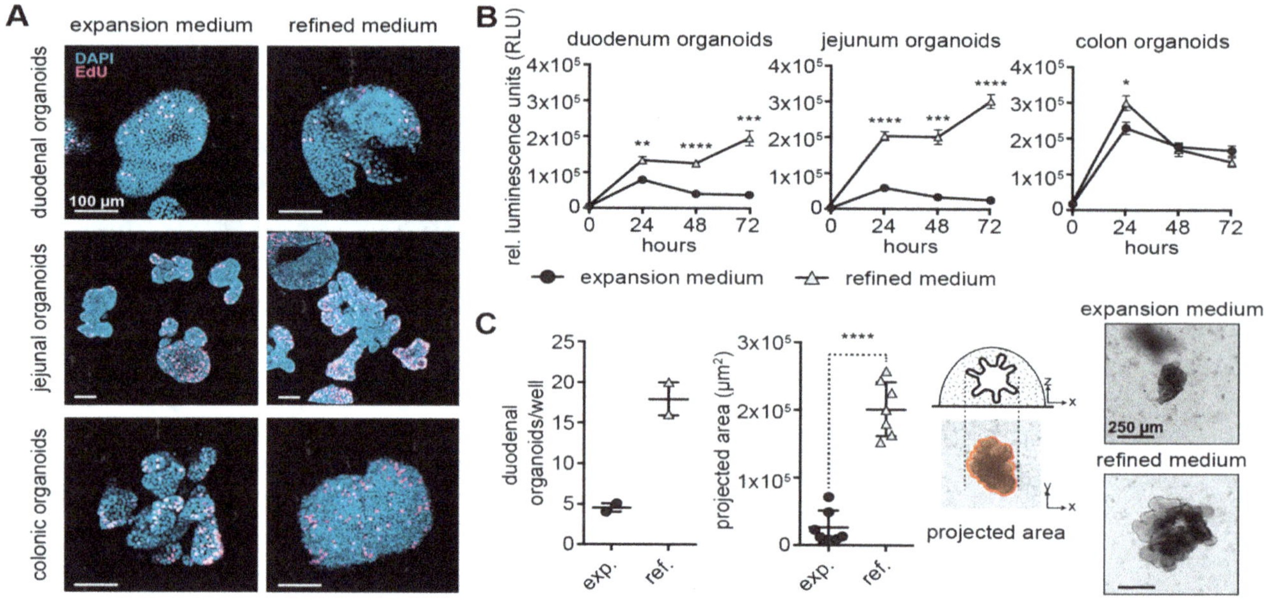

Figure 4. Derivation of organoids is more efficient using refined medium conditions. (**A**) 3D reconstruction of confocal z-stack images from duodenal, jejunal and colonic organoids in expansion and refined medium; staining for EdU after one hour labeling (magenta); organoids counterstained using DAPI (blue); scale bar represents 100 µm. (**B**) Relative luciferase activity of cell viability assay is shown for serial measurements at 0, 24, 48 and 72 h after addition of substrate to duodenal, jejunal and colonic organoids in expansion (circle) and refined medium (triangle); mean is shown; whiskers represent SEM; * $p < 0.05$, ** $p < 0.01$, *** $p < 0.001$ and **** $p < 0.0001$; 50 organoids seeded per replicate; n = 6 replicates. (**C**) Organoid number is shown for freshly isolated duodenal biopsies seeded in expansion and refined medium, in scatter dot plots with mean and range indicated by whiskers, n = 2 wells; projected area of organoids is depicted with mean and SEM defined by whiskers, **** $p < 0.0001$, n = 7 organoids; bright field images of representative organoids are shown, scale bar represents 250 µm.

4. Discussion

Since the first reports on the derivation of organoids from murine and subsequently from human intestine, the establishment of canine intestinal organoids was a matter of time. Particularly as dogs develop GI diseases like cancer, infectious disease and IBD naturally [23], thereby representing a faithful model system for the corresponding human diseases. In early 2017 our group [17], Kingsbury et al. [34], and Powel and Behnke [16] reported on the successful derivation of canine intestinal organoids. We then focused on advancing organoid cultivation to bring them a step closer to the in vivo situation.

In the presented study, we found that expansion medium does not support stem cell commitment in the secretory lineage using transcription profiling. This finding is supported by reports on human intestinal organoids, which usually harbor only a limited number of differentiated cell types, because they are grown in expansion media that allow for the maintenance of stem cells but do not support secretory lineage differentiation [14,17]. This absence of differentiated cells in expansion medium could not be overcome by prolonged cultivation: low- (≤5) and high-passage (≥10) canine intestinal organoids all tested negative for differentiation marker expression. Recently, Chandra et al. published a well-written paper on the derivation of canine intestinal organoids and the presence of goblet and enteroendocrine cells [35]. The formulation of the expansion medium differs in a 5-fold higher concentration of p38-MAPK inhibitor and the addition of 8% fetal bovine serum FBS (Chandra) compared to the medium formulation used herein. p38-MAPK inhibitor in human organoid culture is used to suppress secretory lineage commitment of intestinal stem cells to prevent their depletion and maintain long-term culture [14,36,37]. The addition of 8% FBS is a likely explanation for the presence of secretory cells, potentially via subsequent activation of p38-MAPK signaling. To create a culture system with defined components, the culture medium used in the study presented herein corresponds to a widely used human organoid medium without FBS [14] that was only supplemented with HGF to allow for the cultivation of over 50 passages. These variations in medium additives may explain the differences between both studies.

To further develop canine intestinal organoid culture, we focused on optimizing the growth medium to obtain a more in vivo-like cell composition with stem cells, enterocytes, goblet and enteroendocrine cells. Recently, a newly composed medium was shown to enable stem cells of human intestinal organoids to undergo differentiation, while concomitantly preserving the stem cell pool [25]. To overcome the intrinsic limitations of classical expansion medium, we adapted our canine intestinal expansion medium based on the suggestions of Fujii et al. [25] by (i) withdrawal of nicotinamide, p38-MAPK inhibitor, N2 and EGF; and (ii) addition of IGF1 and FGF2. This medium termed "refined medium" promoted stem cell growth and differentiation of canine intestinal organoids simultaneously for over 6 months of cultivation (data not shown). To assess the benefits and limitations of this refined medium in regard to organoid growth and lineage commitment, canine intestinal organoids grown in expansion, differentiation or refined medium were comparatively characterized.

While duodenal, jejunal and colonic organoids displayed a viable morphology upon culture in expansion and refined medium, differentiation medium induced apoptosis in all three intestinal segments. Interestingly, apoptosis peaked as early as 24 h after transfer to differentiation medium, indicating that programmed cell death was initiated by differentiation-unrelated factors. This assumption is supported by previous reports on the promoting effect of nicotinamide on cell survival in human pluripotent stem cells [38], on proliferation of aged murine organoids [39] and on inducing apoptosis via reduced intracellular NAD concentrations upon its withdrawal [40]. Taken together, differentiation medium induced apoptosis, thus not supporting long-term culture of canine intestinal organoids. This induction of cell death by the medium formulation is unfavorable when differentiation, proliferation or even apoptosis are investigated upon compound treatment for drug development. In addition, differentiation of organoids in expansion medium must be repeated for every individual experiment making them susceptible to fluctuations between media batches, illustrating the need for a better culture system.

Refined medium induced similar levels of differentiation or even outperformed differentiation medium in this regard. Duodenal, jejunal and colonic organoids grown in refined medium revealed enhanced expression of markers indicating the presence of secretory lineage precursors, enteroendocrine and goblet cells. Interestingly, organoids in refined medium expressed comparable or even higher levels of stem cell marker compared to organoids in expansion medium. Organoids irrespective of the intestinal segment or media composition expressed the enterocyte marker *VIL1*. These data confirm findings reported for human intestinal organoids grown in niche-inspired culture media that supported expression of enteroendocrine, goblet and stem cell markers [25] and data reported for villin1 positive enterocytes in murine intestinal organoids [12]. However, expression of the differentiation markers *CHGA*, *MUC2* and *VIL1* in differentiation and refined medium reached the levels of the i.c.i. only to a certain extent. While an in vitro culture model cannot fully reproduce the complex in vivo situation, our data show that growth conditions in refined medium are more similar compared to those in expansion medium.

Since only a limited number of specific antibodies were available for reliable characterization of canine cells, we used two well-established alternative methods, PAS reaction and TEM, for the detection of goblet cells and enteroendocrine cells. These results supported our RT-qPCR data, substantiating that only organoids grown in differentiation and refined medium harbored enteroendocrine and goblet cells. These findings are in accordance with data obtained for human intestinal organoids grown in differentiation [14] or niche-inspired medium [25]. Furthermore, the presence of enterocytes in organoids cultivated in expansion, differentiation and refined medium of all segments was supported by microvilli bearing cells as reported for murine intestinal organoids [12], indicating that enterocytes are present irrespective of culture condition.

So far, our data provided evidence that cultivation of canine intestinal organoids in refined medium is preferable to the use of differentiation medium, since it promotes stem cell growth and differentiation simultaneously, allows long-term growth without fluctuations between media batches and does not induce apoptosis. Therefore, further analysis focused on comparison of organoid growth in expansion and refined medium.

The comparison of proliferative behavior differences between canine intestinal organoids grown in expansion versus refined medium by EdU incorporation assays revealed similar proliferation levels for all organoid types irrespective of the medium used. Yet, viability data of duodenal and jejunal organoids indicate that expansion medium does not support small intestinal organoid cultivation as effective as refined medium. The colonic epithelium is highly populated with various bacteria and has less digestive and absorptive function than small intestinal epithelium possibly resulting in more robustness with respect to cultivation. Therefore, colonic organoids can be expanded in both expansion and refined medium efficiently.

Importantly, generation of organoids from duodenal biopsies was more successful in refined than expansion medium. Not only the number of organoids per well was enhanced, but also their projected area. This feature has an important impact on canine intestinal disease modelling, especially when donor tissue stems from GI enteropathies with severely affected epithelium. The high suitability of refined medium for canine organoid establishment also matches the higher plating efficacy shown for human single-cell dissociated organoids grown in niche-inspired media [25].

5. Conclusions

Taken together, we strongly recommend the use of refined medium for establishment and long-term culture of canine intestinal organoids. We could clearly show that this medium sustains stem cell growth, while simultaneously promoting differentiation of stem cells/immature cells into enterocytes, enteroendocrine and goblet cells. Canine intestinal organoids cultivated in refined medium bear the advantage of an easy to handle, reproducible and stable culture system, thereby representing a physiologically superior in vitro system for disease modelling, drug development, toxicity studies and personalized medicine. Compared to conventional expansion and differentiation media, the refined medium presented herein allows for more accurate assessment of genetic and epigenetic impacts on canine intestinal cell differentiation, bringing organoids a step closer to the in vivo situation.

Supplementary Materials
Figure S1: Cultivation scheme of organoids and images of organoids in differentiation medium and after prolonged cultivation in expansion and refined medium; Figure S2: Gene expression of stem cell marker is increased in colonic organoids; Figure S3: TEM images of microvilli and cell-cell junctions in all conditions and goblet cells in organoids cultivated in differentiation and refined medium; Table S1: qPCR primer specifications used in this study; Table S2: Statistical analysis of duodenal tissue vs. duodenal organoids; Table S3: Statistical analysis of jejunal tissue vs. jejunal organoids; Table S4: Statistical analysis of colonic tissue vs. colonic organoids; Table S5: Statistical analysis of duodenal organoids in different media; Table S6: Statistical analysis of jejunal organoids in different media; Table S7: Statistical analysis of colonic organoids in different media.

Author Contributions: Conceptualization, N.K., B.P., A.M.C.M., A.S., H.S.K., K.S., L.C.P., B.S., M.K., S.B. and I.A.B.; Formal analysis, N.K.; Funding acquisition, L.C.P., B.S., S.B. and I.A.B.; Investigation, N.K., B.P., A.M.C.M., W.T. and I.W.; Methodology, N.K., B.P., A.M.C.M., H.S.K. and K.S.; Project administration, B.P. and I.A.B.; Resources, I.A.B.; Supervision, M.K., S.B. and I.A.B.; Validation, N.K., W.T. and I.W.; Visualization, N.K.; Writing—original draft, N.K. and S.B.; Writing—review & editing, B.P., A.M.C.M., W.T., I.W., A.S., H.S.K., K.S., L.C.P., B.S., M.K. and I.A.B. All authors have read and agreed to the published version of the manuscript.

Acknowledgments: This research was supported using resources of the VetCore Facility (Imaging, VetBioBank and Genomics) of the University of Veterinary Medicine Vienna. Open Access Funding by the University of Veterinary Medicine Vienna.

References

1. Kim, H.P.; Crockett, S.; Shaheen, N.J. The Burden of Gastrointestinal and Liver Disease around the World. In *GI Epidemiology*; Wiley: Oxford, UK, 2014; pp. 1–13.
2. Bray, F.; Ferlay, J.; Soerjomataram, I.; Siegel, R.L.; Torre, L.A.; Jemal, A. Global cancer statistics 2018: GLOBOCAN estimates of incidence and mortality worldwide for 36 cancers in 185 countries. *CA Cancer J. Clin.* **2018**, *68*, 394–424. [CrossRef] [PubMed]
3. Rubin, S.I. Infectious Diseases of the GI Tract. Available online: https://www.msdvetmanual.com/digestive-system/digestive-system-introduction/infectious-diseases-of-the-gi-tract (accessed on 19 September 2019).
4. Albani, S.; Colomb, J.; Prakken, B. Translational Medicine 2.0: From Clinical Diagnosis–Based to Molecular-Targeted Therapies in the Era of Globalization. *Clin. Pharmacol. Ther.* **2010**, *87*, 642–645. [CrossRef]
5. Albani, S.; Prakken, B. The advancement of translational medicine—From regional challenges to global solutions. *Nat. Med.* **2009**, *15*, 1006–1009. [CrossRef]
6. Scherzer, M.; Kramer, N.; Unger, C.; Walzl, A.; Walter, S.; Stadler, M.; Hengstschläger, M.; Dolznig, H. Preclinical Cancer Models with the Potential to Predict Clinical Response. In *Drug Discovery in Cancer Epigenetics*; Elsevier BV: Oxford, UK, 2016; pp. 97–122.
7. Browning, T.H.; Trier, J.S. Organ culture of mucosal biopsies of human small intestine. *J. Clin. Investig.* **1969**, *48*, 1423–1432. [CrossRef] [PubMed]
8. Kedinger, M.; Haffen, K.; Simon-Assmann, P. Intestinal tissue and cell cultures. *Differentiation* **1987**, *36*, 71–85. [CrossRef] [PubMed]
9. Evans, G.S.; Flint, N.; Somers, A.S.; Eyden, B.; Potten, C.S. The development of a method for the preparation of rat intestinal epithelial cell primary cultures. *J. Cell Sci.* **1992**, *101*.
10. Ootani, A.; Li, X.; Sangiorgi, E.; Ho, Q.T.; Ueno, H.; Toda, S.; Sugihara, H.; Fujimoto, K.; Weissman, I.L.; Capecchi, M.R.; et al. Sustained in vitro intestinal epithelial culture within a Wnt-dependent stem cell niche. *Nat. Med.* **2009**, *15*, 701–706. [CrossRef]
11. Wang, X.; Yamamoto, Y.; Wilson, L.H.; Zhang, T.; Howitt, B.; Farrow, M.A.; Kern, F.; Ning, G.; Hong, Y.; Khor, C.C.; et al. Cloning and variation of ground state intestinal stem cells. *Nature* **2015**, *522*, 173–178. [CrossRef]
12. Sato, T.; Vries, R.G.; Snippert, H.J.; Van De Wetering, M.; Barker, N.; Stange, D.; Van Es, J.H.; Abo, A.; Kujala, P.; Peters, P.J.; et al. Single Lgr5 stem cells build crypt-villus structures in vitro without a mesenchymal niche. *Nature* **2009**, *459*, 262–265. [CrossRef]
13. Date, S.; Sato, T. Mini-Gut Organoids: Reconstitution of the Stem Cell Niche. *Annu. Rev. Cell Dev. Boil.* **2015**, *31*, 269–289. [CrossRef]
14. Sato, T.; Stange, D.; Ferrante, M.; Vries, R.G.; Van Es, J.H.; Brink, S.V.D.; Van Houdt, W.J.; Pronk, A.; Van Gorp, J.M.; Siersema, P.D.; et al. Long-term Expansion of Epithelial Organoids From Human Colon, Adenoma, Adenocarcinoma, and Barrett's Epithelium. *Gastroenterology* **2011**, *141*, 1762–1772. [CrossRef] [PubMed]
15. Wang, Y.; Kim, R.; Gunasekara, D.B.; Reed, M.I.; Disalvo, M.; Nguyen, D.L.; Bultman, S.J.; Sims, C.E.; Magness, S.T.; Allbritton, N.L. Formation of Human Colonic Crypt Array by Application of Chemical Gradients Across a Shaped Epithelial Monolayer. *Cell. Mol. Gastroenterol. Hepatol.* **2017**, *5*, 113–130. [CrossRef] [PubMed]
16. Powell, R.H.; Behnke, M. WRN conditioned media is sufficient for in vitro propagation of intestinal organoids from large farm and small companion animals. *Boil. Open* **2017**, *6*, 698–705. [CrossRef] [PubMed]
17. 2017 ACVIM Forum Research Abstract Program. *J. Veter- Intern. Med.* **2017**, *31*, 1225–1361. [CrossRef]
18. Nantasanti, S.; Spee, B.; Kruitwagen, H.S.; Chen, C.; Geijsen, N.; Oosterhoff, L.A.; Van Wolferen, M.E.; Peláez, N.; Fieten, H.; Wubbolts, R.W.; et al. Disease Modeling and Gene Therapy of Copper Storage Disease in Canine Hepatic Organoids. *Stem Cell Rep.* **2015**, *5*, 895–907. [CrossRef]
19. Kruitwagen, H.S.; Oosterhoff, L.A.; Vernooij, I.G.; Schrall, I.M.; Van Wolferen, M.E.; Bannink, F.; Roesch, C.; Van Uden, L.; Molenaar, M.; Helms, J.B.; et al. Long-Term Adult Feline Liver Organoid Cultures for Disease Modeling of Hepatic Steatosis. *Stem Cell Rep.* **2017**, *8*, 822–830. [CrossRef]

20. Derricott, H.; Luu, L.; Fong, W.Y.; Hartley, C.S.; Johnston, L.; Armstrong, S.D.; Randle, N.; Duckworth, C.A.; Campbell, B.J.; Wastling, J.M.; et al. Developing a 3D intestinal epithelium model for livestock species. *Cell Tissue Res.* **2018**, *375*, 409–424. [CrossRef]
21. Hamilton, C.A.; Young, R.; Jayaraman, S.; Sehgal, A.; Paxton, E.; Thomson, S.; Katzer, F.; Hope, J.; Innes, E.A.; Morrison, L.; et al. Development of in vitro enteroids derived from bovine small intestinal crypts. *Veter- Res.* **2018**, *49*, 54. [CrossRef]
22. Pierzchalska, M.; Grabacka, M.; Michalik, M.; Zyla, K.; Pierzchalski, P. Prostaglandin E2 supports growth of chicken embryo intestinal organoids in Matrigel matrix. *Biotechniques* **2012**, *52*, 307–315. [CrossRef]
23. Cerquetella, M.; Spaterna, A.; Laus, F.; Tesei, B.; Rossi, G.; Antonelli, E.; Villanacci, V.; Bassotti, G. Inflammatory bowel disease in the dog: Differences and similarities with humans. *World J. Gastroenterol.* **2010**, *16*, 1050–1056. [CrossRef] [PubMed]
24. Shearin, A.L.; Ostrander, E.A. Leading the way: Canine models of genomics and disease. *Dis. Model. Mech.* **2010**, *3*, 27–34. [CrossRef] [PubMed]
25. Fujii, M.; Matano, M.; Toshimitsu, K.; Takano, A.; Mikami, Y.; Nishikori, S.; Sugimoto, S.; Sato, T. Human Intestinal Organoids Maintain Self-Renewal Capacity and Cellular Diversity in Niche-Inspired Culture Condition. *Cell Stem Cell* **2018**, *23*, 787–793.e6. [CrossRef] [PubMed]
26. Pfaffl, M.W. A new mathematical model for relative quantification in real-time RT-PCR. *Nucleic Acids Res.* **2001**, *29*, 45. [CrossRef] [PubMed]
27. Grün, D.; Muraro, M.J.; Boisset, J.-C.; Wiebrands, K.; Lyubimova, A.; Dharmadhikari, G.; Born, M.V.D.; Van Es, J.; Jansen, E.; Clevers, H.; et al. De Novo Prediction of Stem Cell Identity using Single-Cell Transcriptome Data. *Cell Stem Cell* **2016**, *19*, 266–277. [CrossRef]
28. Audie, J.P.; Janin, A.; Porchet, N.; Copin, M.C.; Gosselin, B.; Aubert, J.P. Expression of human mucin genes in respiratory, digestive, and reproductive tracts ascertained by in situ hybridization. *J. Histochem. Cytochem.* **1993**, *41*, 1479–1485. [CrossRef]
29. Kim, Y.S.; Ho, S.B. Intestinal Goblet Cells and Mucins in Health and Disease: Recent Insights and Progress. *Curr. Gastroenterol. Rep.* **2010**, *12*, 319–330. [CrossRef]
30. Barker, N.; Van Es, J.H.; Kuipers, J.; Kujala, P.; Born, M.V.D.; Cozijnsen, M.; Haegebarth, A.; Korving, J.; Begthel, H.; Peters, P.J.; et al. Identification of stem cells in small intestine and colon by marker gene Lgr5. *Nature* **2007**, *449*, 1003–1007. [CrossRef]
31. Munoz, J.; Stange, D.; Schepers, A.G.; Van De Wetering, M.; Koo, B.-K.; Itzkovitz, S.; Volckmann, R.; Kung, K.S.; Koster, J.; Radulescu, S.; et al. The Lgr5 intestinal stem cell signature: Robust expression of proposed quiescent '+4' cell markers. *EMBO J.* **2012**, *31*, 3079–3091. [CrossRef]
32. Jenny, M.; Uhl, C.; Roche, C.; Duluc, I.; Guillermin, V.; Guillemot, F.; Jensen, J.; Kedinger, M.; Gradwohl, G. Neurogenin3 is differentially required for endocrine cell fate specification in the intestinal and gastric epithelium. *EMBO J.* **2002**, *21*, 6338–6347. [CrossRef]
33. Lopez-Diaz, L.; Jain, R.N.; Keeley, T.M.; VanDussen, K.L.; Brunkan, C.S.; Gumucio, D.L.; Samuelson, L.C. Intestinal Neurogenin 3 directs differentiation of a bipotential secretory progenitor to endocrine cell rather than goblet cell fate. *Dev. Boil.* **2007**, *309*, 298–305. [CrossRef]
34. Kingsbury, D.; Sun, L.; Qi, Y.; Fredericks, J.; Wang, Q.; Wannemuehler, M.; Jergens, A.; Allenspach, K. DDW Abstract: Optimizing the Development and Characterization of Canine Small Intestinal Crypt Organoids as a Research Model. *Gastroenterology* **2017**, *152* (Suppl. 1). [CrossRef]
35. Chandra, L.; Borcherding, D.C.; Kingsbury, D.; Atherly, T.; Ambrosini, Y.M.; Bourgois-Mochel, A.; Yuan, W.; Kimber, M.; Qi, Y.; Wang, Q.; et al. Derivation of adult canine intestinal organoids for translational research in gastroenterology. *BMC Boil.* **2019**, *17*, 33. [CrossRef] [PubMed]
36. Kozuka, K.; He, Y.; Koo-McCoy, S.; Kumaraswamy, P.; Nie, B.; Shaw, K.; Chan, P.; Leadbetter, M.; He, L.; Lewis, J.G.; et al. Development and Characterization of a Human and Mouse Intestinal Epithelial Cell Monolayer Platform. *Stem Cell Rep.* **2017**, *9*, 1976–1990. [CrossRef] [PubMed]
37. Wallach, T.E.; Bayrer, J.R. Intestinal Organoids. *J. Pediatr. Gastroenterol. Nutr.* **2017**, *64*, 180–185. [CrossRef]
38. Meng, Y.; Ren, Z.; Xu, F.; Zhou, X.; Song, C.; Wang, V.Y.-F.; Liu, W.; Lu, L.; Thomson, J.A.; Chen, G. Nicotinamide Promotes Cell Survival and Differentiation as Kinase Inhibitor in Human Pluripotent Stem Cells. *Stem Cell Rep.* **2018**, *11*, 1347–1356. [CrossRef]

39. Uchida, R.; Saito, Y.; Nogami, K.; Kajiyama, Y.; Suzuki, Y.; Kawase, Y.; Nakaoka, T.; Muramatsu, T.; Kimura, M.; Saito, H. Epigenetic silencing of Lgr5 induces senescence of intestinal epithelial organoids during the process of aging. *NPJ Aging Mech. Dis.* **2018**, *4*, 1. [CrossRef]
40. Heikal, A.A. Intracellular coenzymes as natural biomarkers for metabolic activities and mitochondrial anomalies. *Biomark. Med.* **2010**, *4*, 241–263. [CrossRef]

12

Vertically-Aligned Functionalized Silicon Micropillars for 3D Culture of Human Pluripotent Stem Cell-Derived Cortical Progenitors

Alessandro Cutarelli [1,†], Simone Ghio [2,†], Jacopo Zasso [1], Alessandra Speccher [3], Giorgina Scarduelli [4], Michela Roccuzzo [4], Michele Crivellari [2], Nicola Maria Pugno [5,6,7], Simona Casarosa [3], Maurizio Boscardin [2,*,‡] and Luciano Conti [1,*,‡]

1 Laboratory of Stem Cell Biology, Department of Cellular, Computational and Integrative Biology-CIBIO, University of Trento, 38123 Trento, Italy; alessandro.cutarelli@unitn.it (A.C.); Jacopo.Zasso@unitn.it (J.Z.)
2 Fondazione Bruno Kessler-Center for Material and Microsystem, 38123 Trento, Italy; simoneghio1@gmail.com (S.G.); crivella@fbk.eu (M.C.)
3 Laboratory of Neural Development and Regeneration, Department of Cellular, Computational and Integrative Biology-CIBIO, University of Trento, 38123 Trento, Italy; a.speccher.1@unitn.it (A.S.); simona.casarosa@unitn.it (S.C.)
4 Advanced Imaging Facility, Department of Cellular, Computational and Integrative Biology-CIBIO, University of Trento, 38123 Trento, Italy; Giorgina.Scarduelli@unitn.it (G.S.); michela.roccuzzo@unitn.it (M.R.)
5 Laboratory of Bio-Inspired and Graphene Nanomechanics, Department of Civil, Environmental and Mechanical Engineering, University of Trento, 38123 Trento, Italy; nicola.pugno@unitn.it
6 School of Engineering and Materials Science, Queen Mary University of London, London E1 4NS, UK
7 Ket-Lab, Edoardo Amaldi Foundation, via del Politecnico snc, I-00133 Roma, Italy
* Correspondence: boscardi@fbk.eu (M.B.); Luciano.Conti@unitn.it (L.C.)
† A.C. and S.G. contribute equally to this work and share first authorship.
‡ M.B. and L.C. contribute equally to this work and share last authorship.

Abstract: Silicon is a promising material for tissue engineering since it allows to produce micropatterned scaffolding structures resembling biological tissues. Using specific fabrication methods, it is possible to build aligned 3D network-like structures. In the present study, we exploited vertically-aligned silicon micropillar arrays as culture systems for human iPSC-derived cortical progenitors. In particular, our aim was to mimic the radially-oriented cortical radial glia fibres that during embryonic development play key roles in controlling the expansion, radial migration and differentiation of cortical progenitors, which are, in turn, pivotal to the establishment of the correct multilayered cerebral cortex structure. Here we show that silicon vertical micropillar arrays efficiently promote expansion and stemness preservation of human cortical progenitors when compared to standard monolayer growth conditions. Furthermore, the vertically-oriented micropillars allow the radial migration distinctive of cortical progenitors in vivo. These results indicate that vertical silicon micropillar arrays can offer an optimal system for human cortical progenitors' growth and migration. Furthermore, similar structures present an attractive platform for cortical tissue engineering.

Keywords: human cortical progenitors; 3D culture; silicon pillars; cell growth; hiPSC-derived neural progenitors; cerebral cortex

1. Introduction

In biology, developing tissues' microarchitecture is fundamental to allow correct cell differentiation and organization into appropriate structures that relate to specific physiological functions [1,2].

Standard in vitro cell culture models are mainly set in reductionist monolayer settings, conditions that intrinsically lack any structural architecture. This condition often represents a poor proxy to extrapolate cell growth in vivo, thus substantially affecting cell performance and biological assays outcomes [3,4]. Indeed, with respect to whole tissues, monolayer cultured cells are usually more responsive to toxic or therapeutic agents [5,6]. Additionally, cell culture on rigid surfaces can enhance cell proliferation, but might impact cell differentiation due to the partial cell interactions [7].

In this view, more appropriate three-dimensional (3D) cell culture environments can allow more physiological cell-to-cell contact, assisting cell growth and allowing better modelling of developmental processes [8–10]. A 3D environment also facilitates cells to organize into tissue-like structures that better mimic the in vivo function of cells, thus enhancing physiological relevance and predictive accuracy [11–14]. In the last few years, there has been a gradual development and adoption of technologies that enable tissue-like 3D cultures. Both scaffold-free organoid-based technologies and natural or synthetic scaffold-based culture systems have been developed [7]. In particular, since different tissue types show definite assemblies associated with their functional organization, scaffold-based methods allow assisted mimicking of complex tissue geometrical topographies, such as cerebral cortex, thus facilitating effective biofabrication of in vitro 3D tissue-like models [15–17]. With this purpose, silicon-based micro-fabricated culture substrates with well-defined continuous and discontinuous topographies, including the development of surfaces patterned with grooves, nanopillars or nanowires for the study of neural guidance and polarity, have been extensively developed in order to create scaffolds for a variety of applications [18–24].

In the present study, we exploited vertically-aligned silicon micropillar arrays to reproduce the developing cerebral cortex 3D architecture, where effective control over neural columns width, resembling the mammalian neocortex, is required for a large spectrum of applications. In the developing cerebral cortex, cortical progenitors are oriented from the ventricular to the pial surface with an apical-basal polarity. They divide to form radial glia and neuroblasts, the latter can migrate using the apico-basal oriented radial glia fascicles of the subventricular zone (SVZ) as a scaffold, thus forming the different cortical layers [25,26]. Here we focused on mimicking the radially-oriented cortical radial glia fibers, as they are the key players in controlling the expansion, radial migration and differentiation of cortical progenitors, thus allowing for the establishment of the correct multilayered cerebral cortex structure. We show that the scaffold material and structures are compatible with human cortical progenitors' maintenance. Immunofluorescence imaging analysis and RT-PCR results reveal that silicon vertical micropillar arrays efficiently promote the expansion and stemness preservation of human cortical progenitors, when compared to monolayer growth conditions. Furthermore, the precise orientation of the micropillars allows the radial migration, movement that is distinctive of cortical progenitors in vivo. These results indicate that vertically-aligned silicon micropillar arrays can offer an optimal system for human cortical progenitor growth and migration, and a potentially interesting platform for cortical tissue engineering.

2. Materials and Methods

2.1. Vertically-Aligned Silicon Micropillar Array Fabrication

Silicon slides containing vertically-aligned micropillar arrays have been realized on silicon surface through a CMOS like process at the Micro-Nano Facility of Fondazione Bruno Kessler (FBK). The silicon slides are composed of a central zone containing a matrix of vertically-aligned micropillars (height: 250–600 μm). Slides with different micropillar diameters have been obtained, with a diameter of 10 or 15 μm that leads to an aspect ratio that varies from 1:25 to 1:17. These significant proportions are particularly hard to achieve for pillar-like structures. The classical MEMS devices usually have aspect ratio of 1:12, for this reason the Bosch process used to realize the surfaces has been tuned in order to be able to reach higher aspect ratio.

The slide surface used for cellular seeding and culturing is 14 mm long and 5 mm wide these sizes were specifically designed to allow each slide to be accommodated in a well of a 12-well cell culture plate. The design foresees a range of distances between the micropillars, from 20 to 50 μm to evaluate the advantages of smaller and larger spaces between the micropillars for cellular deposition and growth. To realize the structures, the silicon wafer was first oxidized, patterned through soft lithography, then the silicon dioxide (SiO_2) was patterned through dry etching, and finally a deep reactive ion etching (DRIE) process was used to realize the micropillar-like structures. The DRIE process consists of a two-step process: first an isotropic plasma etch, second the deposition of a passivation layer that protects the lateral part of the structure. This two-step process is called Bosch process and it is repeated several times to realize nearly vertical wall. The time lapse of each of the two steps defines the roughness of the vertical wall as the process feed rate while the maximal depth that can be reached is defined by the power of the radio frequency bias that accelerates the ions toward the surface.

Since the fabrication process does not damage the mask, the same stamp can be employed for many subsequent cycles in a very reproducible manner.

During slides production, the micropillars on the open edge of the structures are systematically damaged during the DRIE process. This effect is connected with the presence of a blank space between slides, this void space increments the effective power of the ions impacting on the surfaces and then destroy or modify the shape of the first micropillars row. These defects have been eliminated in the cutting step. The cutting diamond disk was placed over the border, in order to cut away the defected micropillars and leaving a sharp edge.

2.2. Cell Cultures

Human cortical progenitors used in this study were differentiated from a commercial control hiPSC line (Gibco, Thermo Fisher Scientific, Monza, Italy) as previously described [27,28]. Briefly, hiPSCs commitment to a neural lineage and subsequently to the dorsal telencephalic lineage was performed by using N2B27 supplemented with human recombinant Noggin (500 ng/mL, Peprotech, London, UK), SB431542 (20 mM, Santa Cruz Biotechnologies, Heidelberg, Germany) and Fibroblast Growth Factor-2 (4 ng/mL, Peprotech). Cells were then detached and seeded on poly-ornithine and laminin-coated plastic dishes (Sigma-Aldrich, Milan, Italy) in medium supplemented with 10 μM Rock inhibitor Y27632 (Sigma-Aldrich). At day 10, neural rosettes containing cortical progenitor cells were manually collected and plated on poly-ornithine/laminin-treated culture dishes in N2B27 medium containing epidermal growth factor (10 ng/mL), fibroblast growth factor-2 (10 ng/mL) and brain-derived neurothrophic factor (20 ng/mL). Confluent cultures were passed as small multicellular clumps at a ratio of 1:3 using trypsin and amplified until passages 8–10. Cryopreserved stocks of human cortical progenitors were prepared from confluent cultures by trypsinization and resuspension in freezing medium (10% DMSO and 90% foetal calf serum).

Neuronal differentiation of hCPs was induced as previously described. Briefly, hCPs were plated at high density (10^5 cells per cm^2) on laminin-treated silicon vertical micropillar arrays. Cells were allowed to grow for four days and then switched in N2B27 medium without growth factors and cultured for 35 days. The medium was changed every four days.

To visualize the cells for fluorescent imaging analyses and time-lapse experiments, hiPSC-derived cortical progenitors were transduced with a lentiviral vector carrying eGFP cDNA under Cytomegalovirus (CMV) promoter and containing a puromycin selection cassette [29].

Procedure for generation and expansion of mouse NS cells (i.e., radial glia-like neural progenitors) were previously described [30–32]. In this study we used the LC1 NS cell line grown in standard conditions in expansion medium composed of Euromed-N medium (Euroclone) supplemented with 1% N2 (Thermo Fisher Scientific) and 20 ng/mL human recombinant epidermal growth factor (20 ng/mL) and fibroblast growth factor-2 (20 ng/mL).

Human cortical progenitors and mouse NS cells were seeded on silicon slides pre-coated with mouse laminin (Thermo Fisher Scientific).

2.3. Scanning Electron Microscopy (SEM)

Cultures were fixed for 30 min at 4 °C in 25% Glutaraldehyde and 0.1 M cacodylic acid in distilled water (pH 7.2). Then, samples were washed three times with 0.1 M cacodylic acid and dehydrated by the exposure to 50%, 70%, 90% and (2×) 100% *v/v* ethanol concentration, 10 min each. After being air-dried under an air flow, samples were gold coated by evaporation of a thin gold layer on top of the sample surface (thickness 6 nm, 1.5 nm Cr adhesion layer). Silicon micropillar-based devices deprived of cells did not require any treatment prior to SEM image acquisition. SEM micrographs were acquired by using a TESCAN VEGA III scanning electron microscope (Tescan Analytics, Fuveau, France) (operating voltage 4 kV, working distance 18 mm, stage tilting angle 45°).

2.4. Cell Growth/Viability Assay

Analysis of cell growth/viability of all cell types employed in this work (hCPs and mouse NS cells) was performed by MTT assay (Sigma-Aldrich). Briefly, MTT powder was dissolved into culture medium at a final concentration of 1.5 mg/mL. For culture incubation with MTT solution, cell medium was removed, cultures rinsed twice with PBS (Thermo Fisher Scientific) and incubated with MTT solution for 1h at 37 °C. Following incubation, MTT solution was removed, cells were air-dried and violet MTT precipitates dissolved with isopropanol. The absorbance was read at 570 nm wavelength with a microplate reader (Tecan Infinite M200PRO, Tecan Italia, Milan, Italy).

2.5. Immunocytochemistry

To process samples for immunofluorescence analyses, cultures were fixed in 4% paraformaldehyde for 30 min at room temperature (RT), permeabilized in PBS containing 0.5% Triton X-100 for 15 min at RT and then blocked in blocking solution (PBS containing 0.3% Triton X-100 and 5% FCS) for 1 h at RT. Samples were next incubated overnight at 4 °C with primary antibody diluted in antibody solution (PBS containing 0.2% Triton X-100 and 2% FCS), then washed three times with PBS and incubated for 2 h at RT with secondary antibodies. Samples were then counterstained with 1 μg/mL Hoechst 33,258 (Thermo Fisher Scientific) and further rinsed with PBS before proceeding with visualization. Fluorescent signals and Z-Stack of eGFP^{+ve} human cortical progenitors (12 slices of 7.7 μm each, shown at 7 fps) were detected using a Leica DMi8 microscope equipped with an Andor Zyla 4.2 PLUS, monochromatic, sCMOS sensor, 4.2 megapixel camera. Acquired images were processed with the open-source Fiji software (v2.0.0, open source under the GNU General Public License, Madison, WI, USA) [33].

Antibodies used in this study: primary mouse monoclonal anti-NESTIN antibody (R&D Systems, Minneapolis, MN, USA, 1:300), primary mouse monoclonal β3-TUBULIN antibody (Promega, Milan, Italy, 1:1000), primary rabbit polyclonal anti-SOX2 antibody (Millipore, Milan, Italy, 1:200), primary rabbit polyclonal anti-MAP2 antibody (Santa Cruz, Heidelberg, Germany, 1:200), primary mouse monoclonal anti-TBR2 antibody (ABCAM, Cambridge, UK, 1:500), primary mouse monoclonal anti CUX1 (ABCAM, 1:200), AlexaFluor-488 or -568 conjugated secondary antibodies (Thermo Fisher Scientific, 1:500).

2.6. Time Lapse Analysis

Time-lapse movies of live GFP-expressing cells migrating along micropillars were acquired with a Zeiss Axio Observer Z1 inverted microscope equipped with the Apotome 2 module for structured illumination and a 2.83 Megapixel AxioCam 503 mono D (all from Zeiss Italia, Castiglione Olona, Italy). Time-lapses were acquired as z-stacks (10 μm z-step) using a plan-apochromatic 10×/0.3 objective, with a frame interval of 30 min for 12.5 h. The movies shown are maximum intensity projections. Optimal focus selection was performed by manual extraction of each focus z-slices from original z-stack time-lapses to select the best focused z position for each time point, then adjusted for brightness and contrast and saved as 7 fps AVI files using Fiji software [33].

2.7. RNA Isolation and Quantitative RT-PCR (qRT-PCR)

Total RNA was isolated by using TRIzol Reagent (Thermo Fisher Scientific) following the manufacturer's protocol, then retro-transcribed with iScript cDNA Synthesis Kit (BioRad, Segrate, Italy). cDNA was used to verify the expression of specific target genes by qRT-PCR (quantitative RT PCR), using the SsoAdvanced Universal SYBR Green Supermix Kit. Specific primers sets were used (RT-qPCR data were analyzed according to the comparative ΔΔCt method and normalized by using β-Actin housekeeping gene. Sequence of primers used in this study:

Nestin forward 5′-GGAGAAGGACCAAGAACTG-3′, reverse 5′-ACCTCCTCTGTGGCATTC-3′; β3-tubulin forward 5′-TCAGCGTCTACTACAACGAGGC-3′, reverse 5′-GCCTGAAGAGATGTCCAAAGGC-3′; β-Actin forward 5′-GACAGGATGCAGAAGGAGATTACTG-3′, reverse 5′-CTCAGGAGGAGCAATGATCTTGAT-3′.

2.8. Statistical Analysis

Analyses were performed using either a two-sided unpaired Student's *t*-test or a one-way analysis of variance with a Dunnett's post hoc test. Experiments were repeated three times in triplicate and values were considered statistically significant for $p < 0.05$ (*), $p < 0.01$ (**), $p < 0.001$ (***), $p < 0.0001$ (****).

3. Results and Discussion

3.1. Generation of Vertically-Aligned Silicon Micropillar Array Structures

Biomimetic cortical-like 3D platforms have been created based on different approaches, such as cell spheroids, organoids and engineered constructs based on hydrogels [34–36]. These structures present advantages based on cell self-assembly or enabling spatial self-organization. Nonetheless fine control over the 3D micro-architecture, phenotype and reproducibility have been reported to be challenging [37]. To overcome these limitations, we sought to develop a novel scaffold-based approach mimicking the structural organization of the developing cerebral cortex by ideally controlling network topography.

Mammalian cerebral cortex is a complex layered structure organized in columns generated during developmental stages by means of cortical progenitors (CPs) and neuroblasts that move radially along radial glia fibers serving as scaffolds for directed columnar migration [25,26]. Our aim here is to fabricate a 3D culture platform that mimics in vitro these cerebral radial structures, being able to optimally support the outer growth (from the bottom to the top) of CPs such as that occurring in vivo during cortical neurogenesis. With this aim we designed micropatterned structures containing topographically ordered micropillar arrays. Micropatterned substrates with different geometries have been reported in the form of discontinuous micro-grooved configurations and discontinuous geometries like nanopillar arrays made from a several materials comprising polymers, such as PS, PLGA and PDMS together with hard materials, such as silicon and quartz [18,38–41]. In particular, in the neural field pillars and cone geometries at micron scale have been shown to control the outgrowth of neuronal processes, guiding neurite outgrowth alignment and cell growth [20,38,42,43].

Here, we produced a novel 3D neural cell culture platform based on silicon substrates displaying arrays of micropillars fabricated by lithographic patterning (see Methods) of crystalline silicon (Si) wafers. Such technique enables the fabrication of discontinuous micropillars exhibiting at the same time anisotropic geometry. The silicon structures consist of a rectangular seeding surface of 0.7 cm^2 containing vertically arranged micropillars (Figure 1A,B).

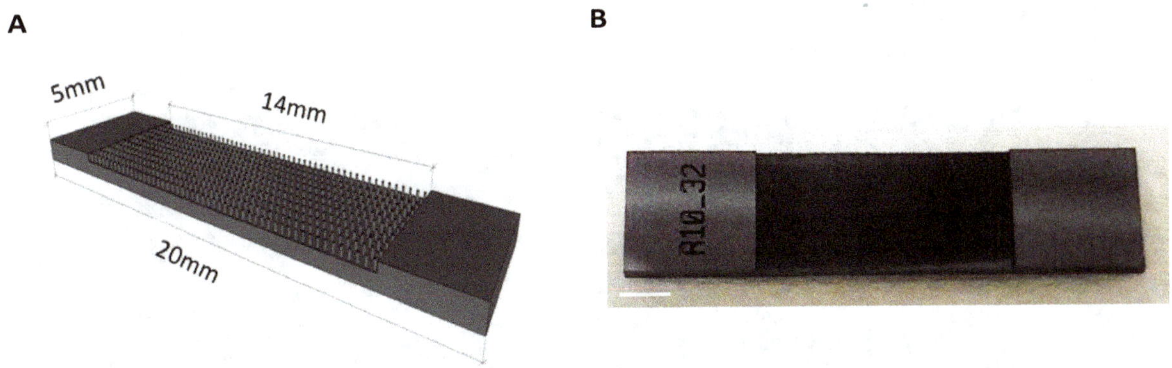

Figure 1. Design, size and morphology of 3D silicon micropillar array slides. (**A**) Schematic structure of a silicon slide containing vertically-aligned micropillars (micropillars are not in scale). (**B**) The image shows a 3D silicon slide. Scale bar: 1 mm.

Silicon-based nanostructures exhibit conceivable applications in several fields. As examples, plasmonic nanostructures based on array of silicon nanopillars could be used for surface enhanced Raman spectroscopy (SERS) or to control the wettability of a silicon surface [44–47]. Silicon micropillar arrays can be assembled by lithographic techniques allowing tight control over the size and density of the micropillars, differently from randomly generated rough surfaces as those presented in several other works [17,43,48,49]. Accordingly, they allow greatest control over the topographic structure of the system, thus warranting high reproducibility and robustness of the experiments. Additionally, differently from other materials, silicon surfaces hold particular technological impacts and potential.

The size of the slide was specifically designated to easily accommodate in a well of a tissue culture 12-multiwell plate (Figure S1A,B). By varying the reaction conditions of growth, we can produce pillars that have 10–15 μm diameter and height that can be regulated in the range of 200–600 μm. For this study we used silicon slides with 250 μm tall micropillars. Different micropillars topographies were successfully fabricated by using different masks. We produced 12 distinctive layouts of silicon micropillar arrays, coded as A1 to S6, with variable micropillar density and topographies (square grid-aligned and hexa grid-staggered micropillar arrays and with variable distances between the micropillars; see Table 1).

Table 1. List of different silicon slide arrays fabricated and tested in this work.

Slide Code	Pillar Diameter (μm)	Distance Among Pillars (μm)	Topography
A1	10	20	Alligned
A2	10	30	Alligned
A3	10	40	Alligned
A4	15	20	Alligned
A5	15	30	Alligned
A6	15	40	Alligned
S1	10	20	Staggered
S2	10	30	Staggered
S3	10	40	Staggered
S4	15	20	Staggered
S5	15	30	Staggered
S6	15	40	Staggered

Additionally, to set up the best conditions for cell growth, we fabricated both oxidized silicon (SiO_2) and silicon nitride (Si_3N_4) vertically-aligned silicon micropillar arrays.

The fabrication accuracy and morphology of the structures at the nanoscale were evaluated by measuring the distance between the micropillars and their diameter by scanning electron microscopy

(SEM). Cross-sectional SEM images were produced to derive the internal structure of the samples (Figure 2A,B) and top-view SEM images to derive the density of the silicon micropillar arrays.

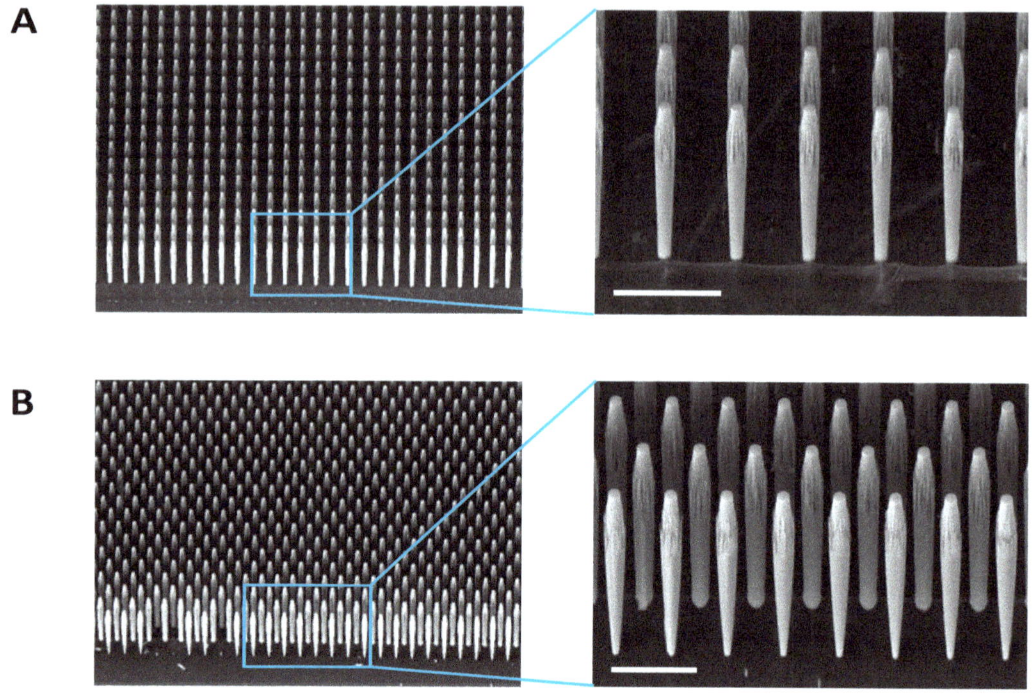

Figure 2. Silicon micropillars can be arranged with different topographies. Silicon micropillar arrays are perfectly vertically-aligned with a height that can be tuned in a 200–600 µm range. Specific spacing, density and morphology of the silicon pillars can be arranged by changing the mask of lithographic process. Low and high magnification of cross-sectional SEM images of aligned (**A**) and staggered (**B**) micropillar arrays. Scale bar: 50 µm.

We found that while the distance between micropillars was quite accurate, the diameter of the micropillars varied depending on the height of the structure. Indeed, the diameters at the bottom of the micropillars were reduced of about 40% (~5–7 µm) with respect to the top of the same micropillar.

This effect is particularly noticeable on the first row of micropillars, and we assumed that this is mainly due to the fact that the chemical-physical processes on which the Bosch process is based depending on the area exposed to the engraving process. Therefore, the etching rate is different in the areas between the micropillars than in the areas between the slides. Therefore, we assume that the first row of micropillars is etched much faster than the others and that this is the cause of the reduction of the diameter of the micropillar. Up to now, we were able to reach structures with a diameter of 15 µm and a height of 600 µm, nevertheless further tests to improve this aspect ratio are ongoing.

3.2. Culturing Human Cortical Progenitors on Vertically-Aligned Silicon Micropillar Arrays

To establish which human cortical progenitors (hCPs) density was the most comparable to the standard 2D monolayer method used as control and to detect potential toxic effects of silicon material, we first performed a cell viability assay. hCPs were seeded on laminin-coated samples A6–S6 with cell density ranging from 2×10^4 to 8×10^4 cells per 3D device (i.e., 2.8×10^4 and 1.1×10^5 cells per cm^2) and cultured for 48 h before being processed in an MTT assay (see Section 2). We found that cell viability was reduced for the lowest cell density (2×10^4 cells) and increased for the highest (8×10^4 cells), compared to the 2D culture. Indeed, 4×10^4 cells per 3D device showed no differences compared to the 2D monolayer culture (Figure 3).

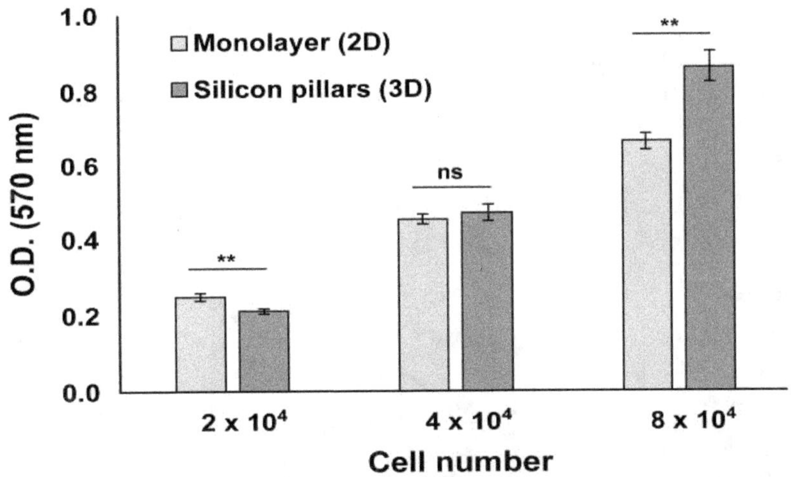

Figure 3. hCPs seeded on silicon micropillar arrays maintain their viability. MTT assay performed on hCPs plated on 3D silicon slide and in standard 2D monolayer shows that hCPs efficiently maintain their viability. Different cell densities were assessed 48 h after seeding. $p < 0.01$ (**), not significant (ns).

We found no significant difference between oxidized silicon or silicon nitride material (not shown), thus indicating that these 3D structures are fully compatible with living cells without affecting cell growth. Moreover, our silicon micropillar arrays might also allow for increased hCPs proliferation activity in long-term cultures by offering a 3D environment in which the cells can also exploit the third (vertical) dimension for growth. Additionally, we did not detect any significant difference between different topographic layouts and silicon types (oxidized silicon or silicon nitride) tested in terms of cell adhesion, viability and growth (not shown).

As further confirmation of these results, an analogous MTT analysis was performed on mouse radial glia-like NS cells plated at the same aforesaid cell densities (Figure S2A). NS cell analysis gave similar results to the hCPs, indicating that 4×10^4 cells per 3D device represents the preferred cell density. This was then selected as standard seeding density in the subsequent experiments in this work. The compatibility of oxidized silicon or silicon nitride for in vitro functional studies on neurons has been already reported, nevertheless, this is the first study to report compatibility with hCPs [50]. Furthermore, contrary to other studies employing silicon uncoated surfaces, we used surface topographies coated with laminin to optimize cell adhesion as also been reported by others [51–53]. In addition, these results demonstrate that our silicon micropillar arrays can be functionalized according to the cells' needs. We then performed SEM analysis in order to monitor the behaviour of the hCPs seeded on the arrays in terms of interaction with the vertical silicon micropillars. We found that cells at high density interact with each other and with the micropillars establishing a uniform network (Figure 4A).

Interestingly, hCPs plated at low density show an increased propensity to adhere to the micropillars rather than to the bottom flat surface of the device, with cells that can be found at different heights of the micropillars (Figure 4B,C). Cells formed 3D neural networks and suspended bridges throughout the micropillar height and along and between the micropillar walls. A similar formation of suspended neural process bridges has also been reported with microtowers, microfibres and 2PP-DLW fabricated microstructures [54–57]. In order to visualize the cells seeded on the silicon micropillar arrays with an inverted microscope, the device has to be set upside down, facing the microscope objectives. To facilitate this operation, we fabricated, by 3D-printing technique, a slide holder support to be placed on a glass bottom dish that allows to visualize the slide with a tilt of 45 degrees, so as to have a 3D visual prospect of the cells (Figure S3). Also, a glass bottom dish was used to place the holder (Figure S2B). To visualize live cultures, we generated eGFP^{+ve} hCPs by infection with lentiviral particles allowing constitutive expression of eGFP cDNA (see Methods). eGFP^{+ve} hCPs were used to live monitor the interactions of hCPs with the device and the single silicon micropillars by time-lapse analysis, and to capture the dynamics of cells movements among the micropillars. We found that hCPs move within the chip and interact with each other and with micropillars. In particular, hCPs can move along the whole micropillar from the bottom to the top (Movie S1) in a manner that resembles the in vivo migration along the radial glia during cortical neurogenesis, and extend processes embracing the micropillars (Movie S2). We also noticed the progressive neurite extension surrounding micropillars at the same height, which might represent the way for the cells to preferentially form a layered architecture.

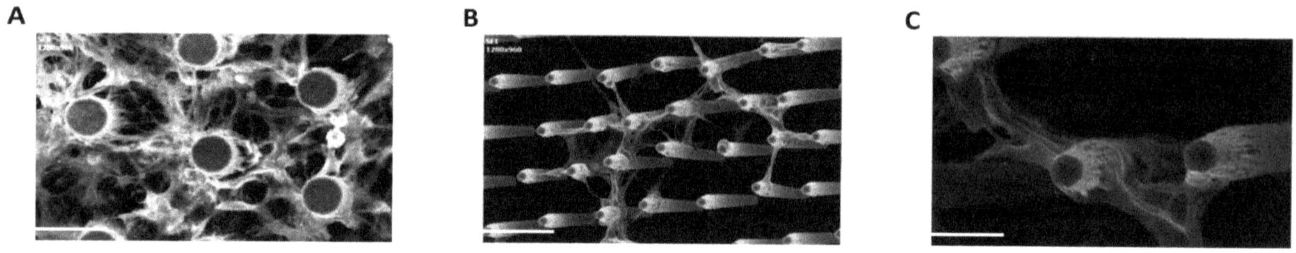

Figure 4. hCPs seeded on silicon slide establish close interactions with micropillars. (**A**) Top-view SEM image showing hCPs seeded at standard density on silicon micropillars. (**B**,**C**) Cross-sectional SEM images showing hCPs seeded at low density on silicon structures to visualize single cells-micropillars interactions. Scale bar: 20 μm (**A**,**B**) and 10 μm (**C**).

3.3. hCPs Seeded on Silicon Micropillar Arrays Proliferate and form 3D Layered Structures

We next assessed the proliferation of hCPs seeded on different silicon 3D devices in order to find the silicon type better supporting hCPs stable long-term growth. To this aim, hCPs were plated and maintained for several days on silicon oxide (SiO_2) and on silicon nitride (Si_3N_4) micropillar arrays, as well as on plastic dishes as standard monolayer growth conditions. Cultures were analysed by MTT-based growth assays at different time points. Resulting growth curves from the three experimental groups show a modest, yet statistically significant, reduction in cell growth within the first four days for both types of 3D silicon devices. After seven days cultures grown on Si_3N_4 devices still exhibit a significant reduction, whereas cultures on SiO_2 device show no difference with respect to the standard monolayer conditions. At the later time point considered, 14 days, a statistically significant difference was detected for the cells seeded on SiO_2 3D silicon device, which resulted in increased growth when compared to the other experimental groups (Figure 5).

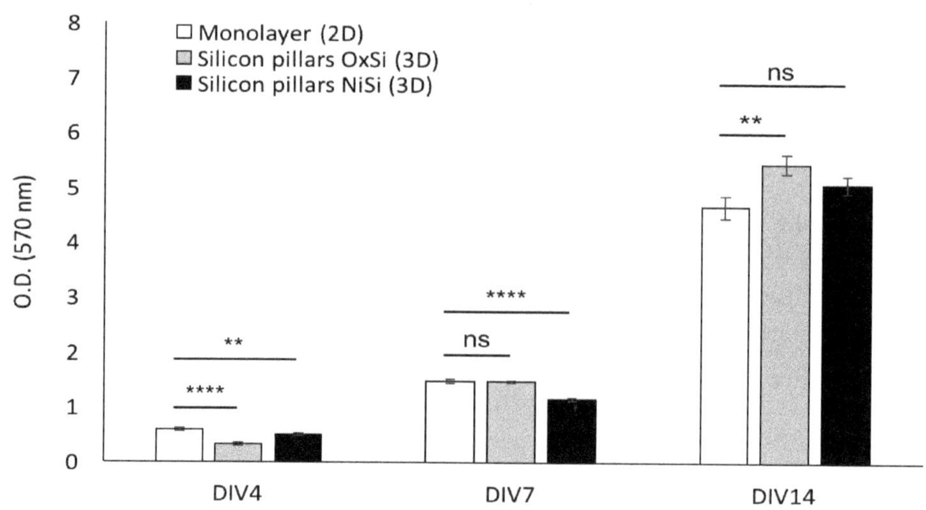

Figure 5. hCPs seeded on silicon slides maintain their proliferation capability. Cell growth analysis (MTT assays) performed on hCPs seeded on oxidized silicon (OxSi) or nitride silicon (NiSi) micropillar arrays. Standard 2D monolayer cultures were used as control. Cultures were assessed at different days in vitro (DIV) after seeding. $p < 0.01$ (**), $p < 0.0001$ (****), not significant (ns).

These results indicate that growth on silicon 3D silicon devices improves hCPs long-term maintenance. Additionally, since oxide silicon 3D arrays resulted to be more compatible and efficient in sustaining hCPs long-term survival and growth with respect with the nitride silicon devices, for the next experiments we employed the former type. Immunofluorescence imaging analysis of hCPs grown on silicon 3D micropillar arrays for two weeks show that cells are well distributed inside the device, filling the whole space among the micropillars (Figure 6A).

Additionally, SEM imaging shows that the cells form a complex three-dimensional multilayered-like structure by growing among the micropillars and interacting with each other, finally establishing regular horizontal layers (Figure 6B,C).

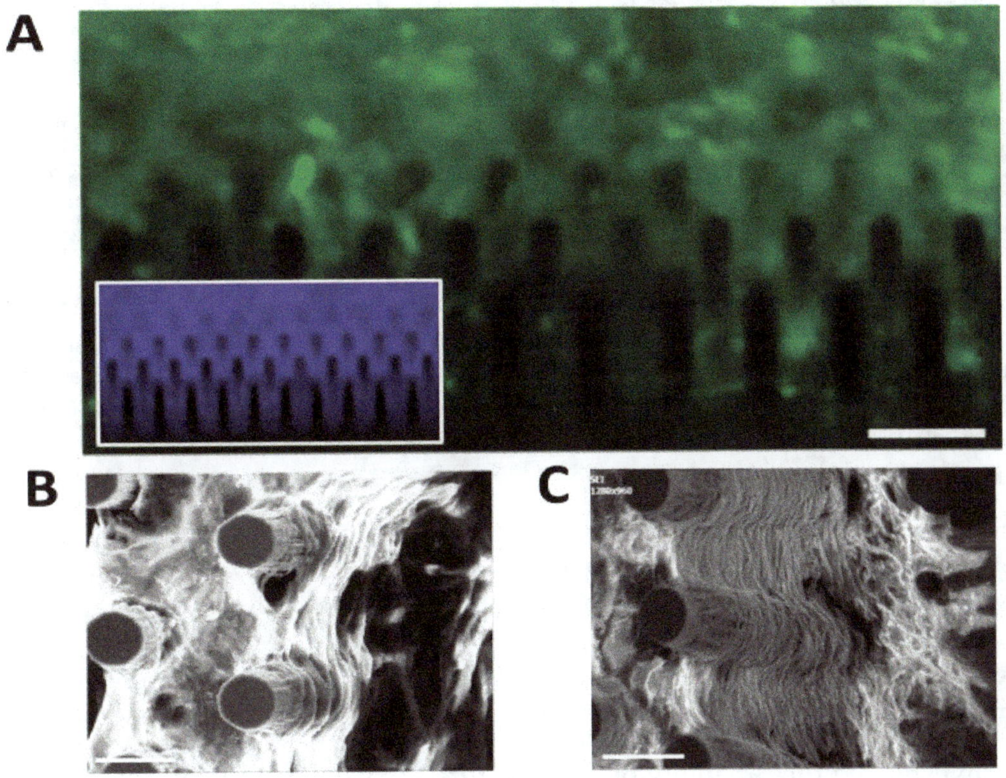

Figure 6. hCPs cultured on silicon micropillar arrays establish layered structures. (**A**) Picture of eGFP^{+ve} hCPs cultured for 14 days on silicon slides showing the establishment of a high-density 3D culture. Inset shows the same culture stained with Hoechst. (**B**,**C**) SEM images of hCPs maintained on silicon pillars device for 14 days. Cultures exhibit the generation of multiple cell layers. Scale bar: 50 μm (**A**) and 20 μm (**B**,**C**).

3.4. hCPs Grown on Silicon Micropillar Arrays Retain Their Multipotency and Regional Identity

We then assessed if growth on silicon 3D arrays might interfere with hCPs (i) multipotency and (ii) preservation of their cortical regional identity. To assess these issues, we first performed an immunostaining analysis for SOX2 and NESTIN, two key neural multipotent markers, on hCP cultures expanded for 14 days on silicon 3D arrays or in standard monolayer conditions. To this respect, we found that silicon 3D cultures show the great majority of cells to co-express SOX2 and NESTIN, comparably to the cultures maintained in standard monolayer conditions previously characterized to efficiently preserve hCPs multipotency (Figure 7A,B and Figure S4).

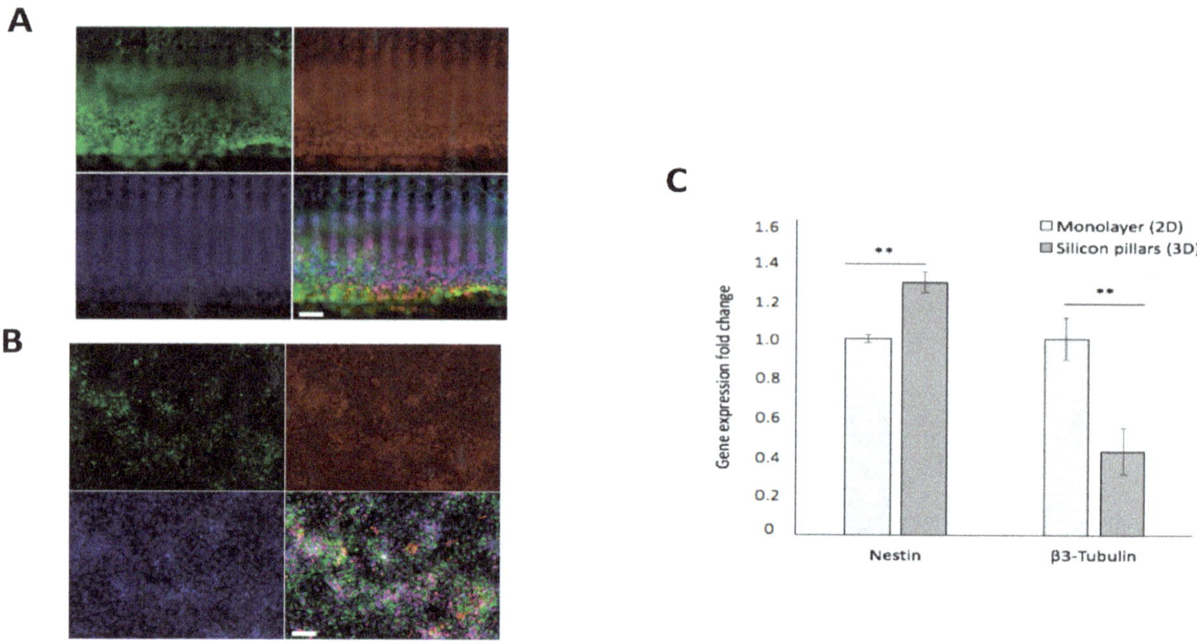

Figure 7. hCPs seeded on silicon slide maintain their neural immature identity. Immunostaining of hCPs for NESTIN (green), SOX2 (red) and nuclear staining with Hoechst (blue) in 3D culture (**A**) and 2D culture (**B**) cultured for 14 days. (**C**) Quantitative RT-PCR assay showing the expression levels of Nestin and β3-Tubulin transcripts in hCPs grown in 2D or 3D cultures. Scale bar: 100 μm (**A**) and 50 μm (**B**). $p < 0.01$ (**), not significant (ns).

To extend this result, we measured the transcript levels of Nestin and of the neuronal marker β3-Tubulin by quantitative RT-PCR (Figure 7C). This assay showed a 30% increase and a 60% decrease on the expression levels of Nestin and β3-Tubulin, respectively, in cultures maintained on silicon 3D arrays with respect to standard monolayer conditions (Figure 7C). These results further confirmed that growth on our silicon 3D devices does not alter hCP's multipotency. Furthermore, growth on silicon 3D arrays further lowers the occurrence of spontaneous neuronal differentiation in the cultures.

We then investigated the capability of hCPs cultured in silicon 3D arrays to retain their original cortical identity. To assess this issue, we first performed an immunostaining analysis for TBR2, a key maker of cortical progenitors, on hCP cultures expanded for 14 days on silicon 3D arrays or in standard monolayer conditions (Figure S5). We found that TBR2 immunoreactivity was maintained in nearly all of the cells in the silicon 3D cultures (Figure 8A; Movie S3). Finally, we tested the ability of hCPs plated in silicon 3D arrays to undergo neuronal maturation giving rise cortical glutamatergic neurons. To this end, hCPs were cultured for 5 days on 3D silicon devices and then exposed for 35 days to differentiative conditions and processed for immunofluorescent analysis for β3-TUBULIN, the mature neuronal marker MAP2 and the cortical neuronal marker CUX1. We found that differentiated cultures exhibited the presence of β3-TUBULIN^{+ve} neuronal cells (Figure 8B, left panel) and the appearance of neurons co-expressing MAP2 and CUX1 (Figure 8B).

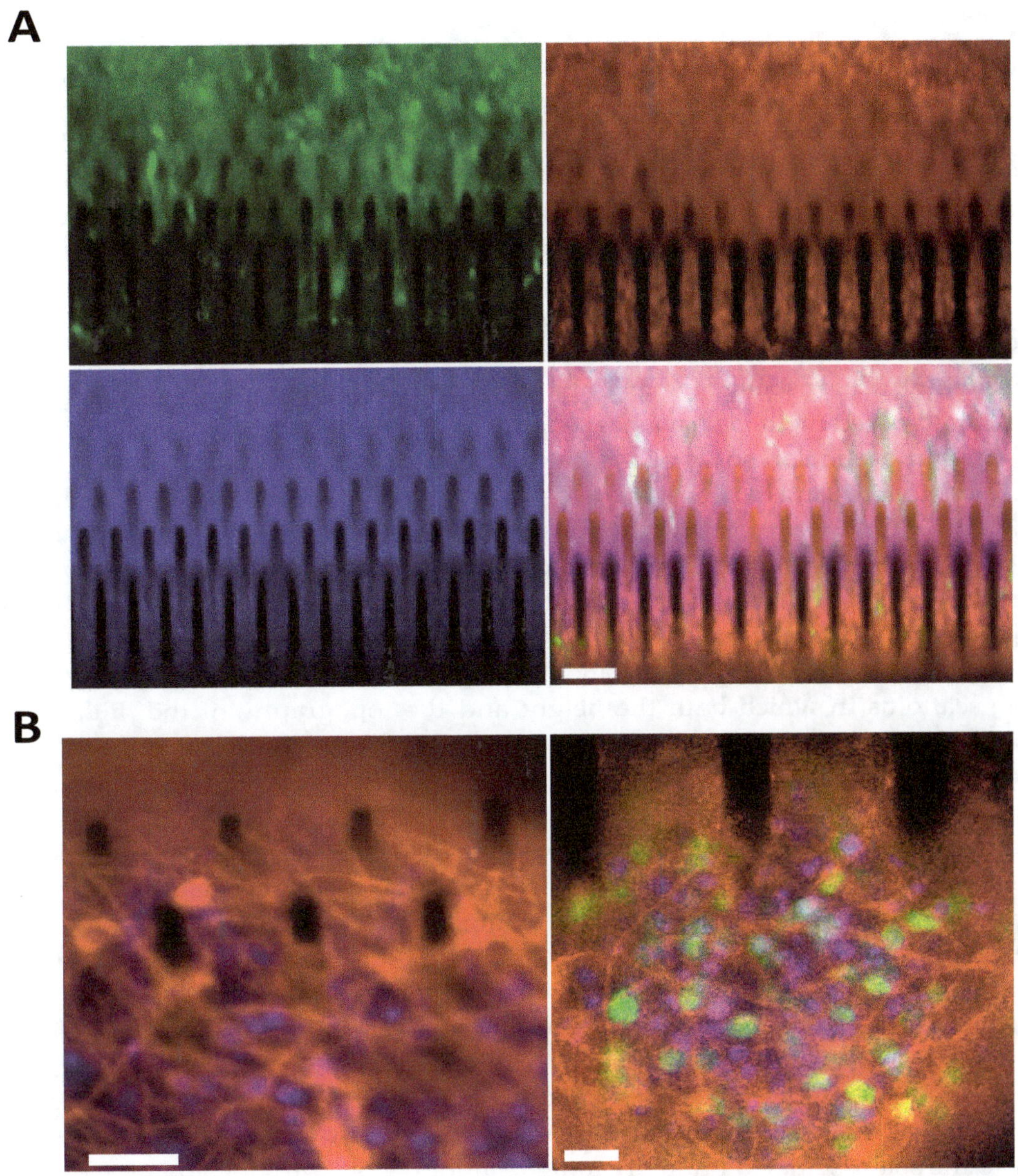

Figure 8. hCPs seeded on silicon micropillar arrays preserve their cortical regional identity and upon exposure to differentiative conditions, generate cortical glutamatergic neurons. (**A**) eGFP^{+ve} hCPs cultured for 14 days on 3D silicon devices preserve the expression of the cortical progenitor marker TBR2 (red). Nuclei are stained with Hoechst (blue). Scale bar: 50 µm. (**B**) hCPs cultured for 5 days on 3D silicon devices and then exposed for 35 days to differentiative conditions maturate into cortical glutamatergic neurons. Left: cultures stained for the pan-neuronal neuronal marker β3-TUBULIN (red). Nuclei are stained with Hoechst (blue). Scale bar: 10 µm. Right: cultures stained for the mature neuronal marker MAP2 (red) and for the cortical neuronal marker CUX1 (green). Nuclei are stained with Hoechst (blue). Scale bar: 10 µm.

These results demonstrate that hCPs grown on silicon micropillar arrays keep their multipotency and cortical identity and are able to generate glutamatergic cortical neurons. However, further experiments are required to further investigate whether the cortical neurons obtained in the 3D environment are able to organize themselves into defined upper and deeper cortical neuronal layers.

4. Conclusions

We have described a novel culture platform for hCPs growth based on 3D vertically-aligned silicon micropillar arrays. This structure mimics the radially-oriented cortical radial glia fibres that during embryonic development are essential to control the expansion, radial migration and differentiation of hCPs.

The silicon micropillar arrays can be arranged using different topographic organizations and report micropillar heights not tested with neural progenitors so far. The structures conceived practically combine the advantages of microscale topographies, without the need of complex fabrication techniques. Importantly, our fabrication process allows the production of biocompatible, 3D devices in a highly versatile and reproducible manner. In fact, the micrometre-sized base confers to micropillars good mechanical stability, while establishing a tight interface with the living hCPs.

Other scaffold-based methods have been reported for the biofabrication of in vitro 3D tissue-like models [15–17]. In particular, some of these systems, especially the ones based on 3D polymeric materials, i.e., microfibres and hydrogel scaffolds, offer a highly useful and strong method for creating large-scale 3D tissue cultures. Additionally, these systems have been shown to be extremely flexible in terms of production, biocompatibility, biodegradability, mechanical properties and functionalization with chemicals or oligopeptides in order to implement their adhesion and/or ECM mimicking properties [16]. The 3D vertically-aligned silicon micropillar arrays here described exhibited a lower flexibility with respect to these microfibers and hydrogel scaffolds, nevertheless they represent highly reproducible scaffolds in which both the height and the topography of the pillars can be finely regulated. Similar pillar-like structures have been previously reported by Limongi and colleagues and characterized for their ability to allow culturing of functional neuronal and glial cells in a 3D manner, allowing the formation of viable and functional neuronal networks [17]. In particular, the pillars there reported were different from ours since they were designed to reach limited height (cylindrical pillars of 10 μm in height and 10 μm in diameter) and to include a patterning in the nanoscale on their sidewall, leading to a spatial modulation in the z direction. Similarly, culturing platform containing arrays of microchannels arranged into ordered 2D arrays (with maximum height of ~3 μm and amplitudes ranging from ~10 to ~1.5 μm) have been reported [15]. These structures were shown to host neuronal cells strongly guiding their axons' growth direction and with additional advantage to enable coupling to devices for active sensing and stimulation at the local scale.

Our future efforts should reveal the actual potential of our 3D vertically-aligned silicon micropillar arrays to extensively support hCPs neuronal maturation in order to generate cortical-like tissue comparable to the self-assembled brain organoid technology but with the advantage of an increased reproducibility intrinsic to the scaffold-assisted process.

Additionally, to further exploit the potential of our 3D vertically-aligned silicon micropillar arrays, in the future we aim at coupling this system with a compartmentalized microfluidic device to reach a complete control of culture environment and reduce media volumes and related costs. The acquired knowledge will certainly pave the path towards the generation of valuable tools to study cortical development in humans and for cortical tissue engineering.

Supplementary Materials

Figure S1: Silicon devices were specifically designed to fit into 12 multiwell plates: Figure S2 Analysis of mouse NS cells viability seeded on 3D silicon micropillars arrays; Figure S3: Silicon slide holder device's picture. The slide lodges on a home-made holder inserted in a bottom glass dish; Figure S4: hCPs seeded on 3D silicon slide preserve their multipotency and create a cell network; Figure S5: hCPs grown in standard 2D conditions exhibit expression of cortical progenitor markers; Video S1: 12.5 hours time-lapse movie performed on eGFP^{+ve} hCPs seeded on silicon micropillars array. A single cell migrating from the bottom to the top of the micropillar is visible; Video S2: 12.5 hours time-lapse movie of eGFP+ve hCPs seeded on 3D silicon slide. Examples of cells extending neurites that interact with pillars are shown; Video S3: Z-stack of eGFP^{+ve} hCPs stained for TBR2 (red) and Hoechst (blue). Cells have been cultured for 14 days on silicon micropillars arrays before the time lapse analysis.

Author Contributions: S.G., M.B. and L.C. designed the pillar architecture. S.G. and M.B. developed the fabrication technology, realized the PDMS molds for pillar imprinting, and with help from M.C. produced the silicon micropillar array slides. A.C. prepared hCPs cultures and performed viability assay, growth curve analyses, morphological and immunohistochemical analyses on hCPs. A.C., J.Z. and A.S. prepared samples for the SEM analyses. J.Z. and A.S. prepared mNS cultures and performed viability assay on mouse NS cells. A.C. and J.Z. performed the time lapse experiments with the help of G.S. and M.R. G.S. designed and produced the slide holder devices. L.C., A.C., S.C., S.G., and M.B. planned the experiments, analysed the data and wrote the main manuscript, with contribution from all authors. N.M.P. and M.B. supervised S.G. activity. L.C. and M.B. conceived and supervised the work. All authors have read and agreed to the published version of the manuscript.

Acknowledgments: The authors wish to thank Francesca Agostinacchio for helpful discussions and Mirco D'Incau for technical assistance for SEM imaging.

References

1. Davenport, R.J. What controls organ regeneration? *Science* **2005**, *309*, 84. [CrossRef] [PubMed]
2. Griffith, L.G.; Swartz, M.A. Capturing complex 3D tissue physiology in vitro. *Nat. Rev. Mol. Cell Biol.* **2006**, *7*, 211–224. [CrossRef] [PubMed]
3. Vergani, L.; Grattarola, M.; Nicolini, C. Modifications of chromatin structure and gene expression following induced alterations of cellular shape. *Int. J. Biochem. Cell Biol.* **2004**, *36*, 1447–1461. [CrossRef] [PubMed]
4. Thomas, C.H.; Collier, J.H.; Sfeir, C.S.; Healy, K.E. Engineering gene expression and protein synthesis by modulation of nuclear shape. *Proc. Natl. Acad. Sci. USA* **2002** *99*, 1972–1977. [CrossRef] [PubMed]
5. Bhadriraju, K.; Chen, C.S. Engineering cellular microenvironments to cell-based drug testing improve cell-based drug testing. *Drug Discov. Today* **2002**, *7*, 612–620. [CrossRef]
6. Sun, T.; Jackson, S.; Haycock, J.W.; MacNeil, S. Culture of skin cells in 3D rather than 2D improves their ability to survive exposure to cytotoxic agents. *J. Biotechnol.* **2006**, *122*, 372–381. [CrossRef]
7. Knight, E.; Przyborski, S. Advances in 3D cell culture technologies enabling tissue-like structures to be created in vitro. *J. Anat.* **2015**, *227*, 746–756. [CrossRef]
8. Abbott, R.D.; Kaplan, D.L. Strategies for improving the physiological relevance of human engineered tissues. *Trends Biotechnol.* **2015**, *33*, 401–407. [CrossRef]
9. LaPlaca, M.C.; Vernekar, V.N.; Shoemaker, J.T.; Cullen, D.K. Three-dimensional neuronal cultures. In *3D Tissue Engineering*; Berthiaume, F., Morgan, J.R., Eds.; Artech House: Boston, MA, USA, 2010; Chapter 11; pp. 187–204.
10. Bercu, M.M.; Arien-Zakay, H.; Stoler, D.; Lecht, S.; Lelkes, P.I.; Samuel, S.; Or, R.; Nagler, A.; Lazarovici, P.; Elchalal, U. Enhanced survival and neurite network formation of human umbilical cord blood neuronal progenitors in three-dimensional collagen constructs. *J. Mol. Neurosci.* **2013**, *51*, 249–261. [CrossRef]
11. Cukierman, E.; Pankov, R.; Yamada, K.M. Cell interactions with three-dimensional matrices. *Curr. Opin. Cell Biol.* **2002** *14*, 633–639. [CrossRef]
12. Mazzoleni, G.; Di Lorenzo, D.; Steimberg, N. Modelling tissues in 3D: The next future of pharmacotoxicology and food research. *Genes Nutr.* **2009** *4*, 13–22. [CrossRef]
13. Vinci, M.; Gowan, S.; Boxall, B.; Patterson, L.; Zimmermann, M.; Court, W.; Lomas, C.; Mendiola, M.; Hardisson, D.; Eccles, S.A. Advances in establishment and analysis of three-dimensional tumor spheroid-based functional assays for target validation and drug evaluation. *BMC Biol.* **2012**, *10*, 29. [CrossRef]
14. Lai, Y.; Cheng, K.; Kisaalita, W. Three-dimensional neuronal cell cultures more accurately model Voltage Gated Calcium Channel functionality in freshly dissected nerve tissue. *PLoS ONE* **2012**, *7*, 45074. [CrossRef]
15. Cavallo, F.; Huang, Y.; Dent, E.W.; Williams, J.C.; Lagally, M.G. Neurite guidance and three-dimensional confinement via compliant semiconductor scaffolds. *ACS Nano* **2014**, *8*, 12219–12227. [CrossRef]

16. McNamara, M.C.; Sharifi, F.; Wrede, A.H.; Kimlinger, D.F.; Thomas, D.G.; Vander Wiel, J.B.; Chen, Y.; Montazami, R.; Hashemi, N.N. Microfibers as physiologically relevant platforms for creation of 3D cell cultures. *Macromol. Biosci.* **2017**, *17*, 1700279. [CrossRef]
17. Limongi, T.; Cesca, F.; Gentile, F.; Marotta, R.; Ruffilli, R.; Barberis, A.; Dal Maschio, M.; Petrini, E.M.; Santoriello, S.; Benfenati, F.; et al. Nanostructured superhydrophobic substrates trigger the development of 3D neuronal networks. *Small* **2013**, *9*, 402–412. [CrossRef]
18. Johansson, F.; Carlberg, P.; Danielsen, N.; Montelius, L.; Kanje, M. Axonal outgrowth on nano-imprinted patterns. *Biomaterials* **2006**, *27*, 1251–1258. [CrossRef]
19. Bucaro, M.A.; Vasquez, Y.; Hatton, B.D.; Aizenberg, J. Fine-tuning the degree of stem cell polarization and alignment on ordered arrays of high-aspect-ratio nanopillars. *ACS Nano* **2012**, *6*, 6222–6230. [CrossRef]
20. Dowell-Mesfin, N.M.; Abdul-Karim, M.A.; Turner, A.M.; Schanz, S.; Craighead, H.G.; Roysam, B.; Turner, J.N.; Shain, W. Topographically modified surfaces affect orientation and growth of hippocampal neurons. *J. Neural Eng.* **2004**, *1*, 78–90. [CrossRef]
21. Kwiat, M.; Elnathan, R.; Pevzner, A.; Peretz, A.; Barak, B.; Peretz, H.; Ducobni, T.; Stein, D.; Mittleman, M.; Ashery, U.; et al. Highly ordered large-scale neuronal networks of individual cells—Toward single cell to 3D nanowire intracellular interfaces. *ACS Appl. Mater. Interfaces* **2012**, *4*, 3542–3549. [CrossRef]
22. Yu, M.; Huang, Y.; Ballweg, J.; Shin, H.; Huang, M.; Savage, D.E.; Lagally, M.G.; Dent, E.W.; Blick, R.H.; Williams, J.C. Semiconductor Nanomembrane Tubes: Three-Dimensional Confinement for Controlled Neurite Outgrowth. *ACS Nano* **2011**, *26*, 2447–2457. [CrossRef] [PubMed]
23. Paul, F.; Huang, Y.; Cangellaris, O.V.; Huang, W.; Dent, E.W.; Gillette, M.U.; Williams, J.C. Toward intelligent synthetic neural circuits: Directing and accelerating neuron cell growth by self-rolled-up silicon nitride microtube array. *ACS Nano* **2014**, *25*, 11108–11117.
24. Mazzini, G.; Carpignano, F.; Surdo, S.; Aredia, F.; Panini, N.; Torchio, M.; Erba, E.; Danova, M.; Scovassi, A.I.; Barillaro, G. 3D Silicon Microstructures: A new tool for evaluating biological aggressiveness of tumor cells. *IEEE Trans. Nanobiosci.* **2015**, *14*, 797–805. [CrossRef] [PubMed]
25. McConnell, S.K. Constructing the cerebral cortex: Neurogenesis and fate determination. *Neuron* **1995**, *15*, 761–768. [CrossRef]
26. Solozobova, V.; Wyvekens, N.; Pruszak, J. Lessons from the embryonic neural stem cell niche for neural lineage differentiation of pluripotent stem cells. *Stem Cell. Rev. Rep.* **2012**, *8*, 813–829. [CrossRef]
27. Boissart, C.; Poulet, A.; Georges, P.; Darville, H.; Julita, E.; Delorme, R.; Bourgeron, T.; Peschanski, M.; Benchoua, A. Differentiation from human pluripotent stem cells of cortical neurons of the superficial layers amenable to psychiatric disease modeling and high-throughput drug screening. *Transl. Psychiatry* **2013**, *3*, 294. [CrossRef]
28. Corti, S.; Faravelli, I.; Cardano, M.; Conti, L. Human pluripotent stem cells as tools for neurodegenerative and neurodevelopmental disease modeling and drug discovery. *Expert Opin. Drug Discov.* **2015**, *10*, 615–629. [CrossRef]
29. Brilli, E.; Reitano, E.; Conti, L.; Conforti, P.; Gulino, R.; Consalez, G.G.; Cesana, E.; Smith, A.; Rossi, F.; Cattaneo, E. Neural stem cells engrafted in the adult brain fuse with endogenous neurons. *Stem Cells Dev.* **2013**, *22*, 538–547. [CrossRef]
30. Conti, L.; Pollard, S.M.; Gorba, T.; Reitano, E.; Toselli, M.; Biella, G.; Sun, Y.; Sanzone, S.; Ying, Q.L.; Cattaneo, E.; et al. Niche-independent symmetrical self-renewal of a mammalian tissue stem cell. *PLoS Biol.* **2005**, *3*, 283. [CrossRef]
31. Pollard, S.; Conti, L.; Smith, A. Exploitation of adherent neural stem cells in basic and applied neurobiology. *Regen. Med.* **2006**, *1*, 111–118. [CrossRef]
32. Pollard, S.M.; Conti, L. Investigating radial glia in vitro. *Prog. Neurobiol.* **2007**, *83*, 53–67. [CrossRef] [PubMed]
33. Schindelin, J.; Arganda-Carreras, I.; Frise, E.; Kaynig, V.; Longair, M.; Pietzsch, T.; Preibisch, S.; Rueden, C.; Saalfeld, S.; Schmid, B.; et al. Fiji: An open-source platform for biological-image analysis. *Nat. Methods* **2012**, *9*, 676–682. [CrossRef] [PubMed]
34. Gjorevski, N.; Ranga, N.; Lutolf, M.P. Bioengineering approaches to guide stem cell-based organogenesis. *Development* **2014**, *141*, 1794–1804. [CrossRef] [PubMed]
35. Kratochvil, M.J.; Seymour, A.J.; Li, T.L.; Sergiu, P.P.; Kuo, C.J.; Heilshorn, S.C. Engineered materials for organoid systems. *Nat. Rev. Mater.* **2019**, *4*, 606–622. [CrossRef]

36. Seliktar, D. Designing cell-compatible hydrogels for biomedical applications. *Science* **2012**, *336*, 1124–1128. [CrossRef]
37. Fang, Y.; Eglen, R.M. Three-dimensional cell cultures in drug discovery and development. *Slas Discov.* **2017**, *22*, 456–472.
38. Simitzi, C.; Efstathopoulos, P.; Kourgiantaki, A.; Ranella, A.; Charalampopoulos, I.; Fotakis, C.; Athanassakis, I.; Stratakis, E.; Gravanis, A. Laser fabricated discontinuous anisotropic microconical substrates as a new model scaffold to control the directionality of neuronal network outgrowth. *Biomaterials* **2015**, *67*, 115–128. [CrossRef]
39. Yao, L.; Wang, S.; Cui, W.; Sherlock, R.; O'Connell, C.; Damodaran, G.; Gorman, A.; Windebank, A.; Pandit, A. Effect of functionalized micropatterned PLGA on guided neurite growth. *Acta Biomater.* **2009**, *5*, 580–588. [CrossRef]
40. Rajnicek, A.; Britland, S.; McCaig, C. Contact guidance of CNS neurites on grooved quartz: Influence of groove dimensions, neuronal age and cell type. *J. Cell Sci.* **1997**, *110*, 2905–2913.
41. Cecchini, M.; Bumma, G.; Serresi, M.; Beltrami, F. PC12 differentiation on biopolymer nanostructures. *Nanotechnology* **2007**, *18*, 505103. [CrossRef]
42. Hanson, J.N.; Motala, M.J.; Heien, M.L.; Gillette, M.; Sweedler, J.; Nuzzo, R.G. Textural guidance cues for controlling process outgrowth of mammalian neurons. *Lab Chip* **2009**, *9*, 122–131. [CrossRef] [PubMed]
43. Mattotti, M.; Micholt, L.; Braeken, D.; Kovacic, D. Characterization of spiral ganglion neurons cultured on silicon micro-pillar substrates for new auditory neuro-electronic interfaces. *J. Neural Eng.* **2015**, *12*, 026001. [CrossRef] [PubMed]
44. Schmidt, M.S.; Hubner, J.; Boisen, A. Large Area Fabrication of Leaning Silicon Nanopillars for Surface Enhanced Raman Spectroscopy. *Adv. Mater.* **2012**, *24*, 11–18. [CrossRef] [PubMed]
45. Magno, G.; Belier, B.; Barbillon, G. Al/Si Nanopillars as Very Sensitive SERS Substrates. *Materials* **2018**, *11*, 1534. [CrossRef]
46. Vereecke, G.; Xu, X.M.; Tsai, W.K.; Yang, H.; Armini, S.; Delande, T.; Doumen, G.; Kentie, F.; Shi, X.; Simms, I.; et al. Partial Wetting of aqueous solutions on high aspect ratio nanopillars with hydrophilic surface finish. *ECS J. Solid State Sci. Technol.* **2014**, *3*, 3095–3100. [CrossRef]
47. Liu, C.; Sun, J.; Li, J.; Xiang, C.; Che, L.; Wang, Z.; Zhou, X. Long-range spontaneous droplet self-propulsion on wettability gradient surfaces. *Sci. Rep.* **2017**, *7*, 7552. [CrossRef]
48. Park, Y.S.; Yoon, S.Y.; Park, J.S.; Lee, J.S. Deflection induced cellular focal adhesion and anisotropic growth on vertically aligned silicon nanowires with differing elasticity. *NPG Asia Mater.* **2016**, *8*, 249. [CrossRef]
49. Tullii, G.; Giona, F.; Lodola, F.; Bonfadini, S.; Bossio, C.; Varo, S.; Desii, A.; Criante, L.; Sala, C.; Pasini, M.; et al. High-aspect-ratio semiconducting polymer pillars for 3D cell cultures. *ACS Appl. Mater. Interfaces* **2019**, *11*, 28125–28137. [CrossRef]
50. Medina Benavente, J.J.; Mogami, H.; Sakurai, T.; Sawada, K. Evaluation of silicon nitride as a substrate for culture of PC12 cells: An interfacial model for functional studies in neurons. *PLoS ONE* **2014**, *9*, 90189. [CrossRef]
51. Liliom, H.; Lajer, P.; Berces, Z.; Csernyus, B.; Szabò, A.; Pinke, D.; Low, P.; Fekete, Z.; Pongracz, A.; Schlett, K. Comparing the effects of uncoated nanostructured surfaces on primary neurons and astrocytes. *J. Biomed. Mater. Res. A* **2019**, *107*, 2350–2359. [CrossRef]
52. Fan, Y.W.; Cui, F.Z.; Hou, S.P.; Xu, Q.Y.; Chen, L.N.; Lee, I.S. Culture of neural cells on silicon wafers with nano-scale surface topograph. *J. Neurosci. Methods* **2002**, *120*, 17–23. [CrossRef]
53. Chen, W.S.; Guo, L.Y.; Tang, C.C.; Tsai, C.K.; Huang, H.H.; Chin, T.Y.; Yang, M.L.; Chen-Yang, Y.W. The effect of laminin surface modification of electrospun silica nanofiber substrate on neuronal tissue engineering. *Nanomaterials* **2018**, *8*, 165. [CrossRef] [PubMed]
54. Turunen, S.; Joki, T.; Hiltunen, M.L.; Ihalainen, T.O.; Narkilahti, S.; Kellomaki, M. Direct laser writing of tubular microtowers for 3D culture of human pluripotent stem cell-derived neuronal cells. *ACS Appl. Mater. Interfaces* **2017**, *9*, 25717–25730. [CrossRef] [PubMed]
55. Sharifi, F.; Patel, B.B.; Dzuilko, A.K.; Montazami, R.; Sakaguchi, D.S.; Hashemi, N. Polycaprolactone microfibrous scaffols to navigate neural stem cells. *Biomacromolecules* **2016**, *17*, 3287–3297. [CrossRef] [PubMed]

56. Timashev, P.S.; Vedunova, M.V.; Guseva, D.; Ponimaskin, E.; Deiwick, A.; Mishchenko, T.A.; Mitroshina, E.V.; Koroleva, A.V.; Pimashkin, A.S.; Mukhina, I.V.; et al. 3D in vitro platform produced by two-photon polymerization for the analysis of neural network formation and function. *Biomed. Phys. Eng. Exp.* **2016**, *2*, 035001. [CrossRef]
57. Koroleva, A.; Gill, A.A.; Ortega, I.; Haycock, J.W.; Schilie, S.; Gittard, S.D.; Chichkov, B.N.; Claeyssens, F. Two-photon polymerization-generated and micromolding-replicated 3D scaffolds for peripheral neural tissue engineering applications. *Biofabrication* **2012**, *4*, 025005. [CrossRef]

Activated Cardiac Fibroblasts Control Contraction of Human Fibrotic Cardiac Microtissues by a β-Adrenoreceptor-Dependent Mechanism

Przemysław Błyszczuk [1,2,*], Christian Zuppinger [3], Ana Costa [4], Daria Nurzynska [5], Franca Di Meglio [5], Mara Stellato [1], Irina Agarkova [6], Godfrey L. Smith [4], Oliver Distler [1] and Gabriela Kania [1,*]

1 Center of Experimental Rheumatology, Department of Rheumatology, University Hospital Zurich, Wagistr. 14, 8952 Schlieren, Switzerland
2 Department of Clinical Immunology, Jagiellonian University Medical College, 30-663 Cracow, Poland
3 Department for BioMedical Research, Department of Cardiology, University Hospital Bern, 3008 Bern, Switzerland
4 Institute of Cardiovascular and Medical Sciences, University of Glasgow, Glasgow G12 8TA, UK
5 Department of Public Health, University of Naples "Federico II", 80131 Naples, Italy
6 InSphero, 8952 Schlieren, Switzerland
* Correspondence: przemyslaw.blyszczuk@uzh.ch (P.B.); gabriela.kania@uzh.ch (G.K.)

Abstract: Cardiac fibrosis represents a serious clinical problem. Development of novel treatment strategies is currently restricted by the lack of the relevant experimental models in a human genetic context. In this study, we fabricated self-aggregating, scaffold-free, 3D cardiac microtissues using human inducible pluripotent stem cell (iPSC)-derived cardiomyocytes and human cardiac fibroblasts. Fibrotic condition was obtained by treatment of cardiac microtissues with profibrotic cytokine transforming growth factor β1 (TGF-β1), preactivation of foetal cardiac fibroblasts with TGF-β1, or by the use of cardiac fibroblasts obtained from heart failure patients. In our model, TGF-β1 effectively induced profibrotic changes in cardiac fibroblasts and in cardiac microtissues. Fibrotic phenotype of cardiac microtissues was inhibited by treatment with TGF-β-receptor type 1 inhibitor SD208 in a dose-dependent manner. We observed that fibrotic cardiac microtissues substantially increased the spontaneous beating rate by shortening the relaxation phase and showed a lower contraction amplitude. Instead, no changes in action potential profile were detected. Furthermore, we demonstrated that contraction of human cardiac microtissues could be modulated by direct electrical stimulation or treatment with the β-adrenergic receptor agonist isoproterenol. However, in the absence of exogenous agonists, the β-adrenoreceptor blocker nadolol decreased beating rate of fibrotic cardiac microtissues by prolonging relaxation time. Thus, our data suggest that in fibrosis, activated cardiac fibroblasts could promote cardiac contraction rate by a direct stimulation of β-adrenoreceptor signalling. In conclusion, a model of fibrotic cardiac microtissues can be used as a high-throughput model for drug testing and to study cellular and molecular mechanisms of cardiac fibrosis.

Keywords: cardiac microtissues; iPSC-derived cardiomyocytes; cardiac fibroblasts; cardiac fibrosis; cardiac rhythm; TGF-β signalling; drug screening; in vitro model

---

1. Introduction

Cardiac fibrosis refers to an excessive accumulation of stromal cells and extracellular matrix (ECM) proteins in the myocardium and represents a common pathophysiological scenario in a broad range of cardiovascular conditions, including myocardial infarction, hypertension, myocarditis, and hypertrophic or dilated cardiomyopathy [1]. Cardiac fibroblasts and myofibroblasts represent the most extensively characterised stromal cell types involved in fibrotic processes in the heart [2]. In the traditional view, resident cardiac fibroblasts become activated and overproduce ECM proteins such as collagen type I and III and fibronectin. Fibrogenesis is regulated by multiple profibrotic inputs, which activate a complex signalling network. In cardiac fibrosis, transforming growth factor β (TGF-β) is recognised as the key profibrotic cytokine activating quiescent cardiac fibroblasts. Constitutive overexpression of TGF-β was shown to induce interstitial cardiac fibrosis and cardiac hypertrophy in transgenic mice [3], whereas targeting TGF-β or its downstream mediators successfully reduced or prevented cardiac fibrosis in various animal models [4–6]. Progressive fibrosis causes tissue stiffening and may affect cardiac rhythm. Clinical data associate fibrotic heart condition with increased resting heart rate [7] and with elevated risk of developing life-threating arrhythmias [8,9]. The underlaying mechanisms remain obscure, however.

In the body, heart rhythm established by pacemaker cells located in the sinoatrial node of the heart is mainly under control of the autonomic nervous system. Sympathetic stimulation increases the rate, whereas parasympathetic stimulation decreases it. Neurotransmitter noradrenaline released by the sympathetic neurons increases heart rate by activating β-adrenergic receptors on cardiomyocytes. Under stress condition, β-adrenoreceptors are further activated by adrenaline—a catecholamine produced by the adrenal glands and released into the circulation. Binding of agonists to β-adrenergic receptors causes the formation of cyclic adenosine monophosphate and activation of protein kinase A that phosphorylates a number of proteins including myosin light chains and thus increases the ability of cardiomyocytes to relax following excitation contraction coupling [10]. In the heart, the increased relaxation is associated with increased rate and strength of contraction.

Because extracardiac stimuli are important regulators of the heart rhythm, it is difficult to study intracardiac mechanisms of cardiac contractility using in vivo systems only. Data from ex vivo and in vitro cardiac models may shed more light on specific aspects of intracardiac rhythm regulation. Disease models using human cells offer a natural advantage over animal models by providing a proper genetic and cellular context. In cardiovascular research, difficulties in obtaining and culturing contractile primary human cardiomyocytes limited the development of in vitro human models for many years. The human induced pluripotent stem cell (iPSC) technology opened new perspectives and became a commonly used, unlimited source of spontaneously beating cardiomyocytes for regenerative medicine, drug discovery and toxicity screening [11].

A lack of the physiologically relevant experimental setting is a major drawback of many in vitro models. In general, traditional two-dimensional (2D) cell culture systems poorly mirror the biomechanical and biochemical microenvironment of the tissue. Three-dimensional (3D) cell culture technologies, instead, much better reflect the organ complexity and represent an attractive alternative to the commonly used 2D models. 3D cell culture systems, called microtissues or organoids, include cellular models on various bioscaffolds, but also scaffold-free spheroids comprising of one or more cell types. In recent years, various types of cardiac microtissues have been successfully used predominantly in regenerative medicine, but also as platforms for drug screening or as disease models [12]. Self-aggregating, scaffold-free cardiac microtissues consisting of human iPSC-derived cardiomyocytes and human primary cardiac fibroblasts cultured in nonadherent plates represent a high-throughput in vitro model of human cardiac tissue. These spontaneously beating cardiac microtissues show typical structures including well-developed myofibrils and recapitulate cardiac functionality including responsiveness to electrical stimulation [13]. In this work, we aimed to address how fibrotic changes affect contractility of human cardiac microtissues. We found that profibrotic condition and, in particular, activation of cardiac fibroblasts increase beating rate of cardiac microtissues, and this effect is at least partially mediated by endogenous activation of β-adrenergic receptors. Furthermore, we demonstrated that human cardiac microtissues represent an easy-to-use, reproducible high-throughput method for small molecule screening.

2. Material and Methods

2.1. Cellular Sources

Frozen human iPSC-derived cardiomyocytes (>99% troponin-positive) were purchased from Cellular Dynamics International, and were used directly upon thawing. Foetal human cardiac fibroblasts were purchased from Sigma (Cell Applications) and cells from passages 10–15 (>99% vimentin-positive, >99% collagen I-positive and <5% positive for filamentous form of α-SMA) were used. Adult cardiac fibroblasts were isolated from cardiac tissue samples obtained from the left atria of hearts from the patients with end-stage heart failure due to ischemic heart disease undergoing heart transplantation and from the waste fragments of donor normal hearts that were trimmed off while adjusting atrium size and shape during transplantation. Specimens were collected without patient identifiers according to the protocol revised by the Ethics Committee of the University of Naples Federico II (approval number 79/18) and in conformity with the principles outlined in the Declaration of Helsinki. Patients characteristic is available in Supplementary Table S1.

2.2. Isolation of Human Adult Cardiac Fibroblasts

Adult cardiac fibroblasts were isolated as described previously [14]. Briefly, cardiac tissue samples were minced and then enzymatically disaggregated by incubation in 0.25% trypsin and 0.1% (*w*/*v*) collagenase II (both Sigma-Aldrich, Basel, Switzerland) for 30 min at 37 °C. The digestion was stopped by adding Hanks' Balanced Salt solution supplemented with 10% foetal bovine serum (FBS, Gibco, Paislay, UK). The tissue was further disaggregated by pipetting; noncardiomyocyte cells were separated from debris and cardiomyocytes by sequential centrifugation and passage through 20 μm cell sieve. Fibroblasts were isolated from cell suspension by immunomagnetic cell sorting through positive selection with anti-fibroblast MicroBeads (MiltenyiBiotec, Bergisch Gladbach, Germany). Following selection, fibroblasts were cultured in Dulbecco's Modified Eagle's Medium (DMEM, Sigma, Basel, Switzerland) supplemented with 10% FBS, penicillin 10,000 U, and streptomycin 10 mg/mL (all from Sigma, Basel, Switzerland) at 37 °C in 5% CO_2. The aCFs from patients with heart failure and unaffected donors were used individually. Cells from passages 4–6 were used.

2.3. Fabrication and Culture of Microtissues

Three types of microtissues containing: (1) iPSC-derived cardiomyocytes and fibroblasts mixed in ratio 4:1, (2) fibroblasts only, and (3) iPSC-derived cardiomyocytes only were generated by self-assembling in 96-well GravityTRAP plates (InSphero, Schlieren, Switzerland), and cultured as one microtissue per well. Accordingly, for generation of one microtissue, 5000 cells suspended in 70 μL maintenance medium were placed in the GravityTRAP plates for three days in the slanted position. Maintenance medium contained 2% FBS (Gibco, Paislay, UK), 50 μM phenylephrine hydrochloride (Sigma), 0.3 μM L-ascorbic acid (Sigma), 50 μM 2-mercaptoethanol (Gibco, Paislay, UK), 50 U/mL penicillin/streptomycin (Gibco, Paislay, UK) in high glucose DMEM (Sigma). Upon self-assembly, microtissues were further cultured in the GravityTRAP plates in the maintenance medium for 10 days. Microtissues were cultured under standard culture conditions (at 37 °C, 5% CO_2 in a humidified incubator). Medium was changed at days 0, 2, 4, and 7. Fibrotic differentiation of cardiac microtissues was induced at day 0 with 10 ng/mL recombinant human TGF-β1 (Peprotech, London, UK) and TGF-βR1 was blocked with 10^{-8}–10^{-5} g/mL SD208 (Tocris, Zug, Switzerland). To address β-adrenoceptor-dependent mechanisms, microtissues were treated with 10 nM isoproterenol (Tocris) and/or 1 μM nadolol (Sigma) 15–90 min prior to video recording. Controls received solvents only.

2.4. Live Imaging and Video Analysis

Randomly chosen live microtissues were analysed using microscopes equipped with a humidified chamber at 37 °C and 5% CO_2. Pictures of microtissues were captured using the Live Cell Imaging System (Olympus, Shinjuku, Japan) and size of microtissues was measured with the Xcellence Pro software (Olympus) as area in pixels. Increase in size was calculated as the size of microtissue at day 10/size of microtissue at day 0. Videos of contracting microtissues were recorded using the AxioObserver Z1 (Zeiss, Hombrechtikon, Switzerland) and the ZEN software. Videos were processed using Fiji software and custom-made macro. Contraction analysis was performed using Fiji software and MUSCLEMOTION macro [15].

2.5. *FluoVolt Measurements*

Voltage experiments were conducted on day 11. The microtissues were washed in serum-free medium consisting of DMEM (Gibco, Paislay, UK) supplemented with 10 mM D-Galactose (Sigma) and 1 mM sodium pyruvate (Sigma). The microtissues were loaded with voltage-sensitive dye: 0.1% FluoVolt and 1% PowerLoad (ThermoFisher, Waltham, WA, USA) in the above listed serum-free medium for 25 min at 37 °C 5% CO_2. The voltage-sensitive dye was removed by washing in serum-free medium, and the multiwell plate was placed in an environmentally controlled stage incubator (37 °C, 5% CO_2, >75% humidity) in the CellOPTIQ® platform (Clyde Biosciences, BioCity Scotland) 30 min before experimentation. The fluorescent signal was recorded with an excitation at 470 nm, and emitted light was collected from the entire microtissue using a 10× Fluor objective and the intensity recorded by a photomultiplier tube at 510–560 nm at 10 kHz. A 15 s recording was taken of each microtissue. Offline analysis was performed using CellOPTIQ®.

2.6. *Electrical Pacing*

Prior to experiment, microtissues were transferred into a glass bottom cell culture dish. Electrical field pacing of microtissues was performed using a modified glass bottom culture dish with inserted platinum wires attached to a MyoPacer (IonOptix, Westwood, MA, USA) set to a frequency of 3 Hz. Microtissues were recorded using the Eclipse TE2000-U microscope (Nikon, Tokyo, Japan) equipped with heating chamber, temperature controller (Ibidi, Graefelfing, Germany) and HeroBlack6 GoPro camera (Back-Bone Gear, Ontario, Canada).

2.7. *Quantitative RT-PCR*

For one sample, 12 microtissues were pooled. Total RNA was isolated using the Quick RNA Micro Kit (Zymo Research, Irvine, CA, USA). cDNAs were amplified using the cDNA Synthesis Kit (Roche). Gene expression was detected using oligonucleotides complementary to transcripts of the analysed genes and the GoTaq qPCR Master Mix (Promega, Dubendorf, Switzerland) using the 7500 Fast Real-Time PCR System (Applied Biosystems, Waltham, MA, USA). Oligonucleotide sequences are available in the Supplementary Table S2. Transcript levels of human RPLP0 were used as endogenous reference, and relative gene expression was analysed using the $2^{-\Delta\Delta Ct}$ method.

2.8. *Immunocytochemistry*

In total, 25–30 microtissues were pooled, fixed overnight with 4% paraformaldehyde and embedded in 2% agarose and in paraffin. Microtissue sections of 3–5 μm were cut (Sophistolab, Muttenz, Switzerland) and deparaffinised following standard methods. Sections were blocked with 10% goat serum (Vector Laboratories, Burlingame, CA, USA) and stained with primary/secondary antibodies: rabbit anti-human α-SMA (Sigma, dilution 1:750)/goat anti-mouse AP (Dako, Glostrup, Denmark, dilution 1:50), rabbit anti-human vimentin (Abcam, Cambridge, UK, dilution 1:250)/goat anti-rabbit (Vector Laboratories, Burlingame, CA, USA, dilution 1:200), rabbit anti-human periostin (Abcam, dilution 1:250)/Polymer anti-rabbit (Histofine, Nichirei Biosciences Inc. Tokyo, Japan), rabbit anti-human fibronectin (Abcam, dilution 1:100)/Polymer anti-rabbit (Histofine), rabbit anti-human Ki67 (Abcam, dilution 1:100)/goat anti-rabbit (Vector Laboratories, dilution 1:200), rabbit anti-human connexin 43 (Abcam, dilution 1:200)/goat anti-rabbit (Vector Laboratories, dilution 1:500), and rabbit anti-human troponin T (Origene Technologies, Rockville, Maryland, USA, dilution 1:2000)/Polymer anti-rabbit (Histofine). Detection was performed with the Bond Polymer HRP Refine Detection kit (Leica, Heerbrugg, Switzerland) according to the manufacturer's guidelines for troponin T, and with DAB or Vector Red (Vector Laboratories) for all other antibodies. Nuclei were counterstained with haematoxylin (J.T.Backer, Gliwice, Poland). Direct Red Sirius Red (Sigma-Aldrich, Basel, Switzerland) staining was used to detect collagen deposition. Stained sections were analysed using the Olympus DP80 microscope and the cellSens Standard imaging software (Olympus, Shinjuku, Japan). Intensity of immunopositive signals and nuclei counterstaining for each section were quantified using the ImageJ software. Staining intensity is presented in arbitrary units (AU) representing immunopositive signal intensity corrected by the counterstained nuclei value.

2.9. Procollagen Type I and IL-6 ELISA

Supernatants from individual microtissues were collected at day 10 and stored at −80 °C. Undiluted supernatants were analysed with the human Procollagen I alpha 1 DuoSet ELISA (R&D Systems, Abingdon, UK) and the human IL-6 DuoSet ELISA (R&D Systems) following manufacturer's protocol. Optical density was measured with the Synergy microplate reader (Biotek, Winooski, Vermont, USA) and the Gen5 software.

2.10. Caspase 3/7 Activity

Supernatants containing one microtissue each were collected at day 10 and immediately measured for apoptosis with the Caspase-Glo 3/7 Assay (Promega, Dubendorf, Switzerland) following the manufacturer's protocol. Luminescence was measured with the Synergy microplate reader (Biotek, Winooski, Vermont, USA) and the Gen5 software.

2.11. Western Blotting

Cellular proteins of foetal human cardiac fibroblasts were extracted with RIPA buffer (Sigma-Aldrich) supplemented with protease inhibitor cocktail (Complete ULTRA Tablets, Roche) and phosphatase inhibitors (PhosphoStop, Roche) from cultivated cells. Protein concentration was quantified by colorimetric bicinchoninic acid assay according to the manufacturer's protocol (Thermo Fisher Scientific). SDS-PAGE electrophoresis and wet-transfer method were used to separate and transfer proteins on nitrocellulose membranes, followed by 45 min incubation in blocking solution (tris buffered saline, Tween-20 (TBST, Thermo Fisher Scientific) containing 5% skim milk powder (Becton Dickinson AG, Allschwil, Swizterland). Membranes were incubated overnight with the following primary antibodies: anti-αSMA (1:1000, A2547, clone 1A4, Sigma-Aldrich), anti-Fibronectin (1:2000, Abcam: ab2413) or GAPDH (1:10,000, clone 14C10, #2118, Cell Signalling). Horseradish peroxidase (HRP)-conjugated secondary antibodies (1:5000) were used for detection with ECL substrate (SuperSignal West Pico Plus, Thermo Fisher Scientific) and development on the Fusion Fx (Vilber, Collegien, France). Densitometric analyses were performed with ImageJ 1.47t. Fold changes were computed after normalization to GAPDH.

2.12. Cell Contraction Assay

To assess the contractile properties of foetal human cardiac fibroblasts, the Contraction Assay Kit (Cell Biolabs, San Diego, CA, USA) was used following the manufacturer's protocol. Cells were cultivated with or without 10 ng/mL TGF-β for 72 h and reseeded in collagen gels for a further 72 h. Each condition was analysed in triplicates or quadruplicates. Images were taken at time 0, 24, 48 and 72 h after reseeding in a collagen gel. Areas of the gels were measured by ImageJ. Percentage of contraction of all conditions was measured compared to the average of unstimulated cells at day 0.

2.13. Statistics

Normally distributed data were analysed by two-tailed unpaired or paired Student's *t*-test and unpaired two-tailed ANOVA followed by uncorrected Fisher's LSD post-hoc test. All analyses were computed using GraphPad Prism 8 software. Differences were considered as statistically significant for $p < 0.05$.

3. Results

3.1. Cardiac Fibroblasts Improve Integrity and Contractility of Human Cardiac Microtissues

Using human iPSC-derived cardiomyocytes (iCMs) and human foetal cardiac fibroblasts (fCFs), we developed a high throughput in vitro model of human cardiac tissue [16]. Cardiac microtissues consisting of 5000 cells each were generated by self-assembly (Figure S1). iCMs assembling without fibroblasts, however, formed loose cell aggregates (Figure 1A, Video S1). Addition of fCFs to iCMs in a ratio of 1:4 (fCFs:iCMs) allowed for formation of compact, spontaneously contracting cardiac microtissues (Figure 1A,B, Video S2) In the next step, we performed transcriptional profiling of microtissues by qPCR for genes characteristic for cardiomyocytes and fibroblasts. Seven genes characteristic for cardiomyocytes were detected in microtissues containing iCMs (iCM and iCM:fCF microtissues), but not in those generated with fCFs only. In contrast, nine out of twelve genes associated with fibroblasts and fibrosis were consistently detected in all types of microtissues (Figure 1C). However, expression levels of most of fibroblastic genes were substantially higher in fCFs than in iCMs microtissues. As expected, the expression profile of iCMs:fCFs microtissues showed high levels of all cardiac and fibroblastic genes (Figure 1C).

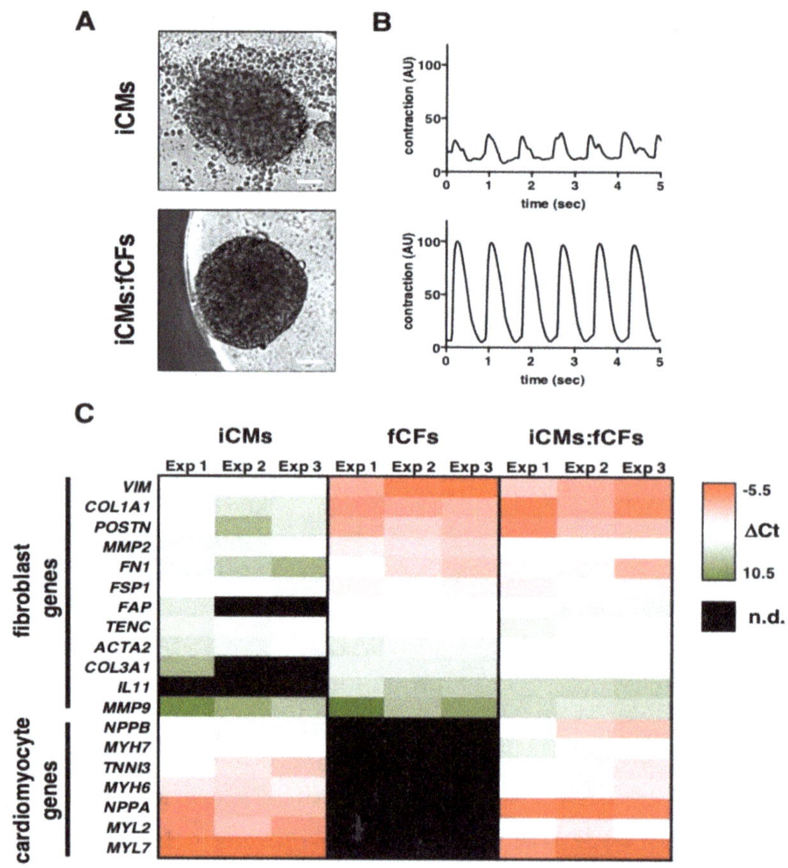

Figure 1. Characteristics of cardiac microtissues. Panel (**A**) illustrates typical morphologies (bar = 50 μm) and panel (**B**) shows typical contraction patterns of cardiac microtissues generated using iPSC-derived cardiomyocytes (iCMs) only (top) or iCMs mixed with foetal cardiac fibroblasts (fCFs) in a ratio of 4:1 (iCMs:fCFs, bottom). Representative contractions are available in the Supplementary Materials (Videos S1 and S2). A heat map in panel (**C**) indicates expression of cardiomyocyte and fibroblast genes in microtissues containing iCMs only (left), fCFs only (center) or iCMs:fCFs (right). Each segment indicates the average (n = 3–5) expression of one experiment. Lower $-\Delta C_t$ values indicate higher relative expression. n.d.—not detected.

3.2. *TGF-β1 Induces Fibrotic Changes in Human Cardiac Microtissues*

TGF-β1 represents a potent inducer of profibrotic changes in cardiac fibroblasts. In the first step, we exposed microtissues to TGF-β1 for 10 days. In the presence of TGF-β1, both fCFs and iCMs:fCFs microtissues significantly increased their size (Figure 2A, Figure S2).

In the next step, we measured procollagen type I secretion, which is an important hallmark of the ongoing fibrotic processes. As expected, fCFs microtissues produced markedly more collagen I upon TGF-β1 stimulation. Under this stimulatory condition, significant increase in procollagen type I production was also observed for iCMs:fCFs microtissues (Figure 2B). Importantly, iCMs microtissues did not produce detectable levels of procollagen I (Figure 2B) pointing to fibroblasts as the main source of collagen I in cardiac microtissues. Increased collagen production in iCMs:fCFs microtissues treated with TGF-β1 was confirmed by picrosirius red staining (Figure 2C). To address whether treatment with TGF-β1 induced apoptosis in this model, we measured caspase 3/7 activity in iCMs:fCFs microtissues. We observed similar caspase 3/7 activity in iCMs:fCFs microtissues treated with and without TGF-β1 (Figure 2D). Additionally, iCMs:fCFs microtissues treated with TGF-β1 produced significantly less IL-6 (Figure 2E).

Cardiac fibrogenesis is a transcriptionally regulated process. Upon TGF-β1 stimulation, genes characteristic for cardiomyocytes were only slightly dysregulated in iCMs microtissues. In iCMs:fCFs microtissues, instead, cardiomyocyte-specific genes were downregulated following TGF-β1 treatment (Figure 2F). Analysis of fibrotic genes showed that most of them were strongly upregulated in fCFs and iCMs:fCFs microtissues in response to TGF-β1 (Figure 2F).

TGF-β1-induced changes in gene expression was followed by the analysis of the respective proteins in cardiac microtissues. To this aim, selected markers were analysed by immunohistochemistry in iCMs:fCFs microtissues cultured with or without TGF-β1. We found significantly less expression of the cardiac cell marker troponin T and gap junction protein connexin 43 in iCMs:fCFs microtissues exposed to TGF-β1. Instead, proteins produced by quiescent (vimentin, fibronectin) or activated (α-SMA, periostin) fibroblasts were more abundant in TGF-β1-treated cardiac microtissues. The number of proliferative (Ki67-positive) cells was also higher in iCMs:fCFs microtissues cultured with TGF-β1 (Figure 3, Figure S3). In conclusion, our results indicate that TGF-β1 effectively induced fibrogenesis in cardiac microtissues.

3.3. *Pharmacological Targeting of TGF-βR1 Signalling Prevents from Fibrotic Changes in Human Cardiac Microtissues*

In the next step, we asked whether the fibrotic cardiac microtissue model is relevant to test anti-fibrotic compounds. To this aim, we used the TGF-βR1 inhibitor SD208 and tested it in the iCMs:fCFs microtissues cultured in the presence or absence of TGF-β. We observed that SD208 inhibitor effectively blocked TGF-β1-mediated microtissue growth (Figure 4A).

Furthermore, we found that SD208 suppressed procollagen type I production in iCMs:fCFs microtissues cultured not only with, but also without TGF-β1 (Figure 4B). Given the effect of the inhibitor in the absence of exogenous TGF-β1, we analysed gene expression in iCMs:fCFs microtissues cultured with and without SD208. Expression of 12 out of 18 analysed genes were significantly changed in response to SD208 treatment (Figure 4C). Similar results were obtained for treatment of iCMs:fCFs microtissues with SD208 in the presence of TGF-β1 (Figure 4D). Next, we analysed dose-dependent responses of microtissues to SD208. In these experiments, iCMs:fCFs microtissues were cultured in the presence of TGF-β1 and serial dilutions of SD208. Our results showed dose-dependent responses of iCMs:fCFs microtissues to SD208 in terms of microtissue size (Figure 4E), collagen type I secretion (Figure 4F) and expression of profibrotic genes (Figure 4G). Taken together, these data suggest that cardiac microtissues represent a useful model for pharmacological studies.

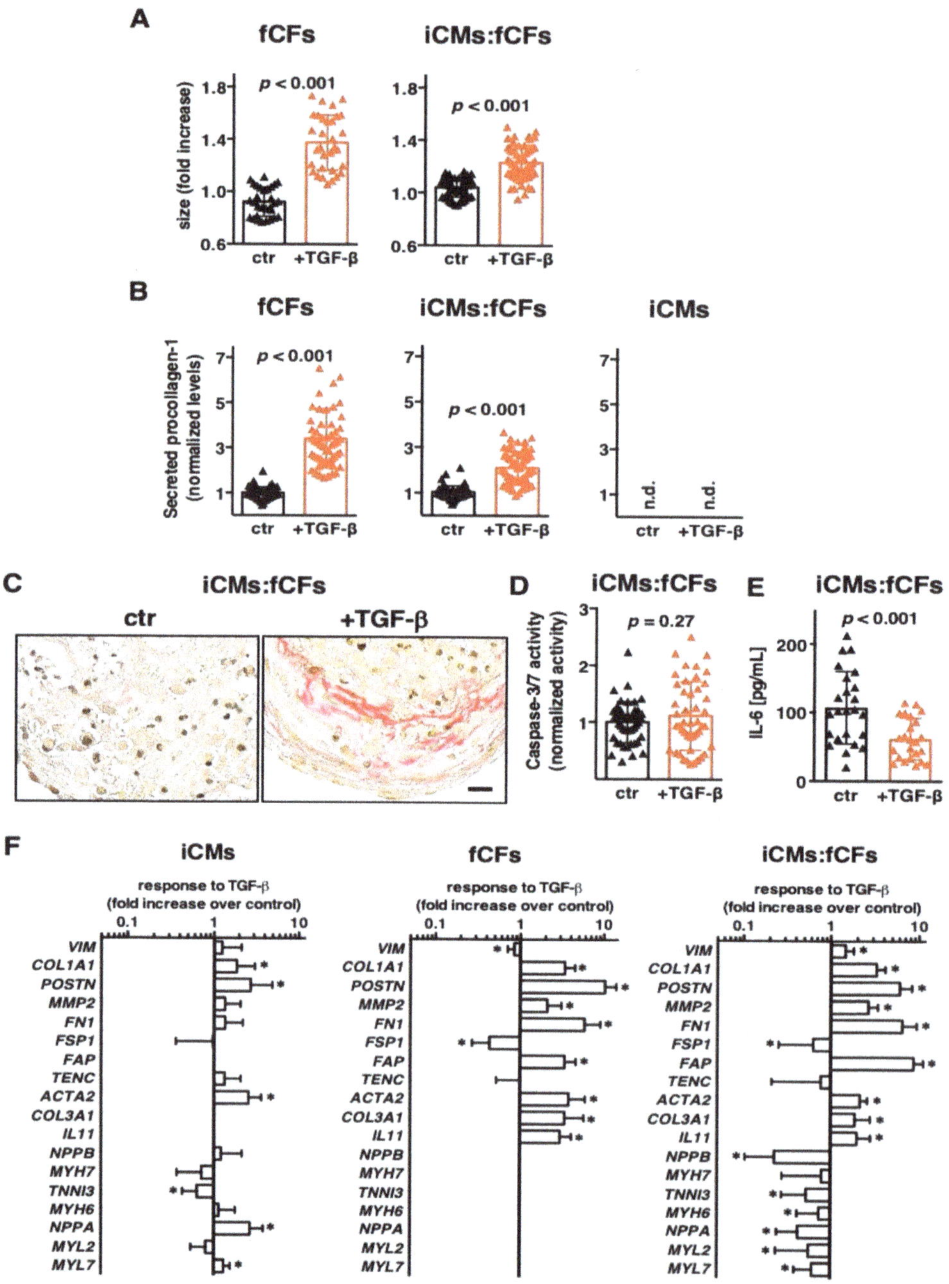

Figure 2. TGF-β1 activates foetal cardiac fibroblasts in microtissues. Panel (**A**) demonstrates changes in size of microtissues generated with fCFs only (fCFs, left) and iCMs mixed with fCFs in ratio 4:1 (iCMs:fCFs, right) cultured in the presence (red) or absence (black) of TGF-β1 (10 ng/mL) for 10 days. Panel (**B**) shows relative levels of procollagen I (measured by ELISA), at day 10 in supernatants of all three microtissue types: fCFs (left), iCMs:fCFs (middle) and iCMs (right). Graphs show cumulative data of 2–5 independent experiments. Each dot represents data of one microtissue. Panel (**C**) illustrates representative picrosirius red staining in iCMs:fCFs microtissues at day 10 (bar = 10 μm). Panel (**D**) shows caspase 3/7 activity measured at day 10 in iCMs:fCFs microtissues. Graphs show cumulative data of 3 independent experiments. Panel (**E**) shows IL-6 levels measured by ELISA, at day 10 in supernatants of iCMs:fCFs microtissues. Graphs show cumulative data of 3 independent experiments. Each triangle represents data of one microtissue. Panel (**F**) summarizes fold changes in gene expression in indicated microtissues in the presence of TGF-β1 (in relation to expression in the absence of TGF-β1). * $p < 0.05$. Graphs show cumulative data of 3–4 independent experiments, $n = 11$–20. For all graphs, p values were calculated with the Student's t-test. n.d.—not detected.

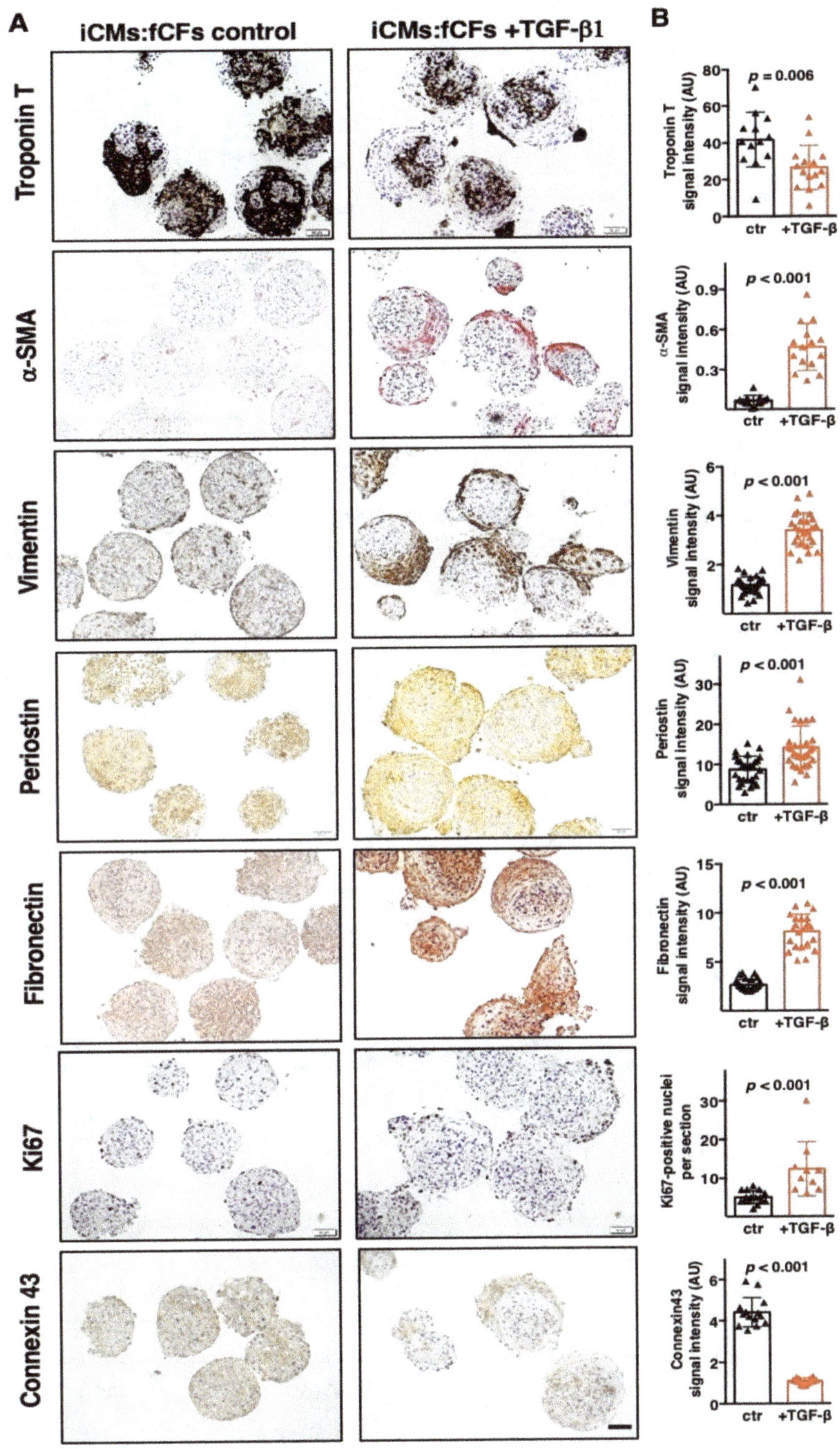

Figure 3. TGF-β1 induces fibrotic phenotype in cardiac microtissues. Immunohistochemistry of iCMs:fCFs microtissues cultured in the presence or absence of TGF-β1 (10 ng/mL) at day 10. Panel (**A**) illustrates representative staining for the indicated proteins at the indicated condition (bar = 50 μm). Higher magnification pictures are available in the Supplementary Materials (Figure S3). Panel (**B**) shows quantification of the respective staining for microtissues cultured in the presence (red) or absence (black) of TGF-β1. Graphs show cumulative data of 3 independent experiments. Each triangle represents data for one microtissue. p values were calculated with the Student's t-test.

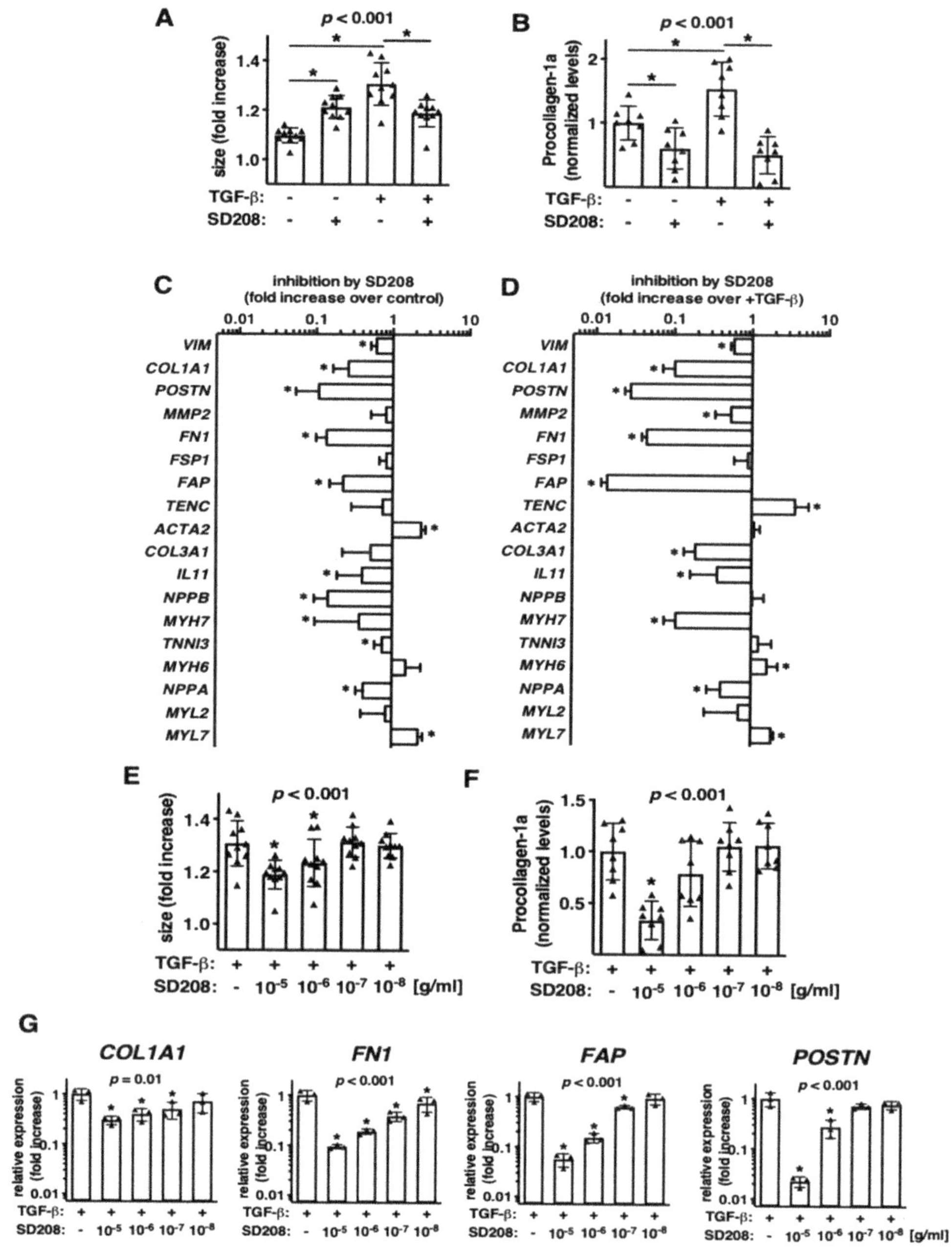

Figure 4. Pharmacological targeting of microtissues with TGF-βR1 inhibitor SD208. (**A–D**) iCMs:fCFs microtissues were cultured in the presence or absence of TGF-β1 (10 ng/mL) and SD208 (10 μg/mL). Panel (**A**) shows changes in size of microtissues and panel (**B**) normalized levels of secreted procollagen I (measured by ELISA in supernatants) at the indicated conditions at day 10. Each dot represents data for one microtissue. p values were calculated with ANOVA followed by uncorrected Fisher's LSD tests for selected groups, * $p < 0.05$. Panels (**C**,**D**) show changes in gene expression in response to treatment with SD208 of microtissues cultured in the absence (**C**) or presence (**D**) of TGF-β1. $n = 6$, p values were calculated with the Student's t-test, * $p < 0.05$. (**E–G**) iCMs:fCFs microtissues were cultured in the presence of TGF-β1 (10 ng/mL) and serial dilutions of SD208. Panel (**E**) shows changes in microtissue size, panel (**F**) normalized levels of secreted procollagen I and panel (**G**) relative expression of selected genes at the indicated conditions at day 10. Each dot represents data for one microtissue. Each triangle represents data for one microtissue. p values were calculated with ANOVA followed by uncorrected Fisher's LSD tests versus +TGF-β group, * $p < 0.05$.

3.4. *Activated Cardiac Fibroblasts Increase the Contraction Rate of Cardiac Microtissues*

Contractility represents the main functional feature of cardiomyocytes. Using high-speed movies and motion-tracking analysis, we examined contractile properties of iCMs:fCFs microtissues at day 10. Fibrotic changes triggered by TGF-β1 substantially deteriorated the contraction pattern of iCMs:fCFs microtissues. We consistently observed significantly increased contraction rate and reduced contraction amplitudes of fibrotic microtissues in all experiments. A more detailed analysis of contraction pointed mainly to the shortening of relaxation phase in iCMs:fCFs microtissues upon treatment with TGF-β1 (Figure 5A, Figures S4 and S5, Video S3).

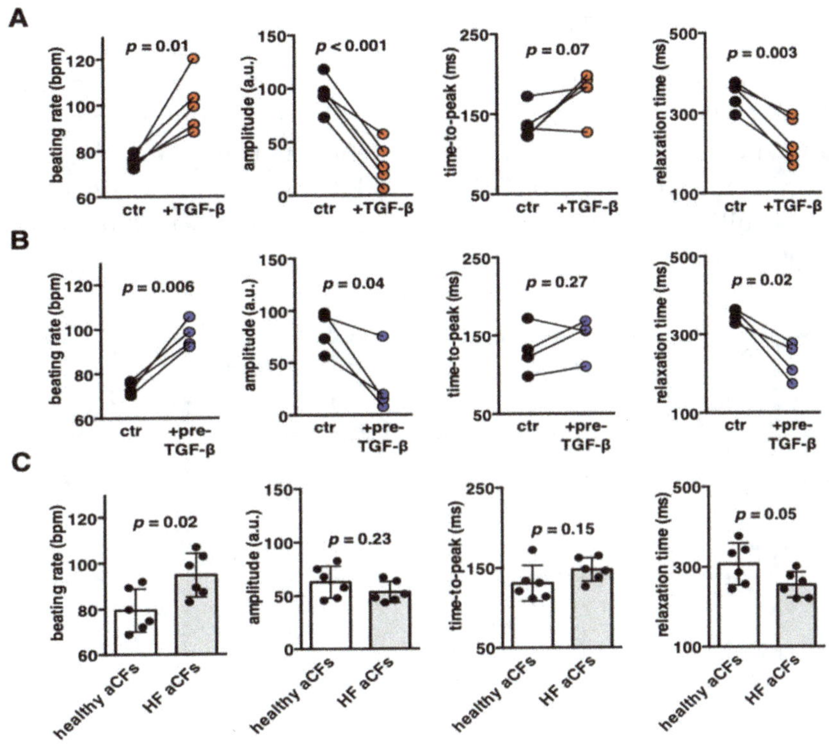

Figure 5. Contractile properties of cardiac microtissues containing foetal or adult cardiac fibroblasts. Panel (A) shows quantification of contraction parameters of iCMs:fCFs microtissues cultured in the presence (red) or absence (black) of TGF-β1 (10 ng/mL) at day 10. Quantification of contraction parameters of microtissues containing fCFs pretreated with TGF-β1 for 3 days prior microtissue formation (blue) or untreated fCFs (black) recorded at day 10 are shown in panel (B). Each dot represents average data of one experiment. Data of individual experiments are available in the Supplementary Materials (Figures S5 and S8). p values were calculated with the paired Student's t-test. Panel (C) shows quantification of contraction parameters of cardiac microtissues containing aCFs from unaffected hearts (white) or heart failure (HF) patients (grey). Each dot represents average data of one experiment (n = 14–18). p values were calculated with the Student's t-test. Representative contraction records are available in the Supplementary Materials (Figures S5, S10B and S11 and Videos S2–S6).

We hypothesized that TGF-β1 affected the contractility of iCMs:fCFs microtissues by activation of fCFs, rather than by a direct effect on iCMs. To validate this hypothesis, we generated cardiac microtissues using fCFs pretreated with TGF-β1 for three days (prior microtissue formation) and analysed their contractility in the absence of exogenous TGF-β1 at day 10. Pretreatment with TGF-β1 effectively activated fCFs as indicated by increased production of collagen, α-SMA and fibronectin in 2D cultures (Figure S6). We observed that iCMs:fCFs microtissues containing activated fCFs displayed increased size and secreted significantly more procollagen I (Figure S7), showed increased contraction rate, reduced contraction amplitudes and shortened relaxation phase (Figure 5B, Figures S4 and S8, Video S4), and successfully reproduced the contraction pattern observed in fibrotic (+TGF-β1) cardiac microtissues.

To further address the relevance of CFs to cardiac microtissue contractility, we compared cardiac microtissues containing adult CFs (aCFs) obtained from unaffected hearts of mostly young individuals (healthy aCFs) to aCFs from patients with heart failure (HF aCFs). Cardiac microtissues containing healthy aCFs and HF aCFs showed similar sizes at day 10 (Figures S9 and S10A). We found that iCMs:aCFs microtissues with HF aCFs showed increased contraction rate and shortened relaxation phase in comparison to cardiac microtissues containing healthy aCFs (Figure 5C, Figures S4 and S10B, Videos S5 and S6). Of note, microtissues containing healthy iCMs:aCFs more often responded to TGF-β1 stimulation by increasing contraction rate than those with HF aCFs (four of six vs. one of six, Figure S11). All these data indicate that contractile properties of cardiac microtissues depend on the activation status of CFs.

3.5. Unaffected Repolarization Phase in Fibrotic Cardiac Microtissues

In cardiomyocytes, mechanical contraction is triggered by electrical excitation. Next, we addressed the effect of external electrical stimulation of cardiac microtissues. Stimulation at 3 Hz induced a beating rate of 178–179 bpm in responsive microtissues, reduced contraction amplitude and substantially shortened contraction duration in both control and TGF-β1-treated iCMs:fCFs microtissues (Figure 6A, Videos S7 and S8), pointing to their full electrical responsiveness. Next, we analysed changes in membrane potential of control and fibrotic iCMs:fCFs microtissues using the relevant fluorescent probe (Figure 6B).

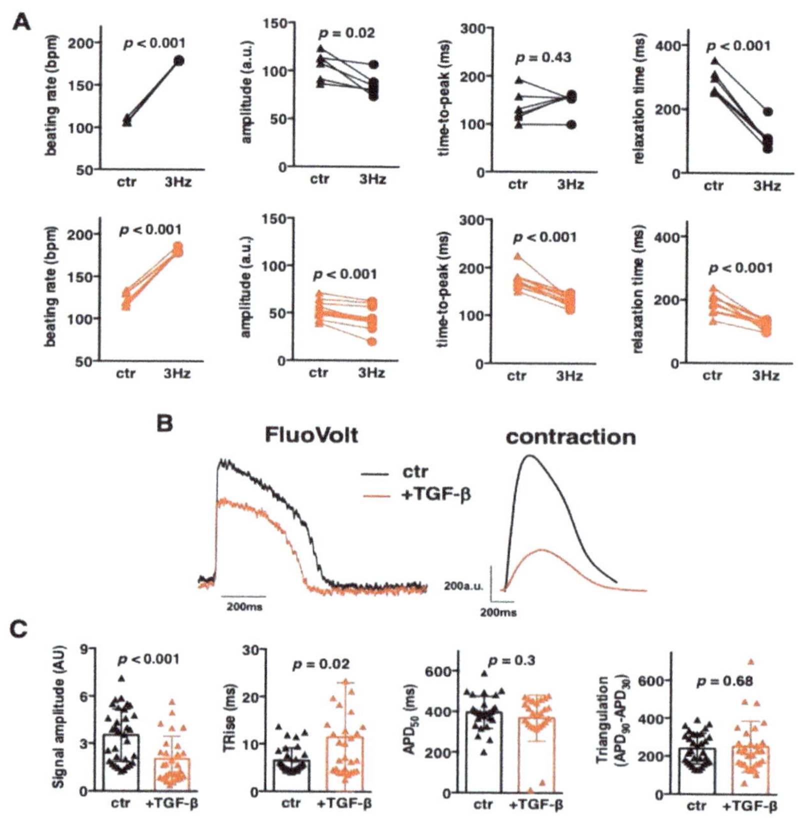

Figure 6. Cardiac microtissue electrophysiology. iCMs:fCFs microtissues were cultured in the presence (red) or absence (black) of TGF-β1 (10 ng/mL) for 10 days. Panel (**A**) shows quantifications of the contraction parameters recorded during spontaneous contractile activity (triangles, ctr) and then upon electrical stimulation with 3 Hz (circles, 3 Hz). Each dot represents data for one microtissue at the indicated condition, lines match data obtained from the same microtissue. p values were calculated with the paired Student's t-test. Panel (**B**) illustrates representative action potentials recorded with FluoVolt probe (left) and the respective contractions (right). Quantifications of action potential parameters are shown in panel (**C**). Graphs show cumulative data of 3 independent experiments. Each dot represents data for one microtissue. p values were calculated with the Student's t-test. APD—action potential duration, TRise—depolarisation phase.

We found significantly lower action potential amplitudes in TGF-β1-treated iCMs:fCFs microtissues in comparison to controls (Figure 6C), suggesting reduced number of electrically active cells in fibrotic microtissues. Instead, there was no difference in the action potential duration, although treatment with TGF-β1 slightly prolonged the depolarization phase (Figure 6C). It seems that electrical activity of iCMs in cardiac microtissues was not significantly affected by profibrotic changes.

3.6. Endogenous β-Adrenergic Receptor Signalling Controls Increased Contraction Rate in Fibrotic Cardiac Microtissues

Action potential triggers cardiomyocyte contraction, but β-adrenoceptor-coupled mechanisms can modulate it. First, we used iCMs:fCFs and iCMs:aCFs microtissues and treated them with β-adrenoceptor agonist isoproterenol. As expected, stimulated microtissues showed increased contraction rate (Figure 7A, Figure S12). Isoproterenol significantly shortened both the contraction and relaxation phases of spontaneously contracting microtissues. In contrast, contraction amplitude remained unaffected. Addition of the β-adrenergic receptor blocker nadolol effectively prevented isoproterenol-induced changes in microtissue contraction (Figure 7A). These data confirmed functional β-adrenoceptors in our cardiac microtissues.

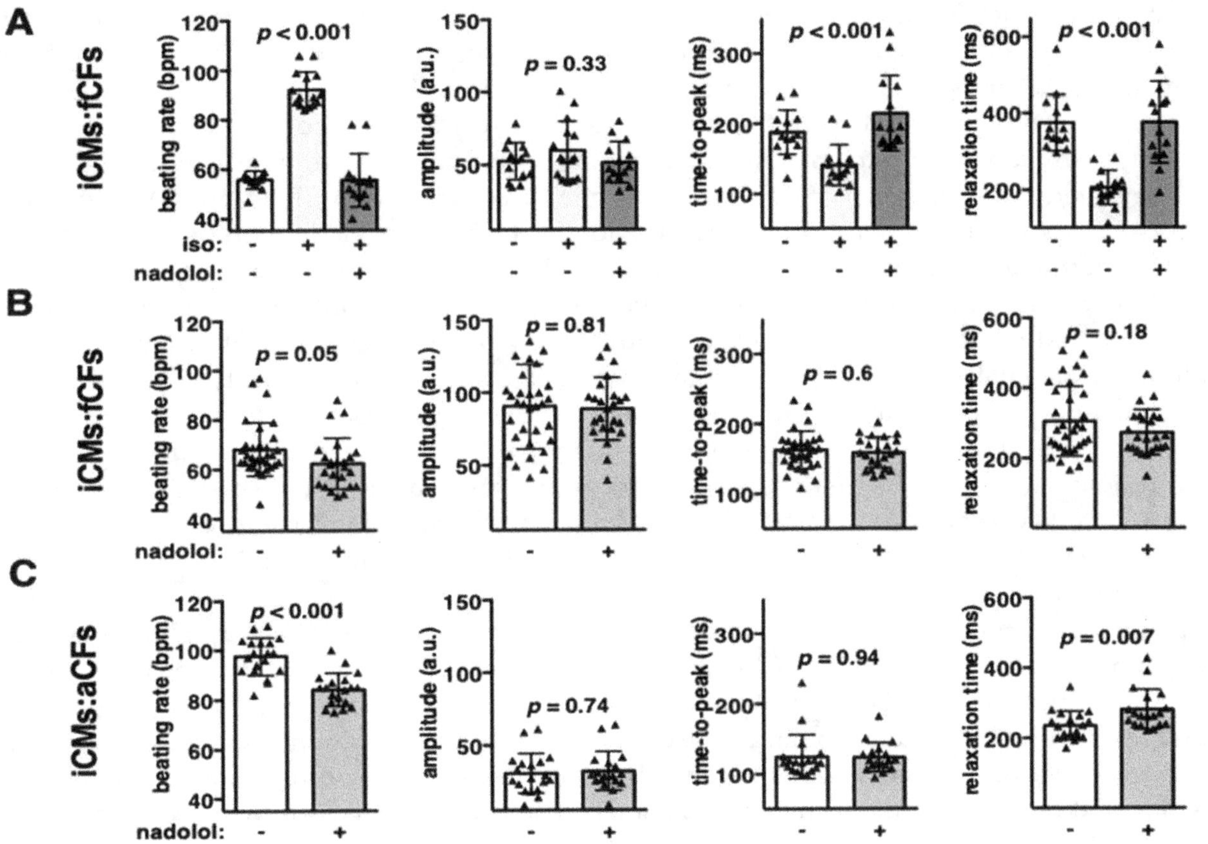

Figure 7. β-adrenergic receptor signalling in cardiac microtissues. Panel (**A**) shows quantification of contraction parameters of iCMs:fCFs microtissues stimulated with β-adrenoceptor agonist isoproterenol (iso, 10 nM) in the presence or absence of β-adrenoceptor blocker nadolol (1 μM). Unstimulated microtissues were used as controls. Graphs in panels (**B**,**C**) show contraction parameters of iCMs:fCFs (**B**) and iCMs:aCFs (**C**) microtissues treated with vehicle or 1 μM nadolol in the absence of isoproterenol. Each triangle represents data for one microtissue. *p* values were calculated with ANOVA for (**A**) and with the Student's *t*-test for (**B**,**C**).

Next, we addressed the relevance of endogenous β-adrenoceptor signalling in cardiac microtissues. To this aim, we analysed the contractility of iCMs:fCFs and iCMs:aCFs microtissues in the presence and absence of nadolol. Treatment with nadolol only slightly reduced contraction of iCMs:fCFs microtissues

(Figure 7B). In the next experiment, we used iCMs:aCFs microtissues showing high spontaneous contraction rate. In this model, nadolol significantly reduced contraction rate, whereas contraction amplitude remained unchanged (Figure 7C). We found that nadolol specifically affected relaxation, but not contraction phase. These results suggest activation of endogenous β-adrenoceptor signalling in iCMs:aCFs microtissues.

4. Discussion

In 3D microtissue models, fibroblasts have been recognized to improve microtissue architecture and biomechanics by providing proper ECM. We confirm that CFs significantly increased the integrity and synchronized contractility of human cardiac microtissues. As recently demonstrated, the use of a 4:1 ratio (iCMs:CFs) results in cardiac microtissues with stable phenotype and beating properties for up to one month [16]. Our data further demonstrated that contractility of these cardiac microtissues was substantially affected under profibrotic condition (i.e., in the presence of TGF-β1). TGF-β1 is a profibrotic cytokine that enhances fibroblast proliferation and production of ECM proteins. In our model, TGF-β1 induced fibrotic phenotype in cardiac microtissues, as indicated by elevated levels of fibrotic genes, secretion of procollagen I and fCFs proliferation. Previously, exogenous TGF-β1 has been shown to induce fibrosis in rat [17,18] or human [19] cardiac microtissue models. Furthermore, fibrotic phenotypes in cardiac microtissues or in biowire were also produced by collagen/fibroblast enrichment [20,21] or by constitutive activation of profibrotic pathways [22].

Our data demonstrated that pretreatment of fCFs with TGF-β1 prior to microtissue formation resulted in a similar effect on microtissue contractility as continuous stimulation with TGF-β1. This result suggested that activation of CFs, rather than a direct effect of TGF-β1 on iCMs, was a main mechanism responsible for deteriorated contraction of fibrotic cardiac microtissues. This idea was further strengthened by the observation that aCFs induced a contraction pattern (high contraction rate with low amplitude) of cardiac microtissues similar to those containing activated fCFs. CFs become activated following heart failure and with aging [23]. We observed that cardiac microtissues containing aCFs obtained from aged heart failure patients showed significantly higher beating rate than microtissues containing aCFs from young, healthy hearts. All these data suggested that activation status of CFs was a key factor determining contraction pattern of cardiac microtissues.

Fibrotic condition in the heart is associated with arrhythmias. Published data showed that activated rodent CFs could cause asynchronous contraction [18] and induce proarrhythmic changes in cardiomyocytes by affecting action potential profile [22,24]. In our model, fibrotic microtissues showed arrhythmic contractions only occasionally and their action potential profile remained unaffected. Instead, the fibrotic condition was associated with reduced action potential signal amplitude, suggesting a lower number of electrically active iCMs in these microtissues. Impaired conduction and contraction of these microtissues might be a consequence of iCMs decoupling and coupling of iCMs with CFs, reduced levels of gap junction protein connexin 43 and/or overproduction of cytokines and ECM by activated CFs. Secreted factors by activated CFs, biomechanical signalling through excessive ECM deposition and remodelling of mechanical junctions are known factors disturbing proper propagation of electrical impulses [25]. Signal conduction abnormalities, such as re-entry, can cause cardiac arrhythmias and may have serious clinical implications for patients with fibrotic hearts [9]. Clinical data suggest that ventricular fibrosis represents a strong predictor of life-threatening ventricular arrhythmia and sudden cardiac death in ischemic and nonischemic heart disorders [8].

In our model, profibrotic condition was consistently associated with increased beating rate of cardiac microtissues. Similar data were obtained in human cardiac microtissues containing mesenchymal stem cells instead of CFs following treatment with TGF-β1 [19]. Physiologically, increased beating rate of the heart is achieved by activation of β-adrenoreceptors. In normal adult cardiomyocytes, β1-adrenoreceptors are dominant, whereas iCMs contractility can be regulated by both β1- and β2-adrenoreceptors [26]. Previous data showed that the beating rate of human iCMs could be increased by treatment with the β-adrenergic agonist isoproterenol [26,27]. In line with these data, we could demonstrate the responsiveness of human cardiac microtissues containing iCMs to isoproterenol. In case of cardiac microtissues showing high beating rate (due to presence of activated aCFs), blockade of β-adrenergic receptors in the absence of exogenous agonists significantly decreased frequency. This finding indicated that activated CFs triggered endogenous β-adrenoreceptor signalling in cardiac microtissues. Mechanistically, stimulation of β-adrenoreceptors increases beating rate by enhancing cardiomyocyte relaxation [10]. Indeed, we observed that the reduced beating rate observed in the presence of β-adrenoreceptor blockers was associated with longer relaxation time. Accordingly, increased contraction rate of fibrotic cardiac microtissues correlated with shortened relaxation time. These results may suggest that activated CFs produce β-adrenergic agonists. Alternatively, fibrotic condition may sensitize β-adrenoreceptors. Cellular response to β-adrenergic stimulations is further modulated by multiple factors, such as cell surface receptor density, activity of the downstream signalling pathway, or by specific costimulations [28]. For example, fibrotic hearts of TGF-β-overexpressing mice showed increased density of β-adrenoreceptors on cardiomyocytes and altered β-adrenergic signalling [3]. Furthermore, collagen receptor β_1-integrin [29], connective tissue growth factor [30] or tissue stiffness [31] have also been shown to regulate β-adrenergic signalling.

A successful preclinical platform has to combine physiological-like relevance with the high-throughput processing. Our fibrotic cardiac microtissue model is established in a 96-well format and certain parameters, such as size, contractile properties, changes in membrane potential, procollagen-I production or apoptosis can be measured for individual microtissues. In particular, recently developed software measuring the contractile activity of the whole microtissue allows for quick and easy analysis of the key properties of beating cardiac microtissues [15]. In our experiments, transcriptomic data were obtained from 10–12 pooled microtissues, but we used a simple RT-PCR to analyse gene expression. Application of the next-generation sequencing technology would allow obtaining transcriptomic data from one microtissue. Thus, this model could be used as a high-throughput platform for pharmacological screening with a timeframe for up to at least 10 days. The proof-of-concept experiments with the TGF-βR1 inhibitor SD208 confirmed that fibrotic cardiac microtissues represent a robust model for drug screening.

Although our human cardiac microtissue-based fibrotic platform combines technical ease, high throughput and reproducibility with physiological-like relevance, it also needs to be discussed in the context of its limitations. iCMs are characterized by their foetal developmental status [32]. Currently, there is no satisfactory alternative to iCMs. The usefulness of adult primary human cardiomyocytes is limited, as these cells contract upon chemical or electrical stimulation only [33]. Despite recent advances in the maturation of iCMs through improved biomechanics and cocultures with other cell types, so far, their maturation status does not reflect adult cardiomyocytes [13,34–36]. Another limitation of our model is the heterogeneity of iCMs. Conventional differentiation of iPSCs into iCMs results in populations of pacemaker, ventricular-like, atrial-like cells as well as nonbeating progenitors. The use of defined types of differentiated cardiomyocytes could further improve the relevance of the cardiac microtissue model. Moreover, addition of other cell types, such as endothelial cells or macrophages, to iCMs and fCFs might result in formation of microtissues that better mimic the structure and function of human cardiac tissue.

Supplementary Materials

Figure S1: Schematic presentation of experimental setup with time lines, Figure S2: Kinetics of microtissue growths, Figure S3: TGF-β1 induces fibrotic phenotypes in cardiac microtissues, Figure S4: Contraction patterns of cardiac microtissues containing foetal or adult cardiac fibroblasts, Figure S5: Effect of exogenous TGF-β1 on contractility of cardiac microtissues - individual experiments, Figure S6: Activation of foetal cardiac fibroblasts with TGF-β1, Figure S7: Effect of activated foetal cardiac fibroblasts on size and procollagen I secretion, Figure S8: Effect of activated foetal cardiac fibroblasts on contractility of cardiac microtissues - individual experiments, Figure S9: Size of cardiac microtissues containing adult cardiac fibroblasts, Figure S10: Size and contractile properties of cardiac microtissues containing adult cardiac fibroblasts, Figure S11: Effect of exogenous TGF-β1 on contractility of healthy iCMs:aCFs and HF iCMs:aCFs microtissues, Figure S12: b-adrenergic receptor stimulation of iCMs:aCFs microtissues; Table S1: Clinical characteristics of patients from which aCFs were obtained, Table S2: Primers used for quantitative RT-PCR; Video S1: Control iCMs only, Video S2: iCMs_fCFs control d10, Video S3: iCMSfCFs TGF-β1, Video S4: iCMsfCFs pre-TGF-β-3d_d10 D10, Video S5: iCMs_healthy aCFs control d10, Video S6: iCMS_HF aCMs control d10, Video S7: iCMs_fCFs control, Video S8: iCMs_fCFs 3Hz.

Author Contributions: Conceptualization: P.B., G.K.; Methodology: P.B., C.Z., A.C., D.N., F.D.M., I.A., M.S., G.K.; Data curation: P.B., C.Z., A.C., M.S., G.K.; Formal analysis: P.B., G.L.S., C.Z., G.K.; Funding acquisition: G.K.; Supervision: O.D., G.K.; Writing—original draft: P.B., G.K.; Writing—review & editing: C.Z., G.L.S., O.D. All authors have read and agreed to the published version of the manuscript.

Acknowledgments: The authors acknowledge financial support of the Swiss Heart Foundation and Vontobel Foundation as well as technical assistance of the Center for Microscopy and Image Analysis of the University of Zurich and Sophistolab AG.

Disclosures: Irina Agarkova is employed by InSphero company, which produces microtissues. Godfrey Smith is a founder, shareholder, executive and honorary Chief Scientific Officer of Clyde Biosciences Ltd. company, which provides an assay service to the pharmaceutical industry using commercial iPSC-derived cardiomyocytes.

References

1. Kong, P.; Christia, P.; Frangogiannis, N.G. The pathogenesis of cardiac fibrosis. *Cell. Mol. Life Sci.* **2014**, *71*, 549–574. [CrossRef]
2. Travers, J.G.; Kamal, F.A.; Robbins, J.; Yutzey, K.E.; Blaxall, B.C. Cardiac fibrosis: The fibroblast awakens. *Circ. Res.* **2016**, *118*, 1021–1040. [CrossRef]
3. Rosenkranz, S.; Flesch, M.; Amann, K.; Haeuseler, C.; Kilter, H.; Seeland, U.; Schluter, K.D.; Bohm, M. Alterations of beta-adrenergic signaling and cardiac hypertrophy in transgenic mice overexpressing TGF-beta(1). *Am. J. Physiol. Heart Circ. Physiol.* **2002**, *283*, H1253–H1262. [CrossRef] [PubMed]
4. Ikeuchi, M.; Tsutsui, H.; Shiomi, T.; Matsusaka, H.; Matsushima, S.; Wen, J.; Kubota, T.; Takeshita, A. Inhibition of TGF-beta signaling exacerbates early cardiac dysfunction but prevents late remodeling after infarction. *Cardiovasc. Res.* **2004**, *64*, 526–535. [CrossRef] [PubMed]
5. Kuwahara, F.; Kai, H.; Tokuda, K.; Kai, M.; Takeshita, A.; Egashira, K.; Imaizumi, T. Transforming growth factor-beta function blocking prevents myocardial fibrosis and diastolic dysfunction in pressure-overloaded rats. *Circulation* **2002**, *106*, 130–135. [CrossRef] [PubMed]
6. Kania, G.; Blyszczuk, P.; Stein, S.; Valaperti, A.; Germano, D.; Dirnhofer, S.; Hunziker, L.; Matter, C.M.; Eriksson, U. Heart-infiltrating prominin-1^+/$CD133^+$ progenitor cells represent the cellular source of transforming growth factor beta-mediated cardiac fibrosis in experimental autoimmune myocarditis. *Circ. Res.* **2009**, *105*, 462–470. [CrossRef]
7. Clements, I.P.; Miller, W.L.; Olson, L.J. Resting heart rate and cardiac function in dilated cardiomyopathy. *Int. J. Cardiol.* **1999**, *72*, 27–37. [CrossRef]
8. Disertori, M.; Mase, M.; Ravelli, F. Myocardial fibrosis predicts ventricular tachyarrhythmias. *Trends Cardiovasc. Med.* **2017**, *27*, 363–372. [CrossRef]
9. Nguyen, T.P.; Qu, Z.; Weiss, J.N. Cardiac fibrosis and arrhythmogenesis: The road to repair is paved with perils. *J. Mol. Cell. Cardiol.* **2014**, *70*, 83–91. [CrossRef]
10. Najafi, A.; Sequeira, V.; Kuster, D.W.; Van der Velden, J. Beta-adrenergic receptor signalling and its functional consequences in the diseased heart. *Eur. J. Clin. Investig.* **2016**, *46*, 362–374. [CrossRef]
11. Karakikes, I.; Ameen, M.; Termglinchan, V.; Wu, J.C. Human induced pluripotent stem cell-derived cardiomyocytes: Insights into molecular, cellular, and functional phenotypes. *Circ. Res.* **2015**, *117*, 80–88. [CrossRef] [PubMed]
12. Zuppinger, C. 3D culture for cardiac cells. *Biochim. Biophys. Acta* **2016**, *1863*, 1873–1881. [CrossRef] [PubMed]
13. Beauchamp, P.; Moritz, W.; Kelm, J.M.; Ullrich, N.D.; Agarkova, I.; Anson, B.D.; Suter, T.M.; Zuppinger, C. Development and characterization of a scaffold-free 3D spheroid model of induced pluripotent stem cell-derived human cardiomyocytes. *Tissue Eng. Part C Methods* **2015**, *21*, 852–861. [CrossRef] [PubMed]
14. Castaldo, C.; Di Meglio, F.; Miraglia, R.; Sacco, A.M.; Romano, V.; Bancone, C.; Della Corte, A.; Montagnani, S.; Nurzynska, D. Cardiac fibroblast-derived extracellular matrix (biomatrix) as a model for the studies of cardiac primitive cell biological properties in normal and pathological adult human heart. *Biomed. Res. Int.* **2013**, *2013*, 352370. [CrossRef] [PubMed]
15. Sala, L.; van Meer, B.J.; Tertoolen, L.G.J.; Bakkers, J.; Bellin, M.; Davis, R.P.; Denning, C.; Dieben, M.A.E.; Eschenhagen, T.; Giacomelli, E.; et al. MUSCLEMOTION: A versatile open software tool to quantify cardiomyocyte and cardiac muscle contraction in vitro and in vivo. *Circ. Res.* **2018**, *122*, e5–e16. [CrossRef]

16. Beauchamp, P.; Jackson, C.B.; Ozhathil, L.C.; Agarkova, I.; Galindo, C.L.; Sawyer, D.B.; Suter, T.M.; Zuppinger, C. 3D co-culture of hiPSC-derived cardiomyocytes with cardiac fibroblasts improves tissue-like features of cardiac spheroids. *Front. Mol. Biosci.* **2020**, *7*, 14. [CrossRef]
17. Figtree, G.A.; Bubb, K.J.; Tang, O.; Kizana, E.; Gentile, C. Vascularized cardiac spheroids as novel 3D in vitro models to study cardiac fibrosis. *Cells Tissues Organs* **2017**, *204*, 191–198. [CrossRef]
18. Sadeghi, A.H.; Shin, S.R.; Deddens, J.C.; Fratta, G.; Mandla, S.; Yazdi, I.K.; Prakash, G.; Antona, S.; Demarchi, D.; Buijsrogge, M.P.; et al. Engineered 3D cardiac fibrotic tissue to study fibrotic remodeling. *Adv. Healthc. Mater.* **2017**, *6*. [CrossRef]
19. Lee, M.O.; Jung, K.B.; Jo, S.J.; Hyun, S.A.; Moon, K.S.; Seo, J.W.; Kim, S.H.; Son, M.Y. Modelling cardiac fibrosis using three-dimensional cardiac microtissues derived from human embryonic stem cells. *J. Biol. Eng.* **2019**, *13*, 15. [CrossRef]
20. Van Spreeuwel, A.C.C.; Bax, N.A.M.; Van Nierop, B.J.; Aartsma-Rus, A.; Goumans, M.T.H.; Bouten, C.V.C. Mimicking cardiac fibrosis in a dish: Fibroblast density rather than collagen density weakens cardiomyocyte function. *J. Cardiovasc. Transl. Res.* **2017**, *10*, 116–127. [CrossRef]
21. Wang, E.Y.; Rafatian, N.; Zhao, Y.; Lee, A.; Lai, B.F.L.; Lu, R.X.; Jekic, D.; Davenport Huyer, L.; Knee-Walden, E.J.; Bhattacharya, S.; et al. Biowire model of interstitial and focal cardiac fibrosis. *ACS Cent. Sci.* **2019**, *5*, 1146–1158. [CrossRef] [PubMed]
22. Kofron, C.M.; Kim, T.Y.; King, M.E.; Xie, A.; Feng, F.; Park, E.; Qu, Z.; Choi, B.R.; Mende, U. Gq-activated fibroblasts induce cardiomyocyte action potential prolongation and automaticity in a three-dimensional microtissue environment. *Am. J. Physiol. Heart Circ. Physiol.* **2017**, *313*, H810–H827. [CrossRef] [PubMed]
23. Nurzynska, D.; Di Meglio, F.; Romano, V.; Miraglia, R.; Sacco, A.M.; Latino, F.; Bancone, C.; Della Corte, A.; Maiello, C.; Amarelli, C.; et al. Cardiac primitive cells become committed to a cardiac fate in adult human heart with chronic ischemic disease but fail to acquire mature phenotype: Genetic and phenotypic study. *Basic Res. Cardiol.* **2013**, *108*, 320. [CrossRef] [PubMed]
24. Trieschmann, J.; Bettin, D.; Haustein, M.; Koster, A.; Molcanyi, M.; Halbach, M.; Hanna, M.; Fouad, M.; Brockmeier, K.; Hescheler, J.; et al. The interaction between adult cardiac fibroblasts and embryonic stem cell-derived cardiomyocytes leads to proarrhythmic changes in in vitro cocultures. *Stem Cells Int.* **2016**, *2016*, 2936126. [CrossRef]
25. Pellman, J.; Zhang, J.; Sheikh, F. Myocyte-fibroblast communication in cardiac fibrosis and arrhythmias: Mechanisms and model systems. *J. Mol. Cell. Cardiol.* **2016**, *94*, 22–31. [CrossRef]
26. Wu, H.; Lee, J.; Vincent, L.G.; Wang, Q.; Gu, M.; Lan, F.; Churko, J.M.; Sallam, K.I.; Matsa, E.; Sharma, A.; et al. Epigenetic regulation of phosphodiesterases 2A and 3A underlies compromised beta-adrenergic signaling in an iPSC model of dilated cardiomyopathy. *Cell Stem Cell* **2015**, *17*, 89–100. [CrossRef]
27. Preininger, M.K.; Jha, R.; Maxwell, J.T.; Wu, Q.; Singh, M.; Wang, B.; Dalal, A.; McEachin, Z.T.; Rossoll, W.; Hales, C.M.; et al. A human pluripotent stem cell model of catecholaminergic polymorphic ventricular tachycardia recapitulates patient-specific drug responses. *Dis. Models Mech.* **2016**, *9*, 927–939. [CrossRef]
28. Ferrara, N.; Komici, K.; Corbi, G.; Pagano, G.; Furgi, G.; Rengo, C.; Femminella, G.D.; Leosco, D.; Bonaduce, D. Beta-adrenergic receptor responsiveness in aging heart and clinical implications. *Front. Physiol.* **2014**, *4*, 396. [CrossRef]
29. Krishnamurthy, P.; Subramanian, V.; Singh, M.; Singh, K. Beta1 integrins modulate beta-adrenergic receptor-stimulated cardiac myocyte apoptosis and myocardial remodeling. *Hypertension* **2007**, *49*, 865–872. [CrossRef]
30. Gravning, J.; Ahmed, M.S.; Qvigstad, E.; Krobert, K.; Edvardsen, T.; Moe, I.T.; Hagelin, E.M.; Sagave, J.; Valen, G.; Levy, F.O.; et al. Connective tissue growth factor/CCN2 attenuates beta-adrenergic receptor responsiveness and cardiotoxicity by induction of G protein-coupled receptor kinase-5 in cardiomyocytes. *Mol. Pharmacol.* **2013**, *84*, 372–383. [CrossRef]
31. Kim, T.J.; Sun, J.; Lu, S.; Zhang, J.; Wang, Y. The regulation of beta-adrenergic receptor-mediated PKA activation by substrate stiffness via microtubule dynamics in human MSCs. *Biomaterials* **2014**, *35*, 8348–8356. [CrossRef] [PubMed]
32. Zuppinger, C.; Gibbons, G.; Dutta-Passecker, P.; Segiser, A.; Most, H.; Suter, T.M. Characterization of cytoskeleton features and maturation status of cultured human iPSC-derived cardiomyocytes. *Eur. J. Histochem.* **2017**, *61*, 2763. [CrossRef] [PubMed]

33. Nguyen, N.; Nguyen, W.; Nguyenton, B.; Ratchada, P.; Page, G.; Miller, P.E.; Ghetti, A.; Abi-Gerges, N. Adult human primary cardiomyocyte-based model for the simultaneous prediction of drug-induced inotropic and pro-arrhythmia risk. *Front. Physiol.* **2017**, *8*, 1073. [CrossRef] [PubMed]
34. Nunes, S.S.; Miklas, J.W.; Liu, J.; Aschar-Sobbi, R.; Xiao, Y.; Zhang, B.; Jiang, J.; Masse, S.; Gagliardi, M.; Hsieh, A.; et al. Biowire: A platform for maturation of human pluripotent stem cell-derived cardiomyocytes. *Nat. Methods* **2013**, *10*, 781–787. [CrossRef]
35. Giacomelli, E.; Bellin, M.; Sala, L.; Van Meer, B.J.; Tertoolen, L.G.; Orlova, V.V.; Mummery, C.L. Three-dimensional cardiac microtissues composed of cardiomyocytes and endothelial cells co-differentiated from human pluripotent stem cells. *Development* **2017**, *144*, 1008–1017. [CrossRef]
36. Ronaldson-Bouchard, K.; Ma, S.P.; Yeager, K.; Chen, T.; Song, L.; Sirabella, D.; Morikawa, K.; Teles, D.; Yazawa, M.; Vunjak-Novakovic, G. Advanced maturation of human cardiac tissue grown from pluripotent stem cells. *Nature* **2018**, *556*, 239–243. [CrossRef]

Permissions

All chapters in this book were first published by MDPI; hereby published with permission under the Creative Commons Attribution License or equivalent. Every chapter published in this book has been scrutinized by our experts. Their significance has been extensively debated. The topics covered herein carry significant findings which will fuel the growth of the discipline. They may even be implemented as practical applications or may be referred to as a beginning point for another development.

The contributors of this book come from diverse backgrounds, making this book a truly international effort. This book will bring forth new frontiers with its revolutionizing research information and detailed analysis of the nascent developments around the world.

We would like to thank all the contributing authors for lending their expertise to make the book truly unique. They have played a crucial role in the development of this book. Without their invaluable contributions this book wouldn't have been possible. They have made vital efforts to compile up to date information on the varied aspects of this subject to make this book a valuable addition to the collection of many professionals and students.

This book was conceptualized with the vision of imparting up-to-date information and advanced data in this field. To ensure the same, a matchless editorial board was set up. Every individual on the board went through rigorous rounds of assessment to prove their worth. After which they invested a large part of their time researching and compiling the most relevant data for our readers.

The editorial board has been involved in producing this book since its inception. They have spent rigorous hours researching and exploring the diverse topics which have resulted in the successful publishing of this book. They have passed on their knowledge of decades through this book. To expedite this challenging task, the publisher supported the team at every step. A small team of assistant editors was also appointed to further simplify the editing procedure and attain best results for the readers.

Apart from the editorial board, the designing team has also invested a significant amount of their time in understanding the subject and creating the most relevant covers. They scrutinized every image to scout for the most suitable representation of the subject and create an appropriate cover for the book.

The publishing team has been an ardent support to the editorial, designing and production team. Their endless efforts to recruit the best for this project, has resulted in the accomplishment of this book. They are a veteran in the field of academics and their pool of knowledge is as vast as their experience in printing. Their expertise and guidance has proved useful at every step. Their uncompromising quality standards have made this book an exceptional effort. Their encouragement from time to time has been an inspiration for everyone.

The publisher and the editorial board hope that this book will prove to be a valuable piece of knowledge for researchers, students, practitioners and scholars across the globe.

List of Contributors

Barbara Breznik
Department of Genetic Toxicology and Cancer Biology, National Institute of Biology, 1000 Ljubljana, Slovenia

Martina Štampar, Metka Filipič and Bojana Žegura
Department of Genetic Toxicology and Cancer Biology, National Institute of Biology, 1000 Ljubljana, Slovenia
Jozef Stefan International Postgraduate School, 1000 Ljubljana, Slovenia

Roman Bruch
Institute of Molecular and Cell Biology, Mannheim University of Applied Sciences, 68163 Mannheim, Germany

Florian Keller, Mathias Hafner and Rüdiger Rudolf
Institute of Molecular and Cell Biology, Mannheim University of Applied Sciences, 68163 Mannheim, Germany
Institute of Medical Technology, Medical Faculty Mannheim of Heidelberg University and Mannheim University of Applied Sciences, 68167 Mannheim, Germany

Richard Schneider and Julia Meier-Hubberten
TIP Oncology, Merck Healthcare KGaA, 64289 Darmstadt, Germany

Enrico Ragni, Carlotta Perucca Orfei, Paola De Luca and Laura de Girolamo
IRCCS Istituto Ortopedico Galeazzi, Laboratorio di Biotecnologie Applicate all'Ortopedia, Via R. Galeazzi 4, I-20161 Milano, Italy

Silvia Palombella, Silvia Lopa and Giuseppe Talò
IRCCS Istituto Ortopedico Galeazzi, Cell and Tissue Engineering Laboratory, Via R. Galeazzi 4, I-20161 Milano, Italy

Matteo Moretti
IRCCS Istituto Ortopedico Galeazzi, Cell and Tissue Engineering Laboratory, Via R. Galeazzi 4, I-20161 Milano, Italy
Regenerative Medicine Technologies Laboratory, Ente Ospedaliero Cantonale (EOC), 6900 Lugano, Switzerland

Robin Demuynck
Cell Death Investigation and Therapy Laboratory, Department of Human Structure and Repair, Ghent University, 9000 Ghent, Belgium
Tissue Engineering and Biomaterials Group, Department of Human Structure and Repair, Ghent University, 9000 Ghent, Belgium

Iuliia Efimova
Cell Death Investigation and Therapy Laboratory, Department of Human Structure and Repair, Ghent University, 9000 Ghent, Belgium
Cancer Research Institute Ghent, 9000 Ghent, Belgium

Abraham Lin
Plasma, Laser Ablation and Surface Modelling Group, University of Antwerp, 2610 Wilrijk, Belgium
Center for Oncological Research, University of Antwerp, 2610 Wilrijk, Belgium

Heidi Declercq
Tissue Engineering and Biomaterials Group, Department of Human Structure and Repair, Ghent University, 9000 Ghent, Belgium
Cancer Research Institute Ghent, 9000 Ghent, Belgium
Tissue Engineering lab, Department of Development and Regeneration, KU Leuven, 8500 Kortrijk, Belgium

Dmitri V. Krysko
Cell Death Investigation and Therapy Laboratory, Department of Human Structure and Repair, Ghent University, 9000 Ghent, Belgium
Cancer Research Institute Ghent, 9000 Ghent, Belgium
Department of Pathophysiology, Sechenov First Moscow State Medical University, 119146 Moscow, Russia

Yu-Ting Yen and May Chien
Drug Development Center, Institute of New Drug Development, China Medical University, Taichung 40402, Taiwan
Integrative Stem Cell Center, China Medical University Hospital, Taichung 40402, Taiwan

Yung-Chih Lai
Integrative Stem Cell Center, China Medical University Hospital, Taichung 40402, Taiwan

Dao-Peng Chen
Kim Forest Enterprise Co., Ltd., Taipei 22175, Taiwan

Cheng-Ming Chuong
Integrative Stem Cell Center, China Medical University Hospital, Taichung 40402, Taiwan
Department of Pathology, Keck School of Medicine, University of Southern California, Los Angeles, CA 90033, USA

Mien-Chie Hung
Center for Molecular Medicine and Graduate Institute of Cancer Biology, China Medical University, Taichung 40402, Taiwan
Cancer Biology Program, The University of Texas Graduate School of Biomedical Sciences, The University of Texas MD Anderson Cancer Center, Houston, TX 77030, USA

Shih-Chieh Hung
Drug Development Center, Institute of New Drug Development, China Medical University, Taichung 40402, Taiwan
Integrative Stem Cell Center, China Medical University Hospital, Taichung 40402, Taiwan
Department of Orthopaedics, China Medical University Hospital, Taichung 40402, Taiwan

Antonio Carlos Ribeiro-Filho and Marluce da Cunha Mantovani
Organoid Development Team, Center of Innovation and Translational Medicine (CIMTRA), University of São Paulo School of Medicine, Sao Paulo 05360-130, Brazil

Débora Levy
Lipids, Oxidation and Cell Biology Team, Laboratory of Immunology (LIM19), Heart Institute (InCor), University of São Paulo School of Medicine, Sao Paulo 05403-900, Brazil

Jorge Luis Maria Ruiz
Life and Nature Science Institute, Federal University of Latin American Integration-UNILA, Foz de Iguaçú, PR 858570-901, Brazil

Sérgio Paulo Bydlowski
Organoid Development Team, Center of Innovation and Translational Medicine (CIMTRA), University of São Paulo School of Medicine, Sao Paulo 05360-130, Brazil
Lipids, Oxidation and Cell Biology Team, Laboratory of Immunology (LIM19), Heart Institute (InCor), University of São Paulo School of Medicine, Sao Paulo 05403-900, Brazil
National Institute of Science and Technology in Regenerative Medicine (INCT-Regenera), CNPq, Rio de Janeiro 21941-902, Brazil

Christina McKee, Christina Brown and G. Rasul Chaudhry
Department of Biological Sciences, Oakland University, Rochester, MI 48309, USA
OU-WB Institute for Stem Cell and Regenerative Medicine, Rochester, MI 48309, USA

Julianna Kobolak, Annamaria Teglasi, Tamas Bellak, Zofia Janstova, Melinda Zana and Istvan Bock
BioTalentum Ltd., H-2100 Gödöllő, Hungary

Kinga Molnar and Lajos Laszlo
Department of Anatomy, Cell and Developmental Biology, Eötvös Loránd University, H-1117 Budapest, Hungary

Andras Dinnyes
BioTalentum Ltd., H-2100 Gödöllő, Hungary
Molecular Animal Biotechnology Laboratory, Szent István University, H-2101 Gödöllő, Hungary
Translational Biomedicine Institute, University of Szeged, H-6720 Szeged, Hungary

Maria C. Ciardulli, Luigi Marino and Carmine Selleri
Department of Medicine, Surgery and Dentistry, University of Salerno, Via S. Allende, 84081 Baronissi (SA), Italy

Joseph Lovecchio and Emanuele Giordano
Department of Electrical, Electronic and Information Engineering "Guglielmo Marconi" (DEI), University of Bologna, Via dell'Università 50, 47522 Cesena (FC), Italy

Nicholas R. Forsyth
Guy Hilton Research Centre, School of Pharmacy and Bioengineering, Keele University, Stoke-on-Trent, Staffordshire ST4 7QB, UK

Nicola Maffulli
Department of Medicine, Surgery and Dentistry, University of Salerno, Via S. Allende, 84081 Baronissi (SA), Italy
Guy Hilton Research Centre, School of Pharmacy and Bioengineering, Keele University, Stoke-on-Trent, Staffordshire ST4 7QB, UK
Centre for Sport and Exercise Medicine, Queen Mary University of London, Barts and The London School of Medicine, London E1 4NL, UK

Giovanna Della Porta
Department of Medicine, Surgery and Dentistry, University of Salerno, Via S. Allende, 84081 Baronissi (SA), Italy
Department of Industrial Engineering, University of Salerno, Via Giovanni Paolo II, 84084 Fisciano (SA), Italy

Zenglai Tan, Aleksandra Rak-Raszewska and Ilya Skovorodkin
Center for Cell Matrix Research, Faculty of Biochemistry and Molecular Medicine, Biocenter Oulu, Laboratory of Developmental Biology, Infotech Oulu, University of Oulu, Aapistie 5A, 90220 Oulu, Finland

Seppo J. Vainio
Center for Cell Matrix Research, Faculty of Biochemistry and Molecular Medicine, Biocenter Oulu, Laboratory of Developmental Biology, Infotech Oulu, University of Oulu, Aapistie 5A, 90220 Oulu, Finland
Borealis Biobank of Northern Finland, Oulu Central Hospital, 90220 Oulu, Finland

Nina Kramer, Barbara Pratscher, Alexander Swoboda, Matthias Kieslinger and Iwan A. Burgener
Division of Small Animal Internal Medicine, Department for Small Animals and Horses, University of Veterinary Medicine, 1210 Vienna, Austria

Andre M. C. Meneses
Division of Small Animal Internal Medicine, Department for Small Animals and Horses, University of Veterinary Medicine, 1210 Vienna, Austria
Department of Clinical Sciences, Faculty of Veterinary Medicine, Utrecht University, 3584 Utrecht, The Netherlands

Hedwig S. Kruitwagen, Kerstin Schneeberger, Louis C. Penning and Bart Spee
Department of Clinical Sciences, Faculty of Veterinary Medicine, Utrecht University, 3584 Utrecht, The Netherlands

Waltraud Tschulenk and Ingrid Walter
Institute of Pathology, Department for Pathobiology, University of Veterinary Medicine, 1210 Vienna, Austria

Sabine Brandt
Research Group Oncology, Equine Surgery, Department of Small Animals and Horses, University of Veterinary Medicine, 1210 Vienna, Austria

Alessandro Cutarelli, Jacopo Zasso and Luciano Conti
Laboratory of Stem Cell Biology, Department of Cellular, Computational and Integrative Biology-CIBIO, University of Trento, 38123 Trento, Italy

Simone Ghio, Michele Crivellari and Maurizio Boscardin
Fondazione Bruno Kessler-Center for Material and Microsystem, 38123 Trento, Italy

Alessandra Speccher and Simona Casarosa
Laboratory of Neural Development and Regeneration, Department of Cellular, Computational and Integrative Biology-CIBIO, University of Trento, 38123 Trento, Italy

Giorgina Scarduelli and Michela Roccuzzo
Advanced Imaging Facility, Department of Cellular, Computational and Integrative Biology-CIBIO, University of Trento, 38123 Trento, Italy

Nicola Maria Pugno
Laboratory of Bio-Inspired and Graphene Nanomechanics, Department of Civil, Environmental and Mechanical Engineering, University of Trento, 38123 Trent don, London E1 4NS, UK
Ket-Lab, Edoardo Amaldi Foundation, via del Politecnico snc, I-00133 Roma, Italy

Mara Stellato, Oliver Distler and Gabriela Kania
Center of Experimental Rheumatology, Department of Rheumatology, University Hospital Zurich, Wagistr. 14, 8952 Schlieren, Switzerland

Przemysław Błyszczuk
Center of Experimental Rheumatology, Department of Rheumatology, University Hospital Zurich, Wagistr. 14, 8952 Schlieren, Switzerland
Department of Clinical Immunology, Jagiellonian University Medical College, 30-663 Cracow, Poland

Christian Zuppinger
Department for BioMedical Research, Department of Cardiology, University Hospital Bern, 3008 Bern, Switzerland

Ana Costa and Godfrey L. Smith
Institute of Cardiovascular and Medical Sciences, University of Glasgow, Glasgow G12 8TA, UK

Daria Nurzynska and Franca Di Meglio
Department of Public Health, University of Naples "Federico II", 80131 Naples, Italy

Irina Agarkova
InSphero, 8952 Schlieren, Switzerland

Index

www.ingramcontent.com/pod-product-compliance
Lightning Source LLC
LaVergne TN
LVHW080951210926
848485LV00004B/820